Geographic Data Mining and Knowledge Discovery

Second Edition

Chapman & Hall/CRC
Data Mining and Knowledge Discovery Series

SERIES EDITOR

Vipin Kumar
University of Minnesota
Department of Computer Science and Engineering
Minneapolis, Minnesota, U.S.A.

AIMS AND SCOPE

This series aims to capture new developments and applications in data mining and knowledge discovery, while summarizing the computational tools and techniques useful in data analysis. This series encourages the integration of mathematical, statistical, and computational methods and techniques through the publication of a broad range of textbooks, reference works, and handbooks. The inclusion of concrete examples and applications is highly encouraged. The scope of the series includes, but is not limited to, titles in the areas of data mining and knowledge discovery methods and applications, modeling, algorithms, theory and foundations, data and knowledge visualization, data mining systems and tools, and privacy and security issues.

PUBLISHED TITLES

UNDERSTANDING COMPLEX DATASETS: Data Mining with Matrix Decompositions
David Skillicorn

COMPUTATIONAL METHODS OF FEATURE SELECTION
Huan Liu and Hiroshi Motoda

CONSTRAINED CLUSTERING: Advances in Algorithms, Theory, and Applications
Sugato Basu, Ian Davidson, and Kiri L. Wagstaff

KNOWLEDGE DISCOVERY FOR COUNTERTERRORISM AND LAW ENFORCEMENT
David Skillicorn

MULTIMEDIA DATA MINING: A Systematic Introduction to Concepts and Theory
Zhongfei Zhang and Ruofei Zhang

NEXT GENERATION OF DATA MINING
Hillol Kargupta, Jiawei Han, Philip S. Yu, Rajeev Motwani, and Vipin Kumar

DATA MINING FOR DESIGN AND MARKETING
Yukio Ohsawa and Katsutoshi Yada

GEOGRAPHIC DATA MINING AND KNOWLEDGE DISCOVERY, Second Edition
Harvey J. Miller and Jiawei Han

Chapman & Hall/CRC
Data Mining and Knowledge Discovery Series

Geographic Data Mining and Knowledge Discovery

Second Edition

Edited by
Harvey J. Miller
Jiawei Han

CRC Press is an imprint of the
Taylor & Francis Group, an **informa** business
A CHAPMAN & HALL BOOK

CRC Press
Taylor & Francis Group
6000 Broken Sound Parkway NW, Suite 300
Boca Raton, FL 33487-2742

Printed in the United States of America on acid-free paper
10 9 8 7 6 5 4 3 2 1

International Standard Book Number-13: 978-1-4200-7397-3 (Hardcover)

Library of Congress Cataloging-in-Publication Data

Geographic data mining and knowledge discovery / editors, Harvey J. Miller and Jiawei Han. -- 2nd ed.
p. cm.
Includes bibliographical references and index.
ISBN 978-1-4200-7397-3 (hard back : alk. paper)
1. Geodatabases. 2. Data mining. I. Miller, Harvey J. II. Han, Jiawei. III. Title.

G70.2.G4365 2009
910.285'6312--dc22 2009010969

Visit the Taylor & Francis Web site at
http://www.taylorandfrancis.com

and the CRC Press Web site at
http://www.crcpress.com

Contents

Acknowledgments

The editors would like to thank the National Center for Geographic Information and Analysis (NCGIA) — Project Varenius for supporting the March 1999 workshop that resulted in the first edition of this book. A special thanks to Randi Cohen (Taylor & Francis Group) who proposed a second edition and worked tirelessly on its behalf with organizational skills that greatly exceed ours.

About The Editors

Harvey J. Miller is professor of geography at the University of Utah. His research and teaching interests include geographic information systems (GIS), spatial analysis, and geocomputational techniques applied to understanding how transportation and communication technologies shape lives, cities, and societies. He has contributed to the theory and methodology underlying *time geography*: a perspective that focuses on personal allocation of time among activities in space and its implications for individual and collective spatial dynamics. He is the author (with Shih-Lung Shaw) of *Geographic Information Systems for Transportation: Principles and Applications* (Oxford University Press) and editor of *Societies and Cities in the Age of Instant Access* (Springer). Professor Miller serves on the editorial boards of *Geographical Analysis, International Journal of Geographical Information Science, Journal of Regional Science, Transportation, URISA Journal,* and *Journal of Transport and Land Use.* He was the North American editor of *International Journal of Geographical Information Science* from 2000 to 2004. Professor Miller has also served as an officer or board member of the North American Regional Science Council, the University Consortium for Geographic Information Science, the Association of American Geographers, and the Transportation Research Board (TRB). He is currently cochair of the TRB Committee on Geographic Information Science and Applications.

Jiawei Han is professor in the department of computer science, University of Illinois at Urbana-Champaign. He has been working on research in data mining, data warehousing, database systems, data mining from spatiotemporal data, multimedia data, stream and radio frequency identification (RFID) data, social network data, and biological data, and has to his credit over 350 journal and conference publications. He has chaired or served on over 100 program committees of international conferences and workshops, including PC cochair of 2005 (IEEE) International Conference on Data Mining (ICDM), American coordinator of 2006 International Conference on Very Large Data Bases (VLDB), and senior PC member for the 2008 ACM SIGKDD International Conference on Knowledge Discovery and Data Mining. He is also serving as the founding editor-in-chief of *ACM Transactions on Knowledge Discovery from Data*. He is an ACM Fellow and has received the 2004 ACM SIGKDD Innovations Award and 2005 IEEE Computer Society Technical Achievement Award. His book *Data Mining: Concepts and Techniques* (2nd ed., Morgan Kaufmann, 2006) has been popularly used as a textbook worldwide.

List of Contributors

Annalisa Appice
Università degli Studi di Bari
Bari, Italy

Yvan Bédard
Laval University
Quebec City, Canada

Eveline Bernier
Laval University
Quebec City, Canada

Arnold P. Boedihardjo
Virginia Tech
Blacksburg, Virginia

Huiping Cao
University of Hong Kong
Hong Kong

Martin Charlton
National University of Ireland
County Kildare, Ireland

Sanjay Chawla
University of Sydney
Sydney, Australia

David W. Cheung
University of Hong Kong
Hong Kong

Urška Demšar
National University of Ireland
County Kildare, Ireland

Rodolphe Devillers
Memorial University of Newfoundland
St. John's, Canada

Matt Duckham
University of Melbourne
Victoria, Australia

A. Stewart Fotheringham
National University of Ireland
County Kildare, Ireland

Mark Gahegan
University of Auckland
Auckland, New Zealand

Marc Gervais
Laval University
Quebec City, Canada

Diansheng Guo
University of South Carolina
Columbia, South Carolina

Otto Huisman
International Institute for
GeoInformation Science
and Earth Observation (ITC)
Enschede, Netherlands

Micheline Kamber
Burnaby, Canada

Menno-Jan Kraak
International Institute for
GeoInformation Science
and Earth Observation (ITC)
Enschede, Netherlands

Antonietta Lanza
Università degli Studi di Bari
Bari, Italy

Patrick Laube
University of Melbourne
Victoria, Australia

Jae-Gil Lee
University of Illinois
Urbana-Champaign
Urbana, Illinois

Brian G. Lees
University of New South Wales
Canberra, Australia

Marie-Andree Levesque
Laval University
Quebec City, Canada

Arend Ligtenberg
Wageningen University and Research
Wageningen, Netherlands

Chang-Tien Lu
Virginia Tech
Blacksburg, Virginia

Jose Macedo
Swiss Federal Institute of Technology
Zurich, Switzerland

Donato Malerba
Università degli Studi di Bari
Bari, Italy

Nikos Mamoulis
University of Hong Kong
Hong Kong

Kyriakos Mouratidis
Singapore Management University
Singapore

Dimitris Papadias
Hong Kong University of Science
and Technology
Clear Water Bay, Hong Kong

Spiros Papadimitriou
IBM T.J. Watson Research Center
Hawthorne, New York

Chiara Renso
KDDLAB-CNR
Pisa, Italy

John F. Roddick
Flinders University
Adelaide, South Australia

Shashi Shekhar
University of Minnesota
Minneapolis, Minnesota

Ranga Raju Vatsavai
Oak Ridge National Laboratory
Oak Ridge, Tennessee

Monica Wachowicz
Technical University of Madrid
Madrid, Spain
and
Wageningen University and Research
Wageningen, Netherlands

May Yuan
University of Oklahoma
Norman, Oklahoma

1 Geographic Data Mining and Knowledge Discovery

An Overview

Harvey J. Miller

Jiawei Han

CONTENTS

1.1 INTRODUCTION

Similar to many research and application fields, geography has moved from a data-poor and computation-poor to a data-rich and computation-rich environment. The scope, coverage, and volume of digital geographic datasets are growing rapidly. Public and private sector agencies are creating, processing, and disseminating digital data on land use, socioeconomic conditions, and infrastructure at very detailed levels of geographic resolution. New high spatial and spectral resolution remote sensing systems and other monitoring devices are gathering vast amounts of geo-referenced digital imagery, video, and sound. Geographic data collection devices linked to location-ware technologies (LATs) such as global positioning system (GPS) receivers allow field researchers to collect unprecedented amounts of data. LATs linked to or embedded in devices such as cell phones, in-vehicle navigation systems, and wireless Internet clients provide location-specific content in exchange for tracking individuals in space and time. Information infrastructure initiatives such as the U.S. National Spatial Data Infrastructure are facilitating data sharing and interoperability. Digital geographic data repositories on the World Wide Web are growing rapidly in both number and scope. The amount of data that geographic information processing systems can handle will continue to increase exponentially through the mid-21st century.

Traditional spatial analytical methods were developed in an era when data collection was expensive and computational power was weak. The increasing volume and diverse nature of digital geographic data easily overwhelm mainstream spatial analysis techniques that are oriented toward teasing scarce information from small and homogenous datasets. Traditional statistical methods, particularly spatial statistics, have high computational burdens. These techniques are confirmatory and require the researcher to have *a priori* hypotheses. Therefore, traditional spatial analytical techniques cannot easily discover new and unexpected patterns, trends, and relationships that can be hidden deep within very large and diverse geographic datasets.

In March 1999, the National Center for Geographic Information and Analysis (NCGIA) — Project Varenius held a workshop on discovering geographic knowledge in data-rich environments in Kirkland, Washington, USA. The workshop brought together a diverse group of stakeholders with interests in developing and applying computational techniques for exploring large, heterogeneous digital geographic datasets. Drawing on papers submitted to that workshop, in 2001 we published *Geographic Data Mining and Knowledge Discovery,* a volume that brought together some of the cutting-edge research in the area of geographic data mining and geographic knowledge discovery in a data-rich environment. There has been much progress in geographic knowledge discovery (GKD) over the past eight years, including the development of new techniques for geographic data warehousing (GDW), spatial data mining, and geo-visualization. In addition, there has been a remarkable rise in the collection and storage of data on spatiotemporal processes and mobile objects, with a consequential rise in knowledge discovery techniques for these data.

The second edition of *Geographic Data Mining and Knowledge Discovery* is a major revision of the first edition. We selected chapters from the first edition and asked authors for updated manuscripts that reflect changes and recent developments in their particular domains. We also solicited new chapters on topics that were not covered well in the first edition but have become more prominent recently. This includes several new chapters on spatiotemporal and mobile objects databases, a topic only briefly mentioned in the 2001 edition.

This chapter introduces geographic data mining and GKD. In this chapter, we provide an overview of knowledge discovery from databases (KDD) and data mining. We identify why geographic data is a nontrivial special case that requires distinctive consideration and techniques. We also review the current state-of-the-art in GKD, including the existing literature and the contributions of the chapters in this volume.

1.2 KNOWLEDGE DISCOVERY AND DATA MINING

In this section, we provide a general overview of knowledge discovery and data mining. We begin with an overview of KDD, highlighting its general objectives and its relationship to the field of statistics and the general scientific process. We then identify the major stages of KDD processing, including data mining. We classify major data-mining tasks and discuss some techniques available for each task. We conclude this section by discussing the relationships between scientific visualization and KDD.

1.2.1 Knowledge Discovery from Databases

Knowledge discovery from databases (KDD) is a response to the enormous volumes of data being collected and stored in operational and scientific databases. Continuing improvements in information technology (IT) and its widespread adoption for process monitoring and control in many domains is creating a wealth of new data. There is often much more information in these databases than the "shallow" information being extracted by traditional analytical and query techniques. KDD leverages investments in IT by searching for deeply hidden information that can be turned into knowledge for strategic decision making and answering fundamental research questions.

KDD is better known through the more popular term "data mining." However, data mining is only one component (albeit a central component) of the larger KDD process. Data mining involves distilling data into *information* or facts about the domain described by the database. KDD is the higher-level process of obtaining information through data mining and distilling this information into *knowledge* (ideas and beliefs about the domain) through interpretation of information and integration with existing knowledge.

KDD is based on a belief that information is hidden in very large databases in the form of *interesting patterns*. These are nonrandom properties and relationships that are valid, novel, useful, and ultimately understandable. *Valid* means that the pattern is general enough to apply to new data; it is not just an anomaly of the current data. *Novel* means that the pattern is nontrivial and unexpected. *Useful* implies that the pattern should lead to some effective action, e.g., successful decision making and scientific investigation. *Ultimately understandable* means that the pattern should be simple and interpretable by humans (Fayyad, Piatetsky-Shapiro and Smyth 1996).

KDD is also based on the belief that traditional database queries and statistical methods cannot reveal interesting patterns in very large databases, largely due to the type of data that increasingly comprise enterprise databases and the novelty of the patterns sought in KDD.

KDD goes beyond the traditional domain of statistics to accommodate data not normally amenable to statistical analysis. Statistics usually involves a small and clean (noiseless) numeric database scientifically sampled from a large population with specific questions in mind. Many statistical models require strict assumptions (such as independence, stationarity of underlying processes, and normality). In contrast, the data being collected and stored in many enterprise databases are noisy, nonnumeric, and possibly incomplete. These data are also collected in an open-ended manner without specific questions in mind (Hand 1998). KDD encompasses principles and techniques from statistics, machine learning, pattern recognition, numeric search, and scientific visualization to accommodate the new data types and data volumes being generated through information technologies.

KDD is more strongly inductive than traditional statistical analysis. The generalization process of statistics is embedded within the broader deductive process of science. Statistical models are confirmatory, requiring the analyst to specify a model *a priori* based on some theory, test these hypotheses, and perhaps revise the theory depending on the results. In contrast, the deeply hidden, interesting patterns being sought in a KDD process are (by definition) difficult or impossible to specify *a priori*, at least with any reasonable degree of completeness. KDD is more concerned about prompting investigators to formulate *new* predictions and hypotheses from data as opposed to testing deductions from theories through a sub-process of induction from a scientific database (Elder and Pregibon 1996; Hand 1998). A guideline is that if the information being sought can only be vaguely described in advance, KDD is more appropriate than statistics (Adriaans and Zantinge 1996).

KDD more naturally fits in the initial stage of the deductive process when the researcher forms or modifies theory based on ordered facts and observations from the real world. In this sense, KDD is to information space as microscopes, remote sensing, and telescopes are to atomic, geographic, and astronomical spaces, respectively. KDD is a tool for exploring domains that are too difficult to perceive with unaided human abilities. For searching through a large information wilderness, the powerful but focused laser beams of statistics cannot compete with the broad but diffuse floodlights of KDD. However, floodlights can cast shadows and KDD cannot compete with statistics in confirmatory power once the pattern is discovered.

1.2.2 Data Warehousing

An infrastructure that often underlies the KDD process is the *data warehouse* (DW). A DW is a repository that integrates data from one or more source databases. The DW phenomenon results from several technological and economic trends, including the decreasing cost of data storage and data processing, and the increasing value of information in business, government, and scientific environments. A DW usually exists to support strategic and scientific decision making based on integrated, shared information, although DWs are also used to save legacy data for liability and other purposes (see Jarke et al. 2000).

The data in a DW are usually read-only historical copies of the operational databases in an enterprise, sometimes in summary form. Consequently, a DW is often several orders of magnitude larger than an operational database (Chaudhuri and Dayal 1997). Rather than just a very large database management system, a DW embodies database design principles very different from operational databases.

Operational database management systems are designed to support *transactional data processing*, that is, data entry, retrieval, and updating. Design principles for transactional database systems attempt to create a database that is internally consistent and recoverable (i.e., can be "rolled-back" to the last known internally consistent state in the event of an error or disruption). These objectives must be met in an environment where multiple users are retrieving and updating data. For example, the normalization process in relational database design decomposes large, "flat" relations along functional dependencies to create smaller, parsimonious relations that logically store a particular item a minimal number of times (ideally, only once; see Silberschatz et al. 1997). Since data are stored a minimal number of times, there is a minimal possibility of two data items about the same real-world entity disagreeing (e.g., if only one item is updated due to user error or an ill-timed system crash).

In contrast to transactional database design, good DW design maximizes the efficiency of *analytical data processing* or data examination for decision making. Since the DW contains read-only copies and summaries of the historical operational databases, consistency and recoverability in a multiuser transactional environment are not issues. The database design principles that maximize analytical efficiency are contrary to those that maximize transactional stability. Acceptable response times when repeatedly retrieving large quantities of data items for analysis require the database to be nonnormalized and connected; examples include the "star" and "snowflake" logical DW schemas (see Chaudhuri and Dayal 1997). The DW is in a sense a buffer between transactional and analytical data processing, allowing efficient analytical data processing without corrupting the source databases (Jarke et al. 2000).

In addition to data mining, a DW often supports *online analytical processing* (OLAP) tools. OLAP tools provide multidimensional summary views of the data in a DW. OLAP tools allow the user to manipulate these views and explore the data underlying the summarized views. Standard OLAP tools include *roll-up* (increasing the level of aggregation), *drill-down* (decreasing the level of aggregation), *slice* and *dice* (selection and projection), and *pivot* (re-orientation of the multidimensional data view) (Chaudhuri and Dayal 1997). OLAP tools are in a sense types of "super-queries": more powerful than standard query language such as SQL but shallower than data-mining techniques because they do not reveal hidden patterns. Nevertheless, OLAP tools can be an important part of the KDD process. For example, OLAP tools can allow the analyst to achieve a synoptic view of the DW that can help specify and direct the application of data-mining techniques (Adriaans and Zantinge 1996).

A powerful and commonly applied OLAP tool for multidimensional data summary is the *data cube*. Given a particular measure (e.g., "sales") and some dimensions of interest (e.g., "item," "store," "week"), a data cube is an operator that returns the power set of all possible aggregations of the measure with respect to the dimensions of interest. These include aggregations over zero dimension (e.g., "total sales"), one dimension (e.g., "total sales by item," "total sales by store," "total sales per week"), two dimensions (e.g., "total sales by item and store") and so on, up to *N* dimensions. The data cube is an *N*-dimensional generalization of the more commonly known SQL aggregation functions and "Group-By" operator. However, the analogous SQL query only generates the zero and one-dimensional aggregations; the data cube operator generates these and the higher dimensional aggregations all at once (Gray et al. 1997).

The power set of aggregations over selected dimensions is called a "data cube" because the logical arrangement of aggregations can be viewed as a hypercube in an *N*-dimensional information space (see Gray et al. 1997, Figure 2). The data cube can be pre-computed and stored in its entirety, computed "on-the-fly" only when requested, or partially pre-computed and stored (see Harinarayan, Rajaman and Ullman 1996). The data cube can support standard OLAP operations including roll-up, drill-down, slice, dice, and pivot on measures computed by different aggregation operators, such as max, min, average, top-10, variance, and so on.

1.2.3 The KDD Process and Data Mining

The KDD process usually consists of several steps, namely, data selection, data pre-processing, data enrichment, data reduction and projection, data mining, and pattern interpretation and reporting. These steps may not necessarily be executed in linear order. Stages may be skipped or revisited. Ideally, KDD should be a human-centered process based on the available data, the desired knowledge, and the intermediate results obtained during the process (see Adriaans and Zantinge 1996; Brachman and Anand 1996; Fayyad, Piatetsky-Shapiro and Smyth 1996; Han and Kamber 2006; Matheus, Chan and Piatetsky-Shapiro 1993).

Data selection refers to determining a subset of the records or variables in a database for knowledge discovery. Particular records or attributes are chosen as foci for concentrating the data-mining activities. Automated data reduction or "focusing" techniques are also available (see Barbara et al. 1997, Reinartz 1999). *Data pre-processing* involves "cleaning" the selected data to remove noise, eliminating duplicate records, and determining strategies for handling missing data fields and domain violations. The pre-processing step may also include *data enrichment* through combining the selected data with other, external data (e.g., census data, market data). *Data reduction and projection* concerns both dimensionality and numerosity reductions to further reduce the number of attributes (or tuples) or transformations to determine equivalent but more efficient representations of the information space. Smaller, less redundant and more efficient representations enhance the effectiveness of the *data-mining* stage that attempts to uncover the information (interesting patterns) in these representations. The *interpretation and reporting* stage involves evaluating, understanding, and communicating the information discovered in the data-mining stage.

Data mining refers to the application of low-level functions for revealing hidden information in a database (Klösgen and Żytkow 1996). The type of knowledge to be mined determines the data-mining function to be applied (Han and Kamber 2006). Table 1.1 provides a possible classification of data-mining tasks and techniques. See Matheus, Chan and Piatetsky-Shapiro (1993) and Fayyad, Piatetsky-Shapiro and

TABLE 1.1
Data-Mining Tasks and Techniques

Knowledge Type	Description	Techniques
Segmentation or clustering	Determining a finite set of implicit groups that describe the data.	Cluster analysis
Classification	Predict the class label that a set of data belongs to based on some training datasets	Bayesian classification Decision tree induction Artificial neural networks Support vector machine (SVM)
Association	Finding relationships among itemsets or association/correlation rules, or predict the value of some attribute based on the value of other attributes	Association rules Bayesian networks
Deviations	Finding data items that exhibit unusual deviations from expectations	Clustering and other data-mining methods Outlier detection Evolution analysis
Trends and regression analysis	Lines and curves summarizing the database, often over time	Regression Sequential pattern extraction
Generalizations	Compact descriptions of the data	Summary rules Attribute-oriented induction

Smyth (1996), as well as several of the chapters in this current volume for other overviews and classifications of data-mining techniques.

Segmentation or *clustering* involves partitioning a selected set of data into meaningful groupings or classes. It usually applies cluster analysis algorithms to examine the relationships between data items and determining a finite set of implicit classes so that the intraclass similarity is maximized and interclass similarity is minimized. The commonly used data-mining technique of *cluster analysis* determines a set of classes and assignments to these classes based on the relative proximity of data items in the information space. Cluster analysis methods for data mining must accommodate the large data volumes and high dimensionalities of interest in data mining; this usually requires statistical approximation or heuristics (see Farnstrom, Lewis and Elkan 2000). *Bayesian classification* methods, such as AutoClass, determine classes and a set of weights or class membership probabilities for data items (see Cheesman and Stutz 1996).

Classification refers to finding rules or methods to assign data items into preexisting classes. Many classification methods have been developed over many years of research in statistics, pattern recognition, machine learning, and data mining, including decision tree induction, naïve Bayesian classification, neural networks, support vector machines, and so on. *Decision* or *classification trees* are hierarchical rule sets that generate an assignment for each data item with respect to a set of known classes. Entropy-based methods such as ID3 and C4.5 (Quinlan 1986, 1992) derive these classification rules from training examples. Statistical methods include the chi-square automatic interaction detector (CHAID) (Kass 1980) and the classification and regression tree (CART) method (Breiman et al. 1984). Artificial neural networks (ANNs) can be used as nonlinear clustering and classification techniques. Unsupervised ANNs such as Kohonen Maps are a type of neural clustering where weighted connectivity after training reflects proximity in information space of the input data (see Flexer 1999). Supervised ANNs such as the well-known feed forward/back propagation architecture require supervised training to determine the appropriate weights (response function) to assign data items into known classes.

Associations are rules that predict the object relationships as well as the value of some attribute based on the value of other attributes (Ester, Kriegel and Sander 1997). *Bayesian networks* are graphical models that maintain probabilistic dependency relationships among a set of variables. These networks encode a set of conditional probabilities as directed acyclic networks with nodes representing variables and arcs extending from cause to effect. We can infer these conditional probabilities from a database using several statistical or computational methods depending on the nature of the data (see Buntine 1996; Heckerman 1997). *Association rules* are a particular type of dependency relationship. An association rule is an expression $X \Rightarrow Y$ ($c\%$, $r\%$) where X and Y are disjoint sets of items from a database, $c\%$ is the *confidence* and $r\%$ is the *support*. Confidence is the proportion of database transactions containing X that also contain Y; in other words, the conditional probability $P(Y|X)$. Support is proportion of database transactions that contain X and Y, i.e., the union of X and Y, $P(X \cup Y)$ (see Hipp, Güntzer and Nakhaeizadeh 2000). Mining association rules is a difficult problem since the number of potential rules is exponential with respect to the number of data items. Algorithms for mining association rules typically use breadth-first or depth-first search with branching rules based on minimum confidence or support thresholds (see Agrawal et al. 1996; Hipp, Güntzer and Nakhaeizadeh 2000).

Deviations are data items that exhibit unexpected deviations or differences from some norm. These cases are either errors that should be corrected/ignored or represent unusual cases that are worthy of additional investigation. Outliers are often a byproduct of other data-mining methods, particularly cluster analysis. However, rather than treating these cases as "noise," special-purpose outlier detection methods search for these unusual cases as signals conveying valuable information (see Breuing et al. 1999).

Trends are lines and curves fitted to the data, including linear and logistic regression analysis, that are very fast and easy to estimate. These methods are often combined with filtering techniques such as stepwise regression. Although the data often violate the stringent regression assumptions, violations are less critical if the estimated model is used for prediction rather than explanation (i.e., estimated parameters are not used to explain the phenomenon). *Sequential pattern extraction* explores time series data looking for temporal correlations or pre-specified patterns (such as curve shapes) in a single temporal data series (see Agrawal and Srikant 1995; Berndt and Clifford 1996).

Generalization and characterization are compact descriptions of the database. As the name implies, *summary rules* are a relatively small set of logical statements that condense the information in the database. The previously discussed classification and association rules are specific types of summary rules. Another type is a *characteristic rule*; this is an assertion that data items belonging to a specified concept have stated properties, where "concept" is some state or idea generalized from particular instances (Klösgen and Żytkow 1996). An example is "all professors in the applied sciences have high salaries." In this example, "professors" and "applied sciences" are high-level concepts (as opposed to low-level measured attributes such as "assistant professor" and "computer science") and "high salaries" is the asserted property (see Han, Cai and Cercone 1993).

A powerful method for finding many types of summary rules is *attribute-oriented induction* (also known as *generalization-based mining*). This strategy performs hierarchical aggregation of data attributes, compressing data into increasingly generalized relations. Data-mining techniques can be applied at each level to extract features or patterns at that level of generalization (Han and Fu 1996). Background knowledge in the form of a *concept hierarchy* provides the logical map for aggregating data attributes. A concept hierarchy is a sequence of mappings from low-level to high-level concepts. It is often expressed as a tree whose leaves correspond to measured attributes in the database with the root representing the null descriptor ("any"). Concept hierarchies can be derived from experts or from data cardinality analysis (Han and Fu 1996).

A potential problem that can arise in a data-mining application is the large number of patterns generated. Typically, only a small proportion of these patterns will encapsulate interesting knowledge. The vast majority may be trivial or irrelevant. A data-mining engine should present only those patterns that are interesting to particular users. *Interestingness measures* are quantitative techniques that separate interesting patterns from trivial ones by assessing the simplicity, certainty, utility, and novelty of the generated patterns (Silberschatz and Tuzhilin 1996; Tan, Kumar and Srivastava 2002). There are many interestingness measures in the literature; see Han and Kamber (2006) for an overview.

1.2.4 Visualization and Knowledge Discovery

KDD is a complex process. The mining metaphor is appropriate — information is buried deeply in a database and extracting it requires skilled application of an intensive and complex suite of extraction and processing tools. Selection, pre-processing, mining, and reporting techniques must be applied in an intelligent and thoughtful manner based on intermediate results and background knowledge. Despite attempts at quantifying concepts such as interestingness, the KDD process is difficult to automate. KDD requires a high-level, most likely human, intelligence at its center (see Brachman and Anand 1996).

Visualization is a powerful strategy for integrating high-level human intelligence and knowledge into the KDD process. The human visual system is extremely effective at recognizing patterns, trends, and anomalies. The visual acuity and pattern spotting capabilities of humans can be exploited in many stages of the KDD process, including OLAP, query formulation, technique selection, and interpretation of results. These capabilities have yet to be surpassed by machine-based approaches.

Keim and Kriegel (1994) and Lee and Ong (1996) describe software systems that incorporate visualization techniques for supporting database querying and data mining. Keim and Kriegel (1994) use visualization to support simple and complex query specification, OLAP, and querying from multiple independent databases. Lee and Ong's (1996) WinViz software uses multidimensional visualization techniques to support OLAP, query formulation, and the interpretation of results from unsupervised (clustering) and supervised (decision tree) segmentation techniques. Fayyad, Grinstein and Wierse (2001) provide a good overview of visualization methods for data mining.

1.3 GEOGRAPHIC DATA MINING AND KNOWLEDGE DISCOVERY

This section of the chapter describes a very important special case of KDD, namely, GKD. We will first discuss why GKD is an important special case that requires careful consideration and specialized tools. We will then discuss GDW and online geographic data repositories, the latter an increasingly important source of digital geo-referenced data and imagery. We then discuss geographic data-mining techniques and the relationships between GKD and *geographic visualization* (GVis), an increasingly active research domain integrating scientific visualization and cartography. We follow this with discussions of current GKD techniques and applications and research frontiers, highlighting the contributions of this current volume.

1.3.1 Why Geographic Knowledge Discovery?

1.3.1.1 Geographic Information in Knowledge Discovery

The digital geographic data explosion is not much different from similar revolutions in marketing, biology, and astronomy. Is there anything special about geographic data that requires unique tools and provides unique research challenges? In this section, we identify and discuss some of the unique properties of geographic data and challenges in GKD.

Geographic measurement frameworks. While many information domains of interest in KDD are high dimensional, these dimensions are relatively independent. Geographic information is not only high dimensional but also has the property that up to four dimensions of the information space are interrelated and provide the measurement framework for all other dimensions. Formal and computational representations of geographic information require the adoption of an implied topological and geometric measurement framework. This framework affects measurement of the geographic attributes and consequently the patterns that can be extracted (see Beguin and Thisse 1979; Miller and Wentz 2003).

The most common framework is the topology and geometry consistent with Euclidean distance. Euclidean space fits in well with our experienced reality and results in maps and cartographic displays that are useful for navigation. However, geographic phenomena often display properties that are consistent with other topologies and geometries. For example, travel time relationships in an urban area usually violate the symmetry and triangular inequality conditions for Euclidean and other distance metrics. Therefore, seeking patterns and trends in transportation systems (such as congestion propagation over space and time) benefits from projecting the data into an information space whose spatial dimensions are nonmetric. In addition, disease patterns in space and time often behave according to topologies and geometries other than Euclidean (see Cliff and Haggett 1998; Miller and Wentz 2003). The useful information implicit in the geographic measurement framework is ignored in many induction and machine learning tools (Gahegan 2000a).

An extensive toolkit of analytical cartographic techniques is available for estimating appropriate distance measures and projecting geographic information into that measurement framework (see Cliff and Haggett 1998; Gatrell 1983; Müller 1982; Tobler 1994). The challenge is to incorporate scalable versions of these tools into GKD. Cartographic transformations can serve a similar role in GKD as data reduction and projection in KDD, i.e., determining effective representations that maximize the likelihood of discovering interesting geographic patterns in a reasonable amount of time.

Spatial dependency and heterogeneity. Measured geographic attributes usually exhibit the properties of *spatial dependency* and *spatial heterogeneity*. Spatial dependency is the tendency of attributes at some locations in space to be related.* These locations are usually proximal in Euclidean space. However, direction, connectivity, and other geographic attributes (e.g., terrain, land cover) can also affect spatial dependency (see Miller and Wentz 2003; Rosenberg 2000). Spatial dependency is similar to but more complex than dependency in other domains (e.g., serial autocorrelation in time series data).

Spatial heterogeneity refers to the nonstationarity of most geographic processes. An intrinsic degree of uniqueness at all geographic locations means that most geographic processes vary by location. Consequently, global parameters estimated from a geographic database do not describe well the geographic phenomenon at any particular location. This is often manifested as apparent parameter drift across space when the model is re-estimated for different geographic subsets.

Spatial dependency and spatial heterogeneity have historically been regarded as nuisances confounding standard statistical techniques that typically require independence and stationarity assumptions. However, these can also be valuable sources of information about the geographic phenomena under investigation. Increasing availability of digital cartographic structures and geoprocessing capabilities has led to many recent breakthroughs in measuring and capturing these properties (see Fotheringham and Rogerson 1993).

Traditional methods for measuring spatial dependency include tests such as Moran's I or Geary's C. The recognition that spatial dependency is also subject to spatial heterogeneity effects has led to the development of *local indicators of spatial analysis* (LISA) statistics that disaggregate spatial dependency measures by

* In spatial analysis, this meaning of spatial dependency is more restrictive than its meaning in the GKD literature. Spatial dependency in GKD is a rule that has a spatial predicate in either the precedent or antecedent. We will use the term "spatial dependency" for both cases with the exact meaning apparent from the context. This should not be too confusing since the GKD concept is a generalization of the concept in spatial analysis.

location. Examples include the Getis and Ord G statistic and local versions of the I and C statistics (see Anselin 1995; Getis and Ord 1992, 1996).

One of the problems in measuring spatial dependency in very large datasets is the computational complexity of spatial dependency measures and tests. In the worst case, spatial autocorrelation statistics are approximately $O(n^2)$ in complexity, since $n(n-1)$ calculations are required to measure spatial dependency in a database with n items (although in practice we can often limit the measurement to local spatial regions). Scalable analytical methods are emerging for estimating and incorporating these dependency structures into spatial models. Pace and Zou (2000) report an $O(n\log(n))$ procedure for calculating a closed form maximum likelihood estimator of nearest neighbor spatial dependency. Another complementary strategy is to exploit parallel computing architectures and cyber-infrastructure. Fortunately, many spatial analytic techniques can be decomposed into parallel and distributed computations due to either task parallelism in the calculations or parallelism in the spatial data (see Armstrong and Marciano 1995; Armstrong, Pavlik and Marciano 1994; Densham and Armstrong 1998; Ding and Densham 1996; Griffith 1990; Guan, Zhang and Clarke 2006).

Spatial analysts have recognized for quite some time that the regression model is misspecified and parameter estimates are biased if spatial dependency effects are not captured. Methods are available for capturing these effects in the structural components, error terms, or both (see Anselin 1993; Bivand 1984). Regression parameter drift across space has also been long recognized. Geographically weighted regression uses location-based kernel density estimation to estimate location-specific regression parameters (see Brunsdon, Fotheringham and Charlton 1996; Fotheringham, Charlton and Brunsdon 1997).

The complexity of spatiotemporal objects and rules. Spatiotemporal objects and relationships tend to be more complex than the objects and relationships in nongeographic databases. Data objects in nongeographic databases can be meaningfully represented as points in information space. Size, shape, and boundary properties of geographic objects often affect geographic processes, sometimes due to measurement artifacts (e.g., recording flow only when it crosses some geographic boundary). Relationships such as distance, direction, and connectivity are more complex with dimensional objects (see Egenhofer and Herring 1994; Okabe and Miller 1996; Peuquet and Ci-Xiang 1987). Transformations among these objects over time are complex but information bearing (Hornsby and Egenhofer 2000). Developing scalable tools for extracting spatiotemporal rules from collections of diverse geographic objects is a major GKD challenge.

In their update of Chapter 2 from the first edition of this book, Roddick and Lees discuss the types and properties of spatiotemporal rules that can describe geographic phenomena. In addition to spatiotemporal analogs of generalization, association, and segmentation rules, there are evolutionary rules that describe changes in spatial entities over time. They also note that the scales and granularities for measuring time in geography can be complex, reducing the effectiveness of simply "dimensioning up" geographic space to include time. Roddick and Lees suggest that geographic phenomena are so complex that GKD may require *meta-mining*, that is, mining large rule sets that have been mined from data to seek more understandable information.

Diverse data types. The range of digital geographic data also presents unique challenges. One aspect of the digital geographic information revolution is that geographic databases are moving beyond the well-structured vector and raster formats. Digital geographic databases and repositories increasingly contain ill-structured data such as imagery and geo-referenced multimedia (see Câmara and Raper 1999). Discovering geographic knowledge from geo-referenced multimedia data is a more complex sibling to the problem of knowledge discovery from multimedia databases and repositories (see Petrushin and Khan 2006).

1.3.1.2 Geographic Knowledge Discovery in Geographic Information Science

There are unique needs and challenges for building GKD into geographic information systems (GIS). Most GIS databases are "dumb." They are at best a very simple representation of geographic knowledge at the level of geometric, topological, and measurement constraints. Knowledge-based GIS is an attempt to capture high-level geographic knowledge by storing basic geographic facts and geographic rules for deducing conclusions from these facts (see Srinivasan and Richards 1993; Yuan 1997). The semantic web and semantic geospatial web attempt to make information understandable to computers to support interoperability, findability, and usability (Bishr 2007; Egenhofer 2002).

GKD is a potentially rich source of geographic facts. A research challenge is building discovered geographic knowledge into geographic databases and models to support information retrieval, interoperability, spatial analysis, and additional knowledge discovery. This is critical; otherwise, the geographic knowledge obtained from the GKD process may be lost to the broader scientific and problem-solving processes.

1.3.1.3 Geographic Knowledge Discovery in Geographic Research

Geographic information has always been the central commodity of geographic research. Throughout much of its history, the field of geography has operated in a data-poor environment. Geographic information was difficult to capture, store, and integrate. Most revolutions in geographic research have been fueled by a technological advancement for geographic data capture, referencing, and handling, including the map, accurate clocks, satellites, GPS, and GIS. The current explosion of digital geographic and geo-referenced data is the most dramatic shift in the information environment for geographic research in history.

Despite the promises of GKD in geographic research, there are some cautions. In Chapter 2, Roddick and Lees note that KDD and data-mining tools were mostly developed for applications such as marketing where the standard of knowledge is "what works" rather than "what is authoritative." The question is how to use GKD as part of a defensible and replicable scientific process. As discussed previously in this chapter, knowledge discovery fits most naturally into the initial stages of hypothesis formulation. Roddick and Lees also suggest a strategy where data mining is used as a tool for gathering evidences that strengthen or refute the null hypotheses consistent with a conceptual model. These null hypotheses are kinds of focusing techniques that constrain the search space in the GKD process. The results will be more acceptable to the scientific community since the likelihood of accepting spurious patterns is reduced.

1.3.2 Geographic Data Warehousing

A change since the publication of the first edition of this book in 2001 is the dramatic rise of the geographic information market, particular with respect to web-mapping services and mobile applications. This has generated a consequent heightened interest in GDW.

A GDW involves complexities that are unique to standard DWs. First is the sheer size. GDWs are potentially much larger than comparable nongeographic DWs. Consequently, there are stricter requirements for scalability. Multidimensional GDW design is more difficult because the spatial dimension can be measured using nongeometric, nongeometric generalized from geometric, and fully geometric scales. Some of the geographic data can be ill structured, for example remotely sensed imagery and other graphics. OLAP tools such as roll-up and drill-down require aggregation of spatial objects and summarizing spatial properties. Spatial data interoperability is critical and particularly challenging because geographic data definitions in legacy databases can vary widely. Metadata management is more complex, particularly with respect to aggregated and fused spatial objects.

In Chapter 3, also an update from the first edition, Bédard and Han provide an overview of fundamental concepts underlying DW and GDW. After discussing key concepts of nonspatial data warehousing, they review the particularities of GDW, which are typically spatiotemporal. They also identify frontiers in GDW research and development.

A *spatial data cube* is the GDW analog to the data cube tool for computing and storing all possible aggregations of some measure in OLAP. The spatial data cube must include standard attribute summaries as well as pointers to spatial objects at varying levels of aggregation. Aggregating spatial objects is nontrivial and often requires background domain knowledge in the form of a geographic concept hierarchy. Strategies for selectively pre-computing measures in the spatial data cube include none, pre-computing rough approximations (e.g., based on minimum bounding rectangles), and selective pre-computation (see Han, Stefanovic and Koperski 2000).

In Chapter 4, Lu, Boedihardjo, and Shekhar update a discussion of the *map cube* from the first edition. The map cube extends the data cube concept to GDW. The map cube operator takes as arguments a base map, associated data files, a geographic aggregation hierarchy, and a set of cartographic preferences. The operator generates an album of maps corresponding to the power set of all possible spatial and nonspatial aggregations. The map collection can be browsed using OLAP tools such as roll-up, drill-down, and pivot using the geographic aggregation hierarchy. They illustrate the map cube through an application to highway traffic data.

GDW incorporates data from multiple sources often collected at different times and using different techniques. An important concern is the quality or the reliability of the data used for GKD. While error and uncertainty in geographic information have been long-standing concerns in the GIScience community, efforts to address these issues have increased substantially since the publication of the first edition of this book in 2001 (Goodchild 2004).

Chapter 5 by Gervais, Bédard, Levesque, Bernier, and DeVillers is a new contribution to the second edition that discusses data quality issues in GKD. The authors identify major concepts regarding quality and risk management with respect to GDW and spatial OLAP. They discuss possible management mechanisms to improve the prevention of inappropriate usages of data. Using this as a foundation, Chapter 5 presents a pragmatic approach of quality and risk management to be applied during the various stages of a spatial data cube design and development. This approach manages the potential risks one may discover during this development process.

1.3.3 Geographic Data Mining

Many of the traditional data-mining tasks discussed previously in this chapter have analogous tasks in the geographic data-mining domain. See Ester, Kriegel and Sander (1997) and Han and Kamber (2006) for overviews. Also, see Roddick and Spiliopoulou (1999) for a useful bibliography of spatiotemporal data-mining research. The volume of geographic data combined with the complexity of spatial data access and spatial analytical operations implies that scalability is particularly critical.

1.3.3.1 Spatial Classification and Capturing Spatial Dependency

Spatial classification builds up classification models based on a relevant set of attributes and attribute values that determine an effective mapping of spatial objects into predefined target classes. Ester, Kriegel and Sander (1997) present a learning algorithm based on ID3 for generating spatial classification rules based on the properties of each spatial object as well as spatial dependency with its neighbors. The user provides a maximum spatial search length for examining spatial dependency relations with each object's neighbors. Adding a rule to the tree requires meeting a minimum information gain threshold.

Geographic data mining involves the application of computational tools to reveal interesting patterns in objects and events distributed in geographic space and across time. These patterns may involve the spatial properties of individual objects and events (e.g., shape, extent) and spatiotemporal relationships among objects and events in addition to the nonspatial attributes of interest in traditional data mining. As noted above, ignoring spatial dependency and spatial heterogeneity effects in geographic data can result in misspecified models and erroneous conclusions. It also ignores a rich source of potential information.

In Chapter 6, also an updated chapter from the first edition, Shekhar, Vatsavai and Chawla discuss the effects of spatial dependency in spatial classification and prediction techniques. They discuss and compare the aspatial techniques of logistic regression and Bayesian classification with the spatial techniques of spatial autoregression and Markov random fields. Theoretical and experimental results suggest that the spatial techniques outperform the traditional methods with respect to accuracy and handling "salt and pepper" noise in the data.

Difficulties in accounting for spatial dependency in geographic data mining include identifying the spatial dependency structure, the potential combinatorial explosion in the size of these structures and scale-dependency of many dependency measures. Further research is required along all of these frontiers. As noted above, researchers report promising results with parallel implementations of the Getis-Ord G statistic. Continued work on implementations of spatial analytical techniques and spatial data-mining tools that exploit parallel and cyber infrastructure environments can complement recent work on parallel processing in standard data mining (see Zaki and Ho 2000).

1.3.3.2 Spatial Segmentation and Clustering

Spatial clustering groups spatial objects such that objects in the same group are similar and objects in different groups are unlike each other. This generates a small set of implicit classes that describe the data. Clustering can be based on combinations of nonspatial attributes, spatial attributes (e.g., shape), and proximity of the objects or events in space, time, and space–time. Spatial clustering has been a very active research area in both the spatial analytic and computer science literatures. Research on the spatial analytic side has focused on theoretical conditions for appropriate clustering in space–time (see O'Kelly 1994; Murray and Estivill-Castro 1998). Research on the computer science side has resulted in several scalable algorithms for clustering very large spatial datasets and methods for finding proximity relationships between clusters and spatial features (Knorr and Ng 1996; Ng and Han 2002).

In Chapter 7, Han, Lee, and Kamber update a review of major spatial clustering methods recently developed in the data-mining literature. The first part of their chapter discusses spatial clustering methods. They classify spatial clustering methods into four categories, namely, partitioning, hierarchical, density-based, and grid-based. Although traditional *partitioning methods* such as k-means and k-medoids are not scalable, scalable versions of these tools are available (also see Ng and Han 2002). *Hierarchical methods* group objects into a tree-like structure that progressively reduces the search space. *Density-based methods* can find arbitrarily shaped clusters by growing from a seed as long as the density in its neighborhood exceeds certain thresholds. *Grid-based methods* divide the information spaces into a finite number of grid cells and cluster objects based on this structure.

The second part of Chapter 7 discusses clustering techniques for trajectory data, that is, data collected on phenomena that changes geographic location frequently with respect to time. As noted above, these data have become more prevalent since the publication of the first edition of this book; this section of the chapter is new material relative to the first edition. Although clustering techniques for trajectory data are not as well developed as purely spatial clustering techniques, there are two major types based on whether they cluster whole trajectories or can discover sub-trajectory clusters. *Probabilistic methods* use a regression mixture model to cluster entire trajectories, while *partition-and-group* methods can discover clusters involving sub-trajectories.

Closely related to clustering techniques are *medoid queries*. A medoid query selects points in a dataset (known as medoids) such that the average Euclidean distance between the remaining points and their closest medoid is minimized. The resulting assignments of points to medoids are clusters of the original spatial data, with the medoids being a compact description of each cluster. Medoids also can be interpreted as facility locations in some problem contexts (see Murray and Estivill-Castro 1998). Mouratidis, Papadias, and Papadimitriou discuss medoids in Chapter 8.

1.3.3.3 Spatial Trends

Spatial trend detection involves finding patterns of change with respect to the neighborhood of some spatial object. Ester, Kriegel and Sander (1997) provide a neighborhood search algorithm for discovering spatial trends. The procedure performs a breadth-first search along defined neighborhood connectivity paths and evaluates a statistical model at each step. If the estimated trend is strong enough, then the neighborhood path is expanded in the next step.

In Chapter 9, a new chapter solicited for the second edition of this book, Fotheringham, Charlton, and Demšar describe the use of geographically weighted regression (GWR) as an exploratory technique. Traditional regression assumes that the relationships between dependent and independent variables are spatially constant across the study area. GWR allows the analyst to model the spatial heterogeneity and seek evidence whether the nonstationarity found is systematic or noise. This allows the analyst to ask additional questions about the structures in the data. GWR is also a technique that benefits greatly from GVis, and Fotheringham, Charlton, and Demšar use GVis analytics to examine some of the interactions in the GWR parameter surfaces and highlight local areas of interest.

1.3.3.4 Spatial Generalization

Geographic phenomena often have complex hierarchical dependencies. Examples include city systems, watersheds, location and travel choices, administrative regions, and transportation/telecommunications systems. *Spatial characterization and generalization* is therefore an important geographic data-mining task. Generalization-based data mining can follow one of two strategies in the geographic case. *Spatial dominant generalization* first spatially aggregates the data based on a user-provided geographic concept hierarchy. A standard attribute-oriented induction method is used at each geographic aggregation level to determine compact descriptions or patterns of each region. The result is a description of the pre-existing regions in the hierarchy using high-level predicates. *Nonspatial dominant generalization* generates aggregated spatial units that share the same high-level description. Attribute-oriented induction is used to aggregate nonspatial attributes into higher-level concepts. At each level in the resulting concept hierarchy, neighboring geographic units are merged if they share the same high-level description. The result is a geographic aggregation hierarchy based on multidimensional information. The extracted aggregation hierarchy for a particular geographic setting could be used to guide the application of confirmatory spatial analytic techniques to the data about that area.

1.3.3.5 Spatial Association

Mining for *spatial association* involves finding rules to predict the value of some attribute based on the value of other attributes, where one or more of the attributes are spatial properties. *Spatial association rules* are association rules that include spatial predicates in the precedent or antecedent. Spatial association rules also have confidence and support measures. Spatial association rules can include a variety of spatial predicates, including topological relations such as "inside" and "disjoint," as well as distance and directional relations. Koperski and Han (1995) provide a detailed discussion of the properties of spatial association rules. They also present a top-down search technique that starts at the highest level of a geographic concept hierarchy (discussed later), using spatial approximations (such as minimum bounding rectangles) to discover rules with large support and confidence. These rules form the basis for additional search at lower levels of the geographic concept hierarchy with more detailed (and computationally intensive) spatial representations.

Chapter 10 by Malerba, Lanza, and Appice discusses INGENS 2.0, a prototype GIS that incorporates spatial data-mining techniques. Malerba and his co-authors reported on INGENS in the first edition of this book; their updated chapter indicates the progress that has been made on this software since 2001. INGENS is a Web-based, open, extensible architecture that integrates spatial data-mining techniques within a GIS environment. The current system incorporates an inductive learning algorithm that generates models of geographic objects from training examples and counter-examples as well as a system that discovers spatial association rules at multiple hierarchical levels. The authors illustrate the system through application to a topographic map repository.

1.3.4 Geovisualization

Earlier in this chapter, we noted the potential for using visualization techniques to integrate human visual pattern acuity and knowledge into the KDD process. Geographic visualization (GVis) is the integration of cartography, GIS, and scientific visualization to explore geographic data and communicate geographic information to private or public audiences (see MacEachren and Kraak 1997). Major GVis tasks include *feature identification*, *feature comparison*, and *feature interpretation* (MacEachren et al. 1999).

GVis is related to GKD since it often involves an iterative, customized process driven by human knowledge. However, the two techniques can greatly complement each other. For example, feature identification tools can allow the user to spot the emergence of spatiotemporal patterns at different levels of spatial aggregation and explore boundaries between spatial classes. Feature identification and comparison GVis tools can also guide spatial query formulation. Feature interpretation can help the user build geographic domain knowledge into the construction of geographic concept hierarchies. MacEachren et al. (1999) discuss these functional objects and a prototype GVis/GKD software system that achieves many of these goals.

MacEachren et al. (1999) suggest that integration between GVis and GKD should be considered at three levels. The conceptual level requires specification of the high-level goals for the GKD process. Operational-level decisions include specification of appropriate geographic data-mining tasks for achieving the high-level goals. Implementation level choices include specific tools and algorithms to meet the operational-level tasks.

In Chapter 11, Gahegan updates his chapter from the first edition and argues that portraying geographic data in a form that a human can understand frees exploratory spatial analysis (ESA) from some of the representational constraints that GIS and geographic data models impose. When GVis techniques fulfill their potential, they are not simply display technologies by which users gain a familiarity with new datasets or look for trends and outliers. Instead, they are environments that facilitate the discovery of new geographical concepts and processes and the formulation of new geographical questions. The visual technologies and supporting science are based on a wide range of scholarly fields, including information visualization, data mining, geography, human perception and cognition, machine learning, and data modeling.

Chapter 12 by Guo is a new chapter solicited for the second edition. In this chapter, Guo introduces an integrated approach to multivariate analysis and GVis. An integrated suite of techniques consists of methods that are visual and computational as well as complementary and competitive. The complementary methods examine data from different perspectives and provide a synoptic view of the complex patterns. The competitive methods validate and crosscheck each other. The integrated approach synthesizes information from different perspectives, but also leverages the power of computational tools to accommodate larger data sets than typical with visual methods alone.

1.3.5 Spatiotemporal and Mobile Objects Databases

Perhaps the most striking change in GKD and data mining since the publication of the first edition of this book in 2001 is the rise of spatiotemporal and mobile objects databases. The development and deployment of LATs and geosensor networks are creating an explosion of data on dynamic and mobile geographic phenomena, with a consequent increase in the potential to discover new knowledge about dynamic and mobile phenomena.

LATs are devices that can report their geographic location in near-real time. LATs typically exploit one or more georeferencing strategies, including radiolocation methods, GPS, and interpolation from known locations (Grejner-Brzezinska 2004). An emerging LAT is radiofrequency identification (RFID) tags. RFID tags are cheap and light devices attached to objects and transmit data to fixed readers using passive or active methods (Morville 2005).

LATs enable location-based services (LBS) that provide targeted information to individuals based on their geographic location though wireless communication networks and devices such as portable computers, PDAs, mobile phones, and in-vehicle navigation systems (Benson 2001). Services include emergency response, navigation, friend finding, traffic information, fleet management, local news, and concierge services (Spiekermann 2004). LBS are widely expected to be the “killer application” for wireless Internet devices; some predict worldwide deployment levels reaching one billion devices by 2010 (Bennahum 2001; Smyth 2001).

Another technology that can capture data on spatiotemporal and mobile phenomena is *geosensor networks*. These are interconnected, communicating, and georeferenced computing devices that monitor a geographic environment. The geographic scales monitored can range from a single room to an entire city or ecosystem. The devices are typically heterogeneous, ranging from temperature and humidity sensors to video cameras and other imagery capture devices. Geosensor networks can also capture the evolution of the phenomenon or environment over time. Geosensor networks can provide fixed stations for tracking individual vehicles, identify traffic patterns, and determine possible stops for a vehicle as it travels across a given domain in the absence of mobile technologies such as GPS or RFID (Stefanidis 2006; Stefanidis and Nittel 2004).

In the first edition of this book, we included only one chapter dedicated to mining trajectory data (Smyth 2001). Recognizing the growth in mobile technologies and trajectory data, the second edition includes five new chapters on knowledge discovery from spatiotemporal and mobile objects databases.

In Chapter 13, Yuan proposes spatiotemporal constructs and a conceptual framework to lead knowledge discovery about geographic dynamics beyond what is directly recorded in spatiotemporal databases. Recognizing the central role of data representation in GKD, the framework develops geographic constructs at a higher level of conceptualization than location and geometry. For example, higher-level background knowledge about the phenomena in question can enhance the interpretation of an observed spatiotemporal pattern. Yuan's premise is that activities, events, and processes are general spatiotemporal constructs of geographic dynamics. Therefore, knowledge discovery about geographic dynamics ultimately aims to synthesize information about activities, events, or processes, and through this synthesis to obtain patterns and rules about their behaviors, interactions, and effects.

Chapter 14 by Wachowicz, Macedo, Renso, and Ligtenberg also addresses the issue of higher-level concepts to support spatiotemporal knowledge discovery. The authors note that although discovering spatiotemporal patterns in large databases is relatively easy, establishing their relevance and explaining their causes are very difficult. Solving these problems requires viewing knowledge discovery as a multitier process, with more sophisticated reasoning modes used to help us understand what makes patterns structurally and meaningfully different from another. Chapter 14 proposes a multitier ontological framework consisting of domain, application, and data ontology tiers. Their approach integrates knowledge representation and data representation in the knowledge discovery process.

In Chapter 15, Cao, Mamoulis, and Cheung focus on discovering knowledge about periodic movements from trajectory data. Discovering periodic patterns (that is, objects following approximately the same routes over regular time intervals) is a difficult problem since these patterns are often not explicitly defined but rather must be discovered from the data. In addition, the objects are not expected to follow the exact patterns but similar ones, making the knowledge discovery process more challenging. Therefore, an effective method needs to discover not only the patterns themselves, but also a description of how they can vary. The authors discuss three algorithms for discovering period motion: an effective but computationally burdensome bottom-up approach and two faster top-down approaches.

Chapter 16 by Laube and Duckham discusses the idea of decentralized spatiotemporal data mining using geosensor networks. In this approach, each sensor-based computing node only possesses local knowledge of its immediate neighborhood. Global knowledge emerges through cooperation and information exchange among network nodes. Laube and Duckham discuss four strategies for decentralized spatial data mining and illustrate their approach using spatial clustering algorithms.

In the final chapter of the book, Kraak and Huisman discuss the *space–time cube* (STC) an interactive environment for the analysis and visualization of spatiotemporal data. Using Hägerstrand's time geographic framework as a conceptual foundation, they illustrate the STC using two examples from the domain of human movement and activities. The first example examines individual movement and the degree to which knowledge can be discovered by linking attribute data to space–time movement data, and demonstrates how the STC can be deployed to query and investigate (individual-level) dynamic processes. The second example draws on the geometry of the STC as an environment for data mining through space–time query and analysis. These two examples provide the basis of a broader discussion regarding the common elements of various disciplines and research areas concerned with moving object databases, dynamics, geocomputation, and GVis.

1.4 CONCLUSION

Due to explosive growth and wide availability of geo-referenced data in recent years, traditional spatial analysis tools are far from adequate at handling the huge volumes of data and the growing complexity of spatial analysis tasks. Geographic data mining and knowledge discovery represent important directions in the development of a new generation of spatial analysis tools in data-rich environment. In this chapter, we introduce knowledge discovery from databases and data mining, with special reference to the applications of these theories and techniques to geo-referenced data.

As shown in this chapter, geographic knowledge discovery is an important and interesting special case of knowledge discovery from databases. Much progress has been made recently in GKD techniques, including heterogeneous spatial data integration, spatial or map data cube construction, spatial dependency and/or association analysis, spatial clustering methods, spatial classification and spatial trend analysis, spatial generalization methods, and GVis tools. Application of data mining and knowledge discovery techniques to spatiotemporal and mobile objects databases is also a rapidly emerging subfield of GKD. However, according to our view, geographic data mining and knowledge discovery is a promising but young discipline, facing many challenging research problems. We hope this book will introduce some recent works in this direction and motivate researchers to contribute to developing new methods and applications in this promising field.

REFERENCES

Adriaans P. and Zantinge, D. (1996) *Data Mining*, Harlow, U.K.: Addison-Wesley.

Agrawal, R., Mannila, H., Srikant, R., Toivonen, H. and Verkamo, A. I. (1996) "Fast discovery of association rules," in U.M. Fayyad, G. Piatetsky-Shapiro, P. Smyth and R. Ulthurusamy (Eds.). *Advances in Knowledge Discovery and Data Mining*, Cambridge, MA: MIT Press, 307–328.

Agrawal, R. and Srikant, R. (1995) "Mining sequential patterns," *Proceedings, 11th International Conference on Data Engineering*, Los Alamitos, CA: IEEE Computer Society Press, 3–14.

Anselin, L. (1993) "Discrete space autoregressive models," in M.F. Goodchild, B.O. Parks and L.T. Steyaert (Eds.). *Environmental Modeling with GIS*, New York: Oxford University Press, 454–469.

Anselin, L. (1995) "Local indicators of spatial association — LISA," *Geographical Analysis*, 27, 93–115.

Armstrong, M. P. and Marciano, R. (1995) "Massively parallel processing of spatial statistics," *International Journal of Geographical Information Systems*, 9, 169–189.

Armstrong, M. P., Pavlik, C.E. and Marciano, R. (1994) "Experiments in the measurement of spatial association using a parallel supercomputer," *Geographical Systems*, 1, 267–288.

Barbara, D., DuMouchel, W., Faloutos, C., Haas, P.J., Hellerstein, J.H., Ioannidis, Y., Jagadish, H.V., Johnson, T., Ng, R., Poosala, V., Ross, K.A. and Servcik, K.C. (1997) "The New Jersey data reduction report," *Bulletin of the Technical Committee on Data Engineering*, 20(4), 3–45.

Beguin, H. and Thisse, J.-F. (1979) "An axiomatic approach to geographical space," *Geographical Analysis*, 11, 325–341.

Bennahum, D.S. (2001) "Be here now," *Wired*, 9.11, 159–163.

Benson, J. (2001) "LBS technology delivers information where and when it's needed," *Business Geographics*, 9(2), 20–22.

Berndt, D.J. and Clifford, J. (1996) "Finding patterns in time series: A dynamic programming approach," in U.M. Fayyad, G. Piatetsky-Shapiro, P. Smyth and R. Ulthurusamy (Eds.) *Advances in Knowledge Discovery and Data Mining*, Cambridge, MA: MIT Press, 229–248.

Bishr, Y. (2007) "Overcoming the semantic and other barriers to GIS interoperability: Seven years on," in P. Fisher (Ed.) *Classics from IJGIS: Twenty Years of the International Journal of Geographical Information Science and Systems*, London: Taylor & Francis, 447–452.

Bivand, R.S. (1984) "Regression modeling with spatial dependence: An application of some class selection and estimation techniques," *Geographical Analysis*, 16, 25–37.

Brachman, R.J. and Anand, T. (1996) "The process of knowledge-discovery in databases: A human-centered approach," in U.M. Fayyad, G. Piatetsky-Shapiro, P. Smyth and R. Ulthurusamy (Eds.) *Advances in Knowledge Discovery and Data Mining*, Cambridge, MA: MIT Press 37–57.

Breiman, L., Friedman, J.H., Olshen, R.A. and Stone, C.J. (1984) *Classification and Regression Trees*, Belmont, CA: Wadsworth.

Breunig, M.M., Kriegel, H.-P., Ng, R.T. and Sander, J. (1999) "OPTICS-OF: Identifying local outliers," in J.M. Żytkow and J. Rauch (Eds.) *Principles of Data Mining and Knowledge Discovery*, Lecture Notes in Artificial Intelligence 1704, 262–270.

Brunsdon, C., Fotheringham, A.S. and Charlton, M.E. (1996) "Geographically weighted regression: A method for exploring spatial nonstationarity," *Geographical Analysis*, 28 281–298.

Buntine, W. (1996) "Graphical models for discovering knowledge," U.M. Fayyad, G. Piatetsky-Shapiro, P. Smyth and R. Ulthurusamy (Eds.) *Advances in Knowledge Discovery and Data Mining*, Cambridge, MA: MIT Press, 59–82.

Câmara, A.S. and Raper, J. (Eds.) (1999) *Spatial Multimedia and Virtual Reality*, London: Taylor & Francis.

Chaudhuri, S. and Dayal, U. (1997) "An overview of data warehousing and OLAP technology," *SIGMOD Record*, 26, 65–74.

Cheesman, P. and Stutz, J. (1996) "Bayesian classification (AutoClass): Theory and results," in U.M. Fayyad, G. Piatetsky-Shapiro, P. Smyth and R. Ulthurusamy (Eds.) *Advances in Knowledge Discovery and Data Mining*, Cambridge, MA: MIT Press, 153–180.

Cliff, A.D. and Haggett, P. (1998) "On complex geographical space: Computing frameworks for spatial diffusion processes," in P.A. Longley, S.M. Brooks, R. McDonnell and B. MacMillan (Eds.) *Geocomputation: A Primer*, Chichester, U.K.: John Wiley & Sons, 231–256.

Densham, P.J. and Armstrong, M.P. (1998) "Spatial analysis," in R. Healy, S. Dowers, B. Gittings and M. Mineter (Eds.) *Parallel Processing Algorithms for GIS*, London: Taylor & Francis, 387–413.

Ding, Y. and Densham, P.J. (1996) "Spatial strategies for parallel spatial modeling," *International Journal of Geographical Information Systems,* 10, 669–698.

Egenhofer, M. (2002) "Toward the semantic geospatial web," Geographic Information Systems: Proceedings of the 10th ACM International Symposium on Advances in Geographic Information Systems, New York: ACM Press, 1–4.

Egenhofer, M.J. and Herring, J.R. (1994) "Categorizing binary topological relations between regions, lines and points in geographic databases," in M. Egenhofer, D.M. Mark and J.R. Herring (Eds.) *The 9-Intersection: Formalism and its Use for Natural-Language Spatial Predicates*, National Center for Geographic Information and Analysis Technical Report 94-1, 1–28.

Elder, J. and Pregibon, D. (1996) "A statistical perspective on knowledge discovery," in U.M. Fayyad, G. Piatetsky-Shapiro, P. Smyth and R. Ulthurusamy (Eds.) *Advances in Knowledge Discovery and Data Mining*, Cambridge, MA: MIT Press, 83–113.

Ester, M., Kriegel, H.-P. and Sander, J. (1997) "Spatial data mining: A database approach," M. Scholl and A. Voisard (Eds.) *Advances in Spatial Databases*, Lecture Notes in Computer Science 1262, Berlin: Springer, 47–66.

Farnstrom, F., Lewis, J. and Elkan, C. (2000) "Scalability for clustering algorithms revisited," *SIGKDD Explorations*, 2, 51–57.

Fayyad, U., Grinstein, G. and Wierse, A. (2001) *Information Visualization in Data Mining and Knowledge Discovery*, San Mateo, CA: Morgan Kaufmann.

Fayyad, U.M., Piatetsky-Shapiro, G. and Smyth, P. (1996) "From data mining to knowledge discovery: An overview" in U.M. Fayyad, G. Piatetsky-Shapiro, P. Smyth and R. Ulthurusamy (Eds.) *Advances in Knowledge Discovery and Data Mining*, Cambridge, MA: MIT Press, 1–34.

Flexer, A. (1999) "On the use of self-organizing maps for clustering and visualization," in J.M. Żytkow and J. Rauch (Eds.) *Principles of Data Mining and Knowledge Discovery*, Lecture Notes in Artificial Intelligence 1704, 80–88.

Fotheringham, A.S., Charlton, M. and Brunsdon, C. (1997) "Two techniques for exploring nonstationarity in geographical data," *Geographical Systems*, 4, 59–82.

Fotheringham, A.S. and Rogerson, P.A. (1993) "GIS and spatial analytical problems," International *Journal of Geographical Information Science*, 7, 3–19.

Gahegan, M. (2000) "On the application of inductive machine learning tools to geographical analysis," *Geographical Analysis*, 32, 113–139.

Gatrell, A.C. (1983) *Distance and Space: A Geographical Perspective*, Oxford: Clarendon Press.

Getis, A. and Ord, J.K. (1992) "The analysis of spatial association by use of distance statistics," *Geographical Analysis,* 24, 189–206.

Getis, A. and Ord, J.K. (1996) "Local spatial statistics: An overview," in P. Longley and M. Batty (Eds.) *Spatial Analysis: Modelling in a GIS Environment*, Cambridge, UK: GeoInformation International, 261–277.

Goodchild, M.F. (2004) "A general framework for error analysis in measurement-based GIS," *Journal of Geographical Systems*, 6, 323–324.

Gray, J., Chaudhuri, S., Bosworth, A., Layman, A., Reichart, D., Venkatrao, M., Pellow, F. and Pirahesh, H. (1997) "Data cube: A relational aggregation operator generalizing group-by, cross-tab and sub-totals," *Data Mining and Knowledge Discovery*, 1, 29–53.

Grejner-Brzezinska, D. (2004) "Positioning and tracking approaches and technologies," in H.A. Karimi and A. Hammad (Eds.) *Telegeoinformatics: Location-Based Computing and Services*, Boca Raton, FL: CRC Press, 69–110.

Griffith, D.A. (1990) "Supercomputing and spatial statistics: A reconnaissance," *Professional Geographer*, 42, 481–492.

Guan, Q., Zhang, T. and Clarke, K.C. (2006) "Geocomputation in the grid computing age," in J.D. Carswell and T. Tezuka (Eds.) *Web and Wireless Geographical Information Systems: 6th International Symposium, W2GIS 2006, Hong Kong, China, December 4-5, 2006, Proceedings,* Berlin: Springer Lecture Notes in Computer Science 4295, 237–246.

Han, J., Cai, Y. and Cercone, N. (1993) "Data-driven discovery of quantitative rules in relational databases," *IEEE Transactions on Knowledge and Data Engineering*, 5, 29–40.

Han, J. and Fu, Y. (1996). "Attribute-oriented induction in data mining." In U.M. Fayyad, G., Piatetsky-Shpiro, P., Sayth, and R. Uthurusamy (eds), *Advances in Knowledge Discovery and Data Mining*, AAAI Press/the MIT Press pp. 399–424.

Han, J. and Kamber, M. (2006), *Data Mining: Concepts and Techniques*, 2nd ed., San Mateo, CA: Morgan Kaufmann.

Han, J., Stefanovic, N. and Koperski, K. (2000) "Object-based selective materialization for efficient implementation of spatial data cubes," *IEEE Trans. Knowledge and Data Engineering*, 12(6), 938–958.

Hand, D.J. (1998) "Data mining: Statistics and more?" *American Statistician,* 52, 112–118.

Harinarayan, V., Rajaramna, A. and Ullman, J.D. (1996) "Implementing data cubes efficiently," *SIGMOD Record*, 25, 205–216.

Heckerman, D. (1997) "Bayesian networks for data mining," *Data Mining and Knowledge Discovery*, 1, 79–119.

Hipp, J., Güntzer, U. and Nakhaeizadeh, G. (2000) "Algorithms for association rule mining: A general survey and comparison," *SIGKDD Explorations*, 2, 58–64.

Hornsby, K. and Egenhofer, M.J. (2000) "Identity-based change: A foundation for spatio-temporal knowledge representation," *International Journal of Geographical Information Science*, 14, 207–224.

Jarke, M., Lenzerini, M., Vassiliou, Y. and Vassiliadis, P. (2000) *Fundamentals of Data Warehouses*, Berlin: Springer.

Kass, G.V. (1980) "An exploratory technique for investigating large quantities of categorical data," *Applied Statistics*, 29, 119–127.

Keim, D.A. and Kriegel, H.-P. (1994) "Using visualization to support data mining of large existing databases," in J.P. Lee and G.G. Grinstein (Eds.) *Database Issues for Data Visualization*, Lecture Notes in Computer Science 871, 210–229.

Klösgen, W. and Żytkow, J.M. (1996) "Knowledge discovery in databases terminology," in U.M. Fayyad, G. Piatetsky-Shapiro, P. Smyth and R. Ulthurusamy (Eds.) *Advances in Knowledge Discovery and Data Mining*, Cambridge, MA: MIT Press 573–592.

Knorr, E.M. and Ng, R.T. (1996) "Finding aggregate proximity relationships and commonalties in spatial data mining," *IEEE Transactions on Knowledge and Data Engineering*, 8, 884–897.

Koperski, K. and Han, J. (1995) "Discovery of spatial association rules in geographic information databases," in M. Egenhofer and J. Herring (Eds.) *Advances in Spatial Databases*, Lecture Notes in Computer Science Number 951, Springer-Verlag, 47–66.

Lee, H.-Y. and Ong, H.-L. (1996) "Visualization support for data mining," *IEEE Expert*, 11(5), 69–75.

MacEachren, A.M. and Kraak, M.-J. (1997) "Exploratory cartographic visualization: Advancing the agenda," *Computers and Geosciences*, 23, 335–343.

MacEachren, A.M., Wachowicz, M., Edsall, R., Haug, D. and Masters, R. (1999) "Constructing knowledge from multivariate spatiotemporal data: Integrating geographic visualization with knowledge discovery in database methods," *International Journal of Geographical Information Science*, 13, 311–334.

Matheus, C.J., Chan, P.K. and Piatetsky-Shapiro, G. (1993) "Systems for knowledge discovery in databases," *IEEE Transactions on Knowledge and Data Engineering*, 5, 903–913.

Miller, H.J. and Wentz, E.A. (2003) "Representation and spatial analysis in geographic information systems," *Annals of the Association of American Geographers*, 93, 574–594.

Morville, P. (2005) *Ambient Findability: What We Find Changes Who We Become*, Sebastopol, CA: O'Reilly Media.

Müller, J.-C. (1982) "Non-Euclidean geographic spaces: Mapping functional distances," *Geographical Analysis*, 14, 189–203.

Murray, A.T. and Estivill-Castro, V. (1998) "Cluster discovery techniques for exploratory data analysis," *International Journal of Geographical Information Science*, 12, 431–443.

National Research Council (1999) *Distributed Geolibraries: Spatial Information Resources*, Washington, D.C.: National Academy Press.

Ng, R.T. and Han, J. (2002) "CLARANS: A method for clustering objects for spatial data mining," *IEEE Transactions on Knowledge and Data Engineering*, 14(5), 1003–1016.

Okabe, A. and Miller, H.J. (1996) "Exact computational methods for calculating distances between objects in a cartographic database," *Cartography and Geographic Information Systems*, 23, 180–195.

O'Kelly, M.E. (1994) "Spatial analysis and GIS," in A.S. Fotheringham and P.A. Rogerson (Eds.) *Spatial Analysis and GIS*, London: Taylor & Francis, 65–79.

Openshaw, S., Charlton, M., Wymer, C. and Craft, A. (1987) "A mark 1 geographical analysis machine for automated analysis of point data sets," *International Journal of Geographical Information Systems*, 1, 335–358.

Pace, R.K. and Zou, D. (2000) "Closed-form maximum likelihood estimates of nearest neighbor spatial dependence," *Geographical Analysis*, 32, 154–172.

Petrushin V.A. and Khan, L. (2006) *Multimedia Data Mining and Knowledge Discovery*, New York: Springer-Verlag.

Peuquet, D.J. and Ci-Xiang, Z. (1987) "An algorithm to determine the directional relationship between arbitrarily-shaped polygons in the plane," *Pattern Recognition*, 20, 65–74.

Quinlan, J.R. (1986) "Induction of decision trees," *Machine Learning*, 1, 81–106.

Quinlan, J.R. (1993) *C4.5 Programs for Machine Learning*, San Matel, CA: Morgan Kaufmann.

Reinartz, T. (1999) *Focusing Solutions for Data Mining*, Lecture Notes in Artificial Intelligence 1623, Berlin: Springer.

Roddick, J.F. and Spiliopoulou, M. (1999) "A bibliography of temporal, spatial and spatio-temporal data mining research," *SIGKDD Explorations*, 1, 34–38.

Rosenberg, M.S. (2000) "The bearing correlogram: A new method of analyzing directional spatial autocorrelation," *Geographical Analysis*, 32, 267–278.

Silberschatz, A., Korth, H.F. and Sudarshan, S. (1997) *Database Systems Concepts*, 3rd ed., New York, NY: McGraw-Hill.

Silberschatz, A. and Tuzhilin, A. (1996) "What makes patterns interesting in knowledge discovery systems," *IEEE Transactions on Knowledge and Data Engineering*, 8, 970–974.

Smyth, C.S. (2001) "Mining mobile trajectories," H.J. Miller and J. Han (Eds.) *Geographic Data Mining and Knowledge Discovery*, London: Taylor & Francis, 337–361.

Spiekerman, S. (2004) "General aspects of location-based services," in J. Schiller and A. Voisard (Eds.) *Location-Based Services*, San Francisco, CA: Morgan Kaufmann, 9–26.

Srinivasan, A. and Richards, J.A. (1993) "Analysis of GIS spatial data using knowledge-based-methods," *International Journal of Geographical Information Systems*, 7, 479–500.

Stefanidis, A. (2006) "The emergence of geosensor networks," *Location Intelligence*, 27 February 2006; http://locationintelligence.net.

Stefanidis, A. and Nittel, S. (Eds.) (2004) *GeoSensor Networks*, Boca Raton, FL: CRC Press.

Tan, P.-N., Kumar, V. and Srivastava, J. (2002) "Selecting the right interestingness measure for association patterns," Proc. 2002 ACM SIGKDD Int. Conf. Knowledge Discovery in Databases (KDD'02), Edmonton, Canada, 32–41.

Tobler, W. R. (1994) Bidimensional regression," Geographical Analysis, 13, 1–20.

Yuan, M. (1997) "Use of knowledge acquisition to build wildfire representation in geographic information systems," *International Journal of Geographical Information Systems,* 11, 723–745

Zaki, M.J. and Ho, C.-T. (Eds.) (2000) *Large-Scale Parallel Data Mining*, Lecture Notes in Artificial Intelligence 1759, Berlin: Springer.

2 Spatio-Temporal Data Mining Paradigms and Methodologies

John F. Roddick

Brian G. Lees

CONTENTS

2.1 INTRODUCTION

With some significant exceptions, current applications for data mining are either in those areas for which there are little accepted discovery methodologies or in those areas that are being used within a knowledge discovery process that does not expect authoritative results but finds the discovered rules useful nonetheless. This is in contrast to its application in the fields applicable to spatial or spatio-temporal discovery, which possess a rich history of methodological discovery and result evaluation.

Examples of the former include market basket analysis, which, in its simplest form (see Agrawal, Imielinski and Swami 1993), provides insight into the correspondence between items purchased in a retail trade environment, and Web log analysis (see also Cooley, Mobasher and Srivastava 1997; Viveros, Wright, Elo-Dean and

Duri 1997; Madria, Bhowmick, Ng and Lim 1999), which attempts to derive a broad understanding of sequences of user activity on the Internet. Examples of the latter include time series analysis and signal processing (Weigend and Gershenfeld 1993; Guralnik and Srivastava 1999; Han, Dong and Yin 1999). The rules resulting from investigations in both of these areas may or may not be the result of behavioral or structural conditions but, significantly, it is the rule* itself rather than the underlying reasons behind the rule, which is generally the focus of interest.

An alternative approach originated in the field of medical knowledge discovery, where a procedure in which the results of data mining are embedded within a process that interprets the results as being merely hints toward further properly structured investigation into the reasons behind the rules (Lavrac 1999). This latter approach has been usefully employed by knowledge discovery processes over geographic data (Pei et al. 2006; Jankowski, Andrienko, and Andrienko 2001; Duckham and Worboys 2005). The whole field of geo-visualization has grown out of this approach (Kousa, Maceachren, and Kraak 2006; Bertolotto et al. 2007). Map-centered approaches, where the map becomes a visual index (Jankowski et al. 2001), and other types of geo-visualization are areas of very active research (Adrienko et al. 2007).

A third, and in some cases more useful approach may be appropriate for many of those areas for which spatial and spatio-temporal rules might be mined. This last approach accepts a (null) hypothesis and attempts to refine it (or disprove it) through the modification of the hypothesis because of knowledge discovery. This latter approach is carried out according to the principles of scientific experimentation and induction and has resulted in theories being developed and refined according to repeatable and accepted conventions.

The promises inherent in the development of data-mining techniques and knowledge discovery processes are manifold and include an ability to suggest rich areas of future research in a manner that could yield unexpected correlations and causal relationships. However, the nature of such techniques is that they can also yield spurious and logically and statistically erroneous conjectures.

Regardless of the process of discovery, the form of the input and the nature and allowable interpretation of the resulting rules can also vary significantly for knowledge discovery from geographic/spatio-temporal data, as opposed to that produced by "conventional" data-mining algorithms. For example, the complexity of the rule space requires significant constraints to be placed on the rules that can be generated to avoid either excessive or useless findings. To this end, some structuring of the data (Lees 1996; Duckham and Worboys 2005; Leung et al. 2007; Jones and Purves 2008) will often enhance the generation of more relevant rules.

This chapter presents a discussion of the issues that make the discovery of spatial and spatio-temporal knowledge different with an emphasis on geographical data. We discuss how the new opportunities of data mining can be integrated into a cohesive and, importantly, scientifically credible knowledge discovery process. This is particularly necessary for spatial and spatio-temporal discovery, as the opportunity

* Although each form of data-mining algorithm provides results with different semantics, we will use the term "rule" to describe all forms of mining output.

for meaningless and expensive diversions is high. We discuss the concepts of spatio-temporal knowledge discovery from geographic and other spatial data, the need to re-code temporal data to a more meaningful metric, the ideas behind higher order or meta-mining, as well as scientific theory formation processes and, briefly, the need to acknowledge the "second-hand" nature of much collected data.

2.2 MINING FROM SPATIAL AND SPATIO-TEMPORAL DATA

Current approaches to spatial and spatio-temporal knowledge discovery exhibit a number of important characteristics that will be discussed in order to compare and contrast them with possible future directions. However, space precludes a full survey of the manner in which spatial and spatio-temporal knowledge discovery is currently undertaken. Therefore, readers are directed to a number of other papers with reviews of the area (Bell, Anand and Shapcott 1994; Koperski, Adhikary and Han 1996; Abraham and Roddick 1998, 1999; Adrienko et al. 2007; Jones and Purves, 2008). In addition, a survey of temporal data-mining research is available (Roddick and Spiliopoulou 2002).

2.2.1 Rule Types

As discussed by Abraham and Roddick (1999), the forms that spatio-temporal rules may take are extensions of their static counterparts and at the same time are uniquely different from them. Five main types can be identified:

1. Spatio-Temporal Associations. These are similar in concept to their static counterparts as described by Agrawal et al. (1993). Association rules* are of the form *X* ® *Y* (c%, s%), where the occurrence of *X* is accompanied by the occurrence of *Y* in c% of cases (while *X* and *Y* occur together in a transaction in s% of cases).† Spatio-temporal extensions to this form of rule require the use of spatial and temporal predicates (Koperski and Han 1995; Estivill-Castro and Murray 1998). Moreover, it should be noted that for temporal association rules, the emphasis moves from the data itself to *changes* in the data (Chen, Petrounias and Heathfield 1998; Ye and Keane 1998; Rainsford and Roddick 1999).
2. Spatio-Temporal Generalization. This is a process whereby concept hierarchies are used to aggregate data, thus allowing stronger rules to be located at the expense of specificity. Two types are discussed in the literature (Lu, Han and Ooi 1993): *spatial-data-dominant* generalization proceeds by first ascending spatial hierarchies and then generalizing attributes data by region, while *nonspatial-data-dominant* generalization proceeds by first ascending the aspatial attribute hierarchies. For each of these, different rules may

* For a recent survey on association mining, see the paper by Ceglar and Roddick (2006).

† Note that while support and confidence were introduced in 1993 (Agrawal et al. 1993), considerable research has been undertaken into the nature of "interestingness" in mining rules (see also Silberschatz and Tuzhilin 1996; Dong and Li 1998; Bayardo Jr. and Agrawal 1999; Freitas 1999; Sahar 1999; Geng and Hamilton 2006).

result. For example, the former may give a rule such as "South Australian summers are commonly hot and dry," while the latter gives "Hot, dry summers are often experienced by areas close to large desert systems."

3. Spatio-Temporal Clustering. While the complexity is far higher than its static, nonspatial counterpart, the ideas behind spatio-temporal clustering are similar; that is, either characteristic features of objects in a spatio-temporal region or the spatio-temporal characteristics of a set of objects are sought (Ng and Han 1994; Ng 1996; Guo 2007).
4. Evolution Rules. This form of rule has an explicit temporal and spatial context and describes the manner in which spatial entities change over time. Due to the exponential number of rules that can be generated, it requires the explicit adoption of sets of predicates that are usable and understandable. Example predicates might include the following:*

follows	One cluster of objects traces the same (or similar) spatial route as another cluster later (i.e., spatial coordinates are fixed, time is varying). Other relationships in this class might include the temporal relationships discussed by Allen (1983) and Freksa (1992).
coincides	One cluster of objects traces the same (or similar) spatial path whenever a second cluster undergoes a specified activity (i.e., temporal coordinates are fixed, spatial activity varies). This may also include a causal relationship in which one cluster of objects undergoes some transformation or movement immediately after a second set undergoes some transformation or movement.
parallels	One cluster of objects traces the same (or a similar) spatial pattern but offset in space (i.e., temporal coordinates are fixed, spatial activity varies). This class may include a number of common spatial translations (such as rotation, reflection, etc.).
mutates	One cluster of objects transforms itself into a second cluster. See the work of Hornsby and Egenhofer (1998) and others that examine change in geographic information.

5. Meta-Rules. These are created when rule sets rather than data sets are inspected for trends and coincidental behavior. They describe observations discovered among sets of rules; for example, *the support for suggestion X is increasing*. This form of rule is particularly useful for temporal and spatio-temporal knowledge discovery and is discussed in more detail in Section 2.3.

2.2.2 Spatial vs. Spatio-Temporal Data

There are some fundamental issues for data mining in both spatial and spatio-temporal data. All spatial analyses are sensitive to the length, area, or volume over which a variable is distributed in space. The term used for these elements is

* This is by no means an exhaustive list, but it gives some idea as to what useful predicates may resemble.

"support" (Dungan 2001). Support encompasses more than just scale or granularity; it also includes orientation. This is not a new concept; scaling effects on spatial analyses have been understood since Gehlke and Biehl (1934) and, more recently Openshaw (1984). Often termed the *Modifiable Areal Unit Problem* (MAUP), or the *Ecological Fallacy*, it is an important characteristic of spatial analysis. Put simply, the scale, or support, chosen for data collection determines which spatial phenomena can be identified (Dale and Desrochers 2007). If spatial data is aggregated, then the larger the unit of aggregation the more likely the attributes will be correlated. Also by aggregating into different groups, you can get different results. Most importantly for data mining, attributes that exist in the data at one level of support can vanish at coarser or finer scales, or other orientations (Manley, Flowerdew and Steel 2006; Schuurman et al. 2007). U.S. elections are a perfect example of the problem in operation where the candidate who captures the majority of votes may not win the presidency.

To try to avoid this effect, an understanding of the scale, or support level, of the phenomenon being searched for is important. Searching a database at other levels is unlikely to be successful and re-scaling data has its own problems. Altering the scale of coarse data to fine is nearly always impossible, as the information lost in the prior generalization is usually not recoverable without complex modeling.

The MAUP also has implications for temporal support. An important phenomenon apparent at an hourly scale, such as the diurnal cycle, disappears at scales longer than one day. The seasonal cycle disappears in data aggregated to years, and so on. The difference between support and scale, where the former considers orientation, is less important for temporal data. Strategies for overcoming this problem in data mining and knowledge discovery include changing the scale of measurement, nonstationary modeling, dynamic modeling, conditional simulation, and constrained optimization (Tate and Atkinson 2001).

The dimensioning-up of the spatial dimension to include time was originally seen as a useful way of accommodating spatio-temporal data. However, the nature of time results in the semantics of time in discovered rules needing to be coded according to the relevant process aspect of time in order to make them useful. In most systems development, time is generally considered unidirectional and linear. Thus, the relational concepts (before, during, etc.) are easily understood, communicated, and accommodated. Conversely, space is perceived as bi-directional and, particularly in spatial/geographic applications, commonly nonlinear.

Although both time and space are continuous phenomena, it is common to encode time as discrete and isomorphic with integers, and a larger granularity is often selected (days, hours, etc.). Space, on the other hand, while a specific granularity is sometimes adopted, is often considered as isomorphic with real numbers, and the granularity relative to the domain is generally smaller. Consider, for example, a land titles system in which spatial accuracy to a meter or less is required across cities or states that can be hundreds of kilometers or more wide (an area ratio typically of the order of $1{:}10^{12}$). Time is commonly captured to the day over possibly three centuries (a granularity of $1{:}10^{6}$). A counter-example might be Advanced Very High Resolution Radiometer (AVHRR) data, captured within a couple of seconds, on a grid of 1.1 km, and commonly reported with a spatial resolution of about 4 km.

There is often an agreement that recent events and the current state of the system are of more interest than past events. While one user may focus on a particular location, it is unlikely that all users of a system will focus on a particular geographic region. Indexing schemes are thus able to be oriented to the "now" point in time but not to the "here" point in space.

Thus, when one is trying to extract new relationships from a database, simple *dimensioning-up* strategies work poorly. There have been numerous attempts to deal with time in the context of spatio-temporal data (see Egenhofer and Golledge 1994, for a review) and the importance of recognizing the differences between the spatial and temporal dimensions cannot be overstated, even when examining apparently static phenomena.

Consideration of the temporal characteristics of some typical datasets used in data mining will highlight this. For example, spectral data represents an instant in time. The time slice is a very narrow and constrained sample of the phenomenon being observed. These data are often included in databases with environmental data of various sorts.

In contrast to the spectral data, environmental data typically represent long-term estimates of mean and variance of very dynamic environmental variables. The time scale of this data is quite different from that of the spectral data. Spatial data (in geographic space) has characteristics that differ from both of these. It is reasonable to question whether this gross scale difference means that our accepted data-mining procedures are flawed. This is not the case, however, as the time scale differences between the data types generally match the characteristics we wish to include in most analyses of land cover. For example, the spectral time slice provides discrimination between vegetation types while the environmental data provides long-term conditions that match the time scale of germination, growth, and development of the largest plants. When we are concerned with forecasting, say, crop production, then shorter time scales would be necessary, as is common practice. Very often, too little consideration is given to the appropriate temporal scales necessary. An example might be the monitoring of wetlands in the dry tropics. The extent of these land-cover elements varies considerably through time, both on a seasonal basis and from year to year. In many years, the inter-annual variability in extent is greater than the average annual variability. This means that a spectral image of wetland extent, without a precise annual and seasonal labeling in the database, and without monthly rainfall and evaporation figures, is a meaningless measurement.

The temporal scales used in conjunction with spatial data are often inconsistent and need to be chosen more carefully. As discussed previously, time, as normally implemented in process models, is a simple, progressive step function. This fails to capture the essence of temporal change in both environmental and many cultural processes and is scale dependent. While our normal indices of time are either categorical or linear, "process time" is essentially spatial in character. The mismatch in the data model for time may well underlie the difficulties that many data miners are experiencing in trying to incorporate spatial and temporal attributes into their investigations. For many spatio-temporal data mining exercises, better results will be achieved by a considered recoding of the temporal data. The work of palaeo-climate reconstruction demonstrates this. In order to make sense of deep-ocean cores and ice cores, the results of down-core analyses are usually analyzed using Fourier, or spectral, analysis to decompose the time series data into a series of repeating cycles. Most, if not all, of the cycles thus identified can be associated with the relative positioning of the Earth, the sun, and the major planets in space. Consideration of this makes it clear that our useful assumption that geographic space is static and time invariant is flawed.

The Cartesian description of location defined by latitude, longitude, and elevation is not only an inaccurate representation of reality; it is an encumbrance to understanding the relationship between time and space. Time is a spatial phenomenon. A fuller understanding of this leads to a resolution of some of the conceptual problems that bedevil the literature on spatio-temporal GIS modeling (Egenhofer and Golledge 1994). Properties of time concepts such as continuous/linear, discrete, monotonic, and cyclic time tend to deal only with limited aspects of time and, as such, have limited application. In data mining, the process aspects of time are particularly important.

In order to progress this discussion, it is worth first returning to consider the Cartesian representation of space using latitude, longitude, and elevation. A point on the Earth's surface defined by this schema is not static in space. It is moving, in a complicated but predictable way, through a complex energy environment. This movement, and the dynamics of the energy environment itself, is "time."

There are three main components to the environmental energy field: gravity, radiation, and magnetism. These fluctuate in amplitude and effectiveness. The effectiveness of gravity, radiation, and magnetism is almost entirely due to the relationships between bodies in space. Interaction between these forces and the internal dynamics of a body such as the sun can alter the amplitude of its radiation and magnetism. These feedback relationships sound complex, but are predictable. The most important relationships have already been indexed as clock and calendar time. These are

- The relative positions of a point on the surface of the Earth and the sun, the diurnal cycle. This is a function of the rotation of the Earth, and the tilt of the Earth's axis relative to the plane of the ecliptic (the declination of the sun).
- The orbit of the moon around the Earth.
- The orbit of the Earth around the sun.

Each of these relationships has a very significant relationship with the dynamics of both our natural, cultural, and even economic environments. These dynamic spatial relationships are the basis of the index we call time, but do not include all the important phenomena we now understand to influence our local process environment. Others include

- The solar day, which sweeps a pattern of four solar magnetic sectors past the Earth in about 27 days. Alternating sectors have reverse polarity and the passage of a sector boundary only takes a few hours. This correlates with a fluctuation in the generation of low-pressure systems.

- The lunar cycle. The lunar cycle is a 27.3-day period in the declination of the moon during which it moves north for 13.65 days and south for 13.65 days. This too correlates with certain movements of pressure systems on the Earth.
- The solar year. The sun is not the center of the solar system. Instead, it orbits the barycenter of the solar system, which at times passes through the sun. The orbit is determined by the numerous gravitational forces within the solar system, but tends to be dominated by the orbits of the larger planets, Jupiter and Saturn, at about 22 to 23 years. This orbit appears to affect solar emissions (the sunspot cycle). Notoriously, this cycle correlates with long-term variation in a large number of natural, cultural, and economic indices from cricket scores and pig belly futures to a host of other, more serious, areas.

There are much longer periods that can be discussed, but the above relate to both the Earth's energy environment and time on the sorts of scales we are most concerned with in data mining. Lees (1999) has reviewed these. Time coded as position using these well-understood astrophysical relationships is not an abstract concept. Such a coding correlates with energy variability, which both drives our natural systems and influences many of our cultural systems. This coding also links directly to variations in spectral space. Illumination is a function of season (apparent declination of the sun) and time of day (diurnal cycle) modified by latitude. The simple process of recoding the time stamp on data to a relevant continuous variable, such as solar declination or time of the solar year, rather than indices such as Julian day, provides most "intelligent" data-mining software a considerably better chance of identifying important relationships in spatio-temporal data.

2.2.3 Handling Second-Hand Data

A significant issue for many systems, and one that is particularly applicable to geographical data, is the need to reuse data collected for other purposes (Healey and Delve 2007). While few data collection methodologies are able to take into account the nondeterministic nature of data mining, the expense and in many cases the difficulty in performing data collection specifically for the knowledge discovery process results in heterogeneous data sources, each possibly collected for different purposes, commonly being bought together. This requires that the interpretation of such data must be carefully considered (Jones and Purves 2008). Possible errors that could result might include

- The rules reflecting the heterogeneity of the data sets rather than any differences in the observed phenomena.
- The data sets being temporally incompatible. For instance, the data collection points may render useful comparison impossible. This is also an issue in time series mining in which the scales of the different data sets must first be reconciled (see Berndt and Clifford 1995).

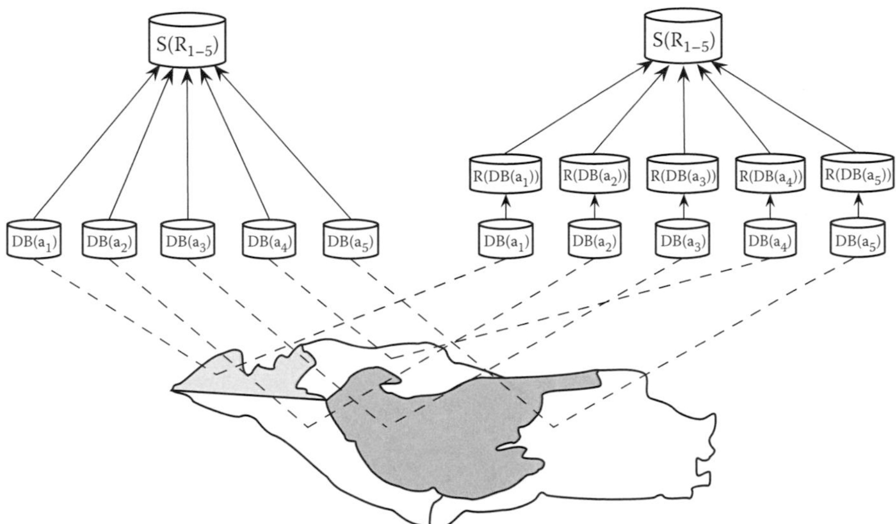

FIGURE 2.1 Mining from data and from rule sets.

- The collection methods being incompatible. For example, the granularities adopted or the aggregation methods of observations may differ. More severe, the implicit semantics of the observations may be different.
- The MAUP discussed earlier.

This puts particular emphasis on either or both of the quality of the data cleaning and the need for the mining process to take account of the allowable interpretations.

2.3 META-MINING AS A DISCOVERY PROCESS PARADIGM

The target of many mining operations has traditionally been the data itself. With the increase in data and the polynomial complexity of many mining algorithms, the direct extraction of useful rules from data becomes difficult. One solution to this, first suggested in the realm of temporal data mining (Abraham and Roddick 1997, 1999; Spiliopoulou and Roddick 2000) is to mine either from summaries of the data or from the results of previous mining exercises as shown in Figure 2.1.

Consider the following results (possibly among hundreds of others) of a mining run on U.K. weather regions as follows:*

SeaArea (Hebrides), Windspeed (High), Humidity (Medium/High)® Forecast (Rain), LandArea (North Scotland)
SeaArea (Hebrides), Windspeed (Medium), Humidity (Medium/High)® Forecast (Rain), LandArea (North Scotland)
SeaArea (Hebrides), Windspeed (Low), Humidity (Medium/High)® Forecast (Fog), LandArea (North Scotland)

* Hebrides, Malin, and Rockall are geographic "shipping" regions to the west of Scotland.

SeaArea (Hebrides), Windspeed (High), Humidity (Low)® Forecast (Windy), LandArea (North Scotland)
SeaArea (Hebrides), Windspeed (Medium), Humidity (Low)® Forecast (Light Winds), LandArea (North Scotland)
SeaArea (Malin), Windspeed (High), Humidity (Medium/High)® Forecast (Rain), LandArea (South Scotland)
SeaArea (Malin), Windspeed (Medium), Humidity (Medium/High)® Forecast (Rain), LandArea (South Scotland)
SeaArea (Malin), Windspeed (Low), Humidity (Medium/High)® Forecast (Fog), LandArea (South Scotland)
SeaArea (Malin), Windspeed (High), Humidity (Low)® Forecast (Windy), LandArea (South Scotland)
SeaArea (Malin), Windspeed (Medium), Humidity (Low)® Forecast (Light Winds), LandArea (South Scotland)
SeaArea (Rockall), Windspeed (High), Humidity (Medium/High)® Forecast (Rain), LandArea (Scotland)
SeaArea (Rockall), Windspeed (Medium), Humidity (Medium/High)® Forecast (Rain), LandArea (Scotland)
SeaArea (Rockall), Windspeed (Low), Humidity (Medium/High)® Forecast (Fog), LandArea (Scotland)
SeaArea (Rockall), Windspeed (High), Humidity (Low)® Forecast (Windy), LandArea (Scotland)
SeaArea (Rockall), Windspeed (Medium), Humidity (Low)® Forecast (Light Winds), LandArea (Scotland)

These rules may be inspected to create higher-level rules such as

SeaArea (West of Scotland), Windspeed (Medium/High), Humidity (Medium/High)® Forecast (Rain), LandArea (Scotland)

or even

SeaArea (West of LandArea), Windspeed (Medium/High), Humidity (Medium/High)® Forecast (Rain)

These higher-level rules can also be produced directly from the source data in a manner similar to the concept ascension algorithms of Cai, Cercone, and Han (1991) and others. However, the source data is not always available or tractable.

Note that the semantics of meta mining must be carefully considered (see Spiliopoulou and Roddick 2000). Each rule generated from data is generated according to an algorithm that, to some extent, removes "irrelevant" data. Association rules, for example, provide a support and confidence rating, which must be taken into account when meta-rules are constructed. Similarly, clusters may use criteria such as lowest entropy to group observations that may mask important outlying facts. Fundamental to all of this is some resolution to the confusion surrounding location, which prevails in many spatial datasets (Overell and Ruger 2008; Leveling 2008).

2.4 PROCESSES FOR THEORY/HYPOTHESIS MANAGEMENT

Analyses into geographic, geo-social, socio-political, and environmental issues commonly require a more formal and in many cases strongly ethically driven approach. For example, environmental science uses a formal scientific experimentation process requiring the formulation and refutation of a credible null hypothesis.

The development of data mining over the past few years has been largely oriented toward the discovery of previously unknown but potentially useful rules that are in some way interesting in themselves. To this end, a large number of algorithms have been proposed to generate rules of various types (association, classification, characterization, etc.) according to the source, structure, and dimensionality of the data and the knowledge sought. In addition, a number of different types of *interestingness* metrics have also been proposed (see Silberschatz and Tuzhilin 1996) that strive to keep the exponential target rule space to within tractable limits.

More recently, in many research forums, a holistic process-centered view of knowledge discovery has been discussed and the interaction between tool and user (and, implicitly, between the rules discovered and the possibly tacit conceptual model) has been stressed. To a great extent, the ambition of totally autonomous data mining has now been abandoned (Roddick 1999). This shift has resulted in the realization that the algorithms used for mining and rule selection need to be put into a process-oriented context (Qui and Zhu 2003; Moran and Bui 2002). This in turn raises the question of which processes might benefit from data-mining research.

One of the motivations for data mining has been the inability of conventional analytical tools to handle, within reasonable time limits, the quantities of data that are now being stored. Data mining is thus being seen as a useful method of providing some measure of automated insight into the data being collected. However, it has become apparent that while some useful rules can be mined and the discipline has had a number of notable successes, the potential for either logical or statistical error is extremely high and the results of much data mining is at best a set of suggested topics for further investigation (Murray and Shyy 2000; Laffan et al. 2005).

2.4.1 The Process of Scientific Induction

The process of investigation (or knowledge discovery) can be considered as having two distinct forms — the process modeling approach in which the real world is modeled in a mathematical manner and from which predictions can be in some way computed, and the pattern matching or inductive approach in which prediction is made based on experience.

It is important to note that the process of rule generation through data mining is wholly the latter (in Figure 2.2, the right-hand side), while scientific induction starts with the latter and aims to translate the process into one that is predominantly the former.

Another view of the scientific induction process can be considered the following. Given a set of observations and an infinitely large hypothesis space, rules (i.e., trends, correlations, clusters, etc.) extracted from the observations constrain the hypothesis space until the space is such that a sufficiently restrictive description of that space can be formed.

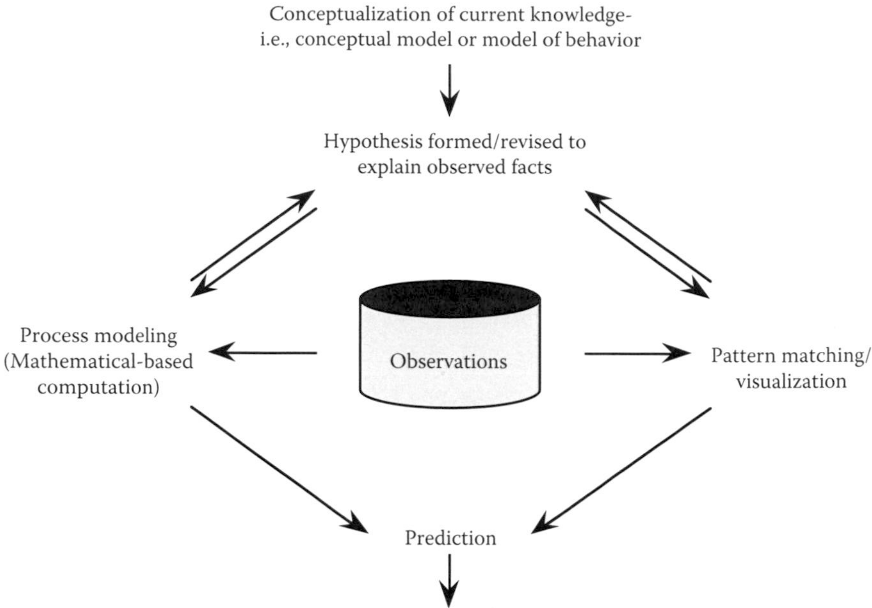

FIGURE 2.2 Investigation paths in discovery.

Experiments are commonly constructed to explore the boundaries between the known regions (i.e., those parts definitely in or out of the solution space). Of course, the intuition and experience of the scientist plays a large part in designing adequate experiments that unambiguously determine whether a region is in or out of the final hypothesis. The *fallacy of induction* comes into play when the hypothesis developed from the observations (or data) resides in a different part of the space from the true solution and yet it is not contradicted by the available data.

The commonly accepted method to reduce the likelihood of a false assumption is to develop alternative hypotheses and prove these are false, and in so doing, to constrain the hypothesis space. An alternative process, which we outline briefly in this chapter, aims to support the development of scientific hypotheses through the accepted scientific methodology of null hypothesis creation and refutation.

To continue with the visual metaphor provided by the hypothesis space described previously, data mining can be considered a process for finding those parts of the hypothesis space that best fit the observations and to return the results in the form of rules. A number of data-mining systems have been developed which aim to describe and search the hypothesis space in a variety of ways.

Unfortunately, the complexity of such a task commonly results in either less than useful answers or high computational overhead. One of the reasons for this is that the search space is exponential to the number of data items, which itself is large. A common constraint, therefore, is to limit the structural complexity of a solution, for example, by restricting the number of predicates or terms in a rule. Data mining also commonly starts with a "clean sheet" approach and while restartable or iterative methods are being researched, little progress has been made to date.

Another significant problem is that the data has often been collected at different times with different schemata, sometimes by different agents and commonly for an alternative purpose to data mining. Data mining is often only vaguely considered (if at all) and thus drawing accurate inferences from the data is often problematic.

2.4.2 Using Data Mining to Support Scientific Induction

An alternative solution is to develop (sets of) hypotheses that will constrain the search space by defining areas within which the search is to take place. Significantly, the hypotheses themselves are not examined; rather, (sets of) null hypotheses are developed which are used instead. Figure 2.3 shows a schematic view of such a data-mining process.

In this model, a user-supplied conceptual model (or an initial hypothesis) provides the starting point from which hypotheses are generated and tested. Generated hypotheses are first tested against known constraints and directed data-mining routines then validate (to a greater or lesser extent) the revised theory. In cases where

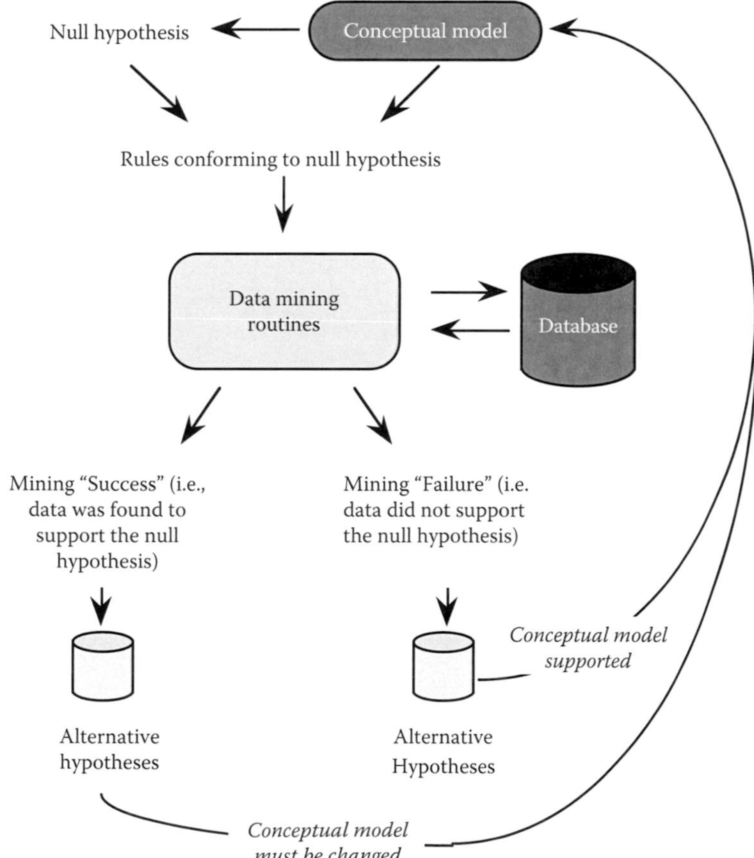

FIGURE 2.3 Data mining for null hypotheses.

the hypothesis is generally supported, weight is added to the confidence of the conceptual model in accordance with the usual notion of scientific induction. In cases where the hypothesis is not supported, either a change to the conceptual model or a need for external input is indicated.

Note that the process provides three aspects of interest:

1. The procedure is able to accept a number of alternative conceptual models and provide a ranking between them based on the available observations. It also allows for modifications to a conceptual model in cases where the rules for such modification are codified.
2. The hypothesis generation component may yield new, hitherto unexplored, insights into accepted conceptual models.
3. The process employs directed mining algorithms and thus represents a reasonably efficient way of exploring large quantities of data, which is essential in the case of mining high-dimensional datasets such as those used in geographic systems.

As this model relies heavily on the accepted process of scientific induction, the process is more acceptable to the general science community.

2.5 CONCLUSION

The ideas outlined in this chapter differ, to some extent, from conventional research directions in spatio-temporal data mining and emphasize that the process into which the mining algorithm will be used can significantly alter the interpretation of the results. Moreover, the knowledge discovery process must take account of this to avoid problems. Johnson-Laird (1993) suggested that *induction should come with a government warning*; this is particularly true of spatio-temporal mining because the scope for error is large.

It should be noted that many of the ideas discussed in this chapter need further examination. For example, while the ideas of using data mining for hypothesis refutation have been discussed in a number of fora, only recently has there been serious investigation of the idea in a scientific setting (Jones and Purves 2008; Adrienko et al. 2007). These are starting to establish a strong framework, but a credible method of hypothesis evolution still needs to be defined. Similarly, in their recent attempt to set a research agenda for geovisual analytics, Adrienko et al. (2007) identified a need for concerted cross-disciplinary efforts in order to make useful progress. They noted that existing methods are still far from enabling the necessary synergy between the power of computational techniques and human analytical expertise.

ACKNOWLEDGMENTS

We are particularly grateful to the NCGIA Varenius Project for organizing the Seattle Workshop on Discovering Knowledge from Geographical Data, which enabled some of these ideas to be explored. We would also particularly like to thank

Tamas Abraham, DSTO, Australia and Jonathan Raper, City University, London, for discussions and comments.

REFERENCES

Abraham, T. and Roddick, J.F. (1997): Discovering meta-rules in mining temporal and spatio-temporal data. *Proc. Eighth International Database Workshop, Data Mining, Data Warehousing and Client/Server Databases (IDW'97)*, Hong Kong, 30-41, Fong, J. (Ed.) Berlin, Springer-Verlag.

Abraham, T. and Roddick, J.F. (1999): Incremental meta-mining from large temporal data sets. In *Advances in Database Technologies, Proc. First International Workshop on Data Warehousing and Data Mining, DWDM'98*. Lecture Notes in Computer Science, **1552**:41–54. Kambayashi, Y., Lee, D.K., Lim, E.-P., Mohania, M. and Masunaga, Y. (Eds.). Berlin, Springer-Verlag.

Abraham, T. and Roddick, J.F. (1999): Survey of spatio-temporal databases. *Geoinformatica*, **3**(1):61–99.

Andrienko, G., Andrienko, N., Jankowski, P. et al. (2007): Geovisual analytics for spatial decision support: Setting the research agenda. *International Journal of Geographical Information Science*, **21**(8):839–857.

Agrawal, R., Imielinski, T. and Swami, A. (1993): Mining association rules between sets of items in large databases. *Proc. ACM SIGMOD International Conference on Management of Data*, Washington D.C., **22**:207–216, ACM Press.

Allen, J.F. (1983): Maintaining knowledge about temporal intervals. *Communications of the ACM*, **26**(11):832–843.

Bayardo Jr., R.J. and Agrawal, R. (1999): Mining the most interesting rules. *Proc. Fifth International Conference on Knowledge Discovery and Data Mining*, San Diego, CA, 145–154, Chaudhuri, S. and Madigan, D. (Eds.), ACM Press.

Bell, D.A., Anand, S.S., and Shapcott, C.M. (1994): Data mining in spatial databases. *Proc. International Workshop on Spatio-Temporal Databases*, Benicassim, Spain.

Berndt, D.J. and Clifford, J. (1995): Finding patterns in time series: A dynamic programming approach. In *Advances in Knowledge Discovery and Data Mining*. 229–248. Fayyad, U.M., Piatetsky-Shapiro, G., Smyth, P. and Uthurusamy, R. (Eds.), AAAI Press/ MIT Press, Cambridge, MA.

Bertolotto, M., Di Martino, S., Ferrucci, F., et al. (2007): Towards a framework for mining and analysing spatio-temporal datasets. *International Journal of Geographical Information Science*, **21**(8):895–906.

Cai, Y., Cercone, N., and Han, J. (1991): Attribute-oriented induction in relational databases. In *Knowledge Discovery in Databases*. 213–228, Piatetsky-Shapiro, G. and Frawley, W.J. (Eds.). AAAI Press/MIT Press, Cambridge, MA.

Ceglar, A. and Roddick, J.F. (2006): Association mining. *ACM Computing Surveys*, **38**(2).

Chen, X., Petrounias, I. and Heathfield, H. (1998): Discovering temporal association rules in temporal databases. *Proc. International Workshop on Issues and Applications of Database Technology (IADT'98)*, 312–319.

Cooley, R., Mobasher, B. and Srivastava, J. (1997): Web mining: information and pattern discovery on the World Wide Web. *Proc. Ninth IEEE International Conference on Tools with Artificial Intelligence*, 558–567, IEEE Computer Society, Los Alamitos, CA.

Dale, M.B, and Desrochers, M. (2007): Measuring information-based complexity across scales using cluster analysis. *Ecological Informatics*, **2**:121–127.

Dong, G. and Li, J. (1998): Interestingness of discovered association rules in terms of neighbourhood-based unexpectedness. *Proc. Second Pacific-Asia Conference on Knowledge*

Discovery and Data Mining: Research and Development, Melbourne, Australia, 72–86, Wu, X., Kotagiri, R. and Korb, K.B. (Eds.). Springer-Verlag, Berlin.

Duckham, M, and Worboys, M. (2005): An algebraic approach to automated geospatial information fusion. *International Journal of Geographical Information Science*, 19(5):537–557.

Dungan, J. (2001): Scaling up and scaling down: The relevance of the support effect on remote sensing in vegetation. In *Modelling Scale in Geographical Information Science.* N.J. Tate and P.M. Atkinson (Eds.). John Wiley & Sons, New York.

Egenhofer, M.J. and Golledge, R.J. (1994): Time in Geographic Space. Report on the Specialist Meeting of Research Initiative, 10: 94-9. National Centre for Geographic Information and Analysis, University of California, Santa Barbara.

Estivill-Castro, V. and Murray, A.T. (1998): Discovering associations in spatial data — an efficient media-based approach. *Proc. Second Pacific-Asia Conference on Research and Development in Knowledge Discovery and Data Mining, PAKDD-98*, 110–121, Springer-Verlag, Berlin.

Freitas, A.A. (1999): On rule interestingness measures. *Knowledge Based Systems* **12**(5–6): 309–315.

Freksa, C. (1992): Temporal reasoning based on semi-intervals. *Artificial Intelligence* **54**: 199–227.

Gehlke, C.E. and Biehl, K. (1934): Certain effects of grouping upon the size of the correlation coefficient in census tract material. *Journal of the American Statistical Association Supplement,* **29:**169–170.

Geng, L. and Hamilton, H.J. (2006): Interestingness measures for data mining: A survey. *ACM Computing Surveys*, 38(3).

Guo, D. (2007): Visual analytics of spatial interaction patterns for pandemic decision support, *International Journal of Geographical Information Science*, **21**(8):859–877.

Guralnik, V. and Srivastava, J. (1999): Event detection from time series data. *Proc. Fifth International Conference on Knowledge Discovery and Data Mining*, San Diego, CA, 33–42, Chaudhuri, S. and Madigan, D. (Eds.). ACM Press.

Han, J., Dong, G., and Yin, Y. (1999): Efficient mining of partial periodic patterns in time series database. *Proc. Fifteenth International Conference on Data Engineering*, Sydney, Australia, 106–115, IEEE Computer Society.

Healey, R.G, and Delve, J. (2007): Integrating GIS and data warehousing in a Web environment: A case study of the U.S. 1880 Census. *International Journal of Geographical Information Science*, **21**(6):603–624.

Hornsby, K. and Egenhofer, M. (1998): Identity-based change operations for composite objects. *Proc. Eighth International Symposium on Spatial Data Handling*, Vancouver, Canada, 202–213, Poiker, T. and Chrisman, N. (Eds.).

Jankowski, P., Andrienko, N., and Andrienko, G., (2001): Map-centred exploratory approach to multiple criteria spatial decision making. *International Journal of Geographical Information Science*, **15**(2):101–127.

Johnson-Laird, P. (1993): *The Computer and the Mind*, 2nd ed., Fontana Press, London.

Jones, C., and Purves, R. (2008): Geographical information retrieval. *International Journal of Geographical Information Science,* **22**(3): in press.

Koperski, K., Adhikary, J. and Han, J. (1996): Knowledge discovery in spatial databases: Progress and challenges. *Proc. ACM SIGMOD Workshop on Research Issues on Data Mining and Knowledge Discovery*, Montreal, Canada, 55–70.

Koperski, K. and Han, J. (1995): Discovery of spatial association rules in geographic information databases. *Proc. Fourth International Symposium on Large Spatial Databases*, Maine, 47–66.

Koua, E.L, Maceachren, A. and Kraak, M.J. (2006): Evaluating the usability of visualization methods in an exploratory geovisualization environment. *International Journal of Geographical Information Science*, **20**(4):425–448.

Laffan, S.W, Nielsen, O. M., Silcock, H., et al. (2005): Sparse grids: A new predictive modelling method for the analysis of geographic data. *International Journal of Geographical Information Science*, **19**(3):267–292.

Lavrac, N. (1999): Selected techniques for data mining in medicine. *Artificial Intelligence in Medicine,* **16**:3–23.

Lees, B.G. (1996): Sampling strategies for machine learning using GIS. In *GIS and Environmental Modelling: Progress and Research Issues*. Goodchild, M.F., Steyart, L., Parks, B. et al. (Eds.). GIS World Inc., Fort Collins, CO.

Lees, B.G. (1999): Cycles, climatic. In *Encyclopedia of Environmental Science*, 105–107. Alexander, D.E. and Fairbridge, R.W. (Eds.). Van Nostrand Reinhold, New York.

Leung, Y., Fung, T., Mi, S. et al. (2007): A rough set approach to the discovery of classification rules in spatial data. *International Journal of Geographical Information Science*, **21**:1033–1058.

Levelling, J. (2008): On metonym recognition for geographic information retrieval. *International Journal of Geographical Information Science,* **22**(3): in press.

Lu, W., Han, J. and Ooi, B.C. (1993): Discovery of general knowledge in large spatial databases. *Proc. 1993 Far East Workshop on GIS (IEGIS 93)*, Singapore, 275–289.

Madria, S.K., Bhowmick, S.S., Ng, W.K. and Lim, E.-P. (1999): Research issues in Web data mining. *Proc. First International Conference on Data Warehousing and Knowledge Discovery, DaWaK '99*, Florence, Italy, Lecture Notes in Computer Science, **1676**:303–312, Mohania, M.K. and Tjoa, A.M. (Eds.). Springer, Berlin.

Manley, D., Flowerdew, R, and Steel, D. (2006): Scales, levels and processes: Studying spatial patterns of British census variables. *Computers Environment and Urban Systems*, **30**(2):143–160.

Moran, C.J. and Bui, E.N. (2002): Spatial data mining for enhanced soil map modeling. International *Jounal of Geographical Information Science*, **16**:533–549.

Murray, A.T., and Shyy, T.K. (2000): Integrating attribute and space characteristics in choropleth display and spatial data mining. *International Journal of Geographical Information Science*, **14**(7):649–667.

Ng, R.T. (1996): Spatial data mining: Discovering knowledge of clusters from maps. *Proc. ACM SIGMOD Workshop on Research Issues on Data Mining and Knowledge Discovery*, Montreal, Canada.

Ng, R.T. and Han, J. (1994): Efficient and effective clustering methods for spatial data mining. *Proc. Twentieth International Conference on Very Large Data Bases*, Santiago, Chile, 144–155, Bocca, J.B., Jarke, M. and Zaniolo, C. (Eds.). Morgan Kaufmann.

Openshaw, S. (1984): *The Modifiable Areal Unit Problem,* Geobooks, Norwich.

Overell, S.E and Ruger, S. (2008): Using co-occurrence models for place name disambiguation. *International Journal of Geographical Information Science,* **22**(3): in press.

Pei, T., Zhu, A.X., Zhou, C.H., et al. (2006): A new approach to the nearest-neighbour method to discover cluster features in overlaid spatial point processes. *International Journal of Geographical Information Science*, **20**(2):153–168.

Qi, F., and Zhu, A.X. (2003): Knowledge discovery from soil maps using inductive learning. *International Journal of Geographical Information Science*, **17**(8):771–795.

Rainsford, C.P. and Roddick, J.F. (1999): Adding temporal semantics to association rules. *Proc. 3rd European Conference on Principles and Practice of Knowledge Discovery in Databases (PKDD'99)*, Prague, Lecture Notes in Artificial Intelligence, **1704**:504–509, Zytkow, J.M. and Rauch, J. (Eds.). Springer, Berlin.

Roddick, J.F. (1999): Data warehousing and data mining: Are we working on the right things? In *Advances in Database Technologies*. Lecture Notes in Computer Science, **1552:**141–144. Kambayashi, Y., Lee, D.K., Lim, E.-P., Masunaga, Y. and Mohania, M. (Eds.). Springer-Verlag, Berlin.

Roddick, J.F. and Spiliopoulou, M. (2002): A survey of temporal knowledge discovery paradigms and methods. *IEEE Transactions on Knowledge and Data Engineering*, **14**(4):750-767.

Sahar, S. (1999): Interestingness via what is not interesting. *Proc. Fifth International Conference on Knowledge Discovery and Data Mining*, San Diego, CA, 332–336, Chaudhuri, S. and Madigan, D. (Eds.).

Schuurman, N., Bell, N., Dunn, J.R., et al. (2007): Deprivation indices, population health and geography: An evaluation of the spatial effectiveness of indices at multiple scales. *Journal of Urban Health — Bulletin of The New York Academy of Medicine* **84**(4):591–603.

Silberschatz, A. and Tuzhilin, A. (1996): What makes patterns interesting in knowledge discovery systems? *IEEE Transactions on Knowledge and Data Engineering* **8**(6):970-974.

Spiliopoulou, M. and Roddick, J.F. (2000): Higher Order Mining: Modeling and Mining the Results of Knowledge Discovery. In *Data Mining II — Proc. Second International Conference on Data Mining Methods and Databases*. 309-320. Ebecken, N. and Brebbia, C.A. (Eds). WIT Press, Cambridge, U.K.

Tate, N.J. and Atkinson, P.M. (Eds.). (2001): *Modeling Scale in Geographical Information Science.* John Wiley & Sons, New York.

Viveros, M.S., Wright, M.A., Elo-Dean, S. and Duri, S.S. (1997): Visitors' behavior: mining web servers. *Proc. First International Conference on the Practical Application of Knowledge Discovery and Data Mining*, 257–269, Practical Application Co., Blackpool, U.K.

Weigend, A.S. and Gershenfeld, N.A. (Eds.) (1993): *Time Series Prediction: Forecasting the Future and Understanding the Past.* Proc. NATO Advanced Research Workshop on Comparative Time Series Analysis **XV**. Santa Fe, NM.

Ye, S. and Keane, J.A. (1998): Mining association rules in temporal databases. *Proc. International Conference on Systems, Man and Cybernetics*, 2803–2808, IEEE, New York.

3 Fundamentals of Spatial Data Warehousing for Geographic Knowledge Discovery

Yvan Bédard

Jiawei Han

CONTENTS

3.1 INTRODUCTION

Recent years have witnessed major changes in the geographic information (GI) market, from interoperable technological offerings to national spatial data infrastructures, Web-mapping services, and mobile applications. The arrival of new major players such as Google, Microsoft, Nokia, and TomTom, for instance, has created tremendous new opportunities and geographic data have become ubiquitous. Thousands of systems are geo-enabled every week, including data warehouses. As a special type of databases, a data warehouse aims at providing organizations with an integrated, homogeneous view of data covering a significant period in order to facilitate decision making. Such a view typically involves data about geographic, administrative, or political places, regions, or networks organized in hierarchies. Data warehouses are separated from transactional databases and are structured to facilitate data analysis. They are built with a relational, object-oriented, multidimensional, or hybrid paradigm although it is with the two latter that they bring the most benefits. Data warehouses are designed as a piece of the overall technological framework of the organization and they are implemented according to very diverse architectures responding to differing users' contexts. In fact, the evolution of spatial data warehouses fits within the general trends of mainstream information technology (IT).

Data warehouses provide these much-needed unified, global, and summarized views of the data dispersed into heterogeneous legacy databases over the years. Organizations invest millions of dollars to build such warehouses in order to efficiently feed the decision-support tools used for strategic decision making, such as dashboards, executive information systems, data mining, report makers, and online analytical processing (OLAP). In fact, data warehouse emerged as the unifying solution to a series of individual circumstances impacting global knowledge discovery.

First, large organizations often have several departmental or application-oriented independent databases that may overlap in content. Usually, such systems work properly for day-to-day operational-level decisions. However, when one needs to obtain aggregated or summarized information integrating data from these different systems, it becomes a long and tedious process that slows down decision making. It then appears easier and much faster to process a homogeneous and unique dataset. However, when several decision makers build their own summarized databases to accelerate the process, incoherencies among these summarized databases rapidly appear, and redundant data extraction/fusion work must be performed. Over the years, this leads to an inefficient, chaotic situation (Inmon, Richard and Hackathorn 1996).

Second, past experiences have shown that fully reengineering the existing systems in order to replace them with a unique corporate system usually leads to failure. It is too expensive and politically difficult. Then, one must find a solution that can cope as much as possible with existing systems but does not seek to replace them. In this regard, data warehouses add value to existing legacy systems rather than attempting to replace them since the unified view of the warehouse is built from an exact or modified copy of the legacy data.

Third, the data structure used today by most decision-support solutions adopts, partly or completely, the multidimensional paradigm. This paradigm is very different from the traditional, normalized relational structure as used by most transaction-oriented, operational-level legacy systems. The problem is that with transactional technologies, it is almost impossible to keep satisfactory response times for both transaction-oriented and analysis-oriented operations within a unique database as soon as this database becomes very large. One must then look for a different solution, which provides short response times for both analytical processing and transaction processing. This has resulted in the concept of the data warehouse, that is, an additional read-only database typically populated with analysis-oriented aggregated or summarized data obtained from the extraction, transformation, and loading (ETL) of the detailed transactional data imported from existing legacy systems. After the ETL process and the new structuring of the resulting data, one typically finds only aggregated data in the warehouse, not the imported detailed legacy data.

Fourth, strategic decision making requires not only different levels of aggregated and summarized data but also direct access to past data as well as present and future data (when possible) to analyze trends over time or predictions. The multidimensional paradigm frequently used in data warehouses efficiently supports such needs.

Finally, decision makers are also hoping for fast answers, simple user interfaces, a high level of flexibility supporting user-driven *ad hoc* exploration of data at different levels of aggregation and different epochs, and automatic analysis capabilities searching for unexpected data patterns.

In other words, the needed solution must support the extraction of useful knowledge from detailed data dispersed in heterogeneous datasets. Such a goal appears reasonable if we consider data warehousing and automatic knowledge discovery as the "common-sense" follow-up to traditional databases. This evolution results from the desire of organizations to further benefit from the major investments initially made into disparate, independent, and heterogeneous departmental systems. Once most operational-level needs are fulfilled by legacy systems, organizations wish to build more global views that support strategic decision making (the frequent bottom-up penetration of innovations). In fact, this evolution is very similar to the situation witnessed in the 1970s where organizations evolved from the management of disparate flat files to the management of integrated databases.

The goal of this chapter is to introduce fundamental concepts underlying spatial data warehousing. It includes four sections. After having presented the "raison d'être" of spatial data warehousing in the previous paragraphs of this section, we introduce key concepts of nonspatial data warehousing in the second section. The third section deals with the particularities of spatial data warehouses, which in fact are typically spatiotemporal. In the last section, we conclude and present future R&D challenges.

3.2 KEY CONCEPTS AND ARCHITECTURES FOR DATA WAREHOUSES

This section provides a global synthesis of the actual state of data warehousing and the related concepts of multidimensional databases, data marts, online analytical processing, and data mining. Specialized terms such as legacy systems, granularity, facts, dimensions, measures, snowflake schema, star schema, fact constellation, hypercube, and N-tiered architectures are also defined.

3.2.1 Data Warehouse

An interesting paradox in the world of databases is that systems used for day-to-day operations store vast amounts of detailed information but are very inefficient for decision support and knowledge discovery. The systems used for day-to-day operations usually perform well for transaction processing where minimum redundancy and maximum integrity checking are key concepts; furthermore, this typically takes place within a context where the systems process large quantities of transactions involving small chunks of detailed data. On the other hand, decision makers need fast answers made of few aggregated data summarizing large units of work, something transactional systems do not achieve today with large databases. This difficulty to combine operational and decision-support databases within a single system gave rise to the dual-system approach typical of data warehouses.

Although the underlying ideas are not new, the term "data warehouse" originated in the early 1990s and rapidly became an explicit concept recognized by the community. It has been defined very similarly by pioneers such as Brackett (1996), Gill and Rao (1996), Inmon, Richard and Hackathorn (1996), and Poe (1995). In general, a data warehouse is an enterprise-oriented, integrated, nonvolatile, read-only collection of data imported from heterogeneous sources and stored at several levels of detail to support decision making. Since this definition has been loosely interpreted and implemented in several projects and, consequently, has not always delivered the promised returns on investments, it is highly important to explain every key characteristic:

- Enterprise-oriented: One of the aims of data warehouses is to become the single and homogeneous source for the data that are of interest to make enterprise-level strategic decision making. Usually, no such homogeneous database exists because system development tends to happen in a bottom-up manner within organizations, resulting in several disparate, specialized systems. Similarly, such single source does not exist because the data stored within operational systems describe detailed transactions (e.g., the amount of cash withdrawn by one person at a given ATM at a precise moment), while enterprise-level strategic decision making requires summarized data (e.g., increase in monetary transactions by our clients in all of our ATMs for the last month), resulting in costly and time-consuming processing to get global information about an enterprise's activities.
- Integrated: This crucial characteristic implies that the data imported from the different source systems must go through a series of transformations so that they evolve from heterogeneous semantics, constraints, formats, and coding into a set of homogeneous results stored in the warehouse. This is the most difficult and time-consuming part of building the warehouse. In a well-regulated application domain (e.g., accounting or finance), this is purely a technical achievement. However, in other fields of activities, severe incompatibilities may exist among different sources, or within the same source through several years due to semantics evolutions, making it impossible to integrate certain data or to produce high quality results. To facilitate this integration process, warehousing technologies offer ETL

capabilities. Such ETL functions include semantics fusion/scission, identification matching, field reformatting, file merging/splitting, field merging/splitting, value recoding, constraints calibration, replacing missing values, measurement scales and units changing, updates filtering, adaptive value calculation, detecting unforeseen or exceptional values, smoothing noisy data, removing outliers, and applying integrity constraints to resolve inconsistencies. These ETL capabilities are sometimes called data cleansing, data scrubbing, data fusion, or data integration. Adherence to standards and to interoperability concepts helps minimize the integration problem. Of particular interest is the so-called "data reduction" process, where one produces a reduced volume of representative data that provides the same or similar analytical results that a complete warehouse would provide (Han and Kamber 2006).

- Nonvolatile: The transactional source systems usually contain only current or near-current data because their out-of-date data are replaced by new values and afterwards are destroyed or archived. On the other hand, warehouses keep these historic (also called "time-variant") data to allow trends analysis and prediction over periods of time (a key component of strategic decision making). Consequently, legacy data are said to be volatile because they are updated continuously (i.e., replaced by most recent values) while, on the other hand, warehouse data are nonvolatile, that is, they are not replaced by new values; they are kept for a long period along with the new values. However, to be more precise, one can specify about nonvolatile data that, "once inserted, [it] cannot be changed, though it might be deleted" (Date 2000). Reasons to delete data are usually not of a transactional nature but of an enterprise-oriented nature such as the decision to keep only the data of the last five years, to remove the data of a division that has been sold by the enterprise, to remove the data of a region of the planet where the enterprise has stopped to do business, etc. Thus, a data warehouse can grow in size (or decrease in rare occasions) but never be rewritten.
- Read-only: The warehouses can import the needed detailed data but they cannot alter the state of the source databases, making sure that the original data always rest within the sources. Such requirement is necessary for technical concerns (e.g., to avoid update loops and inconsistencies) but is mandatory to minimize organizational concerns (such as "Where is the original data?" "Who owns it?" "Who can change it?" and "Do we still need the legacy system?"). Thus, by definition, data warehouses are not allowed to write back into the legacy systems. However, although a data warehouse is conceptually not meant to act as an online transaction processing (OLTP) system oriented toward the entering, storing, updating, integrity checking, securing, and simple querying of data, it is sometimes built to allow direct entry of new data that is of high value for strategic decision making but which does not exist in legacy systems.
- Heterogeneous sources: As previously mentioned, the data warehouse is a new, additional system that does not aim at replacing, in a centralized

approach, the existing operational systems (usually called "legacy systems"). In fact, the implementation of a data warehouse is an attempt to get enterprise-level information while minimizing the impact on existing systems. Consequently, the data warehouse must obtain its data from various sources and massage these data until they provide the desired higher-level information. Usually, the data warehouse imports the raw data from the legacy systems of the organization, but it does not have to be limited to these in-house systems. In all cases, collecting metadata (i.e., data describing the integrated data and integration processes) is necessary to provide the required knowledge about the lineage and quality of the result. Recently, the quality of the produced high-level analytical data has become one of the main concerns of warehouse users; consequently, metadata have become more important and recent projects have introduced risk management approaches in data-warehouse design methods as well as automatic context-sensitive user warnings (see Chapter 5).

- Several levels of detail (also called "granularity" or "abstraction" levels): Decision makers need to get the global picture, but when they see unexpected trends or variations, they need to drill down to get more details to discover the reason for these variations. For example, when sales drop in the company, one must find if it is a general trend for all types of products, for all regions, and for all stores or if this is for a given region, a given store, or a specific category of products (e.g., sport equipment). If it is for a specific category such as sport equipment, one may want to dig further and find out if it is for a certain brand of products since a specific week. Thus, in order to provide fast answers to such multilevel questions, the warehouse must aggregate and summarize data by brand, category, store, region, periods, etc. at different levels of generalization. One such hierarchy could be store-city-region-country, another could be day-week number-quarter-year with a parallel hierarchy date-month-year. The term "granularity" is frequently used to refer to this hierarchical concept. For example, average sales of an individual salesperson is a fine-grained aggregation; average sales by department is coarser; and the sales of the whole company is the most coarse (i.e., a single number). The finest granularity refers to the lowest level of data aggregation to be stored in the database (Date 2000) or, in other words, the most detailed level of information. This may correspond to the imported source data or to a more generalized level if the source data have only served to calculate higher-level aggregations and summarizations before being discarded. Inversely, when talking about the most summarized levels, users sometimes talk about "indicators," especially when the quality of the source data is low. Such indicators give an approximate view of the global picture, which is often sufficient for decision-making purposes.
- Support of decision making: The sum of all the previous characteristics makes a data warehouse the best source of information to support decision making. Data warehouses provide the needed data stored in a structure, which is built specifically to answer global, homogeneous, multilevel, and

TABLE 3.1
Legacy System vs. Data Warehouse

Legacy Systems	Data Warehouse
Built for transactions, day-to-day repetitive operations	Built for analysis, decisions, and exploratory operations
Built for large number of simple queries using few records	Built for *ad hoc* complex queries using millions of records
Original data source with updates	Exact or processed copy of original data, in a read-only mode
Detailed data	Aggregated/summary data
Application-oriented	Enterprise-oriented
Current data	Current + historic data
Normalized data structure and built with the transactional paradigm/concepts	Denormalized data structure and typically built with the multidimensional paradigm/concepts
Top performance for transactions	Top performance for analysis

multiepoch queries from decision makers. This allows for the use of new decision-support tools and new types of data queries, exploration, and analyses, which were too time-consuming in the past.

The characteristics of data warehouses, in comparison to the usual transaction-oriented systems are presented in Table 3.1.

3.2.2 Multidimensional Data Structure

Data warehouses are typically structured using the multidimensional paradigm. Such a structure is preferred by decision-support tools, which dig into the data warehouse (e.g., OLAP, dashboards, and data mining tools). The multidimensional paradigm is built to facilitate the interactive navigation within the database, especially within its different levels of granularity, and to instantly obtain cross-tab information involving several themes of analysis (called *dimensions*) at multiple levels of granularity. It does so with simple functions such as drill-down (i.e., go to finer granularity within a theme), drill-up (i.e., go to coarser granularity), and drill-across (i.e., show another information at the same level of granularity). The term *multidimensional* results from the extension to N-dimensions of the usual matrix representation where the dependant variable is a cell within a 2-D space defined by two axes, one for each independent variable (e.g., purchases could be the cells, while countries and years the axes, giving immediately in the matrix all the purchases per country per year). In the literature, a multidimensional database is usually represented by a cube with three axes (since it is not possible to represent more dimensions), and accordingly the multidimensional database is usually called a *data cube* (or hypercube when N > 3).

The data models of the multidimensional paradigm are based on three fundamental concepts: (1) facts, (2) dimensions, and (3) measures (Rafanelli, 2003, Kimball and Ross 2002). A measure (e.g., total cost, number of items) is the attribute of a fact (e.g., sales), which represents the state of a situation concerning the themes or dimensions of interest (e.g., region, date, product). Thus, one can look at the measures of a fact for a given combination of dimensions (e.g., sales of 123,244 items and $25,000,000 for Canada in 2006 for sport equipment) and say that a measure is the dependent variable while the dimensions are the independent variables. Such an approach appears to be cognitively more compatible with the users' perception, thus facilitating the exploration of the database (i.e., selecting the independent variables first, then seeing what the dependent variable is). "The major reason why multidimensional systems appear intuitive is because they do their business the way we do ours" (Thomsen 2002). One can simply define a multidimensional query by saying "I want to know this (a measure) with regards to that (the dimensions elements)".

Each dimension has members; each member represents a position on the dimensional axis (e.g., January, February, March ...). The members of a single dimension may be structured in a hierarchical manner (e.g., year subdivided into quarters, quarters subdivided into months, months subdivided into weeks, weeks subdivided into days), creating the different levels of granularity of information. Alternative hierarchies can also be defined for the same dimension (e.g., year-month-day vs. year-quarter-week). A hierarchy where every child member has only one parent member is called a strict hierarchy. A nonstrict hierarchy has an M:N relationship between parent members and child members, leading to implementation strategies that create summarizability constraints. A hierarchy can be balanced or not; it is balanced when the number of aggregation levels remains the same whatever the members selected in the dimension.

Such a multidimensional paradigm can be modeled using three data structures: the star schema, the snowflake schema, and the fact constellation. A star schema contains one central fact table made of the measures and one foreign key per dimension to link the fact with the dimension's members (cf. using the primary key of the dimension tables), which are stored in one table per dimension, independent of a member's hierarchical level. A snowflake schema contains one central fact table similar to the star fact table, but the fact table foreign keys are linked to normalized dimension (typically, one table per hierarchical level). A fact constellation contains a set of fact tables, connected by some shared dimension tables. It is not uncommon to see hybrid schemas where some dimensions are normalized and others are not.

Since a data warehouse may consist of a good number of dimensions and each dimension may have multiple levels, there could be a large number of combinations of dimensions and levels, each forming an aggregated multidimensional "cube" (called *cuboids*). For example, a database with 10 dimensions, each having 5 levels of abstraction, will have 6^{10} cuboids. Due to limited storage space, usually only a selected set of higher-level cuboids will be computed as shown by Harinarayan, Rajaraman, and Ullman (1996). There have been many methods developed for efficient computation of multidimensional multilevel aggregates, such as Agarwal et al. (1996), and Beyer and Ramakrishnan (1999). Moreover, if the database contains a large number of dimensions, it is difficult to precompute a substantial number of cuboids due to the explosive number of cuboids. Methods have been developed for efficient high-dimensional OLAP, such as the one by Li, Han, and Gonzalez (2004).

Furthermore, several popular indexing structures, including bitmap index and join index structures have been developed for fast access of multidimensional databases, as shown in Chaudhuri and Dayal (1997). Han and Kamber (2006) give an overview of the computational methods for multidimensional databases.

3.2.3 Data Mart

It is frequent to define *data mart* as a specialized, subject-oriented, highly aggregated mini-warehouse. It is more restricted in scope than the warehouse and can be seen as a departmental or special-purpose sub-warehouse usually dealing with coarser granularity. Typically, the design of data marts relies more on users' analysis needs while a data warehouse relies more on available data. Several data marts can be created in an enterprise. Most of the time, it is built from a subset of the data warehouse, but it may also be built from an enterprise-wide transactional database or from several legacy systems. As opposed to a data warehouse, a data mart does not aim at providing the global picture of an organization. Within the same organization, it is common to see the content of several data marts overlapping. In fact, when an organization builds several data marts without a data warehouse, there is a risk of inconsistencies between data marts and of repeating, at the analytical level, the well-known chaotic problems resulting from silo databases at the transactional level. Figure 3.1 illustrates the distinctions between legacy systems, data warehouses, and data marts, whereas Table 3.2 highlights the differences between data warehouses and data marts.

In the face of the major technical and organizational challenges regarding the building of enterprise-wide warehouses, one may be tempted to build subject-specific data marts without building a data warehouse. This solution involves smaller

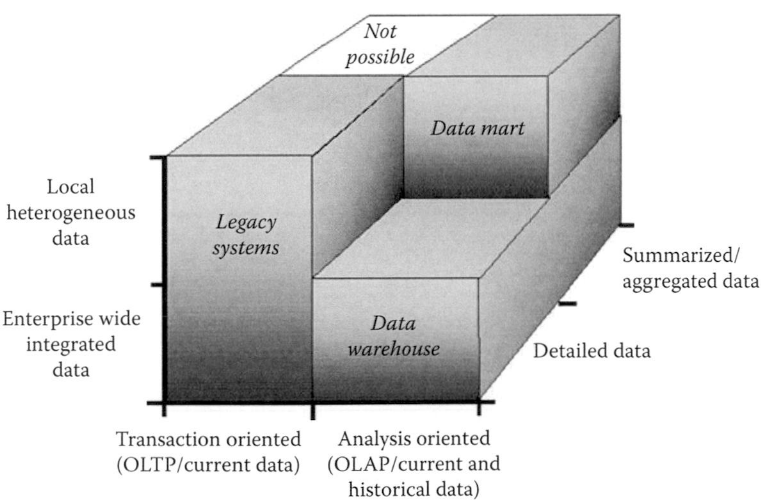

FIGURE 3.1 Comparison between legacy systems, data marts, and data warehouses.

TABLE 3.2
Data Warehouse vs. Data Mart

Data Warehouse	Data Mart
Built for global analysis	Built for more specific analysis
Aggregated/summarized data	Highly aggregated/summarized data
Enterprise-oriented	Subject-oriented
One per organization	Several within an organization
Usually multidimensional data structure	Always multidimensional data structure
Very large database	Large database
Typically populated from legacy systems	Typically populated from warehouse

investments, faster return on investments, and minimum political struggle. However, there is a long-term risk to see several data marts emerging throughout the organization and still have no solution to get the global organizational picture. Nevertheless, this alternative presents several short-term advantages. Thus, it is frequently adopted and may sometimes be the only possible alternative.

3.2.4 Online Analytical Processing (OLAP)

OLAP is a very popular category of decision-support tools that are typically used as clients of the data warehouse and data marts. OLAP provides functions for the rapid, interactive, and easy ad hoc exploration and analysis of data with a multidimensional user interface. Consequently, OLAP functions include the previously defined drill-down, drill-up, and drill-across functions as well as other navigational functions such as filtering, slicing, dicing, and pivoting (see OLAP Council 1995, Thomsen 2002, Wrembel and Koncilia 2006). Users may also be helped by more advanced functions such as focusing on exceptions or locations that need special attention by methods that mark the interesting cells and paths. Sarawagi, Agrawal, and Megiddo (1998) have studied this kind of discovery-driven exploration of data. Also, multifeature databases that incorporate multiple, sophisticated aggregates can be constructed, as shown by Ross, Srivastava, and Chatziantoniou (1998), to further facilitate data exploration and data mining.

OLAP technology provides a high-level user interface that applies the multidimensional paradigm not only to the selection of dimensions and levels within data cubes, but also to the way in which we navigate the different forms of data visualization (Fayyad, Grinstein, and Wierse 2001). Visualization capabilities include, for instance 2D or 3D tables, pie charts, histograms, bar charts, scatter plots, quantile plots, and parallel coordinates where the user can navigate (e.g., drill-down in a bar of a bar chart).

There are several possibilities to build OLAP-capable systems. Each OLAP client can be directly reading the warehouse and it can be used as a simple data exploration tool, or it can have its own data server. Such an OLAP server may structure the data with the relational approach, the multidimensional approach, or a combination of both (based on granularity levels and frequency of the uses of dimensions) (Imhoff, Galemmo, and Geiger 2003). These are then respectively called ROLAP (relational OLAP), MOLAP (multidimensional OLAP) and HOLAP (hybrid OLAP) although most users do not have to care about such distinctions because they are at the underlying implementation techniques Alternatively, one may use specialized SQL servers that support SQL queries over star/snowflake/constellation schemas (Han and Kamber 2006).

One may also encounter so-called "dashboard" applications with capabilities that are similar to OLAP. Although a dashboard can use OLAP components, it is not restricted to present aggregated data from a data cube, and it may also display data from transactional sources (e.g., from a legacy system), Web RSSs, streaming videos, Enterprise Resource Planning (ERP) systems, sophisticated statistical packages, etc. Dashboards wrap different types of data from diverse sources and present them in very simple, pre-defined panoramas and short repetitive sequences of operations to access, day-after-day, the same decision-support data. Strongly influenced by performance management strategies such as balanced scorecards, they are typically used by high-level strategic decision makers who rely on indicators characterizing the phenomena being analyzed. Being easier to use than OLAP, dashboards are not meant to be as flexible or as powerful as OLAP because they support decision processes that are more structured and predictable. They are very popular for top managers but are highly dependent on the proper choice of performance indicators.

3.2.5 Data Mining

Another popular client of the data warehouse server is a category of software packages or built-in functions called data mining. This category of knowledge discovery tools uses different techniques such as neural network, decision trees, genetic algorithms, rule induction, and nearest neighbor to automatically discover hidden patterns or trends in large databases and to make predictions [see Berson and Smith (1997) or Han and Kamber (2006) for a description of popular techniques]. Data mining really shines where it would be too tedious and complex for a human to use OLAP for the manual exploration of data or when there are possibilities to discover highly unexpected patterns. In fact, we use data mining to fully harness the power of the computer and specialized algorithms to help us discover meaningful patterns or trends that would have taken months to find or that we would have never found because of the large volume of data and the complexity of the rules that govern their correlations. Data mining supports the discovery of new associations, classifications, or analytical models by presenting the results with numerical values or visualization tools. Consequently, the line between OLAP and data mining may seem blurred in some technological offerings, but one must keep in mind that data mining is algorithm-driven while OLAP is user-driven and that they are complementary tools. Kim (1997) compares OLTP with DSS, OLAP, and data mining. A good direction for combining the strengths of OLAP and data mining is to study online analytical mining (OLAM) methods, where mining can be performed in an OLAP way, that is, exploring knowledge associated with multidimensional cube spaces by drilling,

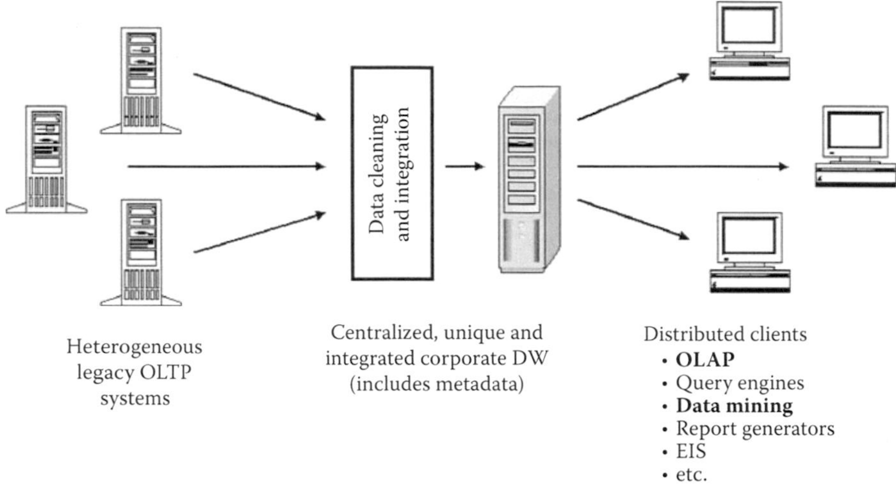

FIGURE 3.2 Generic architecture of a data warehouse.

dicing, pivoting, and using other user-driven data-exploration functions (Han and Kamber 2006).

3.2.6 Data Warehouse Architectures

Data warehouses can be implemented with different architectures depending on technological and organizational needs and constraints (Kimball and Ross 2002). The most typical one is also the simplest, called the *corporated architecture* (Weldon 1997) or the *generic architecture* (Poe 1995). It is represented in Figure 3.2. In such an architecture, the warehouse imports and integrates the desired data directly from the heterogeneous source systems, stores the resulting homogeneous enterprise-wide aggregated/summarized data in its own server, and lets the clients access these data with their own knowledge discovery software package (e.g., OLAP, data mining, query builder, report generator, dashboards). This two-tiered client-server architecture is the most centralized architecture.

There is a frequently used alternative called *federated architecture*. It is a partly decentralized solution and is presented in Figure 3.3. In this example, data are aggregated in the warehouse and other aggregations (at the same or a coarser level of granularity) are implemented in the data marts. This is a standard three-tiered architecture for data warehouses.

While the original concept of the data warehouse suggests that its granularity is very coarse in comparison with that of transaction systems, some organizations decide to keep the integrated detailed data in the warehouses in addition to generating aggregated data. In some cases, for example in the four-tiered architecture shown in Figure 3.4, two distinct warehouses exist. The first stores the integrated data at the granularity level of the source data, while the second warehouse aggregates these data to facilitate data analysis. Such architecture is particularly useful

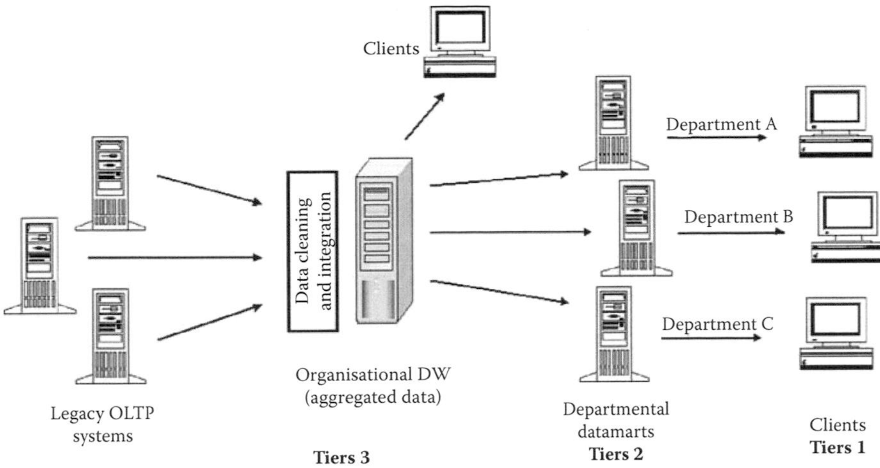

FIGURE 3.3 Standard federated three-tiered architecture of a data warehouse.

when the fusion of detailed source data represents an important effort and that the resulting homogeneous detailed database may have a value of its own besides feeding the second warehouse.

Many more alternatives exist such as the *no-warehouse architecture* (Figure 3.5), which may have two variations to support the data marts: with and without OLAP servers. Similarly, some variations of the previous architectures could also be made without an OLAP server. In this case, a standard Data Base Management Systems (DBMS) supports the star/snowflake schemas and the OLAP client does the mapping between the relational implementation and the multidimensional view offered to the user. In the short term, it results in easier data cube insertion within the organization (such as no software acquisition, smaller learning curve) but on the longer term, it results in increased workloads when building and refreshing data cubes (such as

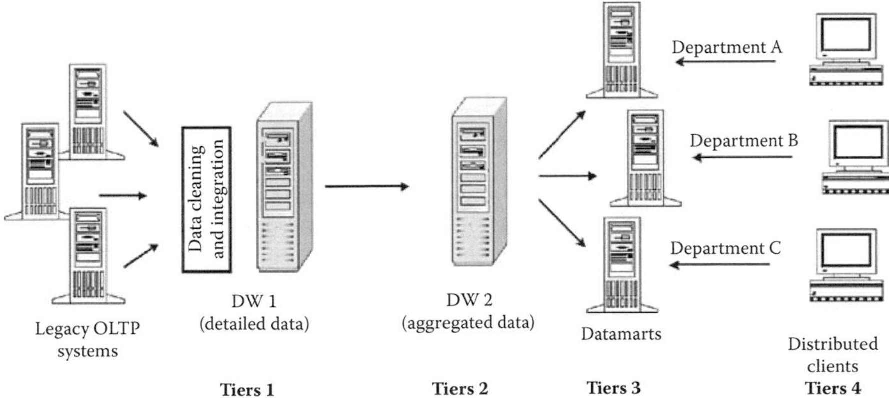

FIGURE 3.4 Multitiered architecture of a data warehouse.

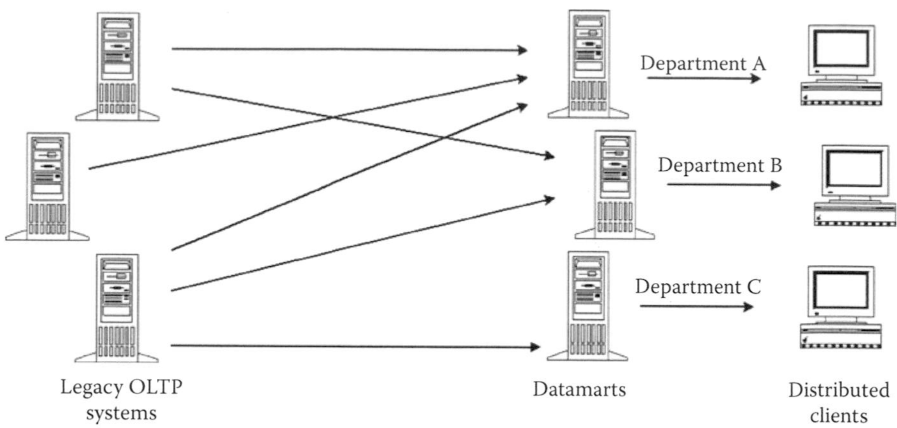

FIGURE 3.5 Data mart architecture without a data warehouse.

workmanship cost, repetitive tasks). Short-term contingencies, used technologies, personnel expertise, cube refreshment frequencies, existing workloads, and long-run objectives must be considered when building the warehouse architecture. Further variations exist when one takes into account the possibility of building virtual data warehouses. In this latter case, integration of data is performed on the fly and not stored persistently, which results in slower response times but smaller data cubes.

Finally, we believe that the data warehouse is a very useful infrastructure for data integration, data aggregation, and multidimensional online data analysis. The advances of new computer technology, parallel processing, and high-performance computing as well as the integration of data mining with data warehousing will make data warehouses more scalable and powerful at serving the needs of large-scale, multidimensional data analysis.

3.3 SPATIAL DATA WAREHOUSING

Spatially enabling data warehouses leads to richer analysis of the positions, shapes, extents, orientations, and geographic distributions of phenomena. Furthermore, maps facilitate the extraction of insights such as spatial relationships (adjacency, connectivity, inclusion, proximity, exclusion, overlay, etc.), spatial distributions (concentrated, scattered, grouped, regular, etc.), and spatial correlations (Bédard, Rivest, and Proulx 2007). When we visualize maps displaying different regions, it becomes easier to compare; when we analyze different maps of the same region, it becomes easier to discover correlations; when we see maps from different epochs, it becomes easier to understand the evolution of a phenomena. Maps facilitate understanding of the structures and relationships contained within spatial datasets better than tables and charts, and when we combine tables, charts, and maps, we increase significantly our potential for geographic knowledge discovery. Maps are natural aids to making the data visible when the spatial distribution of phenomena does not correspond to predefined administrative boundaries. Maps are active instruments to support the end-user's thinking process, leading to a more efficient knowledge discovery process (more alert brain, better visual rhythm, more global perception) (Bédard et al. 2007).

Today's GIS packages have been designed and used mostly for transaction processing. Consequently, GIS is not the most efficient solution for spatial data warehousing and strategic analytical needs of organizations. New solutions have been developed; in most cases, they rely on a coupling of warehousing technologies such as OLAP servers with spatial technologies such as GIS. Research started in the mid-1990s in several universities such as Laval (Bédard et al. 1997, Caron 1998, Rivest, Bédard and Marchand 2001), Simon Fraser (Stefanovic 1997, Han, Stefanovic and Koperski 1998), and Minnesota (Shekhar et al. 2001), and nowadays several researchers and practitioners have become active in spatial data warehousing, spatial OLAP, spatial data mining, and spatial dashboards. Several in-house prototypes have been developed and implemented in government and private organizations, and we have witnessed the arrival of commercial solutions on the market.

This coupling of geospatial technologies with data warehousing technologies has become more common. Some couplings are loose (e.g., import-export-reformatting of data between GIS and OLAP), some are semi-tight (e.g., OLAP-dominant spatial OLAP, GIS-dominant spatial OLAP), while others are tight (e.g., fully integrated spatial OLAP technology) (Rivest et al. 2005, Han and Kamber 2006). See the other chapters in this book and these recent publications for a description of these solutions: Rivest et al. (2005), Han and Kamber (2006), Damiani and Spaccapietra (2006), Bédard, Rivest and Proulx (2007), and Malinowski and Zimanyi (2008). For the remainder of this chapter, we focus on some fundamental spatial extensions of warehousing concepts: spatial data cubes, spatial dimensions, spatial measures, spatial ETL, and spatial OLAP operators (or spatial multidimensional operators).

3.3.1 Spatial Data Cubes

Spatial data cubes are data cubes where some dimension members or some facts are spatially referenced and can be represented on a map. There are two categories of spatial data cubes: feature-based and raster-based (Figure 3.6). Feature-based spatial data cubes include facts that correspond to discrete features having geometry (vectors or cells) or having no geometry (in which case dimension members must have a vector-based or raster-based geometry). Such fact geometry may be specific to this fact (in which case it may be derived from the dimensions) or it may correspond to the geometry of a spatial member. Raster spatial data cubes are made of facts that correspond to regularly subdivided spaces of continuous phenomena, each fine-grained fact being represented by a cell and every fine-grained cell being a fact.

Traditionally, transactional spatial databases consisted of separated thematic and cartographic data (e.g., using a relational database management system and a GIS). Nowadays, it is frequent to have both thematic and cartographic data stored together in a universal server or spatial database engine. Similarly, spatial data cubes can use thematic and cartographic data that are separated into different datasets (e.g., using an OLAP server and a GIS) or they can store natively cartographic data and offer built-in spatial aggregation/summarization operators. Insofar as practical spatial

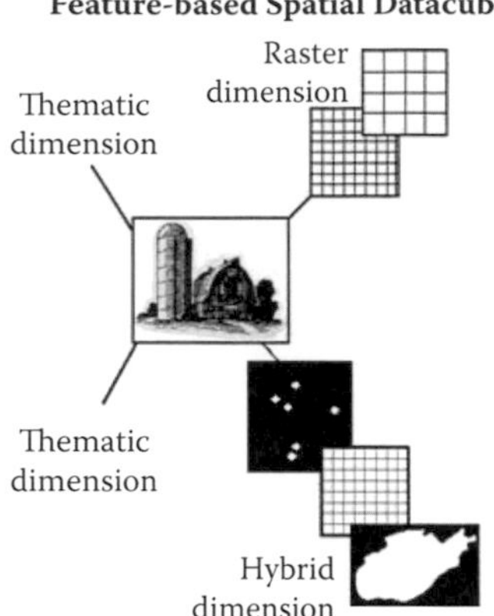

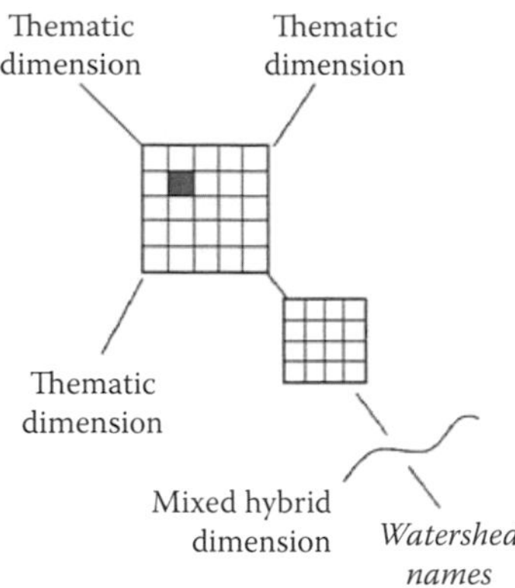

FIGURE 3.6 Feature-based and raster-based spatial data cubes. (Adapted from McHugh, R. 2008. *Etude du potentiel de la structure matricielle pour optimizer l'analyse spatial de données géodécisionnelles*. M.Sc. thesis (draft version), Department of Geomatics Sciences, Laval University, Quebec City, Canada.)

warehousing applications have been based on the coupling of spatial and nonspatial technologies, this is still the only solution commercially available.

3.3.2 Spatial Dimensions

In addition to the usual thematic and temporal dimensions of a data cube, there are spatial dimensions (in the multidimensional sense, not the geometric sense) that can be of three types according to the theory of measurement scales (cf. qualitative = nominal and ordinal scales, quantitative = interval and ratio scales, each scale allowing for richer analysis than the precedent one). These three types of dimension are as follows:

- *Nongeometric spatial dimension* contains only nominal or ordinal location of the dimension members, such as place names (e.g., St. Lawrence River), street addresses (134 Main Street), or hierarchically structured boundaries (e.g., Montreal → Quebec → Canada → North America). Neither shape nor geometry nor cartographic data are used. This is the only type of spatial dimension supported by nonspatial warehousing technology (e.g., OLAP). Caron (1998) has demonstrated the possibilities and limitations of such dimensions; they can only offer a fraction of the analytical richness of the other types of spatial dimensions (Bédard et al. 2007).
- *Geometric spatial dimension* contains a vector-based cartographic representation for every member of every level of a dimension hierarchy to allow the cartographic visualization, spatial drilling, or other spatial operation of the dimension members (Bédard et al. 2007). For example, every

city in North America would be represented by a point, every province/state of Canada, the United States, or Mexico would be represented as polygons, every North American country would also be represented as polygons, as well as North America itself. Similarly, polygons could represent equi-altitude regions in British Columbia, and every generalization, such as regions covering 0 to 500 m, 500 to 1000 m, and so on, would also be represented by a polygonal geometry.

- *Mixed geometric spatial dimension* contains a cartographic representation for some members of the dimension, and nominal/ordinal locators for the other members. For example, this could be all the members of certain levels of a dimension hierarchy (e.g., a point for every city, a polygon for every province/state, but only names for the countries and for North America, that is, no polygon for these latter two levels of the hierarchy). Then, the nongeometric levels can be the finest grained ones (to reduce the map digitizing efforts), the most aggregated ones (when users know exactly where they are), or anywhere in between, in any number and in any sequence. A mixed spatial dimension can also contain a cartographic representation for only some members of the same hierarchy level (e.g., all Canadian cities, but not all Mexican cities). The mixed spatial dimension offers some benefits of the geometric spatial dimension while suffering from some limitations of the nongeometric spatial dimension, all this at varying degrees depending on the type of mixes involved.

Furthermore, spatial dimensions relate to different ways to use geometry to represent a phenomenon: discrete feature-oriented topological vector data vs. continuous phenomena-oriented raster data (McHugh 2008). Depending on the type of geometry used, the users' potential to perform spatial analysis and geographic knowledge discovery changes significantly. As a result, users have the choice among seven categories of spatial dimensions as presented in Figure 3.7.

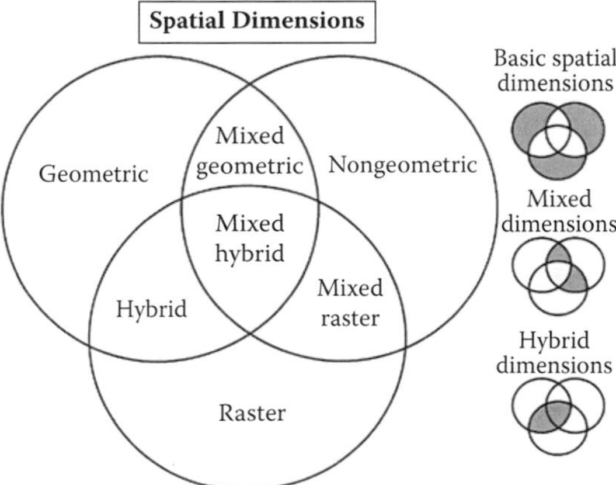

FIGURE 3.7 Raster-based and feature-based spatial data cubes with different examples of spatial dimensions.

The four additional categories of spatial dimensions are presented hereafter:

1. *Raster spatial dimension*: Every level of the dimension hierarchy uses the raster structure, the highest spatial resolution is used for the finest-grained level of the hierarchy. For instance, one could use 100-km cells for North America, 10-km cells for countries, and 1-km cells for province/states.
2. *Hybrid spatial dimension*: Some levels of the dimension hierarchy use the raster structure while other levels use the vector structure. For instance, this can be polygonal geometries for North America and for countries, while the raster structure is used for province/states. Inversely, this could be points for cities, polygonal geometries for provinces/states, 100-km raster cells for countries, and 1000-km cells for North America. All levels must be represented cartographically.
3. *Mixed raster spatial dimension*: Such a dimension contains raster data for some members of the dimension and nominal/ordinal locators for the other members (i.e., no geometry). For instance, this can be all the members of certain levels of a dimension hierarchy (e.g., cells for the province/state level, names for the country and North America levels, that is, no cartographic representation for these latter two levels of the hierarchy). Then, the same mixing possibilities exist as for the mixed geometric spatial dimension. A mixed raster spatial dimension can also contain raster cells for only some members of the same hierarchy level (e.g., all Canadian provinces, but not all American states). The mixed raster dimension offers some benefits of the raster spatial dimension while suffering from some limitations of the nongeometric spatial dimension, all this at varying degrees depending on the type of mixes involved.
4. *Mixed hybrid spatial dimension*: Such a dimension contains raster data for some members of the dimension, vector data for other members, and nominal/ordinal locators for the remaining ones (i.e., no geometry). For instance, this can be all the members of certain levels of a dimension hierarchy (e.g., names for cities, polygons for the province/state level, raster cells for the country level, and a name only for the North America level, that is, no cartographic representation for the finest and most aggregated levels of the hierarchy). Then, the same types of mixing possibilities exist as for the mixed geometric spatial dimension, without restriction. A mixed hybrid spatial dimension can also contain raster cells for some members of the same hierarchy level, polygons for other members, and no geometry for the remaining ones of this level (e.g., all Canadian provinces using raster cells, all American states using polygons, and Mexican states using names). The mixed-hybrid dimension offers some benefits of the raster and the geometric spatial dimensions while suffering from some limitations of the nongeometric spatial dimension, all this at varying degrees depending on the type of mixes involved.

More than one spatial dimension may exist within a spatial data cube (see Figure 3.7).

3.3.3 Spatial Measures

In addition to the nonspatial measures that still exist in a spatial data warehouse, we may distinguish three types of spatial measures (in the multidimensional sense):

1. *Numerical spatial measure*: Single value obtained from spatial data processing (e.g., number of neighbors, spatial density). Such measure contains only a numerical data and is also called *nongeometric spatial measure*.
2. *Geometric spatial measure*: Set of coordinates or pointers to geometric primitives that results from a geometric operation such as spatial union, spatial merge, spatial intersection, or convex hull computation. For example, during the summarization (or roll-up) in a spatial data cube, the regions with the same range of temperature and altitude will be grouped into the same cell, and the measure so formed contains a collection of pointers to those regions.
3. *Complete spatial measure*: Combination of a numerical value and its associated geometry. For example, the number of epidemic clusters with their location.

3.3.4 Spatial ETL

In spite of all these possibilities, it rapidly becomes evident that integrating and aggregating/summarizing spatial data requires additional processing in comparison to nonspatial data. For example, one must make sure that each source dataset is topologically correct before integration and that it respects important spatial integrity constraints, that the overlay of these maps in the warehouse is also topologically correct (e.g., without slivers and gaps) and coherent with regard to updates, that the warehouse maps at the different scales of analysis are consistent, that spatial reference systems and referencing methods are properly transformed, that the geometry of objects is appropriate for each level of granularity, that there is no mismatch problem between the semantic-driven abstraction levels of the dimension hierarchies and the cartographic generalization results (Bédard et al. 2007), that we deal properly with fuzzy spatial boundaries, etc. Consequently, spatial ETL requires an expertise about the very nature of spatial referencing (e.g., spatial reference systems and methods, georeferencing and imaging technologies, geoprocessing, and mapping) and one must not assume this process can be executed 100% automatically. Furthermore, there are issues related to the desire of users to clean and integrate spatial data from different epochs. Trade-offs have to be made and different types of decision-support analyses have to be left out because basic spatial units have been redefined over time, historical data have not been kept, data semantics and codification have been modified over time and are not directly comparable, legacy systems are not documented according to good software engineering practices, spatial reference systems have changed because of the fuzziness in spatial boundaries of certain natural phenomena that are re-observed at different epochs, the spatial precision of measuring technologies has changed, and so on (see Bernier and Bédard 2007 for problems with spatial data, and Kim 1999 for nonspatial data). One must realize that building and refreshing the multisource, multiscale, and multiepoch spatial data warehouse is feasible but requires efforts, strategic trade-offs, and a high level of expertise. In some cases, properly dealing with metadata, quality information, and context-sensitive user warnings becomes a necessity (Levesque et al. 2007).

Spatial ETL technologies are emerging. One may combine a warehouse-oriented nonspatial ETL tool or the built-in functions of OLAP servers with a spatial technology such as a GIS, an open-source spatial library, or a commercial transaction-oriented spatial ETL. One may also look for fully integrated spatial ETL prototypes that are in development in research centers. This new category of spatial ETL tools will include spatial aggregation/summarization operators to facilitate the calculation of aggregated spatial measures.

In spite of these difficulties, it remains possible to develop simple spatial warehousing applications if one keeps the cartographic requirements at a reasonable level. Many applications are running today and succeeded to minimize the above issues. For example, it is the case with administrative data that are highly regulated and are not redefined every five to ten years (e.g., cadastre, municipalities) or with data that have always been collected according to strictly defined procedures of known quality (e.g., topographic databases). With such datasets, the problems are minimal. However, with databases about natural phenomena or with databases that do not keep track of historical data, we must face some of the above-mentioned issues and choose to develop nontemporal data warehouses, semi-temporal data warehouses (historical data exist for given epochs but the data are not comparable over time), warehouses displaying nonmatching maps of different scales, and warehouses of varying data quality. It is our experience that a majority of the efforts to build spatial data warehouses goes to spatial ETL, and that the quality of existing legacy spatial data has an important impact on the design and building of the warehouse.

3.3.5 Spatial OLAP Operators

Spatial data cubes can be explored and analyzed using spatial OLAP operators. Spatial operations allow the navigation within the data cube concerning the spatial dimensions while keeping the same level of thematic and temporal granularities (Bédard et al. 2007). The SOLAP operators are executed directly on the maps and behave the same way as nonspatial operators. Basic operators include spatial drill-down, spatial roll-up, spatial drill-across, and spatial slice and dice, while the most advanced operators include spatial open, spatial close, views synchronization, etc. A detailed definition with examples is provided in Rivest et al. (2005). A recent survey of the commercial technologies proposed to develop spatial OLAP applications (Proulx, Rivest, and Bédard 2007) has shown that only the most tightly integrated SOLAP technologies properly support the basic spatial drill operations; the loosely coupled technologies use traditional "zoom" or "select layer" functions of the traditional transactional paradigm to simulate a change of abstraction level in a spatial dimension.

3.4 DISCUSSION AND CONCLUSION

We have presented an overview of the fundamental concepts of spatial data warehousing in the context of geographic knowledge discovery (GKD). Such spatial data warehouses have become important components of an organization infrastructure. They meet the needs of data integration and summarization from dispersed heterogeneous databases and facilitate interactive data exploration and GKD for power analysts and strategic decision makers. Spatial data warehouses provide fast, flexible, and multidimensional ways to explore spatial data when using the appropriate client technology (e.g., SOLAP, spatial dashboards). Several applications have been developed in many countries. However, as it is still the case with transactional geospatial systems, there remain challenges to populate efficiently such warehouses. Recent research is making significant progress along this direction as there are several universities, government agencies, and private organizations now involved in spatial data warehousing, SOLAP, and spatial dashboard. Today's research issues include the following:

- Spatial data cube interoperability (Sboui et al. 2007, 2008)
- Spatially enabling OLAP servers and ETL tools for spatial data cube (e.g., open-source GeoMondrian and GeoKettle projects by Dube and Badard)
- Developing spatial aggregation/summarization operators for spatial data cubes
- Improving spatial data cube and SOLAP design methods and modeling formalism
- Developing Web services for spatial warehousing (Badard et al. 2008)
- Quasi real-time spatial warehousing
- Mobile wireless spatial warehousing and on the fly creation of spatial mini-cubes (Badard et al. 2008, Dube 2008)
- Formal method to select the best quality legacy sources and ETL processes
- Formal methods and legal issues to properly manage the risk of warehouse data misuse (see Chapter 5)
- Improving the coupling of spatial and nonspatial technologies for spatial warehousing
- Improving the client tools that exploit the spatial warehouses
- Enriching spatial integrity constraints for aggregative spatial measures
- Improving spatial data mining approaches and methods
- Improving the coupling between spatial data mining and SOLAP
- Facilitating the automatic propagation of legacy data updates toward spatial data warehouses
- Increasing the capacities of raster spatial data cubes for interactive SOLAP analysis (McHugh 2008)
- Integrating spatial data mining algorithms, such as spatial clustering, classification, spatial collation pattern mining, and spatial outlier analysis methods, with spatial OLAP mechanisms
- Handling high dimensional spatial data warehouse and spatial OLAP, where a data cube may contain a large number of dimensions on categorical data (such as regional census data) but other dimensions are spatial data (such as maps or polygons)

- Integrating data mining methods as preprocessing methods for constructing high quality data warehouses, where data integration, data cleansing, clustering, and feature selection can be performed first by data mining methods before a data warehouse is constructed
- Integrating space with time to handle spatial temporal data warehouses and sensor-based or RFID-based spatial data warehouses

As it is the case with information technology in general, this field is evolving rapidly as new concepts are emerging, new experimentations are successful, and a larger community becomes involved. If one looks back five years ago, Googling "spatial data warehouse" or "SOLAP" did not return hundreds of hits, while nowadays this is the case. The scientific community is adopting the datacube paradigm to exploit spatial data and the number of papers is rapidly increasing. In a similar manner, the industry has recently started to offer commercial solutions and their improvements will drive the more general adoption by users in the short term.

ACKNOWLEDGMENTS

This research has been supported by Canada NSERC Industrial Research Chair in Spatial Databases for Decision-Support, by the Canadian GEOIDE Network of Centres of Excellence in Geomatics, and by the U.S. National Science Foundation NSF IIS-05-13678 and BDI-05-15813.

REFERENCES

Agarwal, S., Agrawal, R., Deshpande, P.M., Gupta, A., Naughton, J.F., Ramakrishnan, R., and Sarawagi, S., 1996. On the computation of multidimensional aggregates. In *Proc. 1996 Int. Conf. Very Large Data Bases*, pp. 506–521, Morgan Kaufmann, Mumbai, India, September.

Badard, T., Bédard, Y., Hubert, F., Bernier, E., and E. Dube, 2008. Web Services Oriented Architectures for Mobile SOLAP Applications. *International Journal of Web Engineering and Technology* (accepted). Inderscience Publishers, Geneva, Switzerland, p. 434–464.

Bédard, Y., Larrivée, S., Proulx, M-J., Caron, P-Y., and Létourneau, F., 1997. *Étude de l'état actuel et des besoins de R&D relativement aux architectures et technologies des data warehouses appliquées aux données spatiales.* Research report for the Canadian Defense Research Center in Valcartier, Centre for Research in Geomatics, Laval University, Quebec City, Canada, p. 98.

Bédard, Y., Rivest, S., and Proulx, M.-J. 2007. Spatial on-line analytical processing (SOLAP): concepts, architectures and solutions from a geomatics engineering perspective. In Wrembel R. and Koncilia, C., *Data Warehouses and OLAP: Concepts, Architectures and Solutions*, London, UK: IRM Press, pp. 298–319.

Bernier, E. and Bédard, Y., 2007. A data warehouse strategy for on-demand multiscale mapping. In Mackaness, W., Ruas, A., and Sarjakoski T. (Eds.), *Generalisation of Geographic Information: Cartographic Modelling and Applications*, International Cartographic Association, Elsevier, Oxford, UK. pp. 177–198.

Badard, T. 2008. Personal website. http://geosoa.scg.ulaval.ca/en/index.php?module=pagemaster&PAGE_user_op_view_page & PAGE_id=17. Accessed November 21, 2007.

Berson, A. and Smith, S.J., 1997. *Data Warehousing, Data Mining & OLAP*, McGraw- Hill. New York, NY, USA.

Beyer, K. and Ramakrishnan, R., 1999. Bottom-up computation of sparse and iceberg cubes. *Proc. 1999 ACM-SIGMOD Int. Conf. Management of Data (SIGMOD'99),* Philadelphia, PA, pp. 359–370, June. On-line proceedings, Association for computing Machinery, NewYork, NY, USA, http://www.sigmod.org/sigmod99/eproceedings. Accessed, November 20, 2008.

Brackett M.H., 1996, *The Data Warehouse Challenge: Taming Data Chaos*, John Wiley & Sons, New York.

Caron, P.Y., 1998. *Étude du potentiel de OLAP pour supporter l'analyse spatio-temporelle*. M.Sc. thesis, Centre for Research in Geomatics, Laval University, Quebec City, Canada.

Chaudhuri, S. and Dayal, U., 1997. An overview of data warehousing and OLAP technology. *ACM SIGMOD Record*, 26:65–74. Tucson, Arizona, USA, May 13–15, ACM Press, New York, NY, USA, Association for Computing Machinery.

Damiani, M.L. and Spaccapietra, S., 2006. Spatial data warehouse modeling. In: *Processing and Managing Complex Data for Decision Support*. Darmont, J. and Boussaid, O. (Eds.), Idea Group, Hershey, PA, USA, pp. 1–27.

Date, C.J., 2000. *An Introduction to Database Systems*, 7th ed., Addison-Wesley. Reading, MA, USA.

Dube, E., 2008. *Developpement d'un service web de constitution en temps reel de mini cubes SOLAP pour clients mobiles*. M.Sc. thesis (draft), Centre for Research in Geomatics, Department of Geomatics Sciences, Laval University, Quebec City, Canada.

Fayyad, U.M., Grinstein, G., and Wierse, A., 2001. *Information Visualization in Data Mining and Knowledge Discovery*. Morgan Kaufmann, San Francisco, CA, USA.

Gill, S.H. and Rao, P.C., 1996. *The Official Client/Server Computing Guide to Data Warehouse*. QUE Corporation. Indianapolis, IN, USA.

Han, J. and Kamber, M., 2006. *Data Mining: Concepts and Techniques*, 2nd ed., Morgan Kaufmann, San Francisco, CA, USA.

Han, J., Stefanovic, N., and Koperski, K., 1998. Selective materialization: An efficient method for spatial data cube construction, *Proceedings of the Second Pacific-Asia Conference, PAKDD'98*, pp. 144–158. Melbourne, Australia, April 15–17, *Computer Science*, Vol. 1394, Springer Berlin/Heidelbert.

Harinarayan, V., Rajaraman, A., and Ullman, J.D., 1996. Implementing data cubes efficiently, *Proc. 1996 ACM-SIGMOD Int. Conf. Management of Data*, pp. 205–216, Montreal, Canada, June 4–6, ACM Press, New York, NY, USA. Association for Computing Machinery.

Imhoff, C., Galemmo, N., and Geiger, J.G., 2003. *Mastering Data Warehouse Design: Relational and Dimensional Techniques*, John Wiley, New York.

Inmon, W.H., Richard, D., and Hackathorn, D., 1996. *Using the Data Warehouse*, John Wiley & Sons, New York.

Kim, W., 1997. OLTP versus DSS/OLAP/data mining, *Journal of Object-Oriented Programming*, Nov.–Dec., pp. 68–77.

Kim, W., 1999. I/O problems in preparing data for data warehousing and data mining, Part 1. *Journal of Object-Oriented Programming*, February, pp. 13–17. Sigs Publications, Van Nuys, CA, USA.

Kimball, R. and Ross, M., 2002. *The Data Warehouse Toolkit: The Complete Guide to Dimensional Modeling*, 2nd ed., Wiley, New York.

Levesque, M.-A., Bédard, Y., Gervais, M., and Devillers, R., 2007. Towards a safer use of spatial datacubes: Communicating warnings to users, *Proceedings of the 5th International Symposium on Spatial Data Quality*, Enschede, Netherlands, June 13–15. Online Proceedings http://www.itc.nl/ISSDQ2007/proceedings/index.html. Accessed November 20, 2008.

Li, X., Han, J., and Gonzalez, H., 2004. High-dimensional OLAP: A minimal cubing approach, *Proc. 2004 Intl. Conf. Very Large Data Bases (VLDB'04),* Toronto, Canada, pp. 528–539, August. Lecture notes in *Computer Science*, Springer Berlin/Heidelbert.

McHugh, R., 2008. *Etude du potentiel de la structure matricielle pour optimizer l'analyse spatial de données géodécisionnelles.* M.Sc. thesis (draft version), Department of Geomatics Sciences, Laval University, Quebec City, Canada.

Malinowski, E. and Zimanyi, E., 2008. Advanced Data Warehouse Design: From Conventional to Spatial and Temporal Applications, Springer, Berlin, Heidelbert.

OLAP Council, 1995. OLAP council white paper, http://www.olapcounci/.org/research/resrchly.htm. Accessed November 20, 2008, Delware, USA.

Poe, V., 1995. *Building a Data Warehouse for Decision Support*, Prentice Hall, Englewood Cliffs, NJ.

Proulx, M., Rivest, S., and Bédard, Y., 2007. *Évaluation des produits commerciaux offrant des capacités combinées d'analyse multidimensionnelle et de cartographie*, Research report for the partners of the Canada NSERC Industrial Research Chair in Geospatial Databases for Decision-Support, Centre for Research in Geomatics, Laval University, Quebec City, Canada.

Rafanelli, M., 2003. *Multidimensional Databases: Problems and Solutions*. Idea Group Publishing, London, UK.

Rivest, S., Bédard, Y., and Marchand, P., 2001. Towards better support for spatial decision-making: Defining the characteristics of spatial on-line analytical processing, *Geomatica, Journal of the Canadian Institute of Geomatics*, 55(4), 539–555. Ottawa, Canada

Rivest, S., Bédard, Y., Proulx, M.-J., Nadeau, M., Hubert, F., and Pastor, J., 2005. SOLAP: Merging business intelligence with geospatial technology for interactive spatio-temporal exploration and analysis of data, *Journal of International Society for Photogrammetry and Remote Sensing (ISPRS)*, 60(1), 17–33. Elsevier, Amsterdam, Netherlands.

Ross, K.A., Srivastava, D., and Chatziantoniou, D., 1998. Complex aggregation at multiple granularities, *Proc. Int. Conf. of Extending Database Technology (EDBT'98)*, Valencia, Spain, pp. 263–277, March 23–27, Lecture notes in *Computer Science* Vol. 1377, Springer Berlin/Heidelbert.

Sarawagi, S., Agrawal, R., and Megiddo, N., 1998. Discovery-driven exploration of OLAP data cubes, *Proc. Intl. Conf. of Extending Database Technology (EDBT'98),* Valencia, Spain, pp. 168–182, March 23–27, Lecture notes in *Computer Science* Vol. 1377, Springer Berlin/Heidelbert.

Sboui, T., Bédard, Y., Brodeur, J., Badard, T., 2007. A conceptual framework to support semantic interoperability of geospatial datacubes, *Proceedings of ER/2007 SeCoGIS Workshop*, November 5–9, Auckland, New Zealand, Lecture Notes in Computer Sciences, Springer, pp. 378–387. Springer Berlin/Heidelbert.

Sboui, T., Bédard, Y., Brodeur, J., and Badard, T., 2008. Risk management for the simultaneous use of spatial datacubes: A semantic interoperability perspective. *International Journal of Business Intelligence and Data Mining* (submitted). Interscience Publishers, Geneva, Switzerland.

Shekhar, S., Lu, C.T., Tan, X., Chawla, S., and Vatsavai, R., 2001. Map cube: A visualization tool for spatial data warehouses. In: Miller, H.J. and Han, J. (Eds.), *Geographic Data Mining and Knowledge Discovery*, Taylor & Francis, London, UK.

Stefanovic, N., 1997. Design and Implementation of On-Line Analytical Processing (OLAP) of Spatial Data. M.Sc. thesis, Simon Fraser University, Burnaby, British Columbia, Canada.

Thomsen, E., 2002. *OLAP Solutions: Building Multidimensional Information Systems,* 2nd ed., John Wiley & Sons, New York.

Weldon, J.L., 1997. State of the art: Warehouse cornerstones. *Byte*, p.87, January.

Wrembel, R. and Koncilia, C., 2006. *Data Warehouses and OLAP: Concepts, Architectures and Solutions.* IRM Press (IDEA Group Publishing), London.

Zhou, X., Truffet, D., and Han, J., 1999. Efficient polygon amalgamation methods for spatial OLAP and spatial data mining, Proc. *6th Int. Symp. on Large Spatial Databases (SSD'99),* Hong Kong, pp. 167–187, July 20–23 rd, Lecture Notes in *Computer Science*, vol. 1651, Springer Berlin Heidelbert.

4 Analysis of Spatial Data with Map Cubes: Highway Traffic Data

Chang-Tien Lu

Arnold P. Boedihardjo

Shashi Shekhar

CONTENTS

4.1 INTRODUCTION

A DW is a repository of subject-oriented, integrated, and nonvolatile information, aimed at supporting knowledge workers (executives, managers, and analysts) to make better and faster decisions (Chaudhuri and Dayal 1997; Immon 1993; Immon and Hackathorn 1994; Immon, Welch, and Glassey 1997; Kimball et al. 1998; Widom 1995). Data warehouses contain large amounts of information, which is collected from a variety of independent sources and is often maintained separately from the operational databases. Traditionally, operational databases are optimized for online transaction processing (OLTP), where consistency and recoverability are critical. Transactions are typically small and access a small number of individual records based on the primary key. Operational databases maintain current state information. In contrast, DWs maintain historical, summarized, and consolidated information, and are designed for online analytical processing (OLAP) (Codd 1995; Codd, Codd, and Salley 1993). The data in the warehouse are often modeled as a multidimensional space to facilitate the query engines for OLAP, where queries typically aggregate data across many dimensions in order to detect trends and anomalies (Mumick, Quass, and Mumick 1997). There is a set of numeric measures that are the subjects of analysis in a multidimensional data model. Each of the numeric measures is determined by a set of dimensions. In a traffic DW, for example, a fundamental traffic measure is vehicle flow where dimensions of interest are segment, freeway, day, and week. Given N dimensions, the measures can be aggregated in 2^N different ways. The SQL aggregate functions and the group-by operator only produce one out of 2^N aggregates at a time. A data cube (Gray et al. 1995) is an aggregate operator which computes all 2^N aggregates in one shot.

Spatial DWs contain geographic data, for example, satellite images, and aerial photography (Han, Stefanovic, and Koperski 1998; Microsoft 2000) in addition to nonspatial data. Examples of spatial DWs include the U.S. Census data-set (Ferguson 1999; USCB 2000), Earth Observation System archives of satellite imagery (USGS 1998), Sequoia 2000 (Stonebraker, Frew, and Dozier 1993), and highway traffic measurement archives. The research in spatial DWs has focused on case-studies (ESRI 1998; Microsoft 2000) and on the per-dimension concept hierarchy (Han, Stefanovic, and Koperski 1998). A major difference between conventional and spatial DWs lies in the visualization of the results. Conventional DW OLAP results are often shown as summary tables or spreadsheets of text and numbers, whereas in the case of spatial DW the results may be albums of maps. It is not trivial to convert the alpha-numeric output of a data cube on spatial DWs into an organized collection of maps. Another issue concerns the aggregate operators on geometry data types (e.g., point, line, polygon). Neither existing databases nor the emerging standard for geographic data, OGIS (OGIS 1999), has addressed this issue. In this chapter we present the *map cube*, an operator based on the conventional data cube but extended for spatial DWs. With the *map cube* operator, we visualize the data cube in the spatial domain via the superimposition of a set of alpha-numeric measures on its spatial representation (i.e., map). The unified view of a map with its (nonspatial) alpha-numeric measures can help minimize the complexity associated with viewing multiple and related conceptual entities. For spatial applications, for example, traffic data, which require aggregation on different dimensions and comparison between different categories in each attribute, the map cube operator can provide an effective mode of visualization.

A map cube is an operator which takes a base map, associated data tables, aggregation hierarchy, and cartographic preferences to produce an album of maps. This album of maps is organized using the given aggregation hierarchy. The goal is to support exploration of the map collection via roll-up, drill-down, and other operations on the aggregation hierarchy. We also provide a set of aggregate operators for geometric data types and classify them using a well-known classification scheme for aggregate functions. In summary, the proposed map cube operator generates a set of maps for different categories in the chosen dimensions, thus providing an efficient tool for pattern analysis and cross-dimension comparison. At its foundation, traffic data is composed of temporally aggregated measures (e.g., vehicle speed) associated with a set of static spatial properties. The importance of a particular traffic measure is dependent on its spatial location; hence, the task of traffic analyses will necessitate the inclusion of its corresponding spatial information. For example, a typical incident detection scheme employs information at both the upstream and downstream traffic to determine the occurrence of an incident at a particular location. Given the nature of traffic data, the concept of a map cube can be applied to provide an effective tool for the visualization and analyses of traffic information.

This chapter is organized as follows. Section 4.2 discusses some basic concepts of DWs and geographic information systems (GIS). In Section 4.3, the definition and operation of the map cube are introduced. Section 4.4 introduces an application of the map cube for traffic data. Section 4.5 presents a case study of the map cube application for traffic incident analysis. Finally, Section 4.6 includes a summary and a discussion of future research directions.

4.2 BASIC CONCEPTS

4.2.1 AN EXAMPLE OF GEOGRAPHIC INFORMATION SYSTEM

A GIS (Chrisman 1996; Worboys 1995) is a computer-based information system consisting of an integrated set of programs and procedures which enable the capture, modeling, manipulation, retrieval, analysis, and presentation of geographically referenced data. We would define common concepts in GIS via the entities in Figure 4.1. The purpose is to provide a context for a map cube and to show how it may relate to other concepts in GIS. Figure 4.1 is not designed to capture all the concepts in all popular GIS. A map is often organized as a collection of vector and raster layers as shown in Figure 4.1. Each map has its own visual representation, including graphics, layout, legend, and title.

In a raster layer representation, the space is commonly divided into rectangular grid cells, often called pixels. The locations of a geographic object or conditions can be defined by the row and column of the pixels they occupy. The area that each pixel represents may characterize the spatial resolution available. The raster layer is

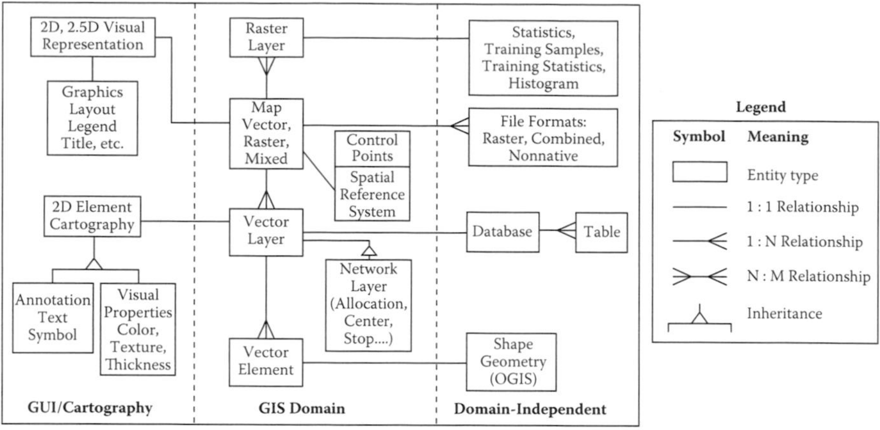

FIGURE 4.1 Concepts in Geographic Information Systems.

a collection of pixels and may represent raw images collected directly from satellites, aerial photography, etc. The raster layer may also represent interpreted images showing a classification of areas.

Information associated with the raster layer representation includes statistics, training samples, training statistics, and histograms for the classified images, among other things. Mean, standard deviation, variance, minimum value, maximum value, variance-covariance matrix, and correlation matrix are some examples of statistical information. Training samples and training statistics are used by supervised classification, which is performed when the analyst has the knowledge about the scene, or has identity and location information about the land cover types, for example, forest, agriculture crops, water, urban, etc. Training samples are collected through a combination of fieldwork, existing maps, or interpretation of high resolution imagery. Based on these training samples, multivariate training statistics are computed from each band for each class of interest. Each pixel is then classified into one of the training classes according to classification decision rules (e.g., minimum distance to means, maximum likelihood, etc.). Vector layers are collections of vector elements. The shape of a vector element may be zero dimensional (point), one dimensional (curves, lines) or two dimensional (surface, polygons). For example, in a vector layer representation, an object type *house* may have attributes referencing further object types: polygon, person name, address, and date. The polygon for the house is stored as a "vector element" in Figure 4.1. A vector layer may have its own cartographic preferences to display the elements. These cartographic preferences include text, symbol, and some visual properties such as color, texture, and thickness. The elements and attributes in the vector layers may be associated with nonspatial attributes managed by a database system which consists of many tables.

A network layer is a special case of a vector layer in Figure 4.1. It is composed of a finite collection of points, the line-segments connecting the points, the location of the points, and the attributes of the points and line-segments. For example, a network layer for transportation applications may store road intersection points and the road segments connecting the intersections.

Maps are also associated with reference systems and control points. A spatial reference system is a coordinate system attached to the surface of the earth, which allows users to locate the map element on the surface of the earth. Different map layers can be geo-registered to a common spatial reference system, thus producing a composite overlay for analysis and display. Control points are common to the base map and the slave map being prepared. The exact locations of the control points are well defined. Examples include intersection of roads and railways or other landmarks or monuments. The control points are used to geo-register newly acquired maps to the well-defined base map at different scales.

4.2.2 Aggregate Functions

A data cube consists of a lattice of cuboids, each of which represents a certain level of hierarchy. Aggregate functions compute statistics for a given set of values within each cuboid. Examples of aggregate functions include sum, average, and centroid. Aggregate functions can be grouped into three categories, namely, distributive, algebraic, and holistic as suggested by Gray et al. (1995). We define these functions in this section and provide some examples from the GIS domain. Table 4.1 shows all of these aggregation functions for different data types.

- **Distributive:** An aggregate function F is called distributive if there exists a function G such that the value of F for an N-dimensional cuboid can be computed by applying a G function to the value of F in an $(N + 1)$-dimensional cuboid. For example, when $N = 1$, $F(M_{ij}) = G(F(C_j)) = G(F(R_i))$, where M_{ij} represents the elements of a two-dimensional matrix, C_j denotes each column of the matrix, and R_i denotes each row of the matrix. Consider the aggregate function Min and Count as shown in Figure 4.2. In the first example, $F = Min$, then $G = Min$, since $Min(M_{ij}) = Min(Min(C_j)) = Min(Min(R_i))$. In the second example, $F = Count$, $G = Sum$, since $Count(M_{ij})$

TABLE 4.1
Aggregation Operations

Data Type	Aggregation Function: Distributive Function	Algebraic Function	Holistic Function
Set of numbers	Count, min, max, sum	Average, standard deviation, MaxN, MinN()	Median, most frequent, rank
Set of points, lines, polygons	Minimal orthogonal bounding box, geometric union, geometric intersection	Centroid, center of mass, center of gravity	Nearest neighbor index, equi-partition

Distributive Aggregate Function: Min

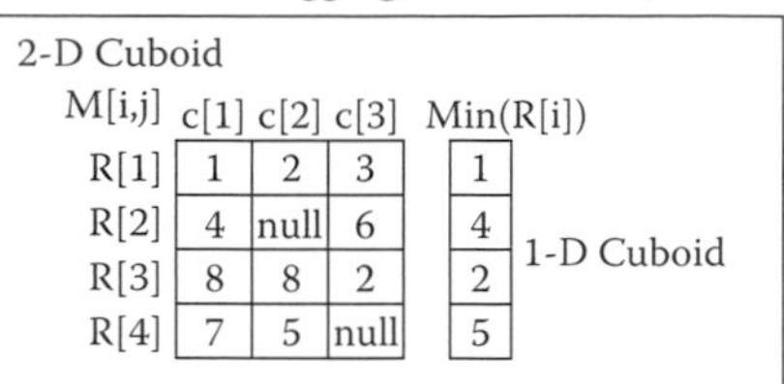

2-D Cuboid
M[i,j] c[1] c[2] c[3] Min(R[i])
R[1] 1 2 3 | 1
R[2] 4 null 6 | 4
R[3] 8 8 2 | 2
R[4] 7 5 null | 5
1-D Cuboid

Min(C[j]) 1 2 2 | 1 0-D Cuboid

1-D Cuboid Min(M[i,j]) = Min(Min of 1-D cuboid)

Distributive Aggregate Function: Count

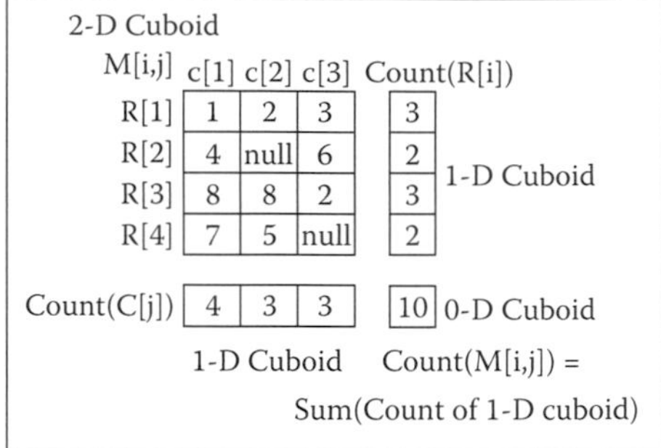

FIGURE 4.2 Computation of distributive aggregate function.

$= Sum(Count(C_j)) = Sum(Count(R_i))$. Other distributive aggregate functions include Max and Sum. Note that "null" valued elements are ignored in computing aggregate functions. Distributive GIS aggregate operations include minimal orthogonal bounding box, geometric union, and geometric intersection. The geometric union is a binary operation that takes two sets of geometric areas and returns the set of regions that are covered by at least one of the original areas. For all of these aggregations, the operator aggregates the computed regions of the subset, and then computes the final result.

- **Algebraic:** An aggregate function F is algebraic if F of an N-dimensional cuboid can be computed using a fixed number of aggregates of the $(N+1)$-dimensional cuboid. Average, variance, standard deviation, MaxN, and MinN are all algebraic. In Figure 4.3, for example, the computations of average and variance for the matrix M are shown. The average of elements

Algebraic Aggregate Function: Average

2-D Cuboid
M[i,j] c[1] c[2] c[3] | Avg (R[i]) | Sum (R[i]) | Count (R[i])
R[1] 1 2 3 | 2 | 6 | 3
R[2] 4 null 6 | 5 | 10 | 2
R[3] 8 8 2 | 6 | 18 | 3
R[4] 7 5 null | 6 | 12 | 2
1-D Cuboid

Avg(C[j]) 5 5 3.6 | 4.6 0-D Cuboid
Sum(C[j]) 20 15 11
Count(C[j]) 4 3 3
1-D Cuboid

$$F(M) = \frac{\sum_{i=1}^{4} Sum(R[i])}{\sum_{i=1}^{4} Count(R[i])} = \frac{\sum_{j=1}^{3} Sum(C[j])}{\sum_{j=1}^{3} Count(C[j])}$$

Algebraic Aggregate Function: Variance

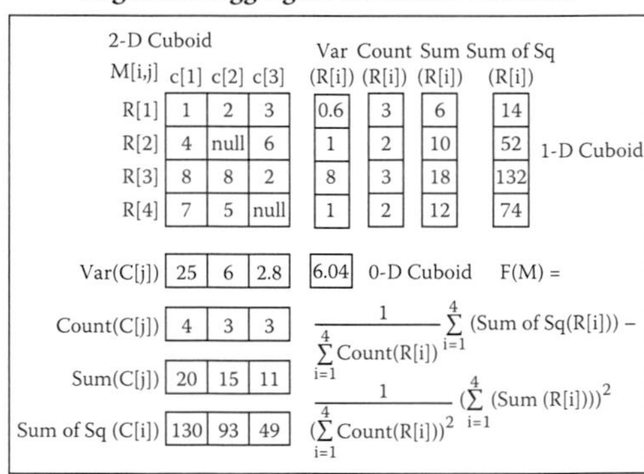

2-D Cuboid
M[i,j] c[1] c[2] c[3] | Var (R[i]) | Count (R[i]) | Sum (R[i]) | Sum of Sq (R[i])
R[1] 1 2 3 | 0.6 | 3 | 6 | 14
R[2] 4 null 6 | 1 | 2 | 10 | 52
R[3] 8 8 2 | 8 | 3 | 18 | 132
R[4] 7 5 null | 1 | 2 | 12 | 74
1-D Cuboid
Var(C[j]) 25 6 2.8 | 6.04 0-D Cuboid
Count(C[j]) 4 3 3
Sum(C[j]) 20 15 11
Sum of Sq (C[i]) 130 93 49
F(M) = 1/(Σ_{i=1}^{4} Count(R[i])) Σ_{i=1}^{4} (Sum of Sq(R[i])) − 1/(Σ_{i=1}^{4} Count(R[i]))^2 (Σ_{i=1}^{4} (Sum (R[i])))^2

FIGURE 4.3 Computation of algebraic aggregate function.

in the two-dimensional matrix M can be computed from sum and count values of the 1-D sub-cubes (e.g., rows or columns). The variance can be derived from count, sum (i.e., $\Sigma_i X_i$), and sum of Sq (i.e., $\Sigma_i X_i^2$), of rows or columns. Similar techniques apply to other algebraic functions. An algebraic aggregate operation in GIS is center. The center of n geometric points $V^i = (V_x^i, V_y^i)$) is defined as $Center = \frac{1}{n}\Sigma V^i, C_x = \frac{\Sigma V_x}{n}, C_y = \frac{\Sigma V_y}{n}$. Both the center and the count are required to compute the result for the next layer. The center of mass and the center of gravity are other examples of algebraic aggregate functions.

- **Holistic:** An aggregate function F is called holistic if the value of F for an N-dimensional cuboid cannot be computed from a constant number of aggregates of the $(N + 1)$-dimensional cuboid. To compute the value of F at each level, we need to access the base data. Examples of holistic functions include median, most frequent, and rank.

Holistic GIS aggregate operations include equi-partition and nearest-neighbor index. Equi-partition of a set of points yields a line L such that there are the same number of point objects on each side of L. Nearest-neighbor index measures the degree of clustering of objects in a spatial field. If a spatial field has the property that like values tend to cluster together, then the field exhibits a high nearest-neighbor index. When new data are added, many of the tuples in the nearest neighbor relationship may change. Therefore, the nearest-neighbor index is holistic. The line of equi-partition could be changed with any new added points. To compute the equi-partition or nearest neighbor-index in all levels of dimensions, we need the base data.

The computation of aggregate functions has graduated difficulty. The distributive function can be computed from the next lower level dimension values. The algebraic function can be computed from a set of aggregates of the next lower level data. The holistic function needs the base data to compute the result in all levels of dimension.

4.2.3 Aggregation Hierarchy

The CUBE operator (Gray et al. 1995) generalizes the histogram, cross-tabulation, roll-up, drill-down, and sub-total constructs. It is the N-dimensional generalization of simple aggregate functions. Figure 4.4 shows the concept for aggregations up to three dimensions. The dimensions are year, company, and region. The measure is sales. The 0-D data cube is a point that shows the total summary. There are three 1-D data cubes: group-by region, group-by company, and group-by year. The three 2-D data cubes are cross tabs, which are a combination of these three dimensions. The 3-D data cube is a cube with three intersecting 2-D cross tabs.

Figure 4.5 shows the tabular forms of the total elements in a 3-D data cube after a CUBE operation. Creating a data cube requires generating a power set of the aggregation columns. The cube operator treats each of the N aggregation attributes as a dimension of the N-space. The aggregate of a particular set of attribute values is a

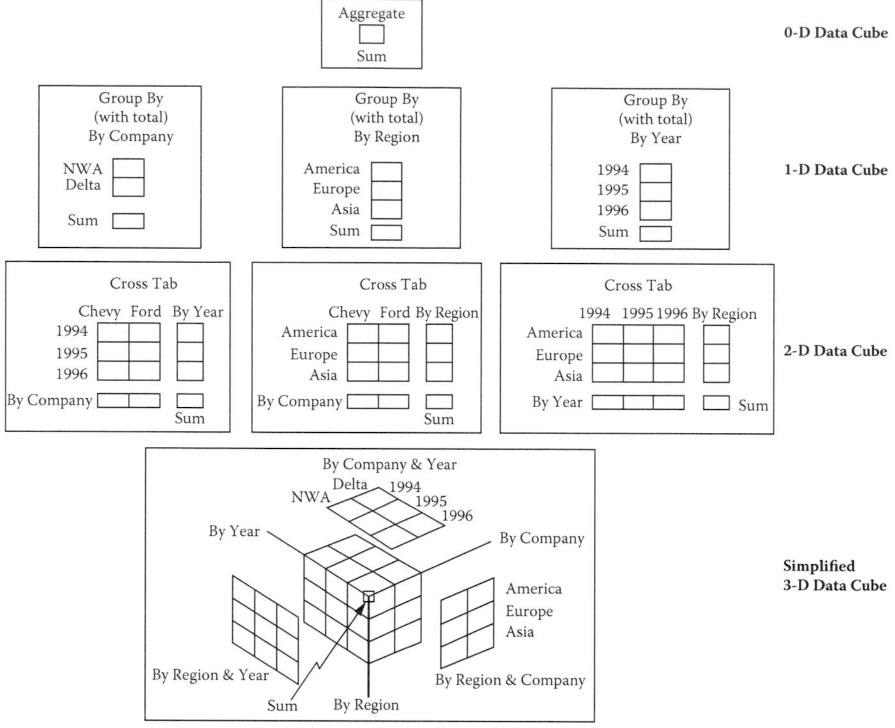

FIGURE 4.4 The 0-d, 1-D, 2-D, and 3-D data cubes (Gray, Bosworth et al. 1995).

point in this space. The set of points forms an *N*-dimensional cube. If there are *N* attributes, there will be $2^N - 1$ super-aggregate values. If the cardinalities of the *N* attributes are C_1, C_2, …, C_N, then the cardinality of the resulting cube operation is $\prod(C_i + 1)$. The extra value in each domain is *ALL*. Each *ALL* value really represents a set over which the aggregate was computed.

A tabular view of the individual sub-space datacubes of Figure 4.4 is shown in Figure 4.6. The union of all the tables in Figure 4.6 yields the resulting table from the data cube operator. The 0-dimensional sub-space cube labeled "Aggregate" in Figure 4.4 is represented by Table "SALES-L2" in Figure 4.6. The one-dimensional sub-space cube labeled "By Company" in Figure 4.5 is represented by Table "SALES-L1-C" in Figure 4.6. The two-dimensional cube labeled "By Company & Year" is represented by Table "SALES-L0-C" in Figure 4.6. Readers can establish the correspondence between the remaining sub-space cubes and tables.

The cube operator can be modeled by a family of SQL queries using *GROUP BY* operators and aggregation functions. Each arrow in Figure 4.6 is represented by an SQL query. In Table 4.2, we provide the corresponding queries for the five arrows labeled Q1, Q2,…,Q5 in Figure 4.6. For example, query Q1 in Table 4.2 aggregates "Sales" by "Year" and "Region," and generates Table "SALES-L0-A" in Figure 4.6.

Cube-query

```
SELECT Company, Year, Region,
       Sum (Sales) AS Sales
FROM SALES
GROUP BY CUBE Company, Year, Region
```

SALES

Company	Year	Region	Sales
NWA	1994	America	20
NWA	1994	Europe	15
NWA	1994	Asia	5
NWA	1995	America	12
NWA	1995	Europe	6
NWA	1995	Asia	18
NWA	1996	America	7
NWA	1996	Europe	20
NWA	1996	Asia	17
Delta	1994	America	15
Delta	1994	Europe	3
Delta	1994	Asia	12
Delta	1995	America	9
Delta	1995	Europe	14
Delta	1995	Asia	20
Delta	1996	America	15
Delta	1996	Europe	8
Delta	1996	Asia	16

Base table

CUBE

Data Cube

Company	Year	Region	Sales
ALL	ALL	ALL	232
NWA	ALL	ALL	120
Delta	ALL	ALL	112
ALL	1994	ALL	70
ALL	1995	ALL	79
ALL	1996	ALL	83
ALL	ALL	America	78
ALL	ALL	Europe	66
ALL	ALL	Asia	88
NWA	1994	ALL	40
NWA	1995	ALL	36
NWA	1996	ALL	44
Delta	1994	ALL	30
Delta	1995	ALL	43
Delta	1996	ALL	39
NWA	ALL	America	39
NWA	ALL	Europe	41
NWA	ALL	Asia	40
Delta	ALL	America	39
Delta	ALL	Europe	25
Delta	ALL	Asia	48
ALL	1994	America	35
ALL	1994	Europe	18
ALL	1994	Asia	17
ALL	1995	America	21
ALL	1995	Europe	20
ALL	1995	Asia	38
ALL	1996	America	22
ALL	1996	Europe	28
ALL	1996	Asia	33
⋮	⋮	⋮	⋮

Resulting table from Cube operator (aka data cube)

FIGURE 4.5 An example of data cube (Gray, Bosworth et al. 1995).

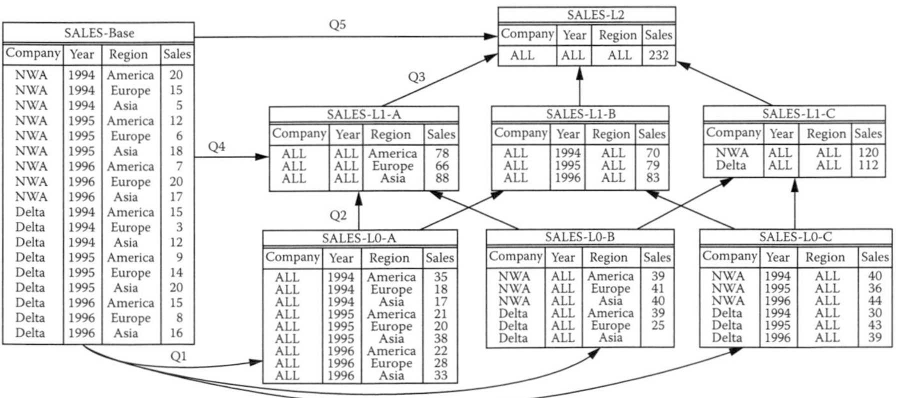

FIGURE 4.6 An example of group-by.

TABLE 4.2
Table of GROUP BY Queries

Q1	SELECT 'ALL', Year, Region, SUM(Sales) FROM SALES-Base GROUP BY Year, Region
Q2	SELECT 'ALL', 'ALL', Region SUM(Sales) FROM SALES-L0-A GROUP BY Region
Q3	SELECT 'ALL', 'ALL', 'ALL' SUM(Sales) FROM SALES-L1-A
Q4	SELECT 'ALL', 'ALL', Region, SUM(Sales) FROM SALES-Base GROUP BY Region
Q5	SELECT 'ALL', 'ALL', 'ALL', SUM(Sales) FROM SALES-Base

The *GROUP BY* clause specifies the grouping attributes, which should also appear in the *SELECT* clause, so that the value resulting from applying each function to a group of tuples appears along with the value of the grouping attributes.

4.2.4 The Role of an Aggregation Hierarchy

To support OLAP, the data cube provides the following operators: roll-up, drill-down, slice and dice, and pivot. We now define these operators.

- Roll-up: Increasing the level of abstraction. This operator generalizes one or more dimensions and aggregates the corresponding measures. For example, Table SALES-L0-A in Figure 4.6 is the roll-up of Table SALES-Base on the company dimension.
- Drill-down: Decreasing the level of abstraction or increasing detail. It specializes in one or a few dimensions and presents low-level aggregations. For example, Table SALES-L0-A in Figure 4.6 is the drill-down of Table SALES-L1-A on the year dimension.
- Slice and dice: Selection and projection. Slicing into one dimension is very much like drilling one level down into that dimension, but the number of entries displayed is limited to that specified in the slice command. A dice operation is like a slice on more than one dimension. On a two-dimensional display, for example, dicing means slicing on both the row and column dimensions. Table 4.3 shows the result of slicing into the value of "America" on the year dimension from the Table SALES-L2 in Figure 4.6.

TABLE 4.3
Slice on the Value America of the Region Dimension

Company	Year	Region	Sales
ALL	ALL	America	78

TABLE 4.4
Dice on Value 1994 of Year Dimension and Value America of Region Dimension

Company	Year	Region	Sales
ALL	1994	America	35

Table 4.4 shows the result of dicing into the value of "1994" on the year dimension and the value of "America" on the region dimension from Table SALES-L2 in Figure 4.6.

- Pivoting: Re-orienting the multidimensional view of data. It presents the measures in different cross-tabular layouts

4.3 MAP CUBE

In this section, we define the map cube operator, which provides an album of maps to browse results of aggregations. We also provide the grammar and the translation rules for the map cube operator.

4.3.1 Definition

We extend the concept of the data cube to the spatial domain by proposing the "map cube" operator as shown in Figure 4.7. A map cube is defined as an operator which takes the input parameters, that is, base map, base table, and cartographic preferences, and generates an album of maps for analysis and comparison. It is built from

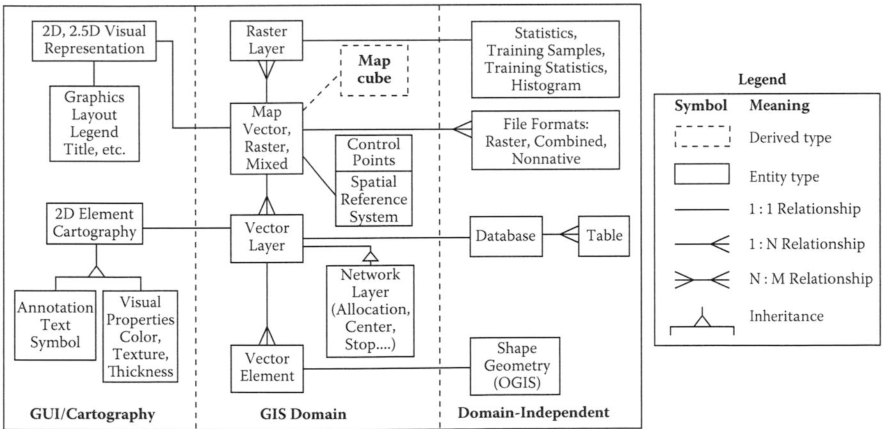

FIGURE 4.7 Concepts in GIS with data cube and map-cube.

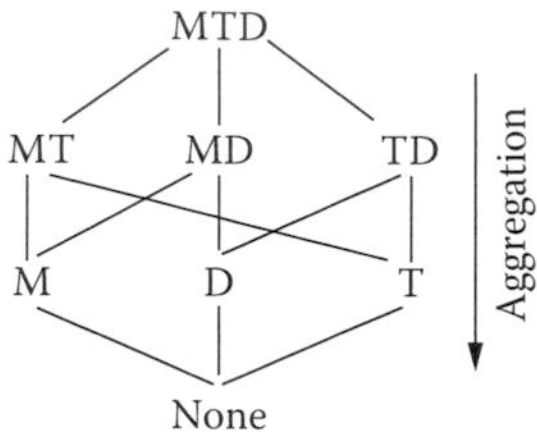

FIGURE 4.8 The Dimension Power-Set Hierarchy.

the requirements of a spatial DW, that is, to aggregate data across many dimensions looking for trends or unusual patterns related to spatial attributes. The basis of a map cube is the hierarchy lattice, either a dimension power-set hierarchy or a concept hierarchy, or a mixture of both. In this chapter, we focus on the dimension power-set hierarchy. Figure 4.8 shows an example of a dimension power-set hierarchy. This example has three attributes: **Maker(M)**, **Type(T)**, and **Dealer(D)**. There are eight possible groupings of all the attributes. Each node in the lattice corresponds to one group-by. Figure 4.9 shows a three-dimensional concept hierarchy. The car makers can be classified as American, European, or Japanese. American car makers include Ford, GM, and Chrysler. The car type dimension can be classified as Sedan and Pickup Trucks. The Sedan can go down one detail level and be classified as Small, Medium, Large, or Luxury.

Let the number of dimensions be m, each dimension be A_i, $i = 1,2, \ldots, m$, and A_m be the geographic location dimension. Then we have $(m - 1)$ different levels of lattice for the dimension power-set hierarchy. Let the level with only one dimension A_i, $i = 1,2, \ldots, m - 1$, be the first level, the level with two dimensions, A_{ij}, where $i \neq j$, $i = 1,2,\ldots m - 1$, let $j = 1,2,\ldots m - 1$ be the second level, and the level with the complete set of dimensions $A_1 A_2 A_3, \ldots, A_{m-1}$ be the $(m - 1)^{\text{th}}$ level. The total number of

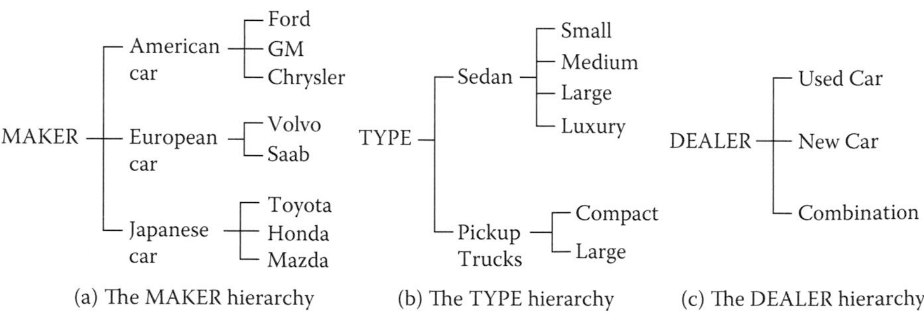

FIGURE 4.9 The Concept Hierarchies of (a) MAKER and (b) VEHICLE TYPE, and (c) DEALER.

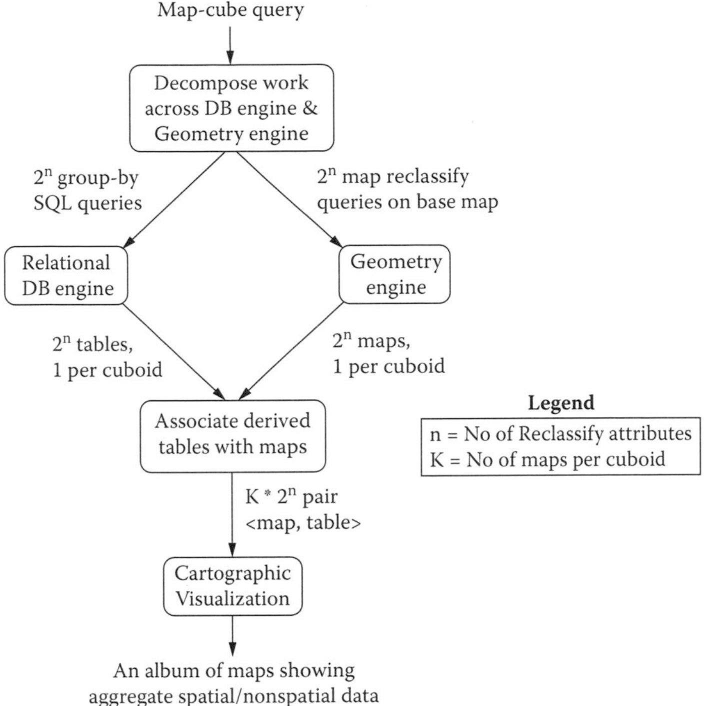

FIGURE 4.10 Steps in Generating a Map Cube.

cuboids is $\sum_{i=1}^{m-1} (i^{m-1})$. Let the cardinality of each dimension A_i be c_i. Then for each cuboid, such as, $A_iA_j \ldots A_k$, we have an album of $c_i \times c_j \times \cdots \times c_k$ maps.

In other words, a map cube is a data cube with cartographic visualization of each dimension to generate an album of related maps for a dimension power-set hierarchy, a concept hierarchy, or a mixed hierarchy. A map cube adds more capability to traditional GIS where maps are often independent (Bédard 1999). The data-cube capability of roll-up, drill-down, and slicing and dicing gets combined with the map view. This can benefit analysis and decision making based on spatial DWs.

4.3.2 Steps in Generating a Map Cube

To generate the map cube, first we issue the map-cube query, as shown in Figure 4.10. This query is decomposed into the DB engine and the geometry engine. The relational database (DB) engine processes the 2^n group-by SQL queries and generates 2^n tables, where n is the number of attributes in the "Reclassify by" clause. The geometry engine processes the map reclassify queries on the base map and generates 2^n maps. These maps and tables are in one-to-one correspondence with each other.

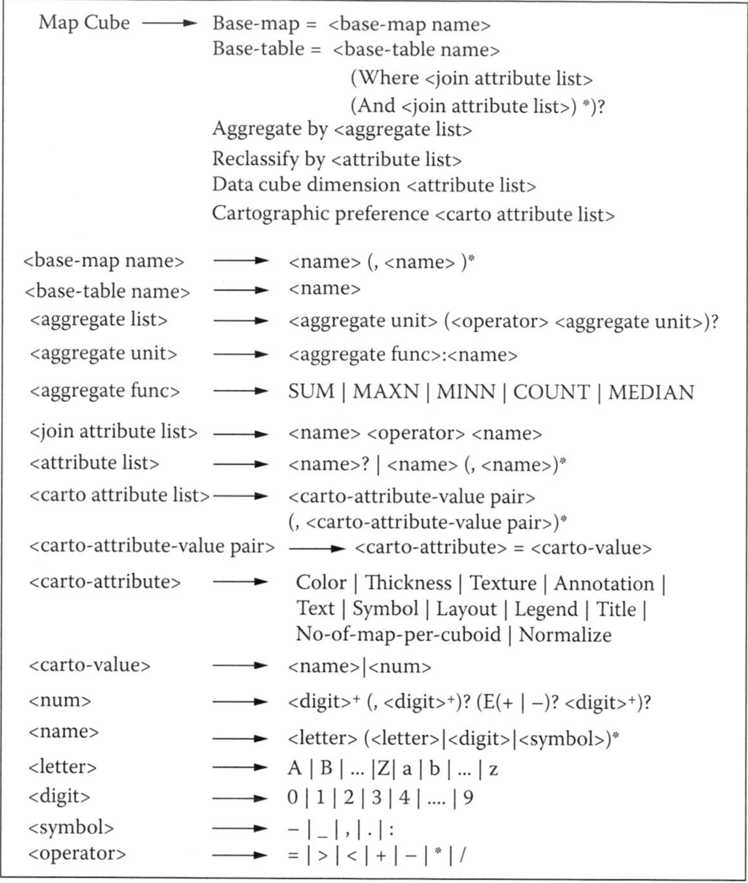

FIGURE 4.11 The grammar for the map cube operator.

Finally, the cartographic preferences are dealt with in the cartographic visualization step, and an album of maps is plotted.

4.3.3 The Grammar for the Map Cube Operator

We have used "yacc" (Levine, Mason, and Brown 1992) like syntax to describe the syntax of a map cube via the grammar and translation rules, as listed in Figure 4.11. Words in angular brackets, for example, < *carto-value* >, denote nonterminating elements of the language. For example, the < *carto-value* > will be translated into either < *name* > or < *num* > . The vertical bar (|) means "or," and the parentheses are used to group sub-expressions. The star (*) denotes zero or more instances of the expression, and the plus (+) denotes one or more instances of the expression. The unary postfix operator (?) means zero or one instance of the expression. These translations will continue until we reach a terminating element. For example, < *letter* > will eventually be translated into one character.

4.4 APPLICATION OF MAP CUBE IN TRANSPORTATION DATA

Transportation and the highway network are critical elements of the public infrastructure system (FHWA 2005; Paniati 2002). A crucial component in sustaining and developing the roadway network is the availability of a traffic information system that allows for the monitoring and analysis of travel behaviors and conditions. It is for these reasons that the transportation roadway is selected for the case study of this chapter. Based on map cube, we have developed an effective traffic visualization system, *AITVS* – *A*dvanced *I*nteractive *T*raffic *V*isualization *S*ystem (http://spatial.nvc.cs.vt.edu/traffic), for observing the summarization of spatiotemporal patterns and travel behavior in loop-detector data. AITVS is designed for browsing the spatiotemporal dimension hierarchy via the integrated roll-up and drill-down operations. AITVS visualization techniques give traffic organizations and personnel powerful tools for extracting practical and insightful information from large collections of roadway data, thus vastly improving their ability to make effective decisions. The identified traffic patterns can assist transportation network management decisions, establish traffic models for researchers and planners, and allow travelers to select commuting routes. Traffic data from Interstate 66 and 95 in metropolitan Washington, D.C. are used to demonstrate the functionalities of AITVS.

Traffic measures and base map: In a traffic DW, the primary measures are volume (number of vehicles passing a given detector), speed (vehicle velocity at a given detector), and occupancy (percentage of time a vehicle is detected within a given detector), and the dimensions are *time* and *space*. Traffic dimensions are hierarchical by nature. For example, the time dimensions can be grouped into "Week", "Month", "Season", or "Year". Similarly, the space dimensions can be grouped into "Station", "County", "Freeway", or "Region". Given the dimensions and hierarchy, the measures can be aggregated into different combinations. For example, for a particular highway and a chosen month, the weekly traffic speeds can be analyzed. For our transportation data, we employed the one-dimensional highway milepost locations as our base map. The mileposts are spatially ordered location identifiers which correspond to the relative distances to an entry point within the highway. For example in I-66, mileposts 48 and 75.25 indicate the start and end of traffic detection, respectively, which provides 27.25 miles of total detection.

AITVS visualization components: Figure 4.12 gives the map cube operations of AITVS which provides six distinct visualization components to comprehensively cover the various performance aspects of a roadway system. Each node of the figure represents a cube operation and a view of the data. Here, S_{HS} is the highway station (logical representation of a group of detectors with identical mileposts), T_{TD} is the time of day, and T_{DW} is the day of week. These nodes are combined to form the following

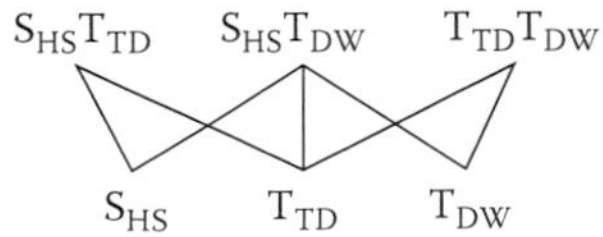

FIGURE 4.12 Map cube dimension lattice as applied to traffic data.

views: $S_{HS}T_{TD}$, $S_{HS}T_{DW}$, and $T_{DW}T_{TD}$. For example, S_{HS} gives the speed of each highway station for all the time and $S_{HS}T_{TD}$ provides the daily traffic speed of each station. The views are partitioned into two visualization sets, one-dimensional and two-dimensional views. The one-dimensional views are composed of *time of day plot* (T_{TD}), *day of week Plot* (T_{DW}), and *highway station plot* (S_{HS}). The two-dimensional views are composed of *highway stations vs. time of day plot* ($S_{HS}T_{TD}$), *highway stations vs. day of week plot* ($S_{HS}T_{DW}$), and *time of day vs. day of week plot* ($T_{DW}T_{TD}$). The following subsections provide the details of each visualization component.

4.4.1 One-Dimensional Map Cube Views

A. Time of Day Plot (T_{TD})

In Figure 4.13(*a*), the x-axis represents the day intervals, Monday to Sunday, and the y-axis shows volume, speed, and occupancy, respectively, from top to bottom. The graph plot represents the I-66 eastbound traffic behavior for the week of November 1, 2004 to November 7, 2004 on station 121 at Milepost 53.2. The graph shows morning and afternoon rush hour patterns on the weekdays (November 1 to November 5) and lighter traffic flow on the weekends. A more detailed view of this behavior can be studied by invoking a drill-down operation to show traffic behavior on a specific day. The user adjustable attributes are the highway station nodes and the date duration.

Figure 4.13(*b*) shows identical value types for its xy-axes as in Figure 4.13(*a*) but exemplifies not only the current traffic trend, but also provides the predicted traffic behavior for Wednesday, June 29, 2005. AITVS calculates the prediction model based on a user-specified time length (e.g., 4,5,6... weeks) and invokes statistical methods to extrapolate the predicted behavior. For Figure 4.13(*b*), the user is able to select his or her choice of highway station node and the span of historical data used to generate the prediction graph. Daily, weekly, and monthly patterns can be analyzed with this component.

B. Day of Week Plot (T_{DW})

In Figure 4.14, the x-axis represents days and the y-axis shows volume, speed, and occupancy, respectively, from top to bottom. The figure shows the I-66 eastbound traffic at 8:30 AM for each day of March 2004 on station 121 where I-66 crosses Route 28 at Milepost 53.2. This plot shows a recurrent weekly pattern that the traffic volume of weekends (February 29, March 7, March 14, March 21, and March 28) is lower than that of Monday to Friday. It is also noticed that as volume and occupancy decreases, speed increases, which reflects another traffic behavior pattern. For this visualization component, the highway station number, date duration, and time can be dynamically selected by the user. Weekly, monthly, and yearly patterns for each station can be analyzed with this component.

C. Highway Station Plot (S_{HS})

In Figure 4.15, the x-axis shows the consecutive mileposts along the route and the y-axis denotes the volume, speed, and occupancy, respectively. The graph represents the I-66 inbound traffic at 8:30 AM on Tuesday, October 26, 2004. The stations at Milepost 60 and mileposts after 65 were malfunctioning with zero volume, speed,

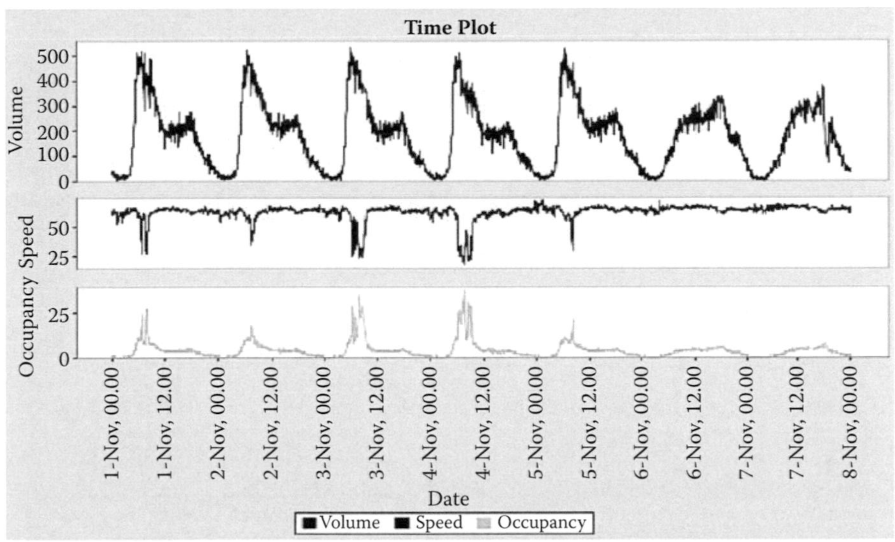

(a) Weekly pattern of volume, speed, and occupancy on a given week

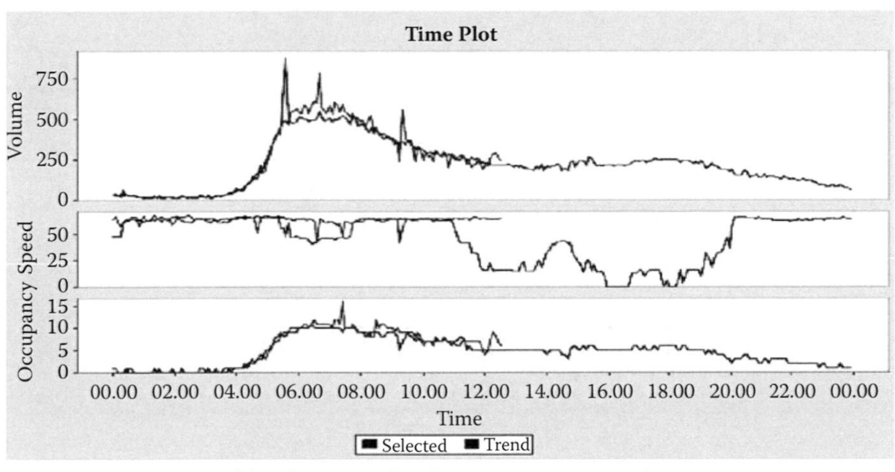

(b) Volume, speed, and occupancy on a given day

FIGURE 4.13 Time of day plots (T_{TD}). (a) Weekly pattern of volume, speed, and occupancy on a given week. (b) Volume, speed, and occupancy on a given day. See color insert after page 148.

and occupancy. Between Mileposts 62 and 64.5, the occupancy graph produces a boundary around its local maximum and, conversely, the speed produces a boundary around its local minimum. These boundaries indicate the morning rush hour in the Oakton/Vienna region. The steady incremental behavior in occupancy from Milepost 47.5 to 64.5 and the simultaneous reduction in speed is a travel behavior pattern associated with congestion. The user-adjustable criteria for this component

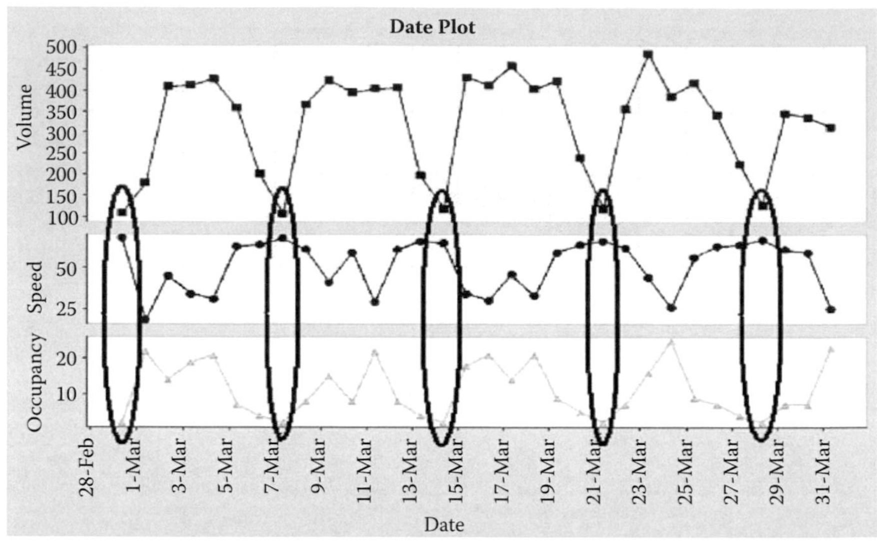

FIGURE 4.14 Day of week plots (T_{DW}). Volume, speed, occupancy in a given month.

are date, time, highway, and traffic direction (i.e., east- or westbound). Monthly patterns along the route span can be analyzed with this component.

4.4.2 Two-Dimensional Map Cube Views

D. Highway Station vs. Time of Day Plot ($S_{HS}T_{TD}$)

In Figure 4.16(*a*), the x-axis denotes the time, the y-axis shows the milepost, and the gray scale represents the speed value. Each row of the graph corresponds to the traffic

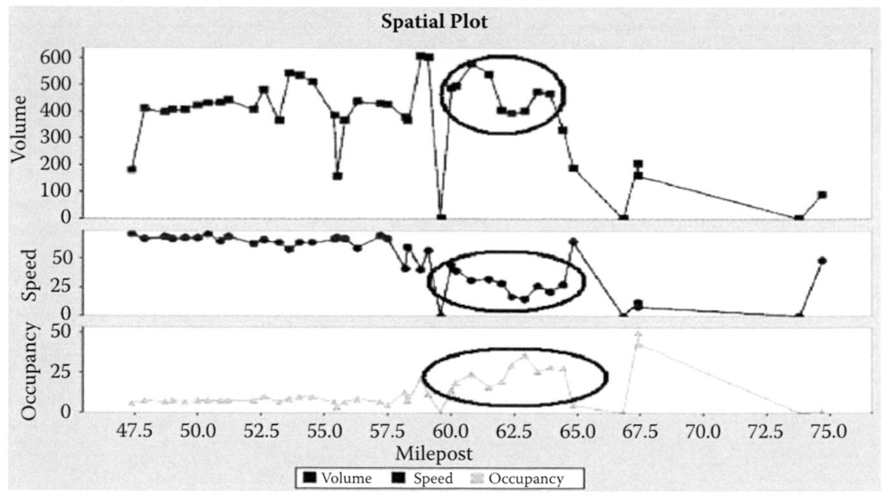

FIGURE 4.15 Highway station plots (S_{HS}). Volume, speed, and occupancy along all mileposts in a highway.

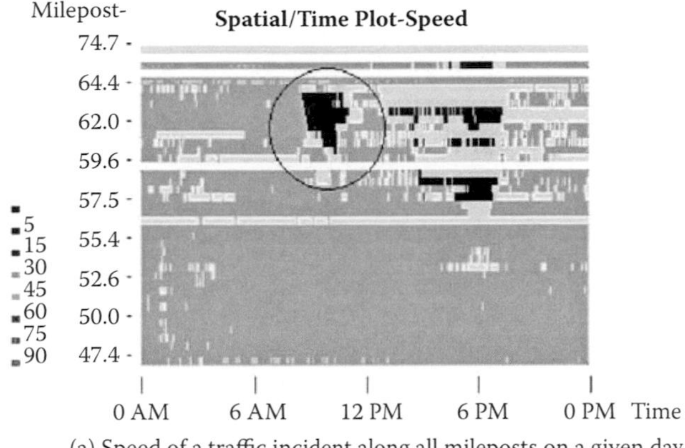

(a) Speed of a traffic incident along all mileposts on a given day

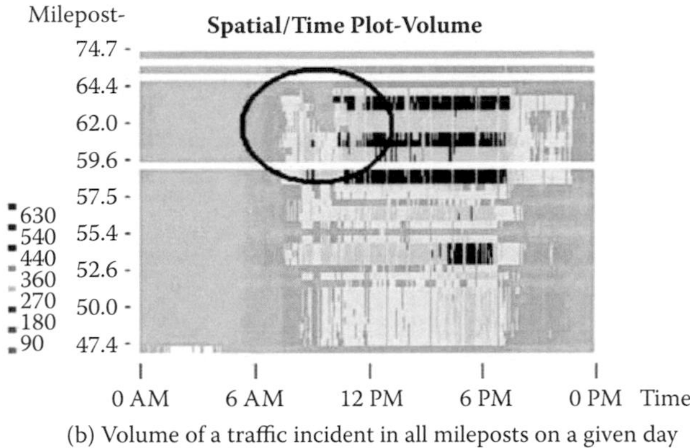

(b) Volume of a traffic incident in all mileposts on a given day

FIGURE 4.16 Highway station vs. time of day plots ($S_{HS}T_{TD}$). (a) Speed of a traffic incident along all mileposts on a given day. (b) Volume of a traffic incident in all mileposts on a given day.

speed of one station on the selected date. This figure depicts the I-66 eastbound traffic on Saturday, November 6, 2004. Notice that in the morning during 9:30 AM to 10:15 AM, traffic congestion occurred for a span of 45 minutes, an unusual event for the regular commuters. We recognize a distinct incident pattern, an upside-down triangle. Figure 4.16(*b*) shows the volume for the same traffic data where we can observe that the volume corresponding to the triangle in Figure 4.16(*b*) is abnormally low. This observation supports the presence of a traffic incident. The user-adjustable properties for this component are date, highway, and traffic direction.

E. Highway Stations vs. Day of Week Plot ($S_{HS}T_{DW}$)

In Figure 4.17(*a*), the x-axis depicts the mileposts along the route and the y-axis shows volume, speed, and occupancy, respectively. The graph represents a series plot of all days of a week for February 2005 at 6:00 PM westbound (outbound), where

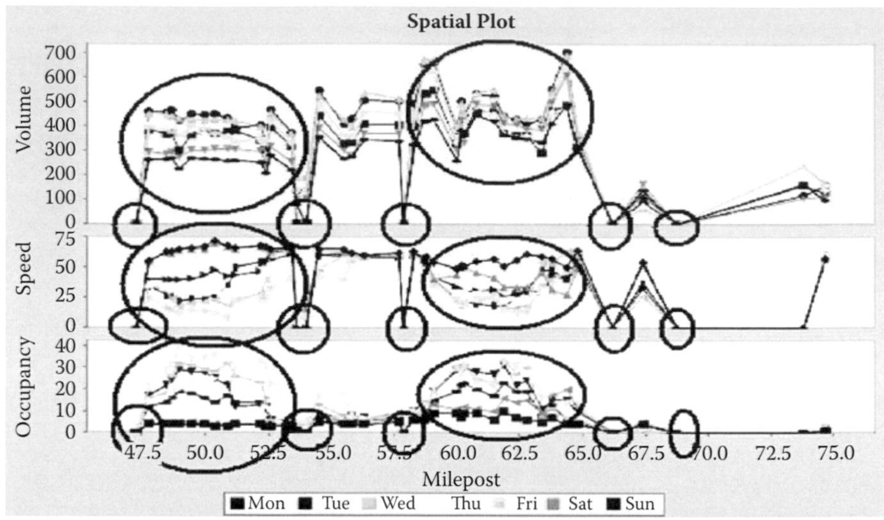

(a) Superimposed graphs of average volume, speed, occupancy of all mileposts for all days of the week

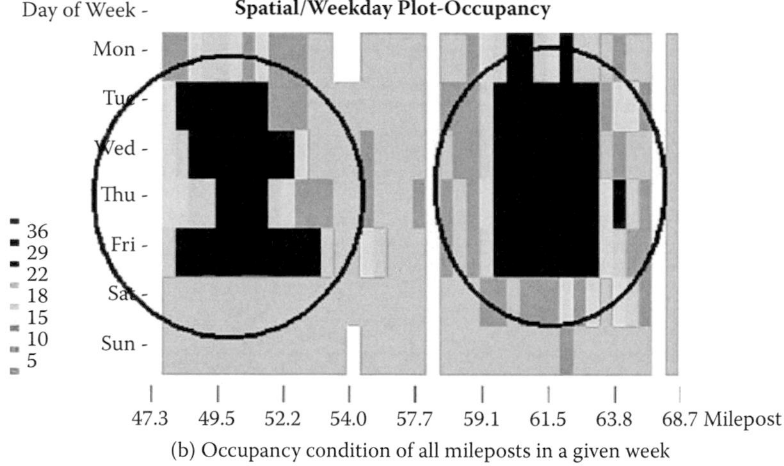

(b) Occupancy condition of all mileposts in a given week

FIGURE 4.17 Highway station vs. day of week plots ($S_{HS}T_{DW}$). (a) Superimposed graphs of average volume, speed, occupancy of all mileposts for all days of the week. (b) Occupancy condition of all mileposts in a given week. See color insert after page 148.

each line represents each average day of the week. From this superimposed plot, we observe that weekdays show heavier traffic activity as compared to the weekends as marked by the ovals. The ovals signify the malfunctioning detectors, which are revealed as having zero values for volume, speed, and occupancy. Figure 4.17(*b*) shows the same westbound traffic data for February 2005 at 6:00 PM, but represented as a gray scale value graph. The x-axis shows the milepost, the y-axis denotes the days of the week, and the colors represent occupancy values. Figure 4.17(*b*) provides a much clearer representation through its gray scale plot where heavy traffic activity is shown as two red clusters. Depending on the subject of analysis, the series [Figure 4.17(*a*)] plot may find more utility than the xy-plot [Figure 4.17(*b*)]. AITVS offers these various modes of visualizations to accommodate different analysis requirements. For both visualization components, the date range, time, highway, and traffic direction are user-adjustable. Average weekly, monthly, and yearly patterns along a particular highway can be evaluated with this component.

F. Time of Day vs. Day of Week Plot ($T_{TD}T_{DW}$)

In Figure 4.18(*a*), the x-axis indicates the time and the y-axis shows volume, speed, and occupancy, respectively. This graph amalgamates values for each day of the week for February 2005 on westbound station 62. An afternoon (4:00 to 7:00 PM) rush hour commute is observed during the weekdays as indicated by the large hump and valleys (circled in blue) on the series graph. Figure 4.18(*b*) depicts the identical traffic data using the color-base value graph, with the x-axis as time, the y-axis as the days of the week, and the colors as speed values. Here, the red region at around 4:00 to 7:00 PM indicates afternoon rush hour, which corresponds to the humps mentioned in Figure 4.18(*a*). The graph also indicates that for an average Friday during February 2005, afternoon rush hour begins early at 1:45 PM. For both visualization components, users may select the highway station node and aggregate date range. This component supports analyses of average traffic patterns for each selected station during a particular period of time.

4.5 CASE STUDY — TRAFFIC INCIDENT ANALYSIS

Traffic planners and engineers can make use of spatial and incident patterns to evaluate current incident response strategies, quantify their economic impact (e.g., loss of productivity), and propose new incident response plans. We utilize the map cube as applied to traffic data to analyze the effects of a vehicular incident on northbound Interstate-95 on Wednesday, February 8, 2006. The incident was of an overturned tractor trailer which occurred at 8:06 AM at Milepost 158.

$S_{HS}T_{TD}$ View: In Figure 4.19(*a*), we give the spatiotemporal speed plot with the incident region circled in blue. We also provide a normal operating situation with the speed spatiotemporal plot of the prior Wednesday, February 1, 2006 in Figure 4.19(*b*). It can be observed from these two sets of figures that the incident caused the average speed between Mileposts 153.8 and 158.6 to fall below 15 mph and the occupancy level to rise above 40%. The horizontal lines are due to malfunctioning detectors or missing data caused by transmission errors.

T_{TD} View: Figure 4.20 provides the traffic metrics of the affected and neighboring stations of the incident. This figure is a drill-down representation of the spatiotemporal speed plot of Figure 4.19. For example, if we were to obtain the horizontal cross-section view of Figure 4.19(*a*) at Milepost 157.44, then we obtain the graph of Figure 4.20(*a*). The boxes highlight the area of interest (i.e., incident). These figures show the traffic behavior at the stations located close to the incident and at the peripheries (i.e., spatial neighborhoods). In Figure 4.20(*a*), traffic begins to form congestion as shown by the sharp increase in occupancy and decrease in speed. Figure 4.20(*b*) shows the worsening congestion as traffic approaches the incident location

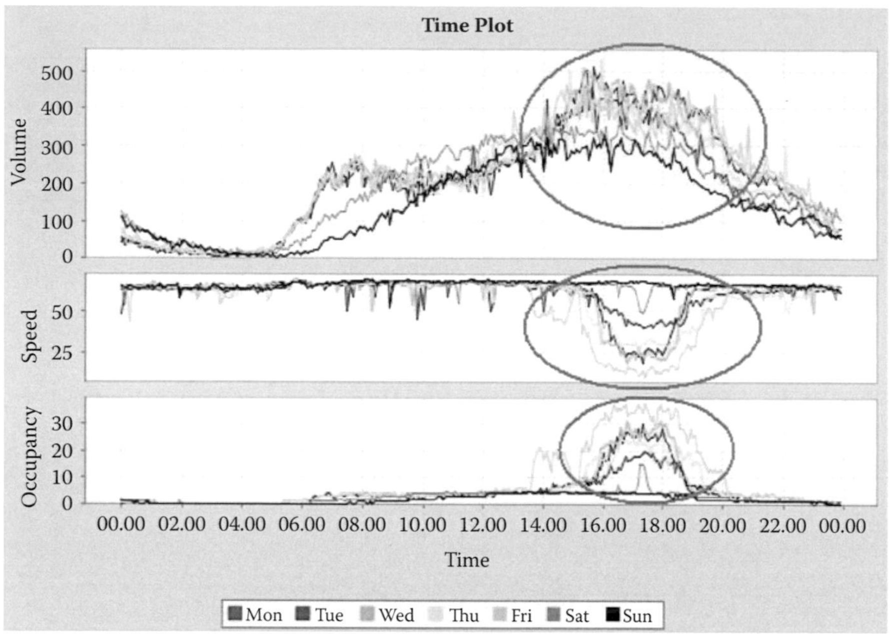

(a) Volume, speed, occupancy conditions of all days in a week on a given time

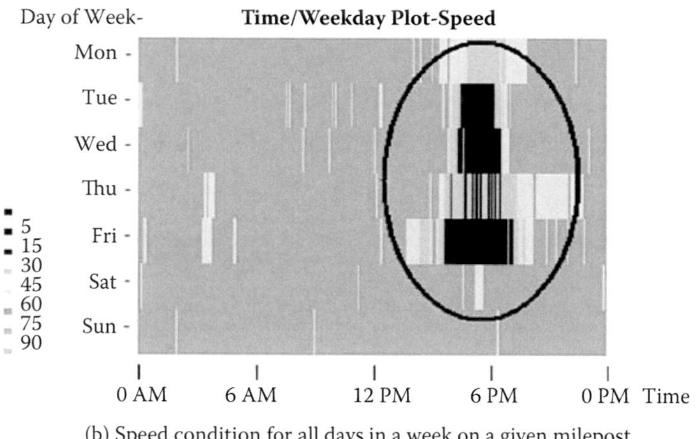

(b) Speed condition for all days in a week on a given milepost

FIGURE 4.18 Time of day vs. day of week plots ($S_{HS}T_{TD}$). (a) Volume, speed, occupancy conditions of all days in a week on a given time. (b) Speed condition for all days in a week on a given milepost. See color insert after page 148.

at Milepost 158.28. This visualization can also provide an estimation on the incident duration, which spans from 8:00 AM to 12:00 PM. In the downstream station [Figure 4.20(*c*)] it is observed that traffic resumes with normal speed and occupancy but reduced volume.

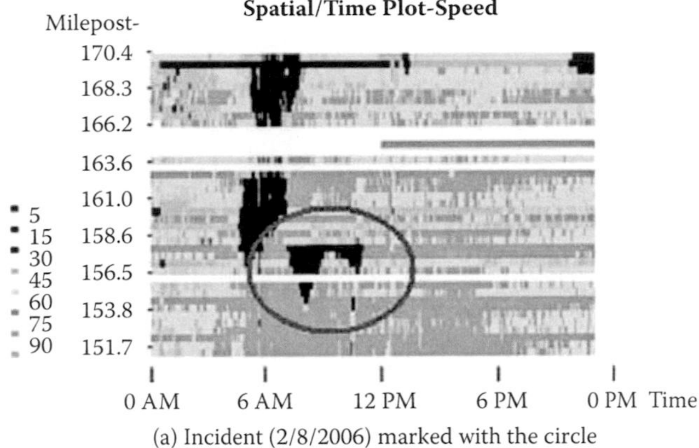

(a) Incident (2/8/2006) marked with the circle

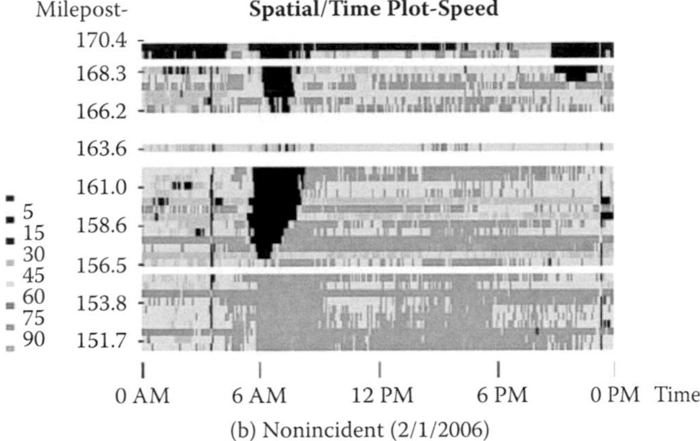

(b) Nonincident (2/1/2006)

FIGURE 4.19 $S_{HS}T_{TD}$ views of incident day against a nonincident day. (a) Incident (2/8/2006) marked in with the circle. (b) Nonincident (2/1/2006).

S_{HS} View: Figure 4.21 shows the spatial plot of the traffic incident Wednesday [Figure 4.21(*a*)] against the nonincident traffic condition of the prior Wednesday [Figure 4.21(*b*)]. Combined, these figures show a detailed view of traffic patterns along northbound I-95 at 8:30 AM. These figures are essentially drilled-down representations of the spatiotemporal plots of Figure 4.19 that depict a specific time and give more fine-grained views of the traffic metric values at each milepost.

$S_{HS}T_{DW}$ View: Figure 4.22 gives an aggregated view, which displays the entire traffic incident week's (2/6/2006 to 2/12/2006) traffic patterns. These visualizations allow a single-view and direct comparison of the incident pattern against the other days. The green lines in Figure 4.22 show the traffic patterns at the day of the incident. As highlighted by the blue boxes, one can easily attain approximate quantifications (e.g., percent volume reduction) relative to another day's traffic.

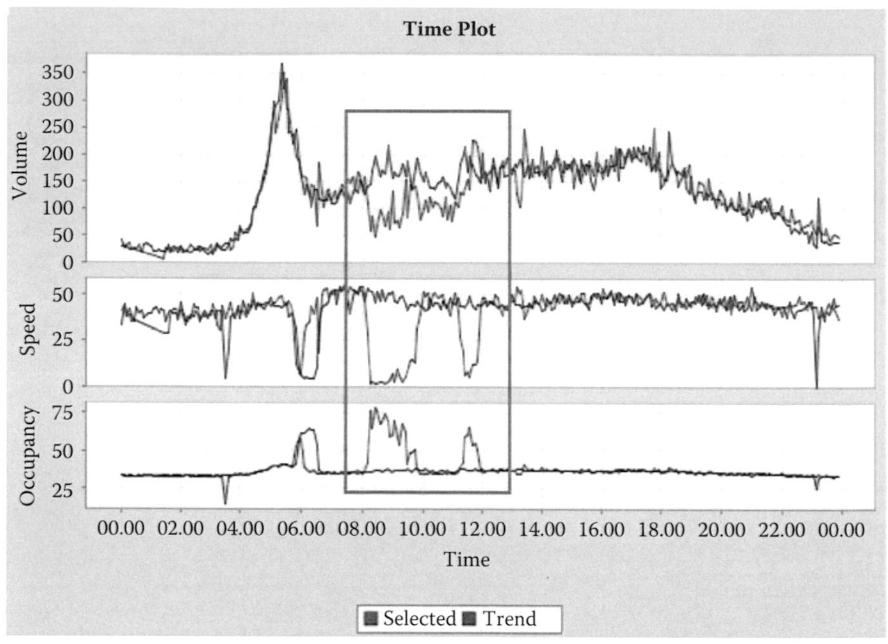

(a) Milepost 157.44 (2/8/2006)

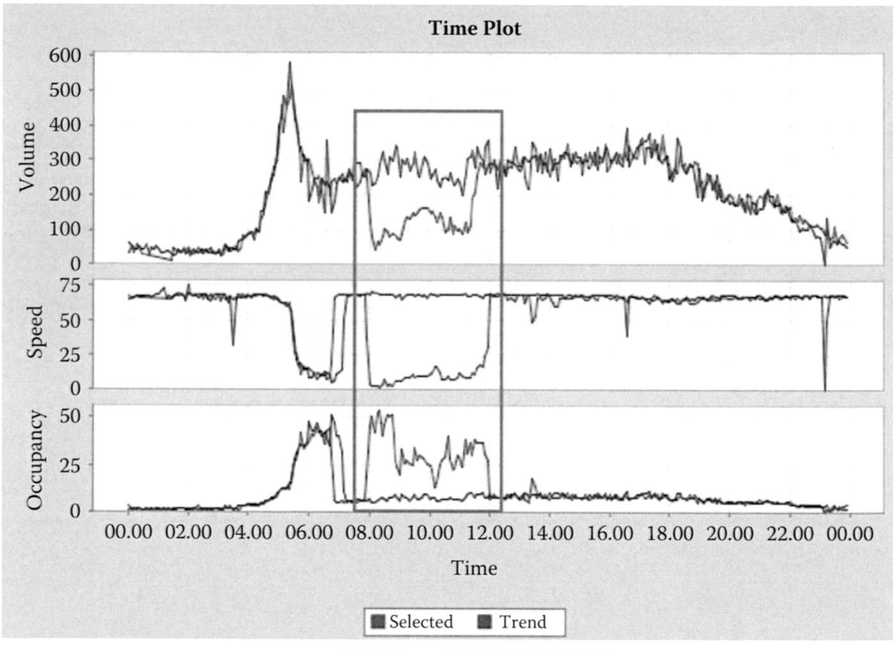

(b) Milepost 158.26 (2/8/2006)

FIGURE 4.20 T_{TD} views for drilled-down analysis of surrounding mileposts. (a) Milepost 157.44 (2/8/2006). (b) Milepost 158.26 (2/8/2006). (c) Milepost 159.63 (2/8/2006). See color insert after page 148.

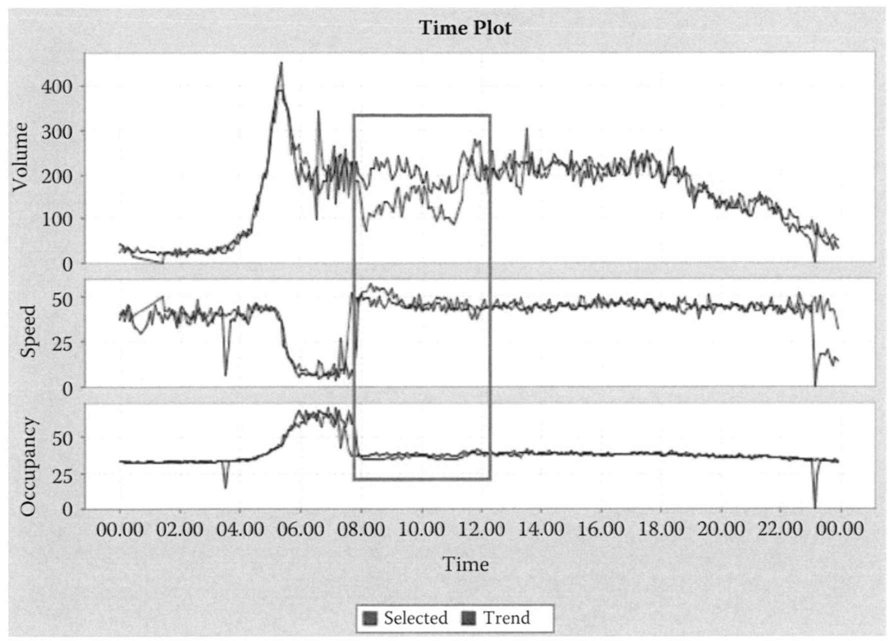

(c) Milepost 159.63 (2/8/2006)

FIGURE 4.20 (Continued).

$T_{TD}T_{DW}$ View: Figure 4.23 is a drilled-down view of Figure 4.22, which shows the effect of the incident at various times of the day against all days of the week. Compared to the nonincident days, it can be observed that the incident at 8:00 AM to 12:00 PM introduced a drastic and quantifiable reduction in speed and volume. Results of the approximate loss in speed, volume, and occupancy can be used to estimate the incident cost (e.g., productivity loss).

T_{DW} View: Figure 4.24 provides the T_{DW} view of the incident on Milepost 158.26 at 8:30 AM. From this figure, rapid observations can be made about the impacts of the incident when compared to the surrounding days' traffic patterns. For example, from the graph it can be seen that the incident caused a tremendous decrease (approximately 66%) in volume against its neighboring days' traffic.

This case study demonstrates the capabilities of map cube to provide a broad and complete analysis of an event under investigation. The visualization module allows users to analyze traffic occurrences at the macro level through roll-up operations, at the micro level through drill-down operations, and a direct comparative analysis through combined visualizations.

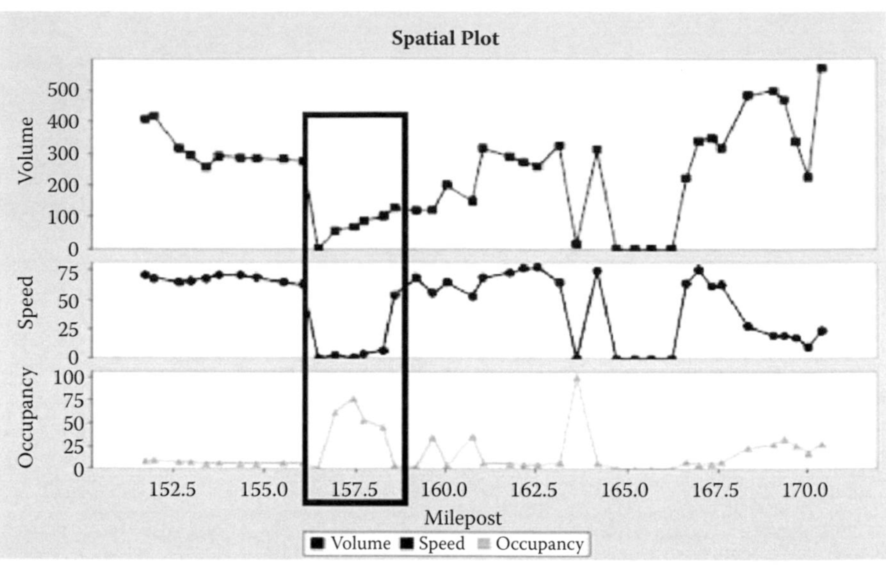

(a) I-95 NB of incident at 8:30 AM (2/8/2006)

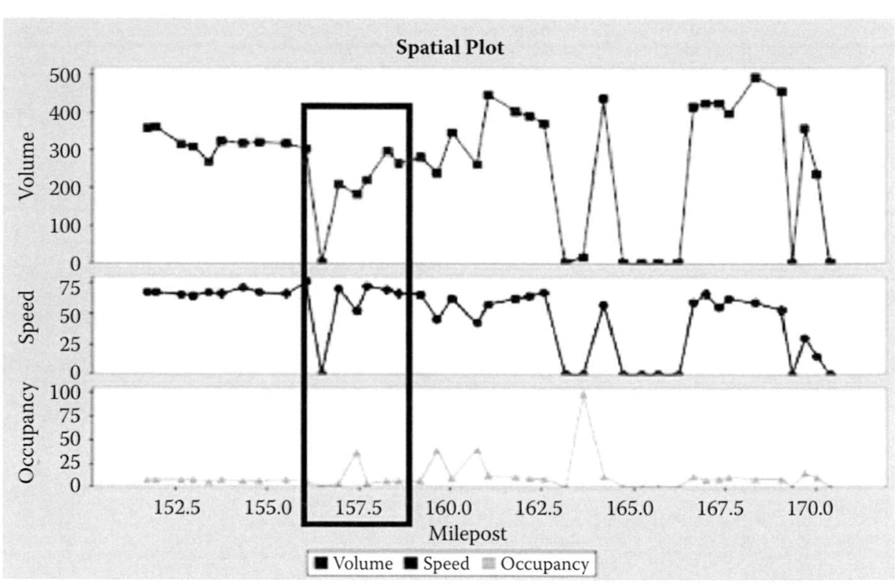

(b) I-95 NB of nonincident (prior Wednesday) at 8:30 AM (2/1/2006)

FIGURE 4.21 S_{HS} views of incident day against nonincident day. (a) I-95 NB of incident at 8:30AM (2/8/2006). (b) I-95 NB of nonincident (prior Wednesday) at 8:30AM (2/1/2006).

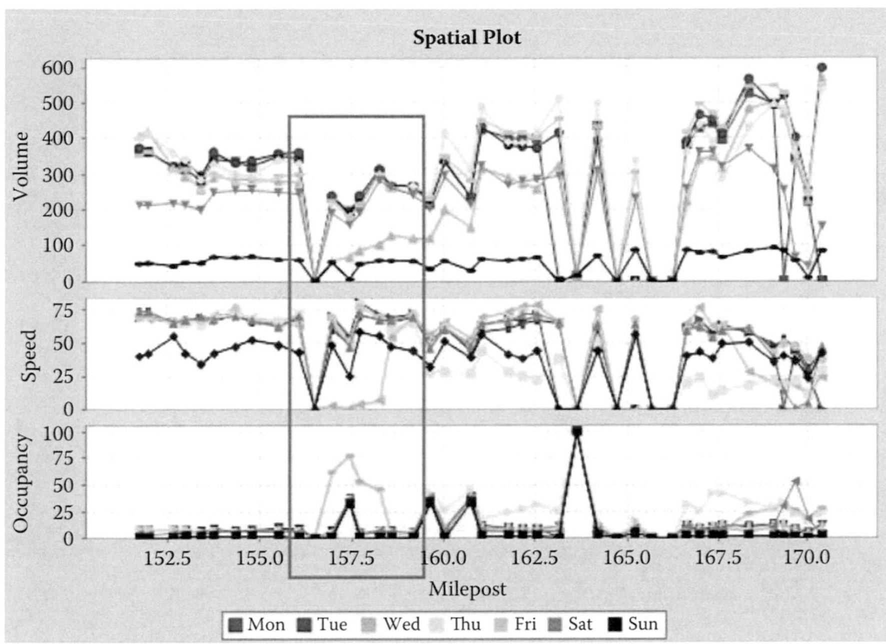

FIGURE 4.22 $S_{HS}T_{DW}$ view of I-95 NB for the week of the incident at 8:30AM (2/6/2006-2/12/2006). See color insert after page 148.

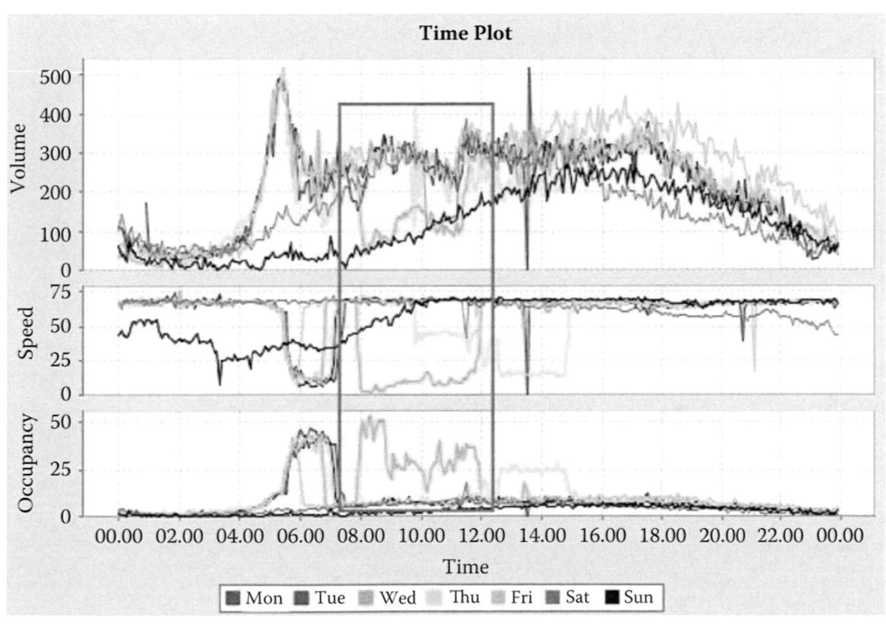

FIGURE 4.23 $T_{TD}T_{DW}$ view of milepost 158.26 for the week of the incident at 8:30AM (2/6/2006-2/12/2006). See color insert after page 148.

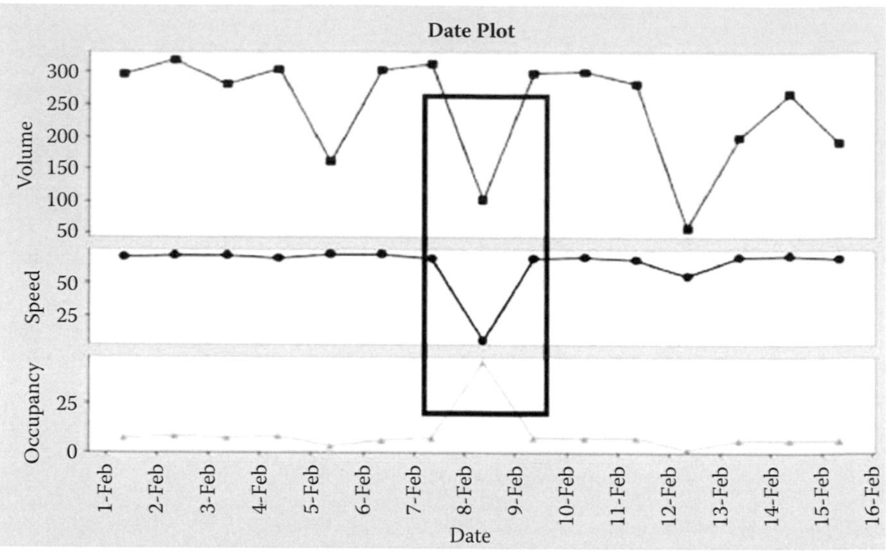

FIGURE 4.24 T_{DW} view of milepost 158.26 from 2/1/2006 to 2/15/2006 at 8:30AM.

4.6 CONCLUSION AND FUTURE WORK

The cube operator generalizes and unifies several common and popular concepts: aggregation, group-by, histogram, roll-up, drill-down, and cross-tab. Maps are the core of the GIS. In this chapter, we extended the concept of the data cube to spatial domain via the proposed "map cube" operator, which is built from the requirements of spatial DWs. We defined the "map cube" operator, provided grammar and translation rules, and developed AITVS as an application of the map cube for traffic data. As demonstrated in the case study, the map cube visualizations can be useful for identifying and analyzing patterns in large traffic data. The employed visualization techniques make the knowledge discovery process much less burdensome and facilitate the usage of the transportation data. Future directions include issues in supporting adaptive user interfaces based on user's expertise and requirements. Additionally, proposals have been made to develop 3-D map representations. Such a representation will provide users with an immersive experience.

We also aim to extend the implementation and application of the map cube to other spatial domains. The map cube can be applied to any spatial application that requires comparisons between different attributes and requires some aggregate functions within each attribute. Therefore, to provide extensions of the map cube to other spatial fields, the aggregate measures will need to be defined and any aggregate functions that are not natively supported by the conventional data cube will need to be developed for the specific domain.

REFERENCES

Bédard, Y. 1999. Visual modeling of spatial databases: towards spatial extensions and uml. *Geomatica* 53(2):169–186.

Chaudhuri, S., and U. Dayal. 1997. An overview of data warehousing and OLAP technology. *SIGMOD Record* 26 (1):65–74.

Chrisman, N. 1996. *Exploring Geographic Information Systems.* New York: John Wiley & Sons.

Codd, E. 1995. Twelve rules for on-line analytic processing. In *Computerworld*, vol. 29, 84–87, 1995.

Codd, E., S. Codd, and C. Salley. 1993. Providing OLAP (On-line Analytical Processing) to user-analysis [http://www.arborsoft.com/essbase/wht_ppr/coddToc.html]. Arbor Software Corporation.

Codd, E., S. Codd, and C. Salley. 1993. *Providing OLAP (On-line Analytical Processing) to user-analysis: an IT mandate*. San Jose, California: Codd & Date, Inc.

ESRI. 1998. White paper: Spatial data warehousing for hospital organizations [http://www.esri.com/library/whitepapers/addl_lit.html]. Environmental Systems Research Institute.

FHWA. 2005. Congestion Mitigation [http://www.fhwa.dot.gov/congestion/index.htm]. Federal Highway Administration.

Gray, J., A. Bosworth, A. Layman, and H. Pirahesh. 1995. Data cube: A relational aggregation operator generalizing group-by, cross-tab, and sub-total. In *The 12th IEEE International Conference on Data Engineering*.

Han, J., N. Stefanovic, and K. Koperski. 1998. Selective materialization: An efficient method for spatial data cube construction. In *Pacific-Asia Conference on Knowledge Discovery and Data Mining (PAKDD'98)*.

Immon, W. 1993. *Building the Data Warehouse*. New York: John Wiley & Sons.

Immon, W., and R. Hackathorn. 1994. *Using the Data Warehouse*. New York: John Wiley & Sons.

Immon, W., J. Welch, and K. Glassey. 1997. *Managing the Data Warehouse*. New York: John Wiley & Sons.

Kimball, R., L. Reeves, M. Ross, and W. Thornthwaite. 1998. *The Data Warehouse Lifecycle Toolkit.* New York: Wiley.

Levine, J., T. Mason, and D. Brown. 1992. *Lex and Yacc.* Edited by D. Dougherty. Cambridge, MA: O'Reilly and Associates.

Microsoft. 2000. Terraserver: A spatial data warehouse [http://terraserver.microsoft.com]. Microsoft.

Mumick, I., D. Quass, and B. Mumick. 1997. Maintenance of data cubes and summary tables in a warehouse. In *1997 ACM SIGMOD International Conference on Mangement of Data*. Tucson, AZ.

OGIS. 1999. Open GIS Simple Features Specification for SQL (Revision 1.1) [http://www.opengis.org/techno/specs.htm]. Open Geospatial Consortium.

Paniati, J. 2002. Traffic Congestion and Sprawl [http://www.fhwa.dot.gov/congestion/congpress.htm]. Federal Highway Administration.

Potok, N. F. 2000. Behind the scenes of census 2000. *The Public Manager*, December 22, 2000 vol. 29, 4:3.

Stonebraker, M., J. Frew, and J. Dozier. 1993. The sequoia 2000 project. In *3rd International Symposium on Large Spatial Databases*.

USCB. 2000. U.S. Census Bureau [http://www.census.gov]. U.S. Census Bureau.

USGS. 1998. National satellite land remote sensing data archive [http://edc.usgs.gov/programs/nslrsda/overview.html]. U.S. Geological Survey.

Widom, J. 1995. Research problems in data warehousing. In *1995 International Conference on Information and Knowledge Management*. Baltimore, MD.

Worboys, M. 1995. *GIS: A Computing Perspective.* London: Taylor & Francis.

5 Data Quality Issues and Geographic Knowledge Discovery

Marc Gervais

Yvan Bédard

Marie-Andree Levesque

Eveline Bernier

Rodolphe Devillers

CONTENTS

5.1 INTRODUCTION

Geographical data warehouses contain data coming from multiple sources potentially collected at different times and using different techniques. One of the most important concerns about geographical data warehouses is the quality or reliability of the data used for knowledge discovery, decision making, and, finally, action. In fact, this is the ultimate objective aimed by using this type of database. On the other hand, with increasing maturity and the proliferation of data warehouses and related applications (e.g., OLAP, data mining, and dashboards), a recent survey indicated that for the second year in a row, data quality has become the first concern for companies using these technologies (Knightsbridge 2006). Similarly, a recent survey of Canadian decision makers using spatial data has identified data quality as the third most important obstacle in increasing the use of spatial data (Environics Research Group 2006). Thus, while data quality has become the number one concern for users of nonspatial data warehouses, it is also recognized as an emerging issue for spatial data (Sonnen 2007, Sanderson 2007) and the quality of spatial datacubes is being investigated seriously within university laboratories. In this context, the concept of data quality is making its way into the realm of geographic knowledge discovery, leading us to think in terms of risks for the users, for the developers, and for the suppliers of data, especially in terms of prevention mechanisms and possible legal consequences.

This chapter first introduces the readers to theoretical concepts regarding quality management and risk management in the context of spatial data warehousing and spatial online analytical processing (SOLAP). Then, it identifies possible management mechanisms to improve the prevention of inappropriate usages of data. Based on this theoretical foundation, this chapter then presents a pragmatic approach of quality and risk management to be applied during the various stages of a spatial datacube design and development. This approach aims at identifying and managing in a more rigorous manner the potential risks one may discover during this development process. Such approach has the merit to (1) be applicable in a real context, (2) be based on recognized quality and risk management models, (3) take into account lessons previously learned, (4) encourage proper documentation and, finally, (5) help clarify the responsibilities for each partner involved in the data warehouse development project. To complete the chapter, associations between these mechanisms and the legal rules governing the relationship between developers and users are presented.

5.2 FUNDAMENTAL CONCEPTS OF SPATIAL DATA QUALITY AND UNCERTAINTY IN A GEOGRAPHIC KNOWLEDGE DISCOVERY CONTEXT

Though data quality has always been an important aspect of geospatial applications, the proliferation of spatial business intelligence (BI) applications in the context of geographic knowledge discovery (GKD) has brought new concerns and raised new issues related to data quality. Their strategic position in organizations is such that these applications may have important impacts on the organization (Ponniah 2001). In order to make informed decisions, decision makers must be aware of the data characteristics and limitations. Otherwise, there is a risk of data misuses or misinterpretations that may cause severe legal, social, and economical impacts on the organization (Devillers et al. 2002). Unfortunately, in the context of GKD and especially with spatial BI applications, several factors increase the risk of data misuses and misinterpretation.

First is the ease with which users interact with the data. As opposed to GIS tools that require specialized knowledge, spatial BI applications are usually based on user interfaces that are easier and do not assume any specific *a priori* knowledge. There is no need to know a query language such as SQL to explore the data or to have specific knowledge about spatial reference methods or internal database structures. By lowering technical skills to operate such applications, they become available to a larger group of users who may not have a complete understanding or knowledge about the spatial, thematic, and temporal characteristics and limitations of the data (Levesque et al. 2007). Also, "the rapidity and ease of data use may lead users to mistakenly feel that data are made-to-order for their decision analysis needs, and hence to deter them from adopting an informed behavior towards data" (Sboui et al. 2008).

Second is the nature of the underlying data warehouses or datacubes. Because GKD applications are often based on data warehousing architectures, the data used have typically undergone several transformations. Building data warehouses or datacubes involves complex data integration and transformation processes (known as ETL procedures, for extract-transform-load) that may affect the meaning of their content (Levesque et al. 2007). Knowing that data sources may also have undergone such processes, it becomes difficult to evaluate the resulting data quality and reliability. Actually, end users of such technologies are rarely aware of these issues, and when they are they rarely receive a robust answer.

Third are the data aggregation methods. GKD and decision makers need aggregated or summarized data to perform their analyses. Hence, aggregation methods must be defined and applied to provide data that will help decision makers and GKD experts to have a global understanding of a phenomenon. This aggregation adds another level of complexity of interpretation (Sboui et al. 2008). Thus, to interpret correctly the data, decision makers must first understand the aggregation method and its impacts on the data.

In short, although spatial BI applications support GKD and the decision-making process, they do not ensure properly informed decisions or quality knowledge. Geospatial data users and decision makers must be aware of data quality in order to reduce the risks of data misuse and misinterpretation (Devillers, Bédard, and Gervais 2004).

5.2.1 Geospatial Data Quality and Uncertainty

In the geospatial literature, the notion of "quality" often mistakenly refers to data precision, uncertainty, or error. Data with good spatial precision are thus often seen as high-quality data. However, the notion of quality goes well beyond the unique concept of spatial precision. In fact, it is usually recognized as including two parts: internal quality and external quality.

Internal quality refers to the respect of data production standards and specification. It is based on the absence of errors in the data and is thus a matter of data producers. According to several standard organizations (such as ISO, ICA, FGDC, and CEN), internal quality is defined using five aspects, also known as the "famous five": (1) positional accuracy, (2) attribute accuracy, (3) temporal accuracy, (4) logical consistency, and (5) completeness (Guptill and Morrison 1995, ISO/TC-211 2002).

Information about internal quality is usually communicated to the users using metadata files transmitted with datasets by data producers (Devillers et al. 2007).

External quality evaluates if a dataset is suited for a specific need and hence refers to the notion of "fitness for use" (Juran, Gryna, and Bingham 1974; Chrisman 1983; Veregin 1999; Morrison 1995; Aalders and Morrison 1998; Aalders 2002; Dassonville et al. 2002; Devillers and Jeansoulin 2006). From a user's point of view, a dataset of quality meets or exceeds his expectations (Kahn and Strong 1998). This second definition has reached an official agreement by standardization organizations (e.g., ISO) and international organizations (e.g., IEEE).

Several researchers break down the concept of quality into sub-classes. Veregin (1999), inspired by the work of Berry (1964) and Sinton (1978), defines three components for geospatial data quality: position, time, and theme. He associates these axes to the notion of precision and resolution (spatial, temporal, and thematic precision, etc.). Bédard and Vallière (1995) propose six aspects that can be used to evaluate spatial data quality:

1. Definition is used to evaluate the nature of the data and the object it describes, i.e., the "what" (semantic, spatial, and temporal definitions).
2. Coverage provides information about the space and the time for which the data is defined, i.e., the "where" and "when".
3. Genealogy is related to the data origin, its acquisition methods, and objectives, i.e., the "how" and "why".
4. Precision is used to evaluate the value of a data and if it is acceptable for the expressed need (semantic, temporal, and spatial precision of the object and its attributes).
5. Legitimacy is associated with the official recognition and the legal extent of a data (*de facto* standards, approved specifications, etc.).
6. Accessibility provides information about the facility with which the user can obtain the data (costs, delivery time, confidentiality, copyrights, etc.).

Uncertainty is another inherent aspect of geospatial data and should be taken into account during their exploration and analysis. In fact, any cartographic representation of a phenomenon is an abstraction of the reality according to a specific goal. Given such abstraction and simplification processes, spatial data are, at different levels, inexact, incomplete, and not actual (Devillers 2004). According to Longley et al. (2001), it is impossible to produce a perfect representation of the reality and thus, this representation is inevitably associated with a certain uncertainty. Hence, there is always a risk associated with the use of spatial data that may be inadequate for some decision-making processes.

Bédard (1987) classifies uncertainty into four categories, which combine to provide the global uncertainty associated with an observed reality:

- (1st order) Conceptual, which relates to the fuzziness in the identification of an observed reality
- (2nd order) Descriptive, which relates to the uncertainty associated with the attributes values of an observed reality

- (3rd order) Locational, which relates to the uncertainty associated with the space and time localization of an observed reality
- (4th order) Meta-uncertainty, which relates to the level to which the previous uncertainties are unknown

Though uncertainty cannot be eliminated in spatial databases, mechanisms can be used to (1) reduce it, and (2) absorb the residuals (Bédard 1987, Hunter 1999). According to Epstein, Hunter, and Agumya (1998), uncertainty may be reduced by acquiring additional information and improving the data quality. According to Bédard (1987), the residual uncertainty is absorbed when an entity, such as the data producer or the distributor, provides a guarantee for the dataset and will cover potential damages resulting from their use for a given purpose or when the user accepts the potential consequences of using the dataset. Absorption can be shared with insurance companies or by contracting professionals with liability insurance. Uncertainty absorption relates to the monetary risk (e.g.,. in case of damages or a legal pursuit) and makes use of different combinations of the previous means depending on local laws and practices. In all cases, good professional practices and legal liability guidelines require using prevention mechanisms.

5.3 EXISTING APPROACHES TO PREVENT USERS FROM SPATIAL DATA MISUSES

Different mechanisms can be used to improve the prevention of inappropriate usages of spatial data. Existing methods are mostly intended to communicate information regarding data quality, characteristics, and limitations to the users. The traditional method consists of transmitting metadata along with spatial datasets. They are usually provided in separate files and contain highly technical information intended for geographic information system (GIS) specialists. However, such information is too cryptic to be understandable by typical users (Timpf, Raubal, and Werner 1996; Harvey 1998; Boin and Hunter 2007) and one is justified to assume the situation worsens with decision makers or data warehouse users who are further away from the technical details of the data acquisition and ETL processes. Furthermore, metadata are rarely integrated with the data, limiting their consultation and analysis as often required for GKD. In fact, it reduces the possibility of easily exploiting this information directly during the analysis process (Devillers et al. 2007). As an alternative to the actual metadata format, some researchers propose different techniques to communicate data quality information based on different colors, textures, opacities, 3D representations, etc. (McGranaghan 1993, Beard 1997, Drecki 2002, Devillers and Beard 2006). Other researchers propose to provide end users with meaningful warnings when they perform illogical GIS operations (e.g., measure a distance without having first set the geographical reference system) (Beard 1989, Hunter and Reinke 2000). This is related to the concept of error-sensitive or error-aware GIS (Unwin 1995, Duckham 2002).

Other researchers have tackled the fitness for use aspect by improving existing tools to select data that will best fit users' needs (Lassoued, Jeansoulin, and

Boucelma 2003), performing risk analysis (Agumya and Hunter 1997), getting opinions from experts (Levesque 2007), and even developing GKD tools to help these experts formulate their opinion by giving them the possibility to integrate, manage, and visualize data quality information at different levels of detail (Devillers 2004; Devillers, Bédard, and Jeansoulin 2005; Devillers et al. 2001; Levesque 2007).

From a data-warehousing point of view, few researchers have tackled the issue of data misuse and misinterpretation. Some have first identified cases where specific online analytical processing (OLAP) operators may lead to inappropriate usages (Lenz and Shoshani 1997, Lenz and Thalheim 2006). Others have suggested restricting the navigation or informing the user when results may be incorrect (Horner, Song, and Chen 2004). Those solutions, however, remain at a theoretical stage and contribute only partially to a global strategy to prevent datacube misuses. They address a subset of the issues related to data warehousing architectures and, above all, they do not consider the spatial aspect of the data. For example, they cannot be used to describe and illustrate the numerous conflicts that must be faced when integrating heterogeneous spatial datasets coming from different producers, or the semantic and geometric aggregations aspects that must be considered for an informed use of datacubes. In fact, most of these solutions are intended for experts in spatial information and data quality rather than the typical users of GKD or BI applications.

5.4 AN APPROACH BASED ON RISK MANAGEMENT TO PREVENT DATA MISUSES IN DATA WAREHOUSING AND GKD CONTEXTS

We suggest using a risk-management approach to face the complexity of the overall data quality issues during the design and feeding of the warehouse datacube. According to ISO/IEC (1999), a risk is defined as a "combination of the probability of occurrence of harm and the severity of that harm." Risk management refers to the reduction of a risk to a level considered acceptable (Morgan 1990, Renn 1998). Our approach is inspired from the risk management approach proposed by ISO/IEC Guide 51 (1999) and considers the notion of "harm" as a data misuse or misinterpretation. Such an approach was proposed by Agumya and Hunter (1999) for transactional geospatial data and is here geared toward multithemes, multiscales, and multiepochs decision-support data underlying GKD applications, and in particular a datacube/SOLAP context. However, the most noticeable difference with the approach proposed by Agumya and Hunter is that the proposed solution takes place during the design process, that is, in a more preventive mode. This key difference relies on the fact that the *raison d'être* and capabilities of datacubes allow us to identify *a priori* the data that will be compared thematically, spatially, temporally, and at different levels of granularity. Several datacubes are typically built from the same data warehouse according to the users' demands and data quality must be analyzed for each application using these cubes. Consequently, the star or snowflake schemas must be designed and populated with data quality in mind to reduce the risks of misuses. As a result, we advocate enriching system development methods (e.g., OMG-MDA or IBM rational unified process) with risk-management processes specific to the prevention of spatial data misuses.

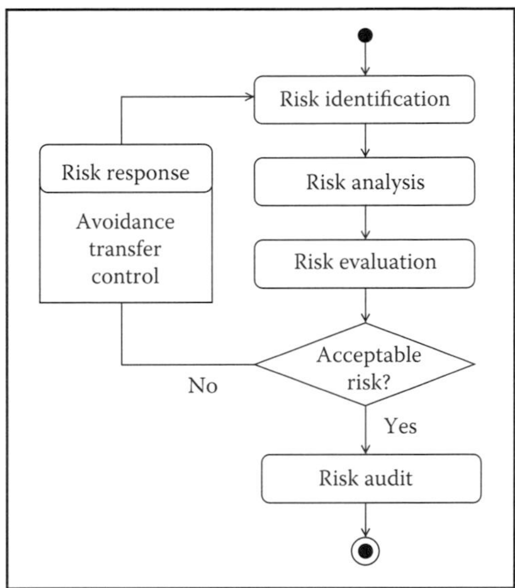

FIGURE 5.1 Proposed steps of the formal risk analysis method.

The proposed approach is a continuous and iterative process that fits with the whole datacube development cycle (needs analysis, design, implementation, feeding). Figure 5.1 shows the different steps proposed: identify risks, analyze them, evaluate potential dangers, prepare responses toward these risks of misuse or misinterpretation, and document the risk-management process as required for quality audits.

Risk identification: This critical step determines the efficiency and quality of the subsequent phases. It aims at finding what could go wrong when using the data, in a way that is as exhaustive as possible. This phase typically involves analyzing (1) the documentation about the data to be integrated (i.e., metadata, data dictionary, source data models), (2) the documentation about the designed datacubes (datacube models, ETL processes, aggregation functions), (3) the material used to train the users, and (4) the existing warnings (e.g., footnotes in reports, tables, and charts, report forewords, restricted accesses, etc.). Once identified, we suggest classifying the risks according to their origin (source) to facilitate further the definition of actions to be undertaken to control them. These categories are (1) data sources (e.g., missing data), (2) ETL procedures (e.g., erroneous aggregation formula), (3) datacube structure (e.g., not satisfying summarizability integrity constraints), and (4) SOLAP functionalities and operators (e.g., adding on the fly a new measure with faulty formula).

Risk analysis: The second step consists of analyzing, for each risk, its probability of occurrence and the severity of the consequences if it occurs. Risk analysis can rely on different techniques such as simulation techniques or probabilistic analysis. We can also look at relevant lessons learned during past projects, consult experts and specialists, etc. These two parameters, that is, the probability of occurrence and the severity of the consequences, are usually evaluated according to an ordinal scale composed of three to five levels (e.g., low, moderate, and high).

TABLE 5.1
General Hierarchy Matrix

		Probability of Occurrence		
		Low	**Moderate**	**High**
Severity level	High	M	H	H
	Moderate	L	M	H
	Low	L	L	M

Note: L = Low, M = Moderate, H = High.

Adapted from Kerzner, H., 2006. *Project Management: A Systems Approach to Planning, Scheduling, and Controlling*, 9th ed., John Wiley & Sons, New York.

Like other risk-based approaches, the risk evaluation step is not a simple task; because it demands that we look in a certain way in the future, it often requires experience, judgment, and sometimes intuition. In addition to these, we also consider that an excellent knowledge and understanding of the datacube users' needs and skills, which represent a legal duty of the datacube producer, are necessary to have the best risk analysis possible.

Risk evaluation: The previous results are combined in a matrix to determine the overall level of danger related to each risk (see Table 5.1).

Risk acceptability: Based on the global level of danger previously defined, we must decide whether a risk is acceptable. This analysis is sensitive and should be done very carefully as it may lead to legal consequences. For instance, in front of an acceptable risk (e.g., a risk with an overall level of danger at low), a datacube producer may accept it as it is without communicating the risk to the end-user. In case of damages for the end-user, the data producer can then be legally declared liable to have chosen to ignore it. If the risk is considered unacceptable, the producer must then choose a response mechanism in order to manage it (see risk response in the following).

At this stage, it is recommended to involve end-users and to select with them the appropriate response mechanisms. When users are involved, they understand the risks and approve the mechanisms proposed; they become directly involved in the uncertainty absorption process. Consequently, the datacube designers/providers are better protected in case of problems related to data quality and legal actions.

Risk response: This step is required to manage the risks that are unacceptable to the users. This is where the datacube producer suggests how to cope with those risks. Several mechanisms can be used, such as

- Avoidance: This mechanism aims at reducing an unacceptable risk by eliminating the source from which it emerges. For example, a data producer may decide not to provide data considered too sensitive or not reliable for the users and their intended usages. Such action is frequent when data are associated with a coefficient of variation above a certain threshold. It is usually applied to moderate to high risks that appear late in the development cycle (Kerzner 2006).

- Transfer: The transfer mechanism is used to move or share a risk with another entity in order to reduce it to a lower level for the datacube producer. For instance, the datacube producer can transfer a risk to a third party (such as an insurance company) who will become liable for the end-user in part or in totality.
- Control: The control mechanism suggests reducing the risk by taking preventive actions. The ISO/IEC Guide 51 standard states that risk reduction must first take place in the design phase, for example by modifying the conceptual model of the datacube or implementing integrity constraints. This is a key step to minimize risks. In addition, Guide 51 suggests producing information for security purposes, that is, "warnings." General warnings can be communicated to datacube users in a user manual while specific ones (i.e., context-sensitive warnings) can be automatically prompted in the SOLAP application when users are facing a risky query.
- We propose warnings according to the ISO 3864-2 (2004) standard for product safety labels. This standard proposes to communicate (1) the danger level of the risk with standardized alert words (e.g., danger, warning, or caution), (2) the nature of the risk with a symbol, (3) the consequence of the risk, and (4) how to avoid the risk. Figure 5.2 shows such a warning message that could be prompted when analyzing the data.

The remaining risks must be treated at the end-user level, for example by providing training, limiting access, building user profiles, etc. Defining the category of controls to apply and when to apply them require an excellent collaboration between the datacube producer and the users.

Risk audit: The last step is to document the previous steps as if preparing for an audit. Ideally, this documentation is made while designing and feeding the datacube as it is mainly during these steps that we think about or discover the potential problems. The documentation about a warning must include the message itself, the

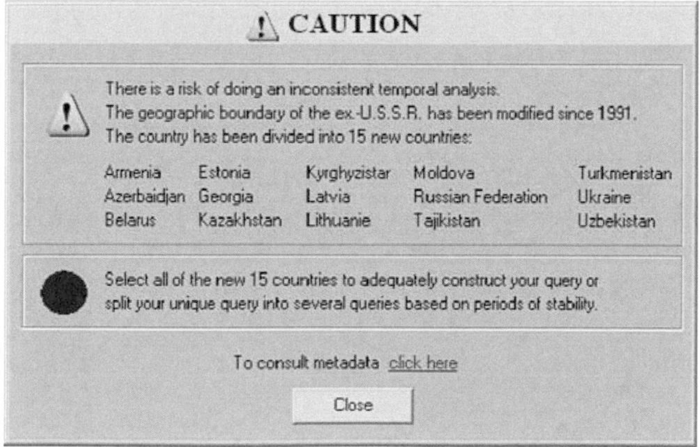

FIGURE 5.2 Example of context-sensitive warning in a SOLAP application.

involved elements, and the triggering elements (e.g., before the SOLAP query or once the results are displayed). This is helped by a series of forms implemented into a Unified Modeling Language (UML)-based CASE (computer-assisted software engineering) tool, and by a dictionary of terms and definitions describing these processes (Levesque 2008). Such documentation is very important from a legal standpoint because (1) this documentation can help prove the designer/producer complied with their legal duties, (2) it is helpful to prepare training material, and (3) it is an important source of information to manage risks in future datacube developments. More generally, it also helps system designers to build systems that are more robust.

5.5 LEGAL ISSUES RELATED TO THE DEVELOPMENT OF SPATIAL DATACUBES

Using a risk management approach to prevent data misuse is not only a matter of satisfying users' requirements, it is also a matter of legal liability principles. This section summarizes the legal principles that apply to the datacube producers and datacube users who are linked by a business relationship. First, we describe legal criteria related to the internal quality of data. Second, we describe the criteria related to the external quality when developing a datacube for a given purpose. Third, we summarize the legal duties of datacube users. Finally, we conclude with the pertinence of using a risk-management approach that involves both datacube producers and users during production of the datacube.

5.5.1 Legal Criteria for Spatial Datacube Producers Related to the Internal Quality of Data

Spatial datacubes are a special category of spatial database. They are not yet sold as commercial datasets per se, they are still designed as *ad hoc* custom services for a given need, and they are populated under the supervision of a professional. In general, it is recognized in many countries (such as in Canada and France) that database production is under the same legal liability regime as information production by agencies (Le Tourneau 2001, 2002; Le Tourneau and Cadiet 2002; Vivant et al. 2002; Dubuisson 2000; Côté et al. 1993). When offering services, the datacube producer must care about internal data quality (quality of the content) and external quality (fitness-for-use and quality of the presentation).

Regarding internal quality, unless specified, database producers are expected to deliver data that are exact, complete, and up-to-date because these are the three most important legal criteria used to assess internal quality (i.e., a subset of the ISO/TC211 data quality indicators). Applying these criteria to spatial datacubes raises some issues (e.g., cascading updates from source databases, completeness of aggregation and summarization of data, exactness of statistical indicators and multiscale generalized maps, time-varying maps, etc.), especially as spatio-temporal data are known to convey inherent uncertainty that cannot be eliminated (Gervais et al. 2007, Gervais 2004, Bédard 1987). Consequently, as it is the case for databases in general (Lucas 2001), one cannot always expect an internally perfect database as it is often impossible to achieve. Rather, it is expected for the data producer to use appropriate means to achieve the required internal quality. Database producers are thus typically facing an obligation of means as opposed to an obligation of results. Obligation of means refers to the obligation of the provider to act carefully to meet the expectations of the client and consequently to use all reasonable means to achieve the desired result, without warranting a perfect result (Baudouin and Jobin 1998). Consequently, it is expected for a datacube producer to formally adopt procedures especially tailored toward ensuring the internal quality of data, but without imposing the production of perfect data. Legally, the emphasis is given to the verification procedures that are used rather than the result obtained. In particular, in a datacube design or an ETL process, the datacube producer should not perform a task without knowing its impact on the resulting values (e.g., measures in the fact table).

5.5.2 Legal Criteria for Spatial Datacube Producers Related to the External Quality of Data

From a legal standpoint, the external quality is directly related to the diffusion and method of presenting data to the users. When the producer cannot guarantee the exactness of the data (as is typically the case with spatial data), there is an obligation to properly inform the users. Such obligation of proper information is in fact the legal mechanism to deal with imperfect products. It is expected that a producer provide all the information necessary to the users so they can properly assess the adequateness of a product concerning their needs. The level of information to provide is directly proportional to the incompetency of the user and to the level of complexity, technicality, and dangerousness of the product when being used. Consequently, from a legal point of view, evaluating the external quality of a spatial datacube becomes the evaluation of the information delivered with the spatial datacube.

Depending on the level of the uncertainty inherent to the datacube or on the level of dangerousness regarding the use of the datacube, one finds three types of such obligations: typical information, advice, and warning. Typical information does not require influencing the decision of the user (Lefebvre 1998) but it must be provided in a language and level of detail compatible with the expected typical users' level of knowledge. For example, providing only the metadata of a datacube could be sufficient if the user has the necessary knowledge to understand the technical terms related to spatial metadata and their impact on the proper use of the data (e.g., a well-trained and experienced user). The obligation of advising becomes important when the producer estimates that the provided datacube is complex and highly technical, or that the users need specific information because they do not have the necessary background to understand the characteristics of the datacubes or the consequences related to the planned usage (Lucas 2001). Such obligations may lead the datacube producer to perform additional research or analysis or to modify the datacube (Le Tourneau 2002). Finally, the obligation to provide warnings is always there (Baudouin and Deslauriers 1998), especially when one estimates that there are potential dangers to using the datacube. Such warnings must be clearly written, complete and up-to-date, and presented to users as soon as a danger is seen as potential (even before a final conclusion). This is an obligation of prevention that may direct the users away from erroneous usages or toward good usages.

Preventing dangerous usages by providing warnings requires identifying and communicating the anticipated risks (Rousseau 1999). A risk-management approach geared toward users' needs and level of tolerance to risks is mandatory. When uncertainty is high, the datacube producer must increase the degree of awareness of users. Considering the higher level of knowledge of the datacube producer, it is expected that he or she will make up for the users' lack of appropriate knowledge. Several court decisions regarding spatial data support this conclusion (e.g., breaking underground infrastructures,* marine charts depth errors,† erroneous transportation costs calculation, ‡ airplane crashes with deaths,§·¶ shipwrecks,**·†† unreasonable fire truck delay, ‡‡ hunting in the wrong area,§§ cross-country skier death,¶¶ and building a house in a forbidden area***).

5.5.3 Legal Criteria for Users of Spatial Datacubes

Users of datacubes also have legal obligations to ensure data are properly used. The most important obligations are those of collaboration with the datacube producer, constancy when defining the needs, and consistency when using the datacube (i.e., in accordance to the conditions emitted by the datacube producer) (Le Tourneau 2002). Collaboration must take place continuously when negotiating, defining the expectations, providing the required documentation and information, identifying the potential risks, designing and populating the datacube, and defining the means to deal with the identified risks of usage.

5.5.4 Legal Pertinence of a Risk-Management Approach

From a legal perspective, using a risk-management approach is necessary to protect both the datacube producers and users, in particular.

- Implementing formally such an approach within the datacube development method indicates the producer's will to take the necessary means to control rigorously the development of the cube and the decisions made during this phase.

---

* *Bell Canada* v. *Québec (Ville)*, [1996) A.Q. 172 (C.S.); *Excavations Nadeau & Fils*. v. *Hydro-Québec*, [1997) A.Q. 1972 (C.S.).

† *Fraser Burrard Diving Ltd.* v. *Lamina Drydock Co. Ltd.*, [1995) B.C.J. 1830 (B.-C.S.C.).

‡ *Côté* v. *Consolidated Bathurst*, [1990) A.Q. 64 (Qué. C.A.).

§ *Aetna Casualty & Surety Co.* v. *Jeppesen & Co.*, 642 F.2d 339 (1981).

¶ *Brocklesby* v. *United States of America*, 767 F.2d. 1288 (9th. Cir., 1985); *Times Mirror Co* v. *Sisk*, 593 P.2d. 924 (Ariz.1978).

** *Algoma Central and Hudson Bay Railway Co.* v. *Manitoba Pool Elevators Ltd.* [1966) S.C.R. 359; *Warwick Shipping Ltd.* v. *Canada* [1983) C.F. 807 (C.A.).

†† *Iron Ore Transport Co.* v. *Canada*, [1960) Ex. C.R. 448.

‡‡ *Bayus* v. *Coquitlam (City)*, [1993) B.C.C.S. 1751; *Bell* v. *Winnipeg (City)*, [1993) M.J. 256.

§§ *R.* v. *Rogue River Outfitters Ltd.* [1996) Y.J. 137 (Y.T.C.).

¶¶ *Rudko* v. *Canada*, [1983) C.F. 915 (C.A.).

*** *Sea Farm Canada* v. *Denton*, [1991) B.C.J. 2317 (B.-C.S.C.).

- Continuously communicating with users allows the datacube producer to better assess their tolerance to risk and to adapt the solutions accordingly. It also increases users' awareness.
- Involving users' collaboration in the complete process helps them to fulfill their duty of collaboration.
- Producing proper documentation helps datacube producers to meet their legal duty for information, advices, and warnings. The documents can be used for users' training or for further reference, and they become tangible proof that the work has been done.

A detailed description of the proposed method is beyond the goal of this chapter; however, it can be found in Levesque (2008). Overall, such an approach helps to clearly share the responsibilities between datacube producers and datacube users with regard to the risks of potential misuses. In addition, it adds rigour in the datacube development cycle, and increased users' satisfaction as well as a higher level of professionalism for the datacube producer.

5.6 CONCLUSION

This chapter focused on spatial datacube quality and, more specifically, on an approach to manage the risks of data misuse. We have synthesized issues related to internal and external data quality and presented how they have impacts on the design, populating, and use of spatial datacubes. This is a very recent concern in the GKD and spatial data warehousing community and indicates a new level of maturity. In particular, we have introduced the basis for adopting a risk management approach while developing datacubes. Such an approach allows reduction of the risks of data misuse, improves the involvement of users in the development of datacubes, and helps identify the responsibilities of the involved participants. Finally, we have made an overview of the legal motivations to adopt such a risk-management approach. Although such an approach cannot prevent all risks of data misuse, it is a means to prevent such risks and to increase users' awareness, leading to spatial datacubes with higher internal and external quality.

REFERENCES

Aalders, H.J.G.L., Morrison, J. 1998. "spatial data quality for GIS." Geographic Information Research: Transatlatic Perspectives (Eds.), Taylor & Francis, London/Bristol, p. 463–475

Aalders, H.J.G.L. 2002. The registration of quality in a GIS, in W. Shi, P. Fisher, and Goodchild M.F. (Eds.), *Spatial Data Quality*, Taylor & Francis, London, pp. 186–199.

Agumya, A. and Hunter, G.J. 1997. Determining the fitness for use of geographic information, *Journal of the International Institute for Aerospace Survey and Earth Science*, 2, 109–113.

Agumya, A. and Hunter, G.J. 1999. A Risk-based approach to Assessing the 'fitness for use' of spatial data, URISA Journal, USA, vol. 11, No. 1, pp. 33–44.

Baudouin, J.-L. and Deslauriers, P. 1998. *La Responsabilité Civile*, Les Éditions Yvon Blais Inc., Cowansville.

Beard, K. 1989. Use error: the neglected error component, *Proc. AUTO-CARTO 9*, Baltimore, MD, pp.808–817.

Beard, K. 1997. Representations of data quality, in Craglia, M. and Couclelis, H. (Eds.), *Geographic Information Research: Bridging the Atlantic*, Taylor & Francis, London, pp. 280–294.

Bédard, Y. 1987. Uncertainties in land information systems databases, *Proceedings of Eighth International Symposium on Computer-Assisted Cartography*, Baltimore, MD, pp. 175–184.

Bédard, Y., Vallière, D. 1995. Qualité des données à référence spatiate dans un contexte gouvernmental, rapport de rechche, Université Laval, Québec, Canada.

Berry, B., "Approaches to regional analysis: a synthesis." 1964 *Annals of the Association of American Geographers*, vol. 54, p. 2–11.

Boin, A.T. and Hunter, G.J. 2007. What communicates quality to the spatial data consumer? *5th International Symposium on Spatial Data Quality*, Enschede, The Netherlands, June 13–15.

Chrisman, N.R. 1983. The role of quality information in the long term functioning of a geographical information system, *Proceedings of International Symposium on Automated Cartography (Auto Carto 6)*, Ottawa, Canada, pp. 303–321.

Côté, R., Jolivet, C., Lebel, G.A., and Beaulieu, B. 1993. La Géomatique, ses enjeux juridiques, Publications du Québec, Québec.

Dassonville, L., Vauglin, F., Jakobsson, A., and Luzet, C. 2002. Quality management, data quality and users, metadata for geographical information, in Shi, W., Fisher, P., and Goodchild, M.F. (Eds.), *Spatial Data Quality*, Taylor & Francis, London, pp. 202–215.

Devillers, R. 2004. Conception d'un système multidimensionnel d'information sur la qualité des données géospatiales. PhD thesis, Sciences Géomatiques, Université Laval, Canada.

Devillers, R. and Beard, K. 2006. Communication and use of spatial data quality information in GIS, in Devillers, R. and Jeansoulin, R. (Eds.), *Fundamentals of Spatial Data Quality*, ISTE Publishing, London, pp. 237–253.

Devillers, R., Bédard, Y., and Gervais, M. 2004. Indicateurs de qualité pour réduire les risques de mauvaise utilisation des données géospatiales, *Revue Internationale de Géomatique*, 14(1), 35–57.

Devillers, R., Bédard, Y., and Jeansoulin, R. 2005. Multidimensional management of geospatial data quality information for its dynamic use within GIS, *Photogrammetric Engineering & Remote Sensing*, 71(2), 205–215.

Devillers, R., Bédard, Y., Jeansoulin, R., and Moulin, B. 2007. Towards spatial data quality information analysis tools for experts assessing the fitness for use of spatial data, *International Journal of Geographical Information Sciences*, 21(3), 261–282.

Devillers, R., Gervais, M., Jeansoulin, R., and Bédard, Y. 2002. Spatial data quality: From metadata to quality indicators and contextual end-user manual, OEEPE/International Society for Photogrammetry and Remote Sensing (ISPRS) Workshop on Spatial Data Quality Management, March 21–22.

Devillers, R. and Jeansoulin, R. (Eds.). 2006. *Fundamentals of Spatial Data Quality*, ISTE, London.

Drecki, I. 2002. Visualisation of uncertainty, in Shi, W., Fisher, P.F., and Goodchild, M.F. (Eds.). *Geographic Data, Spatial Data Quality*, Taylor & Francis, London, pp.140–159.

Dubuisson, B. 2000. Introduction, dans La responsabilité civile liée à l'information et au conseil, Dubuisson, B. and Jadoul, P., Publications des Facultés Universitaires Saint-Louis, Bruxelles, pp. 9–13.

Duckham, M. 2002. A user-oriented perspective of error-sensitive GIS development, *Transactions in GIS*, 6(2), 179–193.

Environics Research Group. 2006. Sondage auprès des décideurs ayant recours à l'information géographique-2006: Sommaire, Technical report for GeoConnection, Natural Resources Canada, October. Also available in English.

Epstein, E. F. Hunter, G.J., and Agumya, A. 1998. Liability insurance and the use of geographical information, *International Journal of Geographical Information Science*, 12(3), 203–214.

Gervais, M. 2004. Pertinence d'un manuel d'instructions au sein d'une stratégie de gestion du risque juridique découlant de la fourniture de données géographiques numériques, PhD Thesis, Université Laval, Québec, Canada and Université Marne-La-Vallée, France.

Gervais, M., Bédard, Y., Jeansoulin, R., and Cervelle, B. 2007. Obligations juridiques potentielles et modèle du producteur raisonnable, Revue Internationale de Géomatique, Éditions Lavoisier, Paris, 17(1), 33–62.

Guptill, S. C. and Morrison, J. L. 1995. Elements of Spatial Data Quality, Elsevier Science, New York.

Harvey, F. 1998. Quality needs more than standards, in Goodchild, M. and Jeansoulin, R. (Eds.), *Data Quality in Geographic Information — From Error to Uncertainty*, Editions Hermes, New York City, pp. 37–42.

Horner, J., Song, II-Y., and Chen, P.P. 2004. An analysis of additivity in OLAP systems, *Proceedings of the 7th ACM International Workshop on Data Warehousing and OLAP*, Washington D.C, pp. 83–91.

Hunter, G. J. 1999. Managing uncertainty in GIS, in Longley, P. A., Goodchild, M.F., Maguire, D.J., and Rhind, D.W. (Eds.), *Geographical Information Systems, Management Issues and Applications*, John Wiley & Sons, New York, pp. 633–641.

Hunter, G.J. and Reinke, K. 2000. Adapting spatial databases to reduce information misuse through illogical operations, *Proceedings Spatial Accuracy Assessment, Land Information Uncertainty in Natural Resources Management,* Amsterdam, The Netherlands, pp. 313–319.

ISO/IEC Guide 51. 1999. Aspects liés à la sécurité — Principes directeurs pour les inclure dans les normes.

ISO-TC/211. 2002. Geographic Information — Quality principles 19113.

ISO 3864–2. 2004. Safety colours and safety signs — Part 2: Design principles for product safety labels.

Juran, J.M., Gryna, F.M.J., and Bingham, R.S. 1974. *Quality Control Handbook*, McGraw-Hill, New York.

Kahn, B.K. and Strong, D.M. 1998. *Product and Service Performance Model for Information Quality: An Update, Conference on Information Quality*, Massachusetts Institute of Technology, Cambridge, MA.

Kerzner, H. 2006. *Project Management: A Systems Approach to Planning, Scheduling, and Controlling*, 9th ed., John Wiley & Sons, New York.

Knightsbridge, 2006. Top 10 Trends in Business Intelligence for 2006. White Paper http://www.tdwi.org. Accessed 01/16/2007.

Lassoued, Y., Jeansoulin, R., and Boucelma, O. 2003. Médiateur de qualité dans les systèmes d'information géographique, SETIT International conference (Sciences Electroniques, Technologies de l'Information et des Télécommunications), Sousse, Tunisia.

Lefebvre, B. 1998. La bonne foi dans la formation des contrats, Les Éditions Yvon Blais Inc., Cowansville.

Lenz, H-J. and Shoshani, A. 1997. Summarizability in OLAP and statistical data bases, *Proceedings of the 9th International Conference on Scientific and Statistical Database Management (SSDB)*, Washington D.C., August 11–13, pp. 132–143.

Lenz, H-J. and Thalheim, B. 2001. OLAP databases and aggregation functions, *Proceedings of the 13th International Conference on Scientific and Statistical Database Management (SSDB)*, pp. 91–101.

Le Tourneau, P. 2001. Responsabilité des vendeurs et fabricants, Droit de l'entreprise, Les Éditions Dalloz, Paris.

Le Tourneau, P. 2002. Contrats informatiques et électroniques, Dalloz reference, Les Éditions Dalloz, Paris.

Le Tourneau, P. and Cadiet, L. 2002. Droit de la responsabilité et des contrats, Les Éditions Dalloz, Paris.

Levesque, J. 2007. Évaluation de la qualité des données géospatiales: Approche top-down et gestion de la métaqualité, M.Sc. Thesis, Université Laval, Québec, Canada.

Levesque, M.A. 2008. Approche générique pour une meilleure identification et gestion des risqué d'usages inappropriés des données géodécisionnelles. MSc thesis draft version, Department of Geomatics Sciences, Laval University, Quebec City, Canada.

Levesque, M.-A., Bédard, Y., Gervais, M., and Devillers, R. 2007. Towards managing the risks of data misuse for spatial datacubes, *Proceedings of the 5th International Symposium on Spatial Data Quality,* June 13–15, Enschede, The Netherlands.

Longley, P.A., Goodchild, M.F., Maguire, D.J., and Rhind, D.W. (Eds.). 2001. *Geographical Information Systems and Science*, John Wiley & Sons, New York.

Lucas, A. 2001. Informatique et droit des obligations, dans Droit de l'informatique et de l'Internet, Thémis Droit privé, Presses Universitaires de France, Paris, p. 441–588.

McGranaghan, M. 1993. A cartographic view of spatial data quality, *Cartographica*, 30, 8–19.

Morgan, M.G. 1990. Choosing and managing technology-induced risks, in Glickman, T.S. and Gough, M. (Eds.), *Readings in Risk*, Resources for the Future, Washington, pp. 5–15.

Morrison, J. L. 1995. Spatial data quality, in Guptill, S.C. and Morrison, J.L. (Eds.), *Elements of Spatial Data Quality*, Elsevier Science, New York.

Ponniah, P. 2001. Data Warehousing Fundamentals: A Comprehensive Guide for IT Professionals, John Wiley & Sons, New York, pp. 291–312.

Renn, O. 1998. Three decades of risk research: Accomplishments and new challenges, *Journal of Risk Research* 1(1), 49–71.

Rousseau, S. 1999. La responsabilité civile de l'analyste financier pour la transmission d'information fausse ou trompeuse sur le marché secondaire des valeurs mobilières, dans La responsabilité civile des courtiers en valeurs mobilières et des gestionnaires de fortune: aspects nouveaux, Les Éditions Yvon Blais Inc., Cowansville, p. 35–62.

Sanderson, M. 2007. Data quality challenges in 2007, *Directions Magazine*, January 18 issue.

Sboui, T., Bédard, Y., Brodeur, J., and Badard, T. 2008. Risk management for the simultaneous use of spatial datacubes: A semantic interoperability perspective, *Annals of Information Systems*, Special Issue on New Trends in Data Warehousing and Data Analysis, Springer, submitted.

Sinton, D.F., 1978. "The inherent structure of information as a constraint in analysis". Harvard Papers on Geographic Information Systems (G. Dulton, Ed), Addison-Wesley, Reading, USA.

Sonnen, D. 2007. Emerging issues: Spatial data quality, *Directions Magazine,* January 4 issue.

Timpf, S., Raubal, M., and Werner, K. 1996. Experiences with metadata, in Kraak, M.-J. and Molenaar, M. (Eds.), *Symposium on Spatial Data Handling, SDH'96, Advances in GIS Research II*, Vol. 2 pp. 12B.31–12B.43.

Unwin, D. 1995. Geographical information systems and the problem of error and uncertainty, *Progress in Human Geography*, 19, 549–558.

Veregin, H. 1999. Data quality parameters, in Longley, P.A., Goodchild, M.F., Maguire, D.J., and Rhind, D.W. (Eds.), *Geographical Information Systems*, Wiley, New York, pp. 177–189.

Vivant, M. et al. 2002. Lamy Droit de l'informatique et des réseaux, Lamy S.A., Paris.

6 Spatial Classification and Prediction Models for Geospatial Data Mining*

Shashi Shekhar

Ranga Raju Vatsavai

Sanjay Chawla

CONTENTS

* This chapter is an extension of our earlier work [43] and IEEE holds the copyright to the figures and tables reproduced in this chapter, © 2002 IEEE. This material is used with permission.

6.1 INTRODUCTION

Widespread use of spatial databases [42], an important subclass of multimedia databases, is leading to an increased interest in mining interesting and useful but implicit spatial patterns [23, 29, 18, 40]. Traditional data mining algorithms [1] often make assumptions (e.g., independent, identical distributions) which violate Tobler's first law of geography: everything is related to everything else but nearby things are more related than distant things [45]. In other words, the values of attributes of nearby spatial objects tend to systematically affect each other. In spatial statistics, an area within statistics devoted to the analysis of spatial data, this is called spatial autocorrelation [12]. Knowledge discovery techniques that ignore spatial autocorrelation typically perform poorly in the presence of spatial data. The simplest way to model spatial dependence is through spatial covariance. Often the spatial dependencies arise due to the inherent characteristics of the phenomena under study, but in particular they arise due to the fact that imaging sensors have better resolution than object size. For example, remote sensing satellites have resolutions ranging from 30 m (e.g., Enhanced Thematic Mapper of Landsat 7 satellite of NASA) to 1 m (e.g., IKONOS satellite from SpaceImaging), while the objects under study (e.g., urban, forest, water) are much bigger than 30 m. As a result, the per-pixel-based classifiers, which do not take spatial context into account, often produce classified images with *salt and pepper* noise. These classifiers also suffer in terms of classification accuracy.

There are two major approaches for incorporating spatial dependence into classification/prediction problems. They are spatial autoregression models (SAR) [3, 24, 26, 36, 37, 25, 28] and Markov random field (MRF) models [15, 11, 21, 44, 8, 48, 27, 2]. Here we want to make a note on the terms spatial dependence and spatial context. These words originated in two different communities. Natural resource analysts and statisticians use spatial dependence to refer to spatial autocorrelation and the image processing community uses spatial context to mean the same. We use spatial context, spatial dependence, and spatial autocorrelation interchangeably to relate to readers of both communities. We also use classification and prediction interchangeably. Natural resource scientists, ecologists and economists have incorporated spatial dependence in spatial data analysis by incorporating spatial autocorrelation into logistic regression models (called SAR). The SAR model states that the class label of a location is partially dependent on the class labels of nearby locations and partially dependent on the feature values. SAR tends to provide better models than logistic regression in terms of achieving higher confidence (R^2). Similarly MRFs is a popular model for incorporating spatial context into image segmentation and land-use classification problems. Over the last decade, several researchers [44, 21, 48, 33] have exploited spatial context in classification using MRF to obtain higher accuracies over their counterparts (i.e., noncontextual classifiers). MRFs provide a uniform framework for integrating spatial context and deriving the probability distribution of interacting objects.

We compare the SAR and MRF models in this chapter using a common probabilistic framework. SAR makes more restrictive assumptions about the probability distributions of feature values as well as the class boundaries. We show that SAR assumes the conditional probability of a feature value given a class label that belongs to an exponential family, for example, Gaussian, binomial, etc. In contrast, MRF models can work with many other probability distributions. SAR also assumes the linear separability of classes in a transformed feature space resulting from the local smoothing of feature values based on autocorrelation parameters. MRFs can be used with nonlinear class boundaries. Readers familiar with classification models which ignore spatial context may find the following analogy helpful. The relationship between SAR and MRF is similar to the relationship between logistic regression and Bayesian classification.

Recent advances in remote sensors have resulted in a huge collection of high temporal, spatial, and spectral resolution images. The high-resolution digital imagery acquired by remote sensing technology has applications in areas such as natural resource monitoring, thematic mapping, flood and fire disaster monitoring, target detection, and urban growth modeling. There is a great demand for accurate land use and land cover classification derived from remotely sensed data in the applications mentioned previously. However, increasing spatial and spectral resolution puts several constraints on supervised classification. The increased spatial resolution invalidates the most widely used assumption of the traditional data mining algorithms (e.g., independent, identical distributions). Often used maximum likelihood estimation requires large amounts of training data for accurate estimation of model parameters and increasing spectral resolution further compounds this problem. We present a novel spatial semisupervised algorithm that addresses spatial autocorrelation as well as parameter estimates with small learning samples.

6.1.1 Outline and Scope of this Chapter

The rest of the chapter is organized as follows. In Section 6.2.1 we introduced two motivating examples which will be used throughout the chapter. In Section 6.2.3 we formally define the location prediction problem. Section 6.3 presents a comparison of classical approaches that do not consider spatial context, namely logistic regression and Bayesian classification. In Section 6.4 we present two modern approaches that model spatial context, namely SAR [24] and MRFs [27]. In Section 6.5 we compare and contrast the SAR and MRF models in a common probabilistic framework and provide experimental results. In Section 6.6 we introduce a spatial semisupervised learning scheme, and provide conclusions and future research directions in Section 6.7.

6.2 ILLUSTRATIVE APPLICATION DOMAINS

6.2.1 Bird Nesting Location Prediction

First we introduce an example to illustrate the different concepts in spatial data mining. We are given data about two wetlands, named Darr and Stubble, on the shores of Lake Erie in Ohio in order to *predict* the spatial distribution of a marsh-breeding bird, the red-winged blackbird (*Agelaius phoeniceus*) [34, 35]. The data were collected from April to June in two successive years, 1995 and 1996.

A uniform grid was imposed on the two wetlands and different types of measurements were recorded at each cell or pixel. In total, values of seven attributes were recorded at each cell. Domain knowledge is crucial in deciding which attributes are important and which are not. For example, *Vegetation Durability* was chosen over *Vegetation Species* because specialized knowledge about the bird-nesting habits of the red-winged blackbird suggested that the choice of nest location is more dependent on plant structure and plant resistance to wind and wave action than on the plant species.

An important goal is to build a model for predicting the location of bird nests in the wetlands. Typically the model is built using a portion of the data, called the learning or training data, and then tested on the remainder of the data, called the testing data. For example, later on we will build a model using the 1995 data on the Darr wetland and then test it on either the 1996 Darr or 1995 Stubble wetland data. In the learning data, all the attributes are used to build the model and in the training data, one value is *hidden*, in our case the location of the nests. Using knowledge gained from the 1995 Darr data and the value of the independent attributes in the test data, we want to predict the location of the nests in Darr 1996 or in Stubble 1995.

In this chapter we focus on three independent attributes, namely *Vegetation Durability*, *Distance to Open Water*, and *Water Depth*. The significance of these three variables was established using classical statistical analysis [35]. The spatial distribution of these variables and the actual nest locations for the Darr wetland in 1995 are shown in Figure 6.1. These maps illustrate the following two important properties inherent in spatial data. The value of attributes that are referenced by spatial location tend to vary gradually over space. While this may seem obvious, classical data-mining techniques, either explicitly or implicitly, assume that the data is *independently* generated. For example, the maps in Figure 6.2 show the spatial distribution of attributes if they were independently generated. One of the authors has applied classical data mining techniques like logistic regression [35] and neural networks [34] to build spatial habitat models. Logistic regression was used because

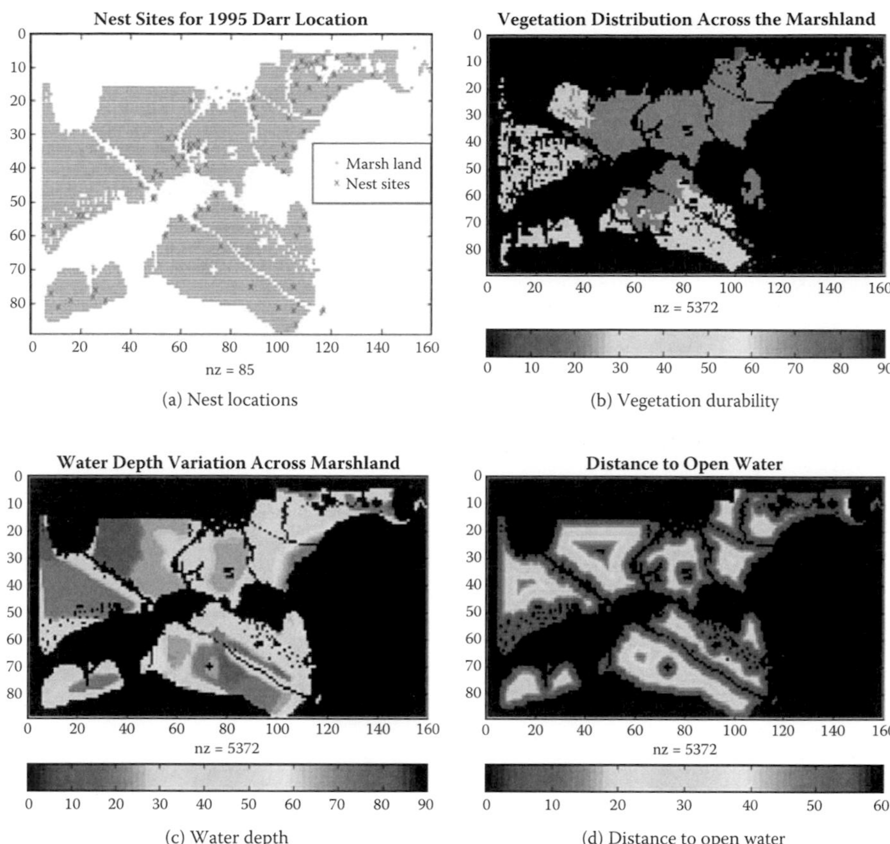

(a) Nest locations

(b) Vegetation durability

(c) Water depth

(d) Distance to open water

FIGURE 6.1 (a) Learning dataset: The geometry of the wetland and the locations of the nests. (b) The spatial distribution of *vegetation durability* over the marshland. (c) The spatial distribution of *water depth*. (d) The spatial distribution of *distance to open water*.

the dependent variable is binary (nest/no-nest) and the logistic function "squashes" the real line onto the unit-interval. The values in the unit-interval can then be interpreted as probabilities. The study concluded that with the use of logistic regression, the nests could be classified at a rate of 24% better than random [34]. The fact that classical data-mining techniques ignore spatial autocorrelation and spatial heterogeneity in the model-building process is one reason why these techniques do a poor job. A second, more subtle, but equally important reason is related to the choice of the objective function to measure classification accuracy. For a two-class problem, the standard way to measure classification accuracy is to calculate the percentage of correctly classified objects. This measure may not be the most suitable in a spatial context. *Spatial accuracy* — how far the predictions are from the actuals — is as important in this application domain due to the effects of discretization of a continuous wetland into discrete pixels, as shown in Figure 6.3. Figure 6.3(a) shows the

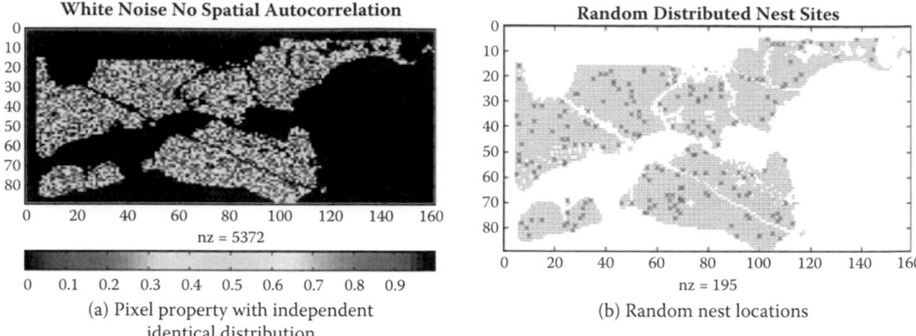

(a) Pixel property with independent identical distribution

(b) Random nest locations

FIGURE 6.2 Spatial distribution satisfying random distribution assumptions of classical regression.

actual locations of nests and Figure 6.3(b) shows the pixels with actual nests. Note the loss of information during the discretization of continuous space into pixels. Many nest locations barely fall within the pixels labeled "A" and are quite close to other blank pixels, which represent "no-nest". Now consider two predictions shown in Figure 6.3(c) and Figure 6.3(d). Domain scientists prefer the prediction that in Figure 6.3(d) over Figure 6.3(c), since predicted nest locations are closer on average to some actual nest locations. The classification accuracy measure cannot distinguish between Figure 6.3(c) and Figure 6.3(d), and a measure of spatial accuracy is needed to capture this preference.

A simple and intuitive measure of spatial accuracy is the average distance to nearest prediction (ADNP) from the actual nest sites, which can be defined as

$$ADNP(A,P) = \frac{1}{K}\sum_{k=1}^{K} d(A_k, A_k.nearest(P)).$$

Here A_k represents the actual nest locations, P is the map layer of predicted nest locations and $A_k.nearest(P)$ denotes the nearest predicted location to A_k. K is the number of actual nest sites.

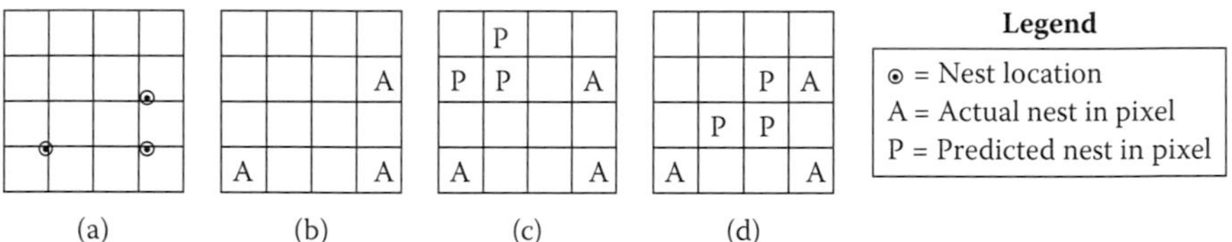

FIGURE 6.3 (a) The actual locations of nests. (b) Pixels with actual nests. (c) Locations predicted by a model. (d) Locations predicted by another model. Prediction (d) is spatially more accurate than (c).

TABLE 6.1
Landsat 7 Spectral Bands

Band Number	Wavelength Interval μm	Spectral Response
1	0.45-0.52	Blue-Green
2	0.52-0.60	Green
3	0.63-0.69	Red
4	0.76-0.90	Near-IR
5	1.55-1.75	Mid-IR
6	10.40-12.50	Thermal-IR
7	2.08-2.35	Mid-IR

6.2.2 Remote Sensing Image Classification

Land management organizations and the public have a need for more current regional land cover information to manage resources and monitor land cover change. Remote sensing, which provides inexpensive, synoptic-scale data with multitemporal coverage, has proven to be very useful in land cover mapping, environmental monitoring, and forest and crop inventory. A common task in analyzing remote sensing imagery is supervised classification, where the objective is to construct a classifier based on a few labeled training samples and then to assign a label (e.g., forest, water, urban) to each pixel (vector, whose elements are spectral measurements) in the entire image. There is a great demand for accurate land use and land cover classification derived from remotely sensed data in various applications. Classified images (thematic information) are also important inputs to geographic information systems (GIS).

A satellite image consists of n-bands or channels. Each band corresponds to measurements in a particular wavelength of the electromagnetic spectrum. Table 6.1 shows Landsat 7 image bands and corresponding wavelengths. Sample raw satellite image and corresponding classified image are shown in Figure 6.4. A typical classification process involves several steps: (1) randomly generate samples (roughly 10 to 100 $\times$ number of bands per class) from satellite image, (2) collect ground truth for each sample (by field visits, or expert knowledge derived from several ancillary sources), (3) divide the labeled samples into training and test datasets, (4) select appropriate classification model and train the classifier using training dataset, (5) evaluate model performance on test dataset, and (6) finally classify the entire image using the classification model generated in step (4). Figure 6.4(b) shows an MLC classification output generated by following these six steps.

6.2.3 Location Prediction: Problem Formulation

The location prediction problem is a generalization of the nest location prediction problem. It captures the essential properties of similar problems from other domains

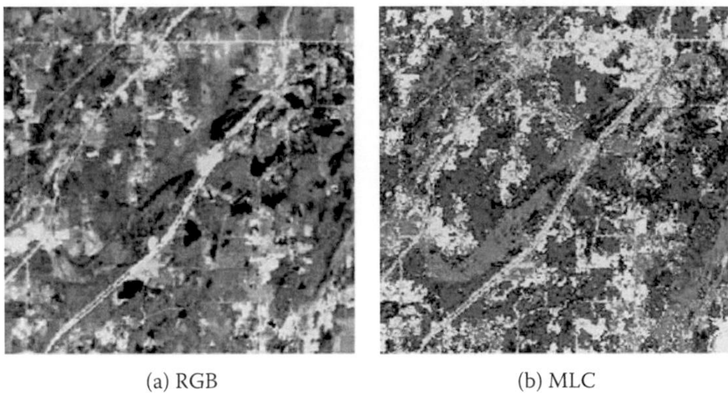

FIGURE 6.4 Sample Red, Green, Blue, (RGB) image (a) and corresponding maximum likelihood classified Maximum Likelihood Classifier (MLC) image (b). See color insert after page 148.

including crime prevention and environmental management. The problem is formally defined as follows:

Given:

- A spatial framework S consisting of sites $\{s_1, \dots, s_n\}$ for an underlying geographic space G.
- A collection of explanatory functions $f_{Xk} : S \rightarrow R^k$, $k = 1, \dots, K$. R^k is the range of possible values for the explanatory functions.
- A dependent class variable $f_L : S \rightarrow L = \{l_1, \dots l_M\}$.
- Value for parameter α, relative importance of spatial accuracy.

Find: Classification model: $\hat{f}_L : R^1 \times \cdots R^k \rightarrow L$.

Objective: Maximize similarity $(map_{si} \in s\,(\hat{f}_L(f_{X_1}, \dots, f_{Xk})(map(f_L)) = (1-\alpha)$ classification_accuracy$(\hat{f}_L, f_L) + (\alpha)$spatial_accuracy$((\hat{f}_L, f_L)$

Constraints:

1. Geographic space S is a multidimensional Euclidean space.[†]
2. The values of the explanatory functions, $f_{X1}, \dots, f_{Xk}$ and the response function f_L may not be independent with respect to those of nearby spatial sites, i.e., spatial autocorrelation exists.
3. The domain R^k of the explanatory functions is the one-dimensional domain of real numbers.
4. The domain of the dependent variable, $L = \{0, 1\}$.

The previous formulation highlights two important aspects of location prediction. It explicitly indicates that (1) the data samples may exhibit spatial autocorrelation

[†] The entire surface of the Earth canot be modeled as a Euclidean sapce but locally the approximation holds true.

and (2) an objective function, i.e., a map similarity measure is a combination of classification accuracy and spatial accuracy. The *similarity* between the dependent variable f_L and the predicted variable $\hat{f}_L$ is a combination of the "traditional classification" accuracy and a representation dependent "spatial classification" accuracy. The regularization term α controls the degree of importance of *spatial accuracy* and is typically domain dependent. As $\alpha \rightarrow 0$, the map similarity measure approaches the traditional classification accuracy measure. Intuitively, α captures the spatial autocorrelation present in spatial data.

The study of the nesting locations of red-winged black birds [34, 35] is an instance of the location prediction problem. The underlying spatial framework is the collection of $5m \times 5m$ pixels in the grid imposed on marshes. Explanatory variables are, for example, water depth, vegetation durability index, distance to open water, map pixels to real numbers. Dependent variables are, for example, nest locations, map pixels to a binary domain. The explanatory and dependent variables exhibit spatial autocorrelation, for example, gradual variation over space, as shown in Figure 6.1. Domain scientists prefer spatially accurate predictions which are closer to actual nests, that is, $\alpha > 0$.

6.3 CLASSIFICATION WITHOUT SPATIAL DEPENDENCE

In this section we briefly review two major statistical techniques that have been commonly used in the classification problem. These are logistic regression modeling and Bayesian classifiers. These models do not consider spatial dependence. Readers familiar with these two models will find it easier to understand the comparison between SAR and MRF.

6.3.1 Logistic Regression Modeling

Given an n—vector **y** of observations and an $n \times m$ matrix $\underline{X}$ of explanatory data, classical linear regression models the relationship between y and $\underline{X}$ as

$$y = X\beta + \varepsilon.$$

Here $X = [1, \underline{X}]$ and $\beta = (\beta_0, \ldots, \beta_m)^t$. The standard assumption on the error vector ε is that each component is generated from an independent, identical, zero-mean and normal distribution, that is, $\varepsilon_i = N(0, \sigma^2)$.

When the dependent variable is binary, as is the case in the "bird-nest" example, the model is transformed via the logistic function and the dependent variable is interpreted as the probability of finding a nest at a given location. Thus, $Pr(l|y) = \frac{e^y}{1+e^y}$. This transformed model is referred to as *logistic* regression [3].

The fundamental limitation of classical regression modeling is that it assumes that the sample observations are independently generated. This may not be true in the case of spatial data. As we have shown in our example application, the explanatory and the independent variables show a moderate to high degree of spatial autocorrelation (see Figure 6.1). The inappropriateness of the independence assumption

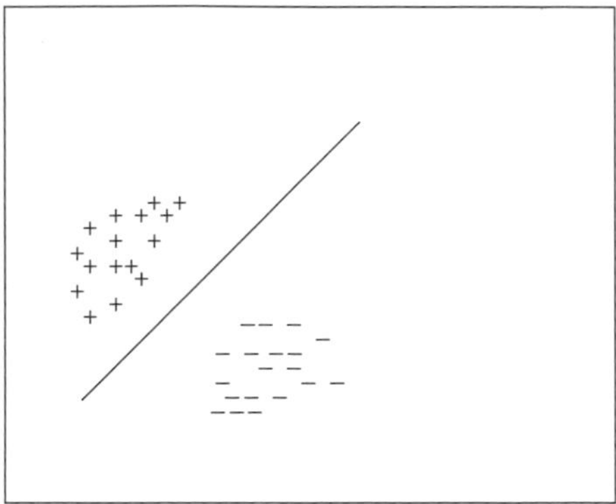

FIGURE 6.5 Two-dimensional feature space, with two classes (+: nest, –: no-nest) that can be separated by a linear surface.

shows up in the residual errors, the ε_i's. When the samples are spatially related, the residual errors reveal a systematic variation over space, that is, they exhibit high spatial autocorrelation. This is a clear indication that the model was unable to capture the spatial relationships existing in the data. Thus, the model may be a poor fit to the geospatial data. Incidentally, the notion of spatial autocorrelation is similar to that of time autocorrelation in time series analysis but is more difficult to model because of the multidimensional nature of space. A statistic that quantifies spatial autocorrelation is introduced in the SAR model.

Also the logistic regression finds a discriminant surface, which is a hyperplane in feature space as shown in Figure 6.5. Formally, a logistic regression based classifier is equivalent to a perceptron [19, 20, 41], which can only separate linearly separable classes.

6.3.2 Bayesian Classification

Bayesian classifiers use Bayes' rule to compute the probability of the class labels given the data:

$$Pr(l_i|X) = \frac{Pr(X|l_i)\Pr(l_i)}{Pr(X)} \tag{6.1}$$

In the case of the location prediction problem, where a single class label is predicted for each location, a decision step can assign the most-likely class chosen by Bayes' rule to be the class for a given location. This solution is often referred to as the maximum *a posteriori* (MAP) estimate.

Given a learning data set, $Pr(l_i)$ can be computed as a ratio of the number of locations s_j with $f_L(s_j) = l_i$ to the total number of locations in S. $Pr(X|l_i)$ also can be estimated directly from the data using the histograms or a kernel

TABLE 6.2
Comparison of Logistic Regression and Bayesian Classification

	Classifier	
Criteria	Logistic regression	Bayesian
Input	$f_{x_1},\ldots,f_{x_k},f_i$	$f_{x_1},\ldots,f_{x_k},f_i$
Intermediate result	β	$Pr(l_i)$, $Pr(X\|l_i)$ using kernel estimate
Output	$Pr(l_i\|X)$ based on β	$Pr(l_i\|X)$ based on $Pr(l_i)$ and $Pr(X\|l_i)$
Decision	Select most-likely class for a given feature value	Select most-likely class for a given feature value
Assumptions		
- $Pr(X\|l_i)$	Exponential family	—
- Class boundaries	Linearly separable in feature space	—
- Autocorrelation in class labels	None	None

density estimate over the counts of locations s_j in S for different values X of features and different class labels l_i. This estimation requires a large training set if the domains of features f_{Xk} allow a large number of distinct values. A possible approach is that when the joint-probability distribution is too complicated to be directly estimated, then a sufficiently large number of samples from the conditional probability distributions can be used to estimate the *statistics* of the full joint probability distribution. $Pr(X)$ need not be estimated separately. It can be derived from estimates of $Pr(X \mid l_i)$ and $Pr(l_i)$ Alternatively it may be left as unknown, since for any given dataset, $Pr(X)$ is a constant that does not affect the assignment of class labels.

Table 6.2 summarizes key properties of logistic regression based classifiers and Bayesian classifiers. Both models are applicable to the location prediction problem if spatial autocorrelation is insignificant. However, they differ in many areas. Logistic regression assumes that the $Pr(X + l_i)$ distribution belongs to an exponential family (e.g., binomial, normal) whereas Bayesian classifiers can work with arbitrary distribution. Logistic regression finds a linear classifier specified by β and is most effective when classes are not linearly separable in feature space, since it allows nonlinear interaction among features in estimating $Pr(X \mid l_i)$. Logistic regression can be used with a relatively small training set since it estimates only $(k + 1)$ parameters, that is, β. Bayesian classifiers usually need a larger training set to estimate $Pr(X \mid l_i)$ due to the potentially large size of feature space. Within the machine learning community, logistic regression is considered as an example of discriminative learning and Bayesian classification as an instance of generative learning [31]. In many domains, parametric probability distributions (e.g., normal [44], Beta) are used with Bayesian classifiers if large training datasets are not available.

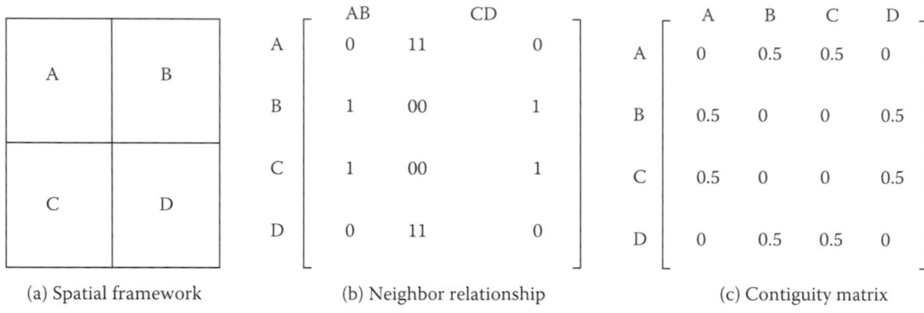

(a) Spatial framework (b) Neighbor relationship (c) Contiguity matrix

FIGURE 6.6 A spatial framework and its four-neighborhood contiguity matrix.

6.4 MODELING SPATIAL DEPENDENCIES

Modeling of spatial dependency (often called context) during the classification process has improved overall classification accuracy in several previous studies. Spatial context can be defined by the correlations between spatially adjacent pixels in a small neighborhood. The spatial relationship among locations in a spatial framework is often modeled via a contiguity matrix. A simple contiguity matrix may represent the neighborhood relationship defined using adjacency, Euclidean distance, etc. Example definitions of a neighborhood using adjacency include four-neighborhood and eight-neighborhood. Given a gridded spatial framework, the four-neighborhood assumes that a pair of locations influence each other if they share an edge. The eight-neighborhood assumes that a pair of locations influence each other if they share either an edge or a vertex.

Figure 6.6(a) shows a gridded spatial framework with four locations, namely A, B, C, and D. A binary matrix representation of a four-neighborhood relationship is shown in Figure 6.6(b). The row normalized representation of this matrix is called a contiguity matrix, as shown in Figure 6.6(c). Other contiguity matrices can be designed to model a neighborhood relationship based on distance. The essential idea is to specify the pairs of locations that influence each other along with the relative intensity of interaction. More general models of spatial relationships using cliques and hypergraphs are available in the literature [48].

6.4.1 SPATIAL AUTOREGRESSION MODEL (SAR)

We now show how spatial dependencies are modeled in the framework of regression analysis. In spatial regression, the spatial dependencies of the error term, or the dependent variable, are directly modeled in the regression equation [3]. If the dependent values y_i' are related to each other, that is, $y_i = f(y_j)\ i \neq j$, then the regression equation can be modified as

$$y = \rho W y + X\beta + \varepsilon. \tag{6.2}$$

Here W is the neighborhood relationship contiguity matrix and ρ is a parameter that reflects the strength of spatial dependencies between the elements of the dependent variable. After the correction term ρWy is introduced, the components of the residual error vector ε are then assumed to be generated from independent and identical standard normal distributions. As in the case of classical regression, the SAR equation has to be transformed via the logistic function for binary dependent variables.

We refer to this equation as the SAR. Notice that when $\rho = 0$, this equation collapses to the classical regression model. The benefits of modeling spatial autocorrelation are many. The residual error will have much lower spatial autocorrelation, that is, systematic variation. With the proper choice of W, the residual error should, at least theoretically, have no systematic variation. If the spatial autocorrelation coefficient is statistically significant, then SAR will quantify the presence of spatial autocorrelation. It will indicate the extent to which variations in the dependent variable (y) are explained by the average of neighboring observation values. Finally, the model will have a better fit, that is, a higher R-squared statistic. We compare SAR with linear regression for predicting nest location in Section 4.

A mixed model extends the general linear model by allowing a more flexible specification of the covariance matrix of ε. The mixed model can be written as

$$y = X\beta + X\gamma + \varepsilon \tag{6.3}$$

where γ is the vector of random-effects parameters. The name *mixed model* comes from the fact that the model contains both fixed-effects parameters, β, and random-effects parameters, γ. The SAR model can be extended to a mixed model that allows for explanatory variables from neighboring observations [25]. The new model (MSAR) is given by

$$y = \alpha Wy + X\beta + WX\gamma + \varepsilon. \tag{6.4}$$

The marginal impact of the explanatory variables from the neighboring observations on the dependent variable y can be encoded as a $k * 1$ parameter vector γ.

6.4.1.1 Solution Procedures

The estimates of ρ and β can be derived using maximum likelihood theory or Bayesian statistics. We have carried out preliminary experiments using the spatial econometrics Matlab package,‡ which implements a Bayesian approach using sampling-based Markov chain Monte Carlo (MCMC) methods [26]. Without any optimization, likelihood-based estimation would require $O(n^3)$ operations. Recently several authors [36, 37, 25, 9] have proposed several efficient techniques to solve SAR. Several techniques have been studied and compared in Reference [22].

6.4.2 Markov Random Field Classifiers

A set of random variables whose interdependency relationship is represented by a undirected graph (i.e., a symmetric neighborhood matrix) is called an MRF [27]. The Markov property specifies that a variable depends only on the neighbors and is independent of all other variables. The location prediction problem can be modeled

‡ We would like to thank James Lesage (http://www.spatial-econometrics.com/) for making the Matlab toolbox available on the Web.

in this framework by assuming that the class label, $f_L(s_i)$, of different locations, s_i, constitutes an MRF. In other words, random variable $f_L(s_i)$ is independent of $f_L(s_j)$ if $W(s_i, s_j) = 0$.

The Bayesian rule can be used to predict $f_L(s_i)$ from feature value vector X and neighborhood class label vector L_M as follows:

$$Pr(l(s_i)|X, L\backslash l(s_i)) = \frac{Pr(X(s_i)|l(s_i), L\backslash l(s_i))Pr(l(s_i)|L\backslash l(s_i))}{Pr(X(s_i))} \tag{6.5}$$

The solution procedure can estimate $Pr(l(s_i)|L\backslash l(s_i))$ from the training data by examining the ratios of the frequencies of class labels to the total number of locations in the spatial framework. $Pr(X(s_i)|l(s_i), L\backslash l(s_i))$ can be estimated using kernel functions from the observed values in the training dataset. For reliable estimates, even larger training datasets are needed relative to those needed for the Bayesian classifiers without spatial context, since we are estimating a more complex distribution. An assumption on $Pr(X(s_i)|l(s_i), L\backslash l(s_i))$ may be useful if a large enough training dataset is not available. A common assumption is the uniformity of influence from all neighbors of a location. Another common assumption is the independence between X and L_N, hypothesizing that all interaction between neighbors is captured via the interaction in the class label variable. Many domains also use specific parametric probability distribution forms, leading to simpler solution procedures. In addition, it is frequently easier to work with the Gibbs distribution specialized by the locally defined MRF through the Hammersley-Clifford theorem [6].

6.4.2.1 Solution Procedures

Solution procedures for the MRF Bayesian classifier include stochastic relaxation [15], iterated conditional modes [4], dynamic programming [13], highest confidence first [11], and graph cut [8]. We have used the graph cut method; more details can be found in Reference [43].

6.5 COMPARISON OF SAR AND MRF BAYESIAN CLASSIFIERS

SAR and MRF Bayesian classifiers both model spatial context and have been used by different communities for classification problems related to spatial datasets. We compare these two approaches to modeling spatial context in this section using a probabilistic framework as well as an experimental framework.

6.5.1 COMPARISON OF SAR AND MRF USING A PROBABILISTIC FRAMEWORK

We use a simple probabilistic framework to compare SAR and MRF in this section. We will assume that classes $L = l_1, l_2, \ldots, l_M$ are discrete and the class label estimate $\hat{f}_L(s_i)$ for location s_i is a random variable. We also assume that feature values (X) are constant since there is no specified generative model. Model parameters for SAR are assumed to be constant, that is, β is a constant vector and ρ is a constant number. Finally, we assume that the spatial framework is a regular grid.

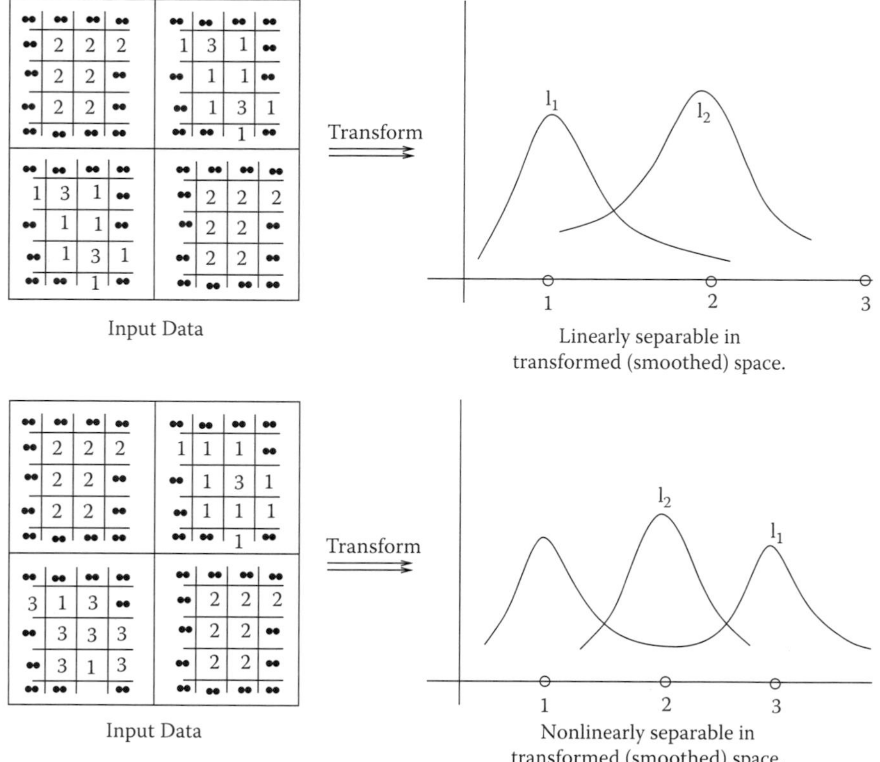

FIGURE 6.7 Spatial datasets with *salt and pepper* spatial patterns.

We first note that the basic SAR model can be rewritten as follows:

$$
\begin{aligned}
y &= X\beta + \rho W y + \varepsilon \\
(I - \rho W) y &= X\beta + \varepsilon \\
y &= (I - \rho W)^{-1} X\beta + (I - \rho W)^{-1} \varepsilon = (QX)\beta + Q\varepsilon
\end{aligned} \tag{6.6}
$$

where $Q = (I - \rho W)^{-1}$ and β, ρ are constants (because we are modeling a particular problem). The effect of transforming feature vector X to QX can be viewed as a spatial smoothing operation. The SAR model is similar to the linear logistic model in the transformed feature space. In other words, the SAR model assumes linear separability of classes in a transformed feature space.

Figure 6.7 shows two datasets with a *salt and pepper* spatial distribution of the feature values. There are two classes, l_1 and l_2, defined on this feature. Feature values close to 2 map to class l_2 and feature values close to 1 or 3 will map to l_1. These classes are not linearly separable in the original feature space. Spatial smoothing can eliminate the *salt and pepper* spatial pattern in the feature values to transform the distribution of the feature values. In the top part of Figure 6.7, there are few values of 3 and smoothing revises them close to 1 since most neighbors have values of 1.

SAR can perform well with this dataset because classes are linearly separable in the transformed space. However, the bottom part of Figure 6.7 show a different spatial dataset where local smoothing does not make the classes linearly separable. Linear classifiers cannot separate these classes even in the transformed feature space.

Although MRF and SAR classification have different formulations, they share a common goal, estimating the posterior probability distribution: $p(l_i|X)$. However, the posterior for the two models is computed differently with different assumptions. For MRF, the posterior is computed using Bayes' rule. On the other hand, in logistic regression, the posterior distribution is directly fit to the data. For logistic regression, the probability of the set of labels L is given by

$$Pr(L|X) = \prod_{i=1}^{N} p(l_i|X) \tag{6.7}$$

One important difference between logistic regression and MRF is that logistic regression assumes no dependence on neighboring classes. Given the logistic model, the probability that the binary label takes its first value l_1 at a location s_i is

$$Pr(l_i|X) = \frac{1}{1+\exp(-Q_i X\beta)} \tag{6.8}$$

where the dependence on the neighboring labels exerts itself through the W matrix, and subscript i denotes the ith row of the matrix Q. Here we have used the fact that y can be rewritten as in Equation (6.6).

To find the local relationship between the MRF formulation and the logistic regression formulation, at point s_i,

$$\begin{aligned} Pr((l_i = 1)|X, L) &= \frac{Pr(X|l_i = 1, L\backslash l_i)Pr(l_i = 1, L\backslash l_i)}{Pr(X|l_i = 1, L\backslash l_i)Pr(l_i = 1, L\backslash l_i) + Pr(X|l_i = 0, L\backslash l_i)Pr(l_i = 0, L\backslash l_i)} \\ &= \frac{1}{1+\exp(-Q_i X\beta)} \end{aligned} \tag{6.9}$$

which implies

$$Q_i X\beta = \ln\left(\frac{Pr(X|l_i = 1, L\backslash l_i)Pr(l_i = 1, L\backslash l_i)}{Pr(X|l_i = 0, L\backslash l_i)Pr(l_i = 0, L\backslash l_i)}\right) \tag{6.10}$$

This last equation shows that the spatial dependence is introduced by the W term through Q_i. More importantly, it also shows that in fitting β we are trying to simultaneously fit the relative importance of the features and the relative frequency $\left(\frac{Pr(l_i=1,L\backslash l_i)}{Pr(l_i=0,L\backslash l_i)}\right)$ of the labels. In contrast, in the MRF formulation, we explicitly *model* the relative frequencies in the class prior term. Finally, the relationship shows that we are making distributional assumptions about the class conditional distributions in logistic regression. Logistic regression and logistic SAR models belong to a more general exponential family. The exponential family is given by

$$Pr(x|l) = e^{A(\theta_1)+B(x,\pi)+\theta_i^T x} \tag{6.11}$$

This exponential family includes many of the common distributions as special cases such as Gaussian, binomial, Bernoulli, Poisson, etc. The parameters θ_l and π control the form of the distribution. Equation (6.10) implies that the class conditional distributions are from the exponential family. Moreover, the distributions $Pr(X|l_i = 1, L\backslash l_i)$ and $Pr(X|l_i = 0, L\backslash l_i)$ are matched in all moments higher than the mean (e.g., covariance, skew, kurtosis, etc.), such that in the difference $ln(Pr(X|l_i = 1, L\backslash l_i)) - ln(Pr(X|l_i = 0, L\backslash l_i))$, the higher-order terms cancel out leaving the linear term $(\theta_i^T x)$ in Equation (6.11) on the left-hand side of Equation (6.10).

6.5.2 Experimental Comparison of SAR and MRF

We have carried out experiments to compare the classical regression, SAR [10], and the MRF-based Bayesian classifiers.

The goals of the experiments were

1. To evaluate the effect of including an SAR term ρWy in the logistic regression equation.
2. To compare the accuracy and performance of a graph-partitioning-based MRF approach with spatial logistic regression on a synthetic image corrupted with Gaussian noise.
3. To compare the accuracy and performance of MRF and spatial logistic regression to predict the location of bird nests.

The experimental setup is shown in Figure 6.8. The bird habitat datasets described as in Section 6.1 are used for the learning portion of the experiments, that is, to predict locations of bird nests, as shown in Figure 6.1. Explanatory variables in these datasets are defined over a spatial grid of approximately 5000 cells. The 1995 data acquired in the Stubble wetland served as the testing datasets. This data is similar to the learning data except for the spatial locations. We have also generated a few synthetic datasets over the same marshlands using original feature values and synthetic class labels for each location to control the shape of class boundaries.

6.5.2.1 Metrics of Comparison for Classification Accuracy

In a binary prediction model (e.g., nests/no-nests), there are two types of error predictions: the false-negative ratio (FNR) and the false-positive ratio (FPR). The classification accuracy of various measures for such a binary prediction model is usually summarized in an error (or confusion) matrix as shown in Table 6.3.

We compared the classification accuracy achieved by classical and spatial logistic regression models on the test data. Receiver operating characteristic (ROC)[14] curves were used to compare classification accuracy. ROC curves plot the relationship between the true-positive rate (TPR) and the false-positive rate (FPR). For each

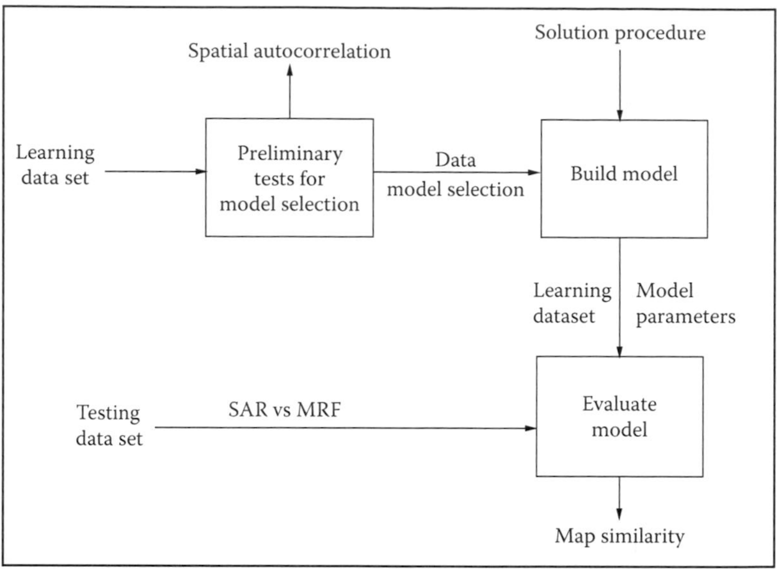

FIGURE 6.8 Experimental method for evaluation of SAR and MRF.

cut-off probability b, $TPR(b)$ measures the ratio of the number of sites where the nest is actually located and was predicted, divided by the number of actual nest sites, that is, $TPR = \frac{AP_n}{APn+AnPnn}$. The FPR measures the ratio of the number of sites where the nest was absent but predicted, divided by the number of sites where the nests were absent, that is, $FPR = \frac{AnnPn}{AnnPn+APnn}$. The ROC curve is the locus of the pair ($TPR(b)$, $FPR(b)$) for each cut-off probability. The higher the curve above the straight line $TPR = FPR$, the better the accuracy of the model. We also calculated the total error (TE), which is $TE = AnPnn + AnnPn$.

6.5.2.2 Metrics of Comparison for Spatial Accuracy

We compared spatial accuracy achieved by SAR and MRF by using ADNP (average distance to nearest prediction), which is defined as

$$ADNP(A,P) = \frac{1}{K}\sum_{k=1}^{K} d(A_k, A_k.nearest(P)).$$

TABLE 6.3
Confusion Matrix Measures

	Predicted Nest (present)	Predicted No-Nest (absence)
Actual nest (present)	APn	AnPnn
Actual no-nest (absence)	AnnPn	APnn

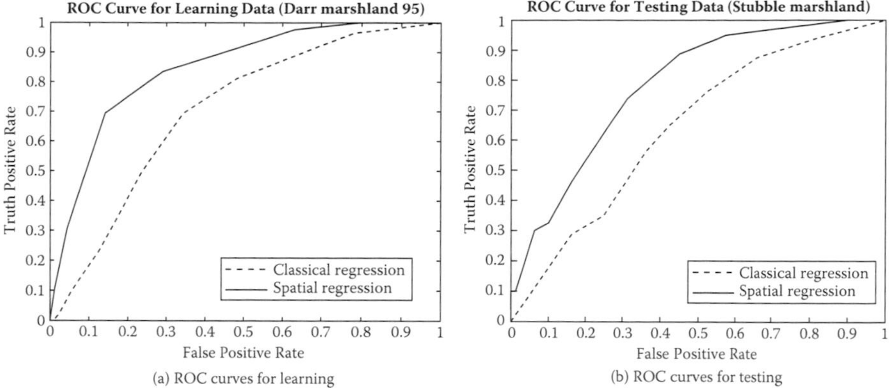

FIGURE 6.9 (a) Comparison of classical regression model with the SAR model on the Darr learning data. (b) Comparison of the models on the testing data.

Here A_k stands for the actual nest locations, P is the map layer of predicted nest locations, and $A_k.nearest(P)$ denotes the nearest predicted location to A_k. K is the number of actual nest sites. The units for ADNP are the number of pixels in the experiment. The results of our experiments are shown in the following subsection.

6.5.3 Evaluation of Classical Accuracy on SAR and Classical Regression Models for Genuine Bird Datasets

Figure 6.9(a) illustrates the ROC curves for SAR and classical regression models built using the real surveyed 1995 Darr learning data and Figure 6.9(b) displays the ROC curve for the real Stubble test data. It is clear that using spatial regression resulted in better predictions at all cut-off probabilities relative to the classical regression model.

6.5.4 Nonlinear Class Boundary Simulation by Synthetic Bird Datasets

We created a set of synthetic bird datasets based on nonlinear generalization. We carried out experiments on these synthetic bird nesting datasets. Table 6.10 is the confusion matrix via the SAR and MRF models for these nonlinear generalization synthetic datasets. From Table 6.10, we can easily calculate the TE for the synthetic testing data. The TE of MRF for the testing data is 563, which is significantly less than that of SAR (665).

6.5.4.1 Spatial Accuracy Results (SAR and MRF Comparision)

The results of spatial accuracy for the nonlinear class boundary simulation via SAR and MRF models are shown in Table 6.4. As can be seen, MRF achieves better spatial accuracy on both learning and test datasets.

We also draw maps, shown in Figure 6.11 and Figure 6.12, to visualize the results of the comparison between the SAR and MRF approaches. Figure 6.11 shows the

		Learning		Testing	
		Predicted Nest	Predicted No-nest	Predicted Nest	Predicted No-nest
SAR	Actual Nest	248	1277	99	399
	Actual No-nest	700	3147	266	1053
MRF	Actual Nest	339	1186	124	303
	Actual No-nest	609	3238	260	1137

FIGURE 6.10 Error matrix of nonlinear generalized synthetic data via MRF and SAR models ($\lambda = 1000$ for MRF, # of predicted nests are identical in both models).

TABLE 6.4
Spatial Accuracies for the Nonlinear Generalized Synthetic Datasets via SAR and MRF

Dataset		SAR	MRF
Learning	Spatial accuracy	2.8[a]	1.5
Testing	Spatial accuracy	2.1	1.7

[a] The result is measured in pixel units, 1 pixel = 5 × 5m.

Note: $\lambda = 1000$ for MRF, # of predicted nests are identical in both models.

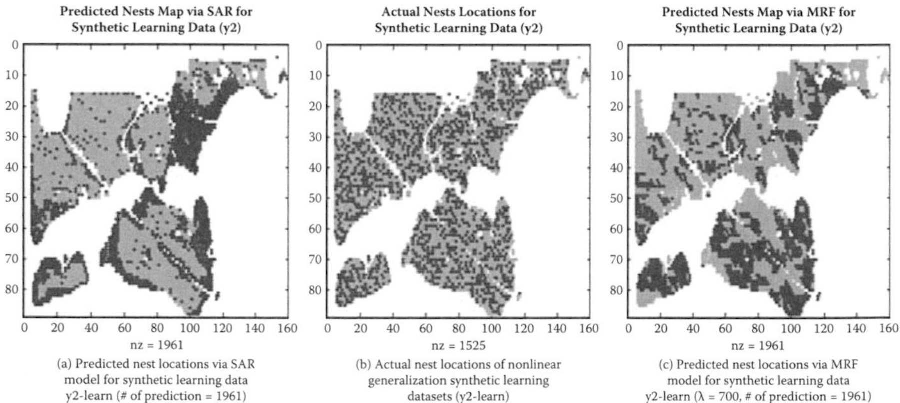

(a) Predicted nest locations via SAR model for synthetic learning data y2-learn (# of prediction = 1961)

(b) Actual nest locations of nonlinear generalization synthetic learning datasets (y2-learn)

(c) Predicted nest locations via MRF model for synthetic learning data y2-learn ($\lambda = 700$, # of prediction = 1961)

FIGURE 6.11 Maps of predicted nest locations via SAR and MRF for nonlinear generalized synthetic learning datasets.

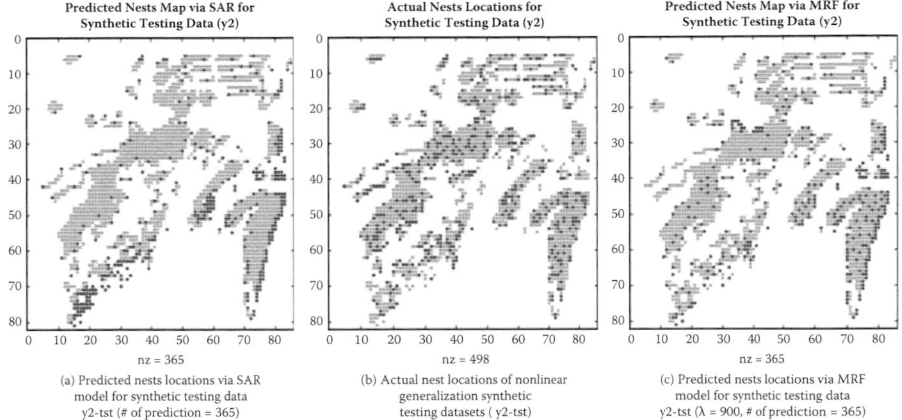

(a) Predicted nests locations via SAR model for synthetic testing data y2-tst (# of prediction = 365)

(b) Actual nest locations of nonlinear generalization synthetic testing datasets (y2-tst)

(c) Predicted nests locations via MRF model for synthetic testing data y2-tst (λ = 900, # of prediction = 365)

FIGURE 6.12 Results of predicted nest sites via SAR and MRF for nonlinear generalization synthetic testing datasets.

actual nonlinear generalized synthetic learning data. Maps of the nest predictions for the nonlinear generalized synthetic learning dataset via the SAR and MRF models are displayed, respectively, in Figure 6.11(a) and Figure 6.11(c).

Figure 6.11(c) illustrates the nest locations predicted by the MRF approach. As can be seen, these are significantly closer to the actual nest locations compared to the locations predicted by the SAR model shown in Figure 6.11(a). The results of the prediction maps via SAR and MRF models for the synthetic testing datasets are shown in Figures 6.12(a) and 6.12(c). The trend of the predicted nests via the MRF approach also shows that the predicted nests are located near the actual nest sites. For our synthetic dataset, the MRF model gives a better prediction than the SAR model for both the learning and testing datasets.

6.5.5 Linearly Separable Synthetic Bird Dataset

We have created another synthetic bird dataset using linearly separable class boundaries and used a similar experimental structure as that used to evaluate the SAR and MRF models. The error matrix of these linear generalized synthetic datasets are shown in Figure 6.13. For both learning and testing, the TEs of the MRF approach are less than but comparable to that of the SAR model.

6.5.5.1 Spatial Accuracy Result of Comparison SAR and MRF Models for Linearly Separable Synthetic Datasets

The spatial accuracy results on this simulated dataset for SAR and MRF models are shown in Table 6.5. As can be seen, MRF achieves better spatial accuracy on learning datasets. For the testing datasets, MRF achieves slightly better spatial accuracy than SAR.

The prediction maps, shown in Figure 6.14 and Figure 6.15 visually illustrate the results of the comparison between the SAR and MRF approaches for the linearly

		Learning		Testing	
		Predicted Nest	Predicted No-nest	Predicted Nest	Predicted No-nest
SAR	Actual Nest	606	1314	160	367
	Actual No-nest	1016	2436	357	943
MRF	Actual Nest	679	1241	167	350
	Actual No-nest	943	2509	354	946

FIGURE 6.13 Error matrix of synthetic data with linearly separable classes via MRF and SAR models. ($\lambda = 700$ for MRF, # of predicted nests are identical in both models).

TABLE 6.5
Spatial Accuracies for the Linear Generalized Synthetic Datasets via SAR and MRF

Dataset		SAR	MRF
Learning	Spatial accuracy	3.31	1.075
Testing	Spatial accuracy	2.96	2.02

Note: $\lambda = 700$ for MRF, # of predicted nests are identical in both models.

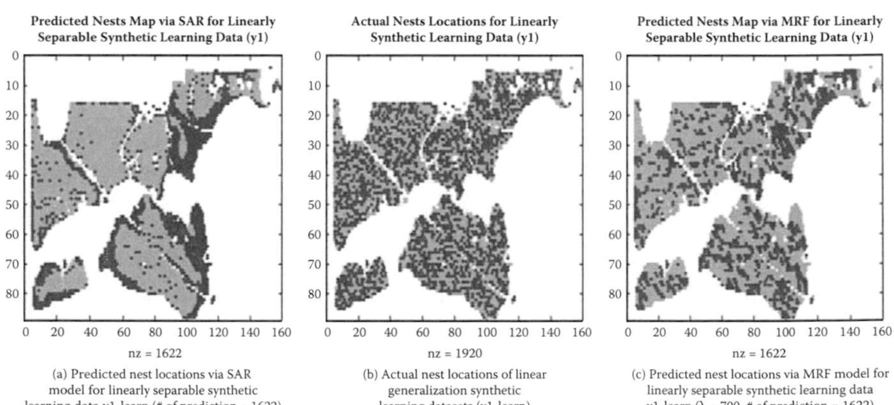

(a) Predicted nest locations via SAR model for linearly separable synthetic learning data y1-learn (# of prediction = 1622)

(b) Actual nest locations of linear generalization synthetic learning datasets (y1-learn)

(c) Predicted nest locations via MRF model for linearly separable synthetic learning data y1-learn ($\lambda = 700$, # of prediction = 1622)

FIGURE 6.14 Maps of predicted nest locations via SAR and MRF for linearly separable synthetic learning datasets.

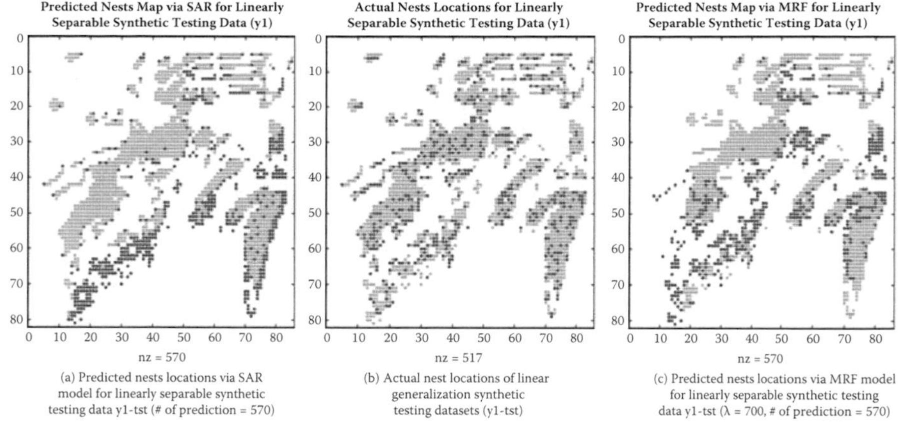

(a) Predicted nests locations via SAR model for linearly separable synthetic testing data y1-tst (# of prediction = 570)

(b) Actual nest locations of linear generalization synthetic testing datasets (y1-tst)

(c) Predicted nests locations via MRF model for linearly separable synthetic testing data y1-tst (λ = 700, # of prediction = 570)

FIGURE 6.15 Results of predicted nest sites via SAR and MRF for linear generalization synthetic testing datasets.

separable synthetic datasets. Figure 6.14 shows the actual linearly separable synthetic learning data. Maps of the nest predictions for this synthetic learning dataset via the SAR and MRF models are displayed, respectively, in Figure 6.14(a) and Figure 6.14(c). As can be seen, the nest locations predicted by the MRF approach illustrated in Figure 6.14(c) are closer to the actual nest locations compared to the locations predicted by the SAR model shown in Figure 6.14(a). The results of the prediction maps via SAR and MRF models for the linearly separable synthetic testing datasets are shown in Figures 6.15(a) and 6.15(c). The trend of the predicted nests via the MRF approach also shows that the predicted nests are located near the actual nest sites. For this linearly separable synthetic dataset, the MRF model gives a better prediction than the SAR model for both the learning and testing datasets.

6.6 SPATIAL SEMISUPERVISED CLASSIFICATION

We mentioned in the Introduction that increasing spatial and spectral resolutions put several constraints on supervised classification. The increased spatial resolution invalidates the most widely used assumption of the traditional data mining algorithms (e.g., independent, identical distributions). Often used maximum likelihood estimation requires large amounts of training data (10 to 100 $\times$ number of dimensions per class) for accurate estimation of model parameters and increasing spectral resolution further compounds this problem.

Recently, semisupervised learning techniques that utilize large unlabeled training samples in conjunction with small labeled training data are becoming popular in machine learning and data mining [30, 16, 32]. This popularity can be attributed to the fact that several of these studies have reported improved classification and prediction accuracies, and that the unlabeled training samples come almost for free. This is also true in the case of remote sensing classification, as collecting samples is almost free, although assigning labels to them is not. Many semisupervised learning algorithms can be found in the recent survey article [49]. In this section, we briefly present a semisupervised learning scheme [46] and its spatial extension via MRF [47].

6.6.1 Semisupervised Learning

In many supervised learning situations, the class labels (y_i)'s are not readily available. However, assuming that the initial parameters Θ^k can be guessed (as in clustering), or can be estimated (as in semisupervised learning), we can easily compute the parameter vector Θ using the expectation maximization (EM) algorithm. In brief, the EM algorithm consists of two steps. The EM algorithm at the first step (called the E-step) maximizes the expectation of the *log-likelihood* function (Equation 6.12), using the current estimate of the parameters and conditioned upon the observed samples.

$$L(\Theta) = \ln(P(X,L) \mid \Theta)) = \sum_{i=1}^{n} \ln(P(x_i \mid l_i)P(l_i)) = \sum_{i=1}^{n} \ln(\alpha_{l_i} p_{l_i}(x_i \mid \theta_{l_i})). \tag{6.12}$$

In the second step of the EM algorithm, called maximization (or the M-step), the new estimates of the parameters are computed. The EM algorithm iterates over these two steps until the convergence is reached. The *log-likelihood* function is guaranteed to increase until a maximum (local or global or saddle point) is reached. For multivariate normal distribution, the expectation $E[.]$, which is denoted by p_{ij}, is nothing but the probability that Gaussian mixture j generated the data point i, and is given by

$$p_{ij} = \frac{\left|\hat{\Sigma}_j\right|^{-1/2} e^{\left\{-\frac{1}{2}(x_i-\hat{\mu}_j)^t \hat{\Sigma}_j^{-1}(x_i-\hat{\mu}_j)\right\}}}{\sum_{l=1}^{M} \left|\hat{\Sigma}_l\right|^{-1/2} e^{\left\{-\frac{1}{2}(x_i-\hat{\mu}_j)^t \hat{\Sigma}_j^{-1}(x_i-\hat{\mu}_j)\right\}}} \tag{6.13}$$

The new estimates (at the kth iteration) of parameters in terms of the old estimates at the M-step are given by the following equations:

$$\hat{\alpha}_j^k = \frac{1}{n}\sum_{i=1}^{n} p_{ij}, \quad \hat{\mu}_j^k = \frac{\sum_{i=1}^{n} x_i p_{ij}}{\sum_{i=1}^{n} p_{ij}}, \quad \sum_j^k = \frac{\sum_{i=1}^{n} p_{ij}\left(x_i - \hat{\mu}_j^k\right)\left(x_i - \hat{\mu}_j^k\right)^t}{\sum_{i=1}^{n} p_{ij}} \tag{6.14}$$

A more-detailed derivation of these equations can be found in Reference [7].

6.6.2 Spatial Extension of Semisupervised Learning

In this section we provide an extension of semisupervised learning algorithm (Section 6.1) to model spatial context via the MRF model. MRF exploits spatial context through the prior probability $p(l_i)$ term in the Bayesian formulation (Section 3.2). Since MRF models spatial context in the *a priori* term, we optimize a *penalized log-likelihood* [17] instead of the *log-likelihood* function [see Equation (6.12)]. The *penalized log-likelihood* can be written as

$$\ln(P(X,L|\Theta)) = -\sum_{C} V_C(l,\beta) - \ln C(\beta) + \sum_{i}\sum_{j} L_{ij} \ln p_j(x_i|\Theta_i) \tag{6.15}$$

where $V_c(l)$ is called clique potential. Then the E-step for a given Θ^k reduces to computing

$$Q(\Theta,\Theta^k) = \sum_{i}\sum_{j} E(L_{ij}|x,\theta^k)\ln p_j(x_i|\theta_i) - \sum E(V_c(L,\beta)|x,\theta^k) - \ln C(\beta) \tag{6.16}$$

However, exact computation of the quantities $E(Vc(L, \beta)|x, \theta^k)$ and $E(L_{ij} \mid x, \theta^k)$ in Equation (6.16) are impossible [38]. Also the maximization of Equation (6.16) with respect to β is also very difficult because computing $z = C(\beta)$ is intractable except for very simple neighborhood models. Several approximate solutions for this problem in unsupervised learning can be found in [38, 39]. An approximate solution for semisupervised learning via an extension of MRF [38] is provided in [47]. The E-step is divided into two parts: first, we compute the complete data *log-likelihood* for all data points; second, for the given neighborhood, we iteratively optimize contextual energy using the iterative conditional modes (ICM) [5] algorithm. Since the estimation of β is difficult [38], we assume that it is given *a priori*, and proceed with the M-step as described in the semisupervised learning algorithm. This basic spatial semisupervised learning scheme is summarized in Table 6.6.

TABLE 6.6
Spatial Semisupervised Learning Algorithm

Inputs: Training data set $D = D^l \cup D^u$, where D^l consists of labeled samples and D^u contains unlabeled samples, s a neighborhood model, and β a homogeneity weight.

Initial Estimates: Build initial classifier (MLC or MAP) from the labeled training samples, D^l. Estimate initial parameter using MLE, to find $\hat{\theta}$.

Loop: While the complete data *log-likelihood* improves [see Equation (6.16)]:

- **E-step:** Use current classifier to estimate the class membership of each unlabeled sample, that is, the probability that each Gaussian mixture component generated the given sample point, p_{ij} [see Equation (6.13)].
- **ICM-step:** Optimize contextual energy using the ICM [5] algorithm.
- **M-step:** Re-estimate the parameter, $\hat{\theta}$, given the estimated Gaussian mixture component membership of each unlabeled sample [see Equation (6.14)].

Output: An MAP-MRF classifier that takes the given sample (feature vector), a neighborhood model, and predicts a class label.

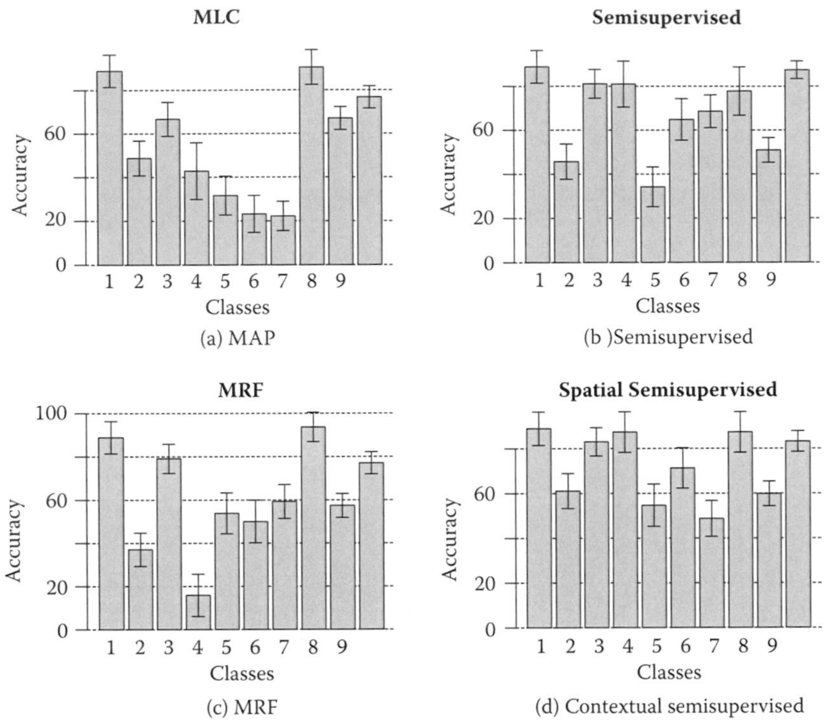

FIGURE 6.16 Comparison of accuracies (10 classes); (a) MLC/MAP (60%), (b) semisupervised (68%), (c) MRF (65%), and (d) spatial semisupervised (72%).

6.6.3 Experimental Results

We used a spring Landsat 7 scene, taken on May 31, 2000 over the town of Cloquet located in Carlton County, Minnesota. For all experiments, we considered the following 10 classes: water, bare soil, agriculture, upland conifer, upland hardwood, lowland conifer, lowland hardwood, wetlands, low-density urban, and high-density urban. For discussion purposes we summarized key results as graphs for easy understanding. We applied all four classifiers, namely, MLC/MAP, MRF, semisupervised, and spatial semisupervised (SSSL). The training data consisted of 20 labeled plots, and 100 unlabeled plots, and the test data consisted of 85 labeled plots. From each plot, we extracted exactly nine feature vectors by centering a 3×3 window on the plot center. The classification performance (accuracy) results are summarized in Figure 6.16. Individual class (labeled 1 to 10) accuracies along with the two-tailed 95% confidence intervals are shown for each classifier. As can be seen, the overall classification accuracy of the spatial semisupervised algorithm is about 72%, as compared to the BC (60%), MAP-MRF (65%), and semisupervised (68%) classifiers on the test dataset. The overall accuracy obtained from the efficient spatial semisupervised algorithm is about 71%. Figure 6.17 shows the classified images generated by each of the four methods for a small area from the northwest corner of the full image.

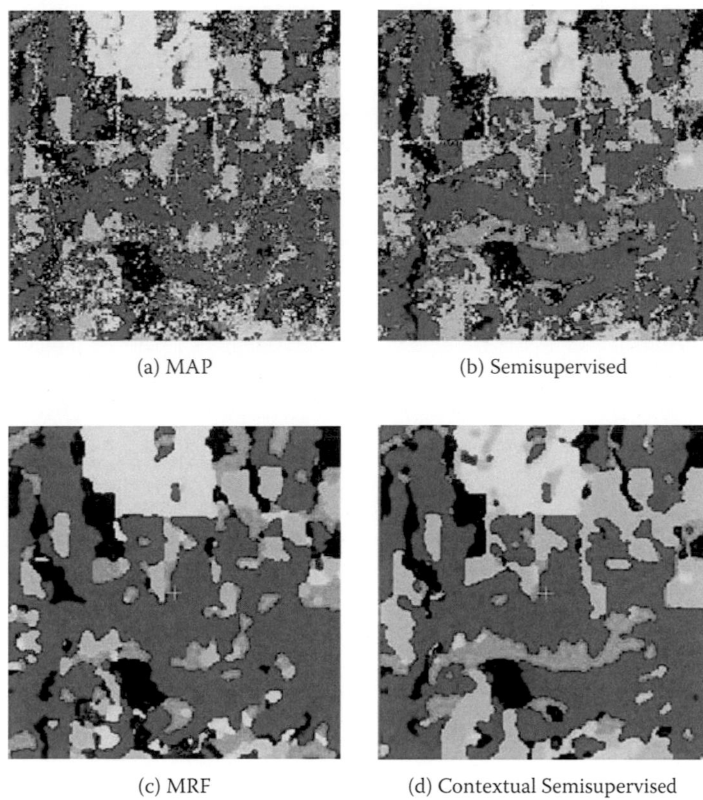

FIGURE 6.17 Small portion from the classified image (Carleton, Minnesota). (a) Bayesian (MAP), (b) semisupervised (EM-MAP), (c) MRF (MAP-MRF), and (d) contextual semisupervised (EM-MAP-MRF). See color insert after page 148.

This figure clearly shows the superior performance of spatial semisupervised learning over the BC (MLC), MAP-MRF, and semisupervised algorithms. In addition to superior classification accuracy, the output generated by the spatial semisupervised learning algorithm is also more preferable (less salt and pepper noise) in various application domains.

6.7 CONCLUSIONS AND FUTURE WORK

In this chapter, we presented two popular classification approaches that model spatial context in the framework of spatial data mining. These two models, SAR and MRF, were compared and contrasted using a common probabilistic framework [43]. Our study shows that the SAR model makes more restrictive assumptions about the distribution of features and class shapes (or decision boundaries) than MRF. We also observed an interesting relationship between classical models that do not consider spatial dependence and modern approaches that explicitly model spatial context. The relationship between SAR and MRF is analogous to the relationship between logistic regression and Bayesian classifiers. We have provided theoretical results using a probabilistic framework and as well as experimental results validating the comparison between SAR and MRF. We also described a spatial semisupervised learning scheme that overcomes small sample problems and also models spatial context via MRF [47]. This approach showed improved accuracy and also eliminated *salt-and-pepper* noise, which is common to noncontextual classification schemes.

In the future we would like to study and compare other models that consider spatial context in the classification decision process. We would also like to extend the graph cut solution procedure for SAR.

6.8 ACKNOWLEDGMENTS

We are particularly grateful to our collaborators Professor Vipin Kumar, Professor Paul Schrater, Professor Chang-Tien Lu, Professor Weili Wu, Professor Uygar Ozesmi, Professor Yan Huang, and Dr. Pusheng Zhang for their various contributions. We also thank Xiaobin Ma, Professor Hui Xiong, Professor Jin Soung Yoo, Dr. Qingsong Lu, Dr. Baris Kazar, Betsy George, and anonymous reviewers for their valuable feedback on early versions of this chapter. We would like to thank Kim Koffolt for improving the readability of this chapter.

Prepared by Oak Ridge National Laboratory, P. O. Box 2008, Oak Ridge, Tennessee 37831-6285, managed by UT-Battelle, LLC for the U.S. Department of Energy under contract no. DEAC05-00OR22725.

REFERENCES

[1] R. Agrawal. Tutorial on database mining. In *Thirteenth ACM Symposium on Principles of Databases Systems*, pp. 75–76, Minneapolis, MN, 1994.

[2] D. Anguelov, B. Taskar, V. Chatalbashev, D. Koller, D. Gupta, G. Heitz, and A. Ng. Discriminative learning of Markov random fields for segmentation of 3D scan data. In *CVPR '05: Proceedings of the 2005 IEEE Computer Society Conference on Computer Vision and Pattern Recognition (CVPR'05) -Volume 2*, pp. 169–176, Washington, D.C., 2005.

[3] L. Anselin. *Spatial Econometrics:* Methods and Models. Kluwer, Dordrecht, Netherlands, 1988.

[4] J. Besag. On the statistical analysis of dirty pictures. *J. Royal Statistical Soc.*, (48):259–302, 1986.

[5] J. Besag. On the statistical analysis of dirty pictures. *Journal of Royal Statistical Society*, 48(3):259–302, 1986.

[6] J.E. Besag. Spatial interaction and statistical analysis of lattice systems. *Journal of Royal Statistical Society, Ser. B*, 36:192–236, 1974.

[7] J. Bilmes. A gentle tutorial on the em algorithm and its application to parameter estimation for Gaussian mixture and hidden markov models. Technical Report, University of Berkeley, ICSI-TR-97-021, 1997.

[8] Y. Boykov, O. Veksler, and R. Zabih. Fast approximate energy minimization via graph cuts. *International Conference on Computer Vision*, September 1999.

[9] M. Celik, B.M. Kazar, S. Shekhar, D. Boley, and D.J. Lilja. Northstar: A parameter estimation method for the spatial autoregression model. AHPCRC Technical Report No: 2005-001, 2007.

[10] S. Chawla, S. Shekhar, W. Wu, and U. Ozesmi. Extending data mining for spatial applications: A case study in predicting nest locations. *2000 ACM SIGMOD Workshop on Research Issues in Data Mining and Knowledge Discovery (DMKD 2000),* Dallas, TX, May 2000.

[11] P.B. Chou, P.R. Cooper, M.J. Swain, C.M. Brown, and L.E. Wixson. Probabilistic network inference for cooperative high and low level vision. In *In Markov Random Field, Theory and Applications*. Academic Press, New York, 1993.

[12] N.A. Cressie. *Statistics for Spatial Data* (Revised Edition). Wiley, New York, 1993.

[13] H. Derin and H. Elliott. Modeling and segmentation of noisy and textured images using Gibbs random fields. *IEEE Transaction on Pattern Analysis and Machine Intelligence*,(9):39–55, 1987.

[14] J.P. Egan. *Signal Detection Theory and ROC Analysis*. AcademicPress, New York, 1975.

[15] S. Geman and D. Geman. Stochastic relaxation, Gibbs distributions and the Bayesian restoration of images. *IEEE Transactions on Pattern Analysis and Machine Intelligence*, (6):721–741, 1984.

[16] S. Goldman and Y. Zhou. Enhancing supervised learning with unlabeled data. In *Proc. 17th International Conf. on Machine Learning*, pp. 327–334. Morgan Kaufmann, San Francisco, CA, 2000.

[17] P.J. Green. On use of the em algorithm for penalized likelihood estimation. *Journal of the Royal Statistical Society, Series B*, 52(3):443– 452, 1990.

[18] C. Greenman. Turning a map into a cake layer of information. *New York Times*, (http://www.nytimes.com/library/tech/00/01/circuits/arctiles/20giss.html) Jan. 20, 2000.

[19] S. Haykin. *Neural Networks—A Comprehensive Foundation*. Prentice Hall, Englewood Cliffs, NJ, 1998.

[20] D.W. Hosmer and S. Lemeshow. *Applied Logistic Regression*. John Wiley & Sons, New York, 2000.

[21] Y. Jhung and P. H. Swain. Bayesian contextual classification based on modified M-estimates and Markov random fields. *IEEE Transaction on Pattern Analysis and Machine Intelligence*, 34(1):67–75, 1996.

[22] B.M. Kazar, S. Shekhar, D.J. Lilja, R.R. Vatsavai, and R.K. Pace. Comparing exact and approximate spatial auto-regression model solutions for spatial data analysis. In *Third International Conference on Geographic Information Science* (GIScience 2004). LNCS, Springer, October 2004.

[23] K. Koperski, J. Adhikary, and J. Han. Spatial data mining: Progress and challenges. In *Workshop on Research Issues on Data Mining and Knowledge Discovery* (*DMKD'96*), pp. 1–10, Montreal, Canada, 1996.

[24] J. LeSage. Regression analysis of spatial data. *The Journal of Regional Analysis and Policy*, 27(2):83–94, 1997.

[25] J.P. LeSage and R.K. Pace. Spatial dependence in data mining. In *Geographic Data Mining and Knowledge Discovery*. Taylor & Francis, London, 2001.

[26] J.P. LeSage. Bayesian estimation of spatial autoregressive models. *International Regional Science Review*, (20):113–129, 1997.

[27] S. Z. Li. *Markov random field modeling in image analysis*. Springer-Verlag New York, 2001.

[28] C. Ma. Spatial autoregression and related spatio-temporal models. *J. Multivariate Analysis*, 88(1):152–162, 2004.

[29] D. Mark. Geographical information science: Critical issues in an emerging cross-disciplinary research domain. In *NSF Workshop*, Feburary 1999.

[30] T. Mitchell. The role of unlabeled data in supervised learning. In *Proceedings of the Sixth International Colloquium on Cognitive Science*, San Sebastian, Spain, 1999.

[31] A. Ng and M. Jordan. On discriminative vs. generative classifiers: A comparison of logistic regression and naive Bayes, A. Y. Ng nad M. I. Jordan, In T. Dietterich, S. Becker and Z. Ghahramani (Eds.), *Advances in Neural Information Processing systems* (NIPS) 14, 2002.

[32] K. Nigam, A. K. McCallum, S. Thrun, and T. M. Mitchell. Text classification from labeled and unlabeled documents using EM. *Machine Learning*, 39(2/3):103–134, 2000.

[33] R. Nishii and S. Eguchi. Image classification based on Markov random field models with Jeffreys divergence. *J. Multivar. Anal.*, 97(9):1997–2008, 2006.

[34] S. Ozesmi and U. Ozesmi. An artificial neural network approach to spatial habitat modeling with interspecific interaction. *Ecological Modelling,* (116):15–31, 1999.

[35] U. Ozesmi and W. Mitsch. A spatial habitat model for the marsh-breeding red-winged black-bird (agelaius phoeniceus l.) In coastal Lake Erie wetlands. *Ecological Modelling*, (101):139–152, 1997.

[36] R. Pace and R. Barry. Quick computation of regressions with a spatially autoregressive dependent variable. *Geographic Analysis*, 29:232–247, 1997.

[37] R. Pace and R. Barry. Sparse spatial autoregressions. *Statistics and Probability Letters*, (33):291–297, 1997.

[38] W. Qian and D.M. Titterington. Estimation of parameters in hidden Markov models. *Philosophical Transactions of the Royal Statistical Society, Series A*, 337:407–428, 1991.

[39] W. Qian and D.M. Titterington. Stochastic relaxations and em algorithms for Markov random fields. *Journal of Statistical Computation and Simulation*, 41, 1991.

[40] J. F. Roddick and M. Spiliopoulou. A bibliography of temporal, spatial and spatio-temporal data mining research. ACM *Special Interest Group on Knowledge. Discovery in Data Mining (SIGKDD) Explorations, 1999.*

[41] W.S. Sarle. Neural networks and statistical models. In *Proceeding of 9th Annual SAS User Group Conference*. SAS Institue, 1994.

[42] S. Shekhar, S. Chawla, S. Ravada, A. Fetterer, X. Liu, and C.T. Lu. Spatial databases: Accomplishments and research needs. *IEEE Transactions on Knowledge and Data Engineering*, 11(1), Jan–Feb 1999.

[43] S. Shekhar, P. Schrater, R. Vatsavai, W. Wu, and S. Chawla. Spatial contextual classification and prediction models for mining geospatial data. *IEEE Transaction on Multimedia,* 4(2):174–188, 2002.

[44] A. H. Solberg, T. Taxt, and A. K. Jain. A Markov random field model for classification of multisource satellite imagery. *IEEE Transaction on Geoscience and Remote Sensing*, 34(1):100–113, 1996.

[45] W.R. Tobler. *Cellular Geography, Philosophy in Geography*. Gale and Olsson, Eds., Reidel, Dordrecht, 1979.

[46] R. R. Vatsavai, S. Shekhar, Tobler, W. R. and T. E. Burk. *Cellular Geography, Philosophy in Geography*. Gate and Olsson, Eds., Reidel, Dordrecht, 1979, pp. 379–386.

[47] R. R. Vatsavai, S. Shekhar, and T. E. Burk. A spatial semisupervised learning method for classification of multispectral remote sensing imagery. In *Seventh International Workshop on Multimedia Data Mining, MDM/KDD,* 2006.

[48] C. E. Warrender and M. F. Augusteijn. Fusion of image classifications using Bayesian techniques with Markov rand fields. *International Journal of Remote Sensing,* 20(10):1987–2002, 1999.

[49] X. Zhu. semisupervised learning literature survey. University of Wisconsin-Madison, Technical Report No: 1530, 2007.

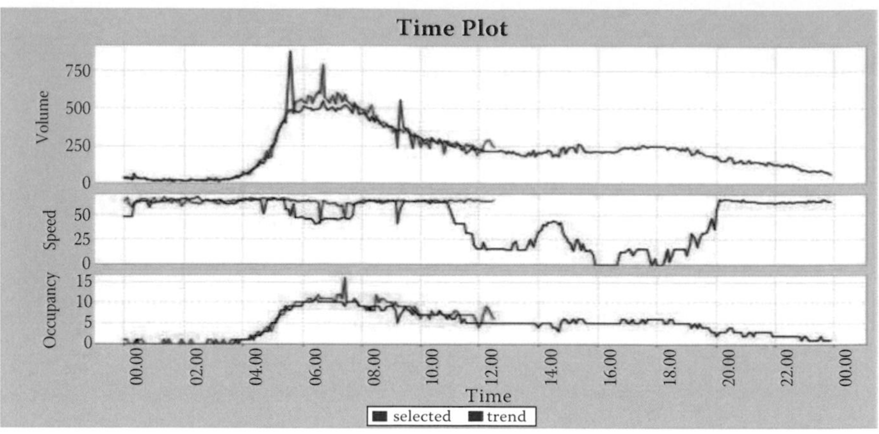

(b) Volume, speed, and occupancy on a given day (Blue - predicted traffic; Red - current traffic)

COLOR FIGURE 4.13 Time of day plots (T_{TD}). (b) Volume, speed, and occupancy on a given day (Blue - predicted traffic; Red - current traffic).

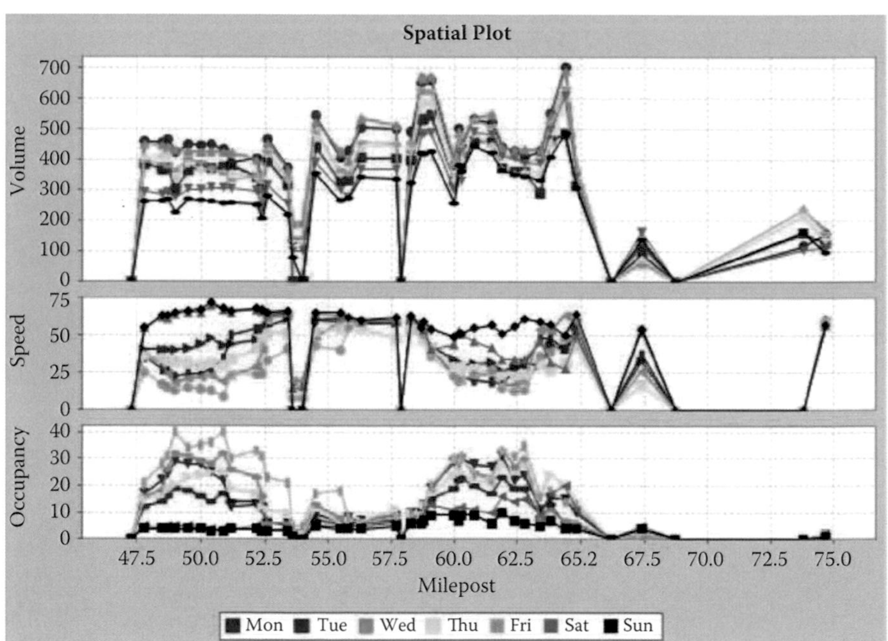

COLOR FIGURE 4.17 Highway station vs. day of week plots ($S_{HS}T_{DW}$). (a) Superimposed graphs of average volume, speed, occupancy of all mileposts for all days of the week.

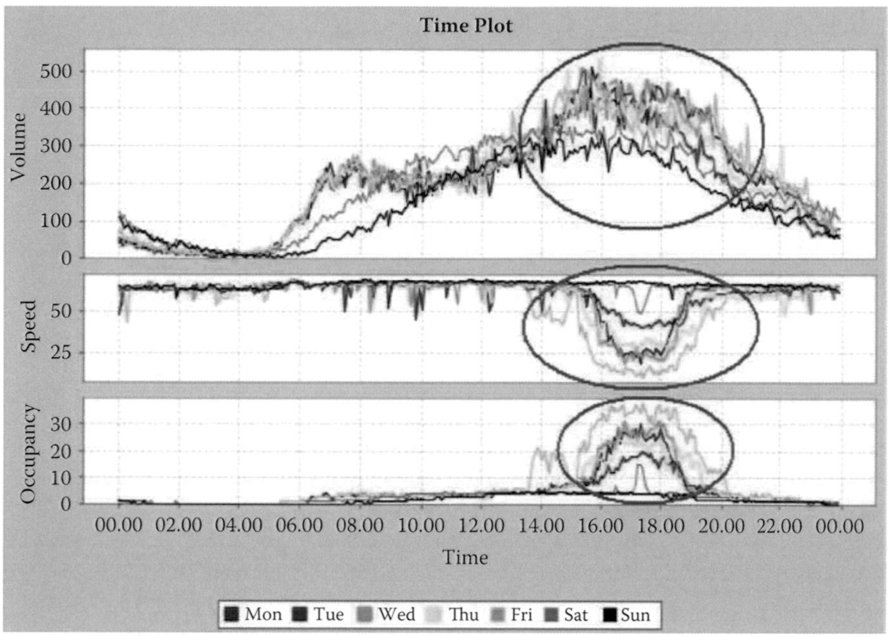

COLOR FIGURE 4.18 Time of day vs. day of week plots ($S_{HS}T_{TD}$). (a) Volume, speed, occupancy conditions of all days in a week on a given time. (b) Speed condition for all days in a week on a given milepost.

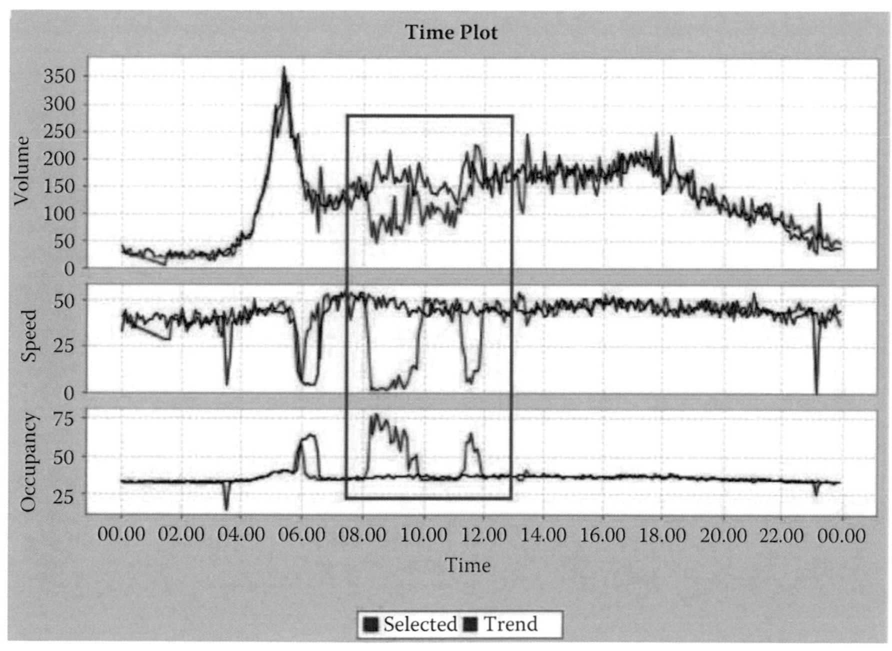

(a) Milepost 157.44 (2/8/2006)

COLOR FIGURE 4.20 T_{TD} views for drilled-down analysis of surrounding mileposts. (a) Milepost 157.44 (2/8/2006). (b) Milepost 158.26 (2/8/2006). (c) Milepost 159.63 (2/8/2006).

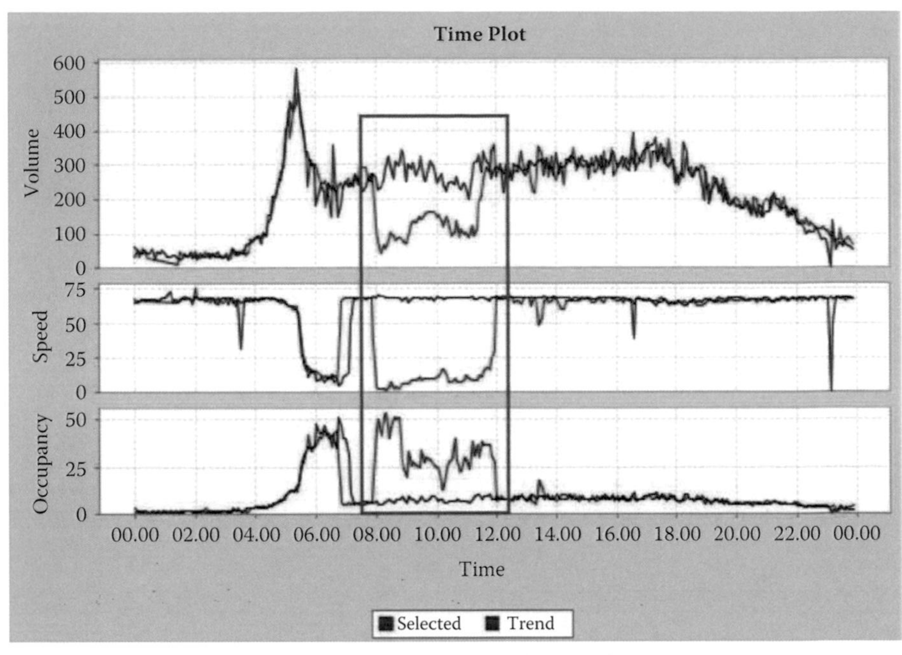

(b) Milepost 158.26 (2/8/2006)

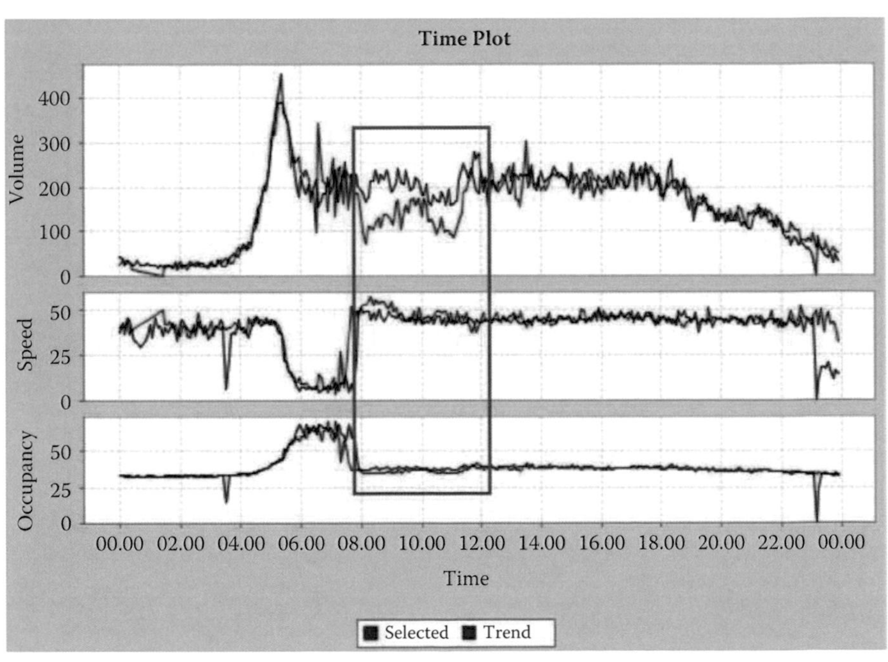

(c) Milepost 159.63 (2/8/2006)

COLOR FIGURE 4.20 (Continued).

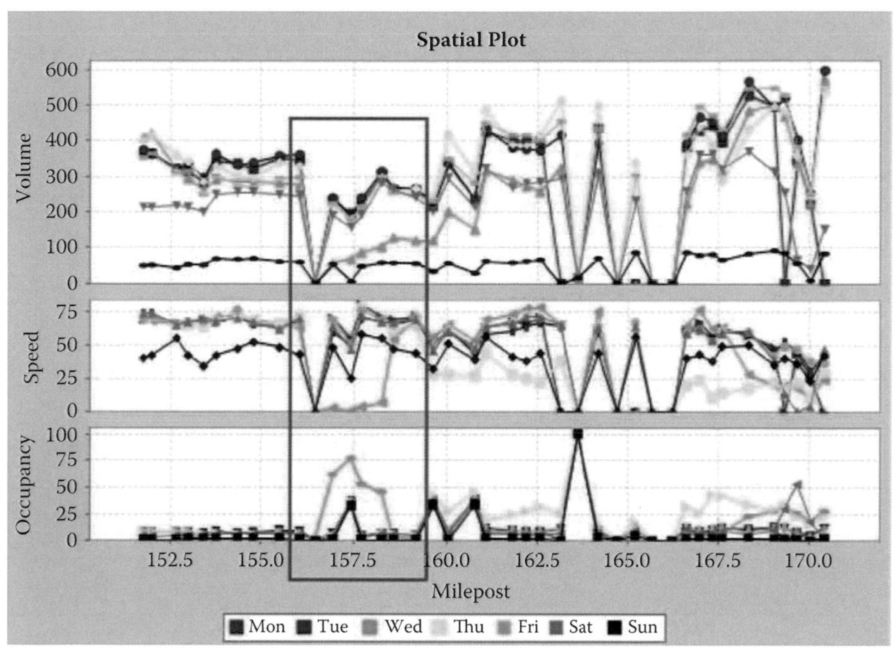

COLOR FIGURE 4.22 $S_{HS}T_{DW}$ view of I-95 NB for the week of the incident at 8:30AM (2/6/2006-2/12/2006).

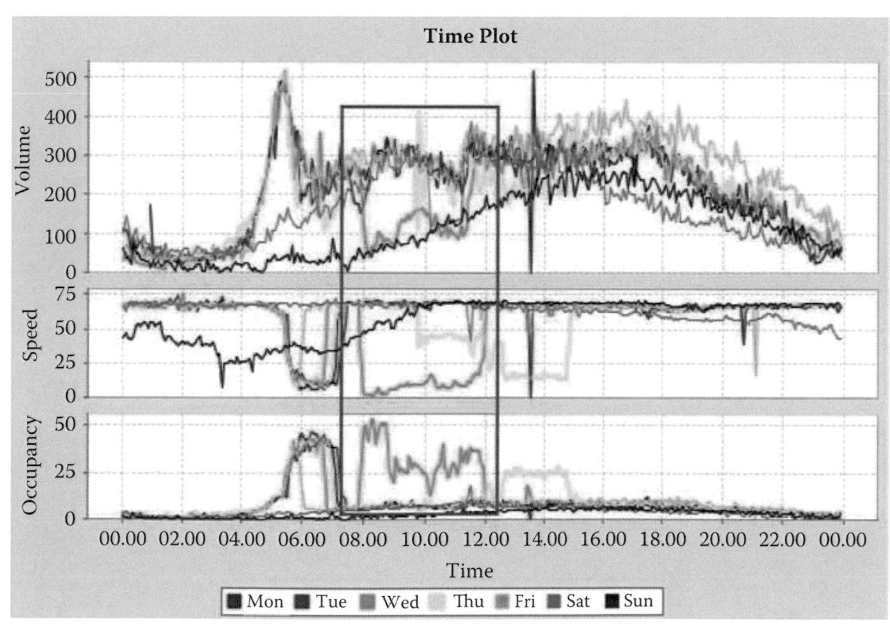

COLOR FIGURE 4.23 $T_{TD}T_{DW}$ view of milepost 158.26 for the week of the incident at 8:30AM (2/6/2006-2/12/2006)

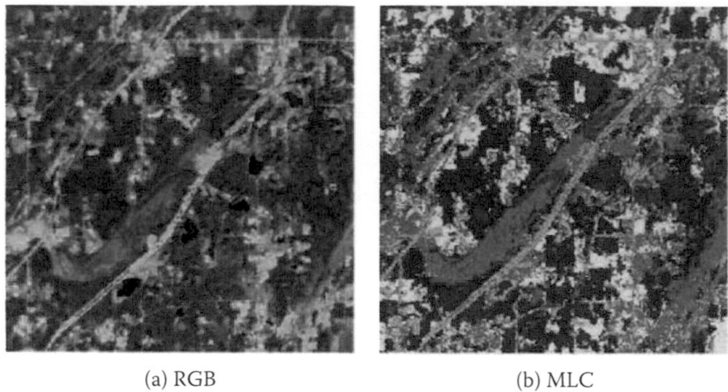

(a) RGB (b) MLC

COLOR FIGURE 6.4 Sample RGB image (a) and corresponding maximum likelihood classified (MLC) image (b).

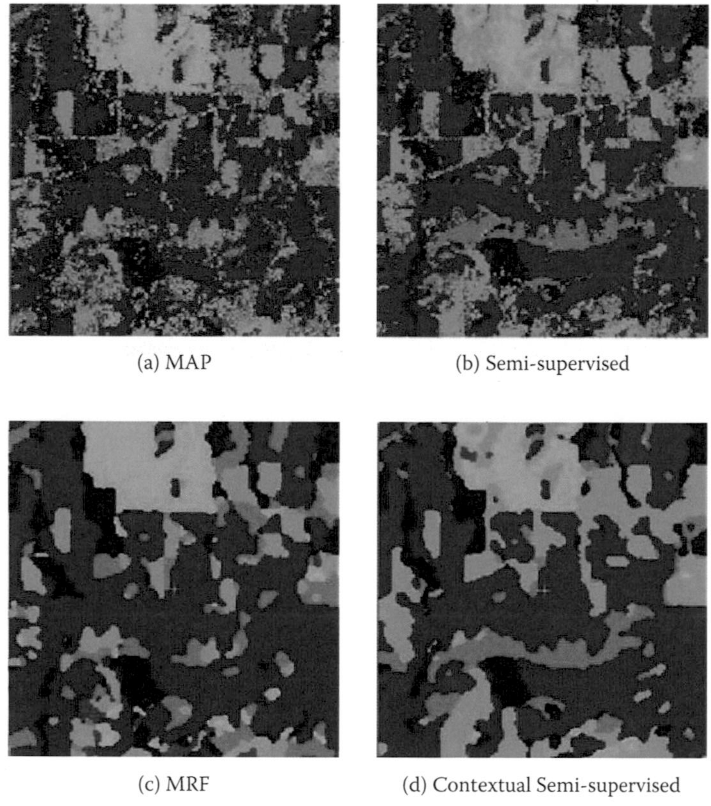

(a) MAP (b) Semi-supervised

(c) MRF (d) Contextual Semi-supervised

COLOR FIGURE 6.17 Small portion from the classified image (Carleton, Minnesota). (a) Bayesian (MAP), (b) semisupervised (EM-MAP), (c) MRF (MAP-MRF), and (d) contextual semisupervised (EM-MAP-MRF).

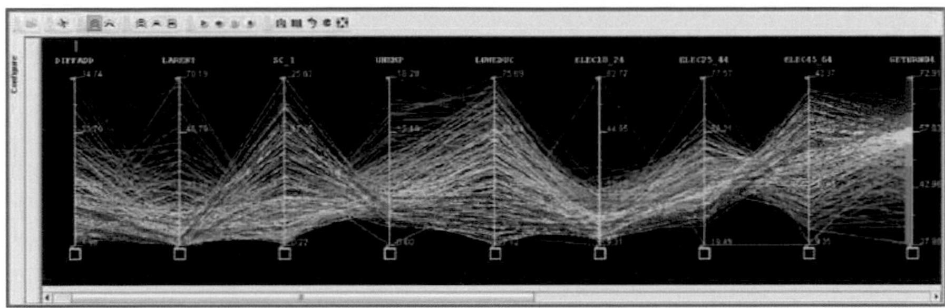

COLOR FIGURE 9.5 Rescaled parallel coordinates plot showing relationship between variables in the regression colored by quintile. Red: high turnout; yellow: middle turnout; green: low turnout.

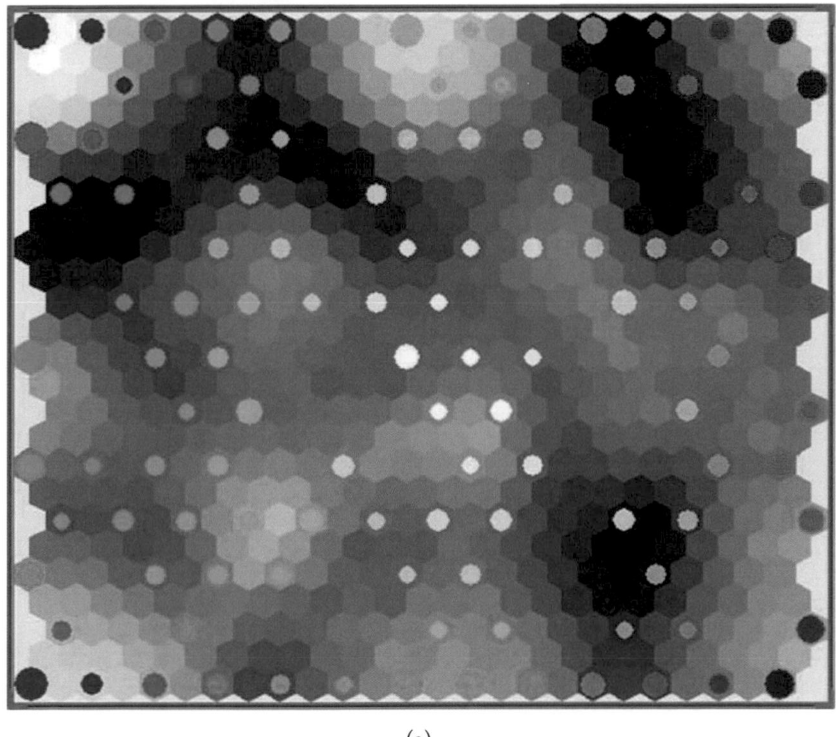

(a)

COLOR FIGURE 9.10 Visualizing the structure of the GWR space with (a) a self-organizing map (SOM) and (b) a map.

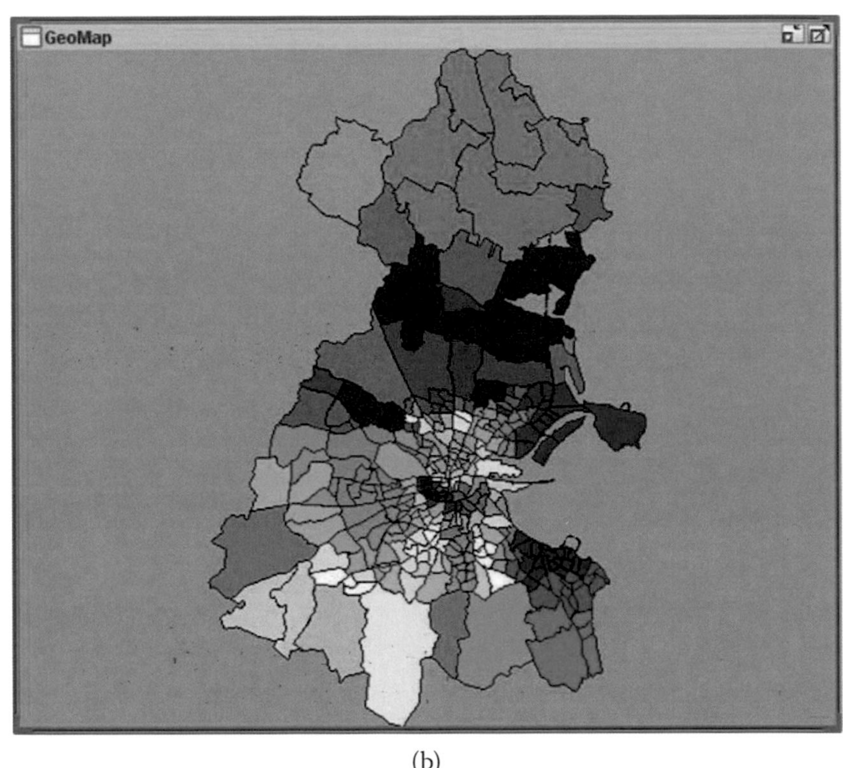

(b)

COLOR FIGURE 9.10 (Continued)

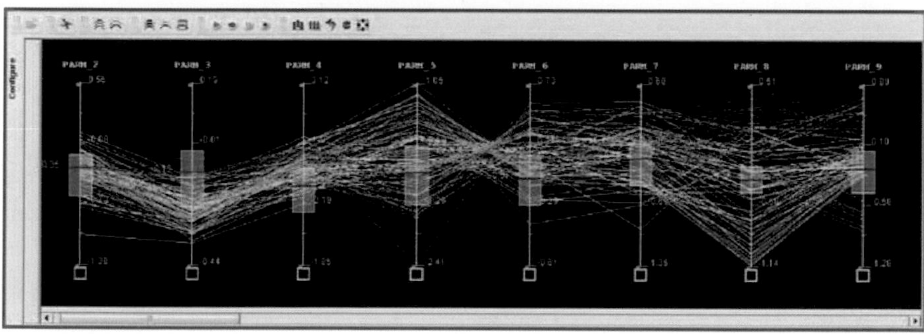

COLOR FIGURE 9.11 (a) The map and (b) the parallel coordinates plot (PCP) showing the selection of the two clusters from the SOM (violet and yellow, respectively) that represent two areas in the center of Dublin. River Liffey (running through the center in the east-west direction) divides the center in the north yellow area and south violet areas where the social processes behind voting mechanisms have different characteristics. This can be clearly seen by comparing the violet and yellow trajectories in the PCP.

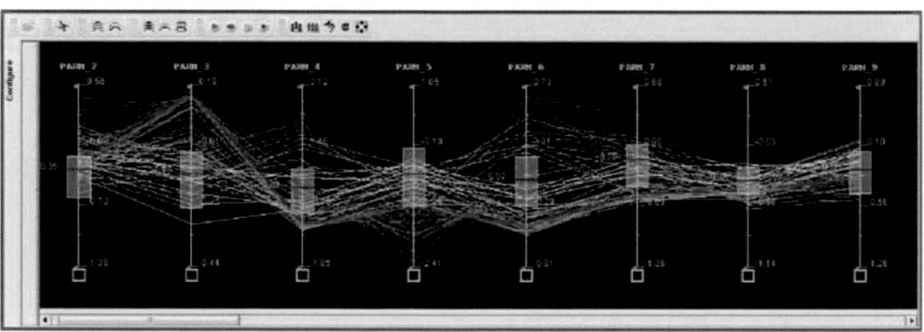

COLOR FIGURE 9.12 (a) The map and (b) the PCP showing the selection of the two clusters in SOM (red and blue) that represent two areas north of city center. Blue EDs are located around Howth and are the most affluent areas in Dublin. Red EDs, on the other hand, represent one of the more problematic areas in Dublin (Ballymun). The two different trajectories in the PCP again capture the differences between areas with different social situations.

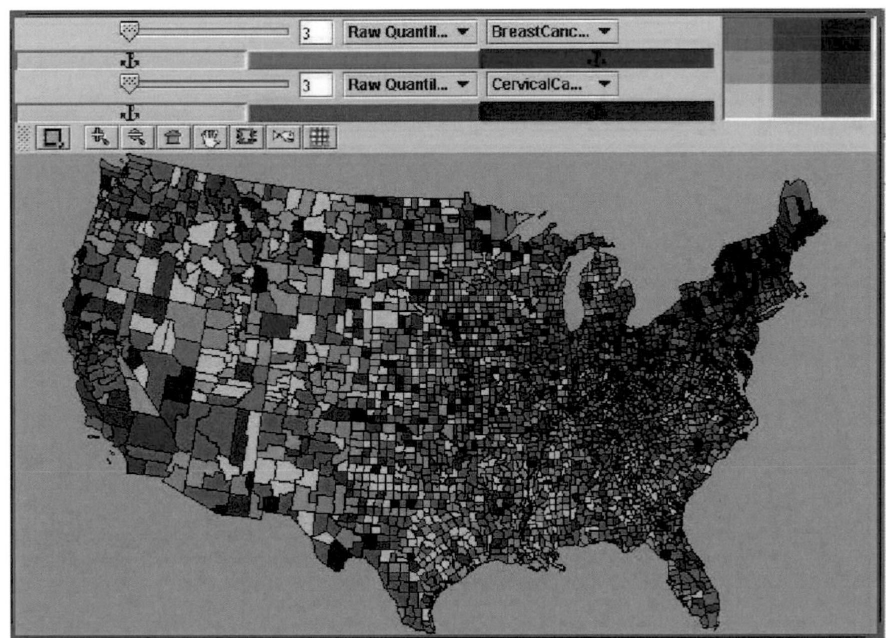

COLOR FIGURE 11.3 An example of a map-based visualization technique. A bivariate choropleth map shows cancer incidence by county for the conterminous United States. Redness increases with the rate for cervical cancer, blueness with the rate for breast cancer. Thus, counties colored light gray have low rates for both cancers, dark purple indicates places with high rates for both. See text for further details.

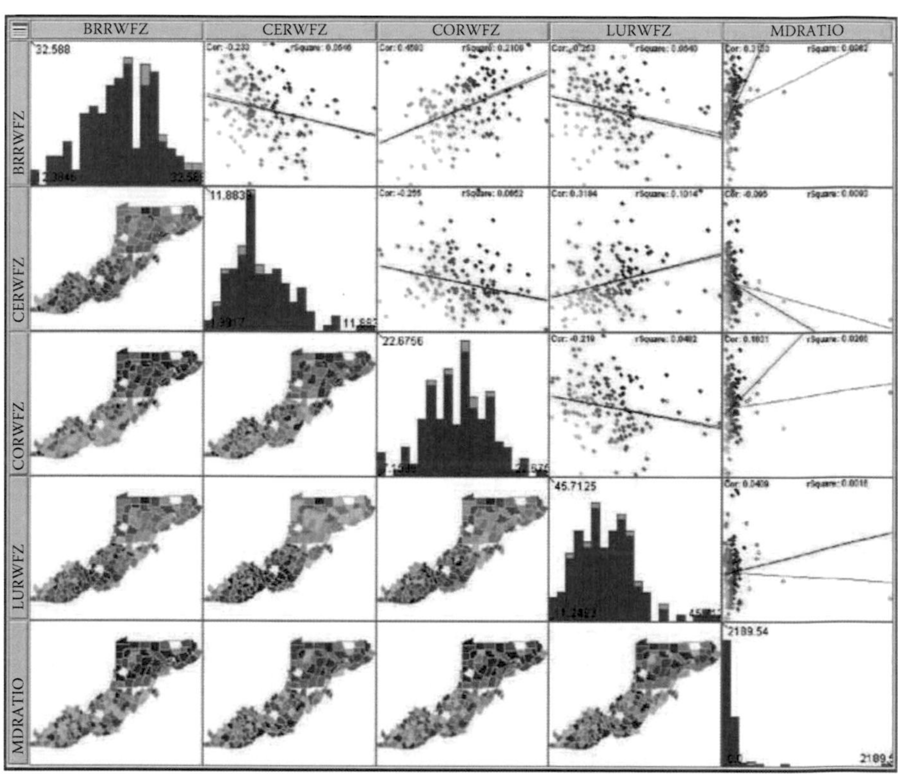

COLOR FIGURE 11.4 A matrix of maps and scatterplots, relating incidence rates for different kinds of cancer (first four rows and columns) to the ratio of doctors per 1000 population, for the Appalachian region of the United States (MD ratio, the right hand column and bottom row). This shows every pair of bivariate displays as both a scatterplot (above and right) and a map (below and left). The on-diagonal elements of the matrix simply show the distribution of values for each variable in isolation. A bivariate color scheme is used throughout.

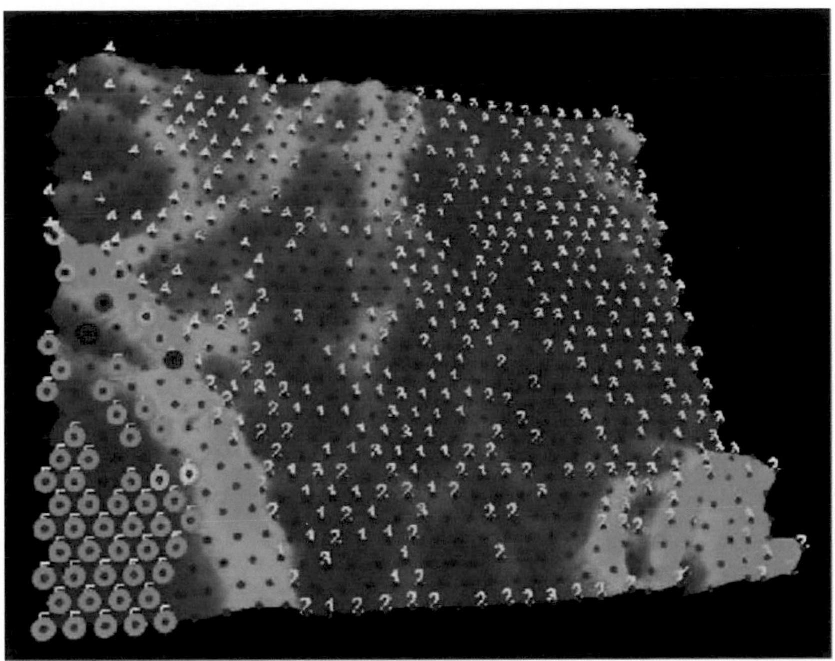

COLOR FIGURE 11.5 An example of a compositional landscape, created using a self-organizing map (SOM) that projects a high-dimensional dataset into a form that can be visualized in 2 or 3 dimensions — shown here as a surface. See text for a full explanation of how to read this figure.

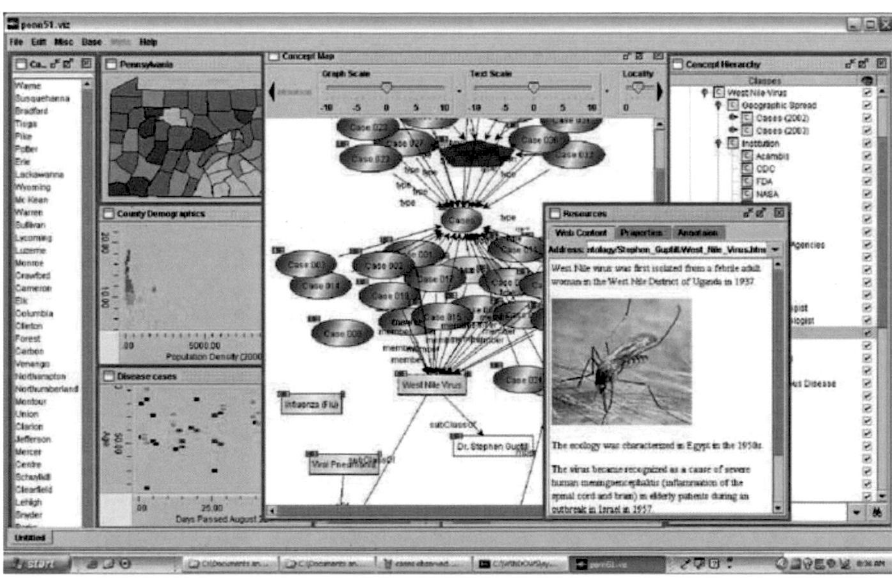

COLOR FIGURE 11.9 A screenshot of the *Improvise* visualization system, used to investigate symptoms and spread of vector-borne diseases. The display shows a mixture of many visualization techniques, to emphasize the point that techniques can be easily combined, to provide alternative views onto the same data in a coordinated manner. See text for further details.

Clustering with SOM

Multivariate Mapping

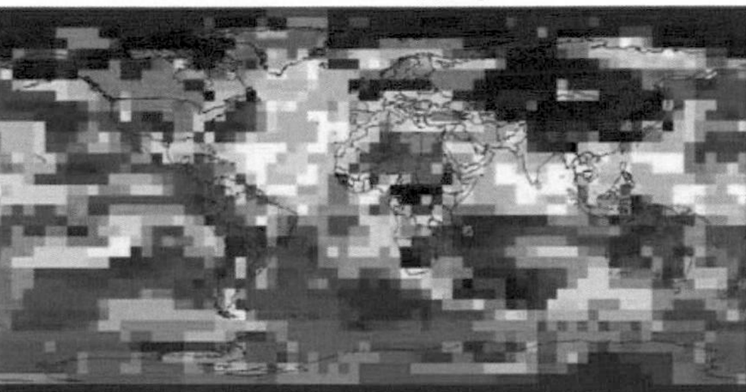

Multivariate Visualization of Clusters

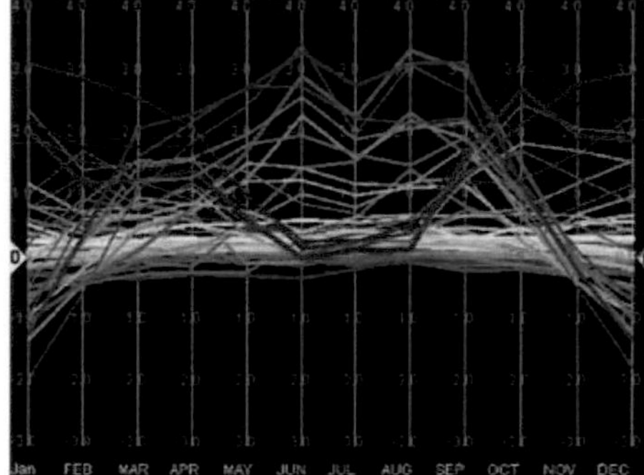

Multivariate Visualization of Data Items

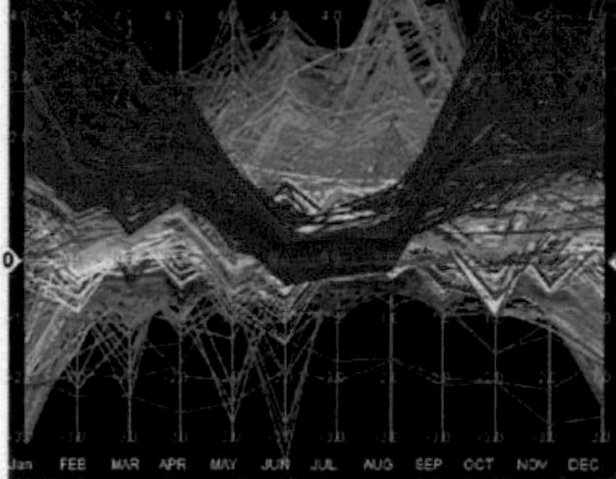

COLOR FIGURE 12.3 This is the overview of multivariate (seasonal) and spatial patterns of global climate (temperature) change. The multivariate mapping component (top right) is the central view, while other views can assist the understanding of the map. From such an integrated overview, one can easily perceive the major patterns in the data even without user interactions.

5 Regions (*minimum region size* >200)

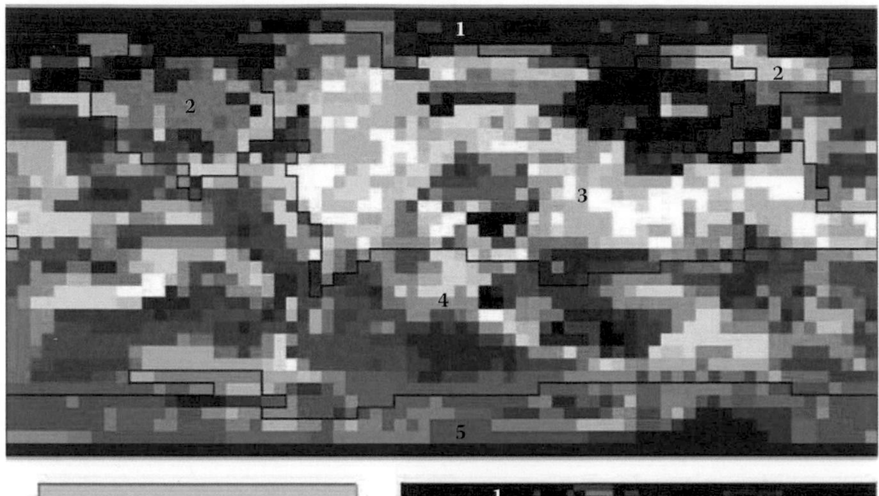

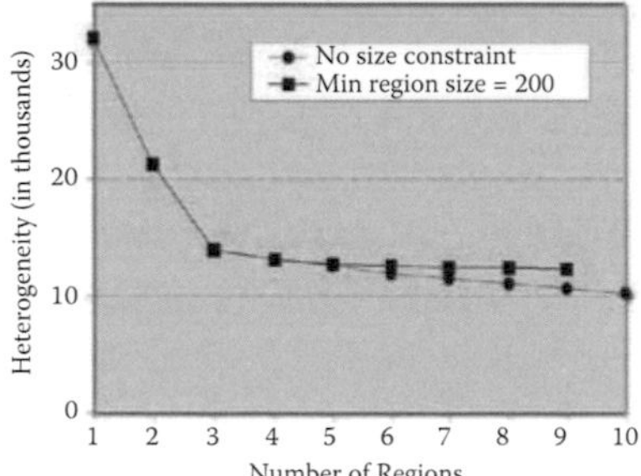

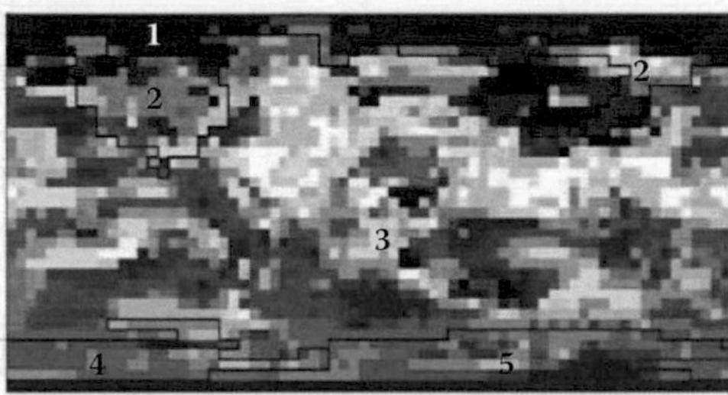

5 Regions (*without region size constraint*)

COLOR FIGURE 12.5 Two regionalization results. The small map shows five regions derived without a size constraint while the top (larger) map shows five regions under a size constraint (i.e., a region must be larger than 200 grid cells).

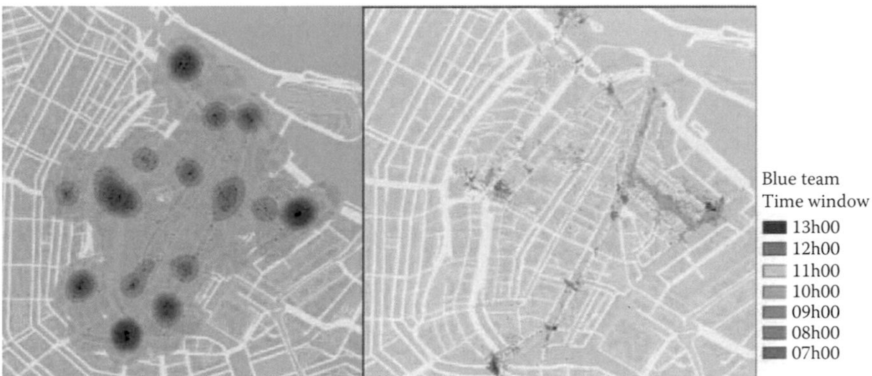

COLOR FIGURE 14.3 The visualization of (a) spatial clusters of stops and (b) the temporal clusters of moves of a group of 50 people (Blue Team) moving around the city of Amsterdam. (Data Source: Waag Society, Netherlands.)

7 An Overview of Clustering Methods in Geographic Data Analysis

Jiawei Han

Jae-Gil Lee

Micheline Kamber

CONTENTS

7.1 INTRODUCTION

Clustering is the process of grouping a set of physical or abstract objects into classes of *similar* objects. A **cluster** is a collection of data objects that are *similar* to one another within the same cluster and are *dissimilar* to the objects in other clusters. Although classification is an effective means for distinguishing groups or classes of objects, it often requires costly collection and labeling of a large set of training tuples or patterns, which the classifier uses to model each group. In contrast, clustering does not require such labeling at all.

Cluster analysis has been widely used in numerous applications, including market research, pattern recognition, data analysis, and image processing. In business, clustering can help marketers discover distinct groups in their customer bases and characterize customer groups based on purchasing patterns. In biology, it can be used to derive plant and animal taxonomies, categorize genes with similar functionality, and gain insight into structures inherent in populations. Clustering may also help in the identification of areas of similar land use in an earth observation database; in the identification of groups of houses in a city according to house type, value, and geographical location; and in the identification of groups of automobile insurance policy holders with a high average claim cost. It can also be used to help classify documents on the Web for information discovery.

Clustering is also called **data segmentation** in some applications because clustering partitions large data sets into groups according to their *similarity*. As a data mining function, cluster analysis can be used as a stand-alone tool to gain insight into the distribution of data, to observe the characteristics of each cluster, and to focus on a particular set of clusters for further analysis. Alternatively, it may serve as a preprocessing step for other algorithms, such as characterization, attribute subset selection, and classification, which would then operate on the detected clusters and the selected attributes or features.

For the geographical community, data mining promises many new and valuable tools for geographic data analysis. Spatial clustering is one of these tools. Owing to the huge amounts of geographic data collected in databases, spatial clustering has become a highly active topic in geographic data analysis. The immense applications of spatial clustering has resulted in tremendous growth of the field, making it worth a dedicated and comprehensive overview.

There are various forms of geographic data. The most popular form is *point* data. The location of a spatial object or an event is represented as a point. Examples of point data include the locations of buildings, the centers of roads or lands, and the locations of outbreaks of epidemics. On the other hand, recent improvements in satellites and tracking facilities have made it possible to collect a large amount of *trajectory* data of moving objects. Examples include vehicle position data, hurricane track data, and animal movement data.

Our survey of spatial clustering proceeds as follows. First, we study traditional clustering methods developed for point data. Second, we study recent clustering methods developed for trajectory data. For each part, we first introduce interesting applications to facilitate understanding of the reader, and then we present representative clustering methods.

7.2 CLUSTERING OF POINT DATA

In this section, we present an overview of clustering methods developed for point data. First, we introduce some interesting applications of point data clustering in Section 7.2.1. Next, we provide a categorization of major clustering methods in Section 7.2.2. Each application introduced represents a possible usage scenario of each category. Then, the major clustering methods are explained in Sections 7.2.3 through 7.2.5.

7.2.1 Applications

7.2.1.1 Classical Example: John Snow's Analysis

In the 1850s, Dr. John Snow performed pioneering data analysis to prove his hypothesis that cholera was spread by the drinking of water infected with cholera bacteria. This analysis, if performed today, would come under the realms of GIS (geographic information system) and data mining. Some have claimed his Broad Street map as being the first example of GIS, even though it was performed with a pen and paper.

The cholera outbreak occurred in the Soho District of London, in and around Broad Street. As shown in Figure 7.1, Dr. Snow plotted each cholera case on a map and also plotted houses with multiple cases. The cases were not distributed uniformly, but rather were distributed in a tight cluster around a water pump located on Broad Street (now Broadwick Street). Dr. Snow disabled the water pump by removing its handle. The Broad Street cholera outbreak stopped almost literally overnight.

7.2.1.2 Geographic Customer Segmentation

People living within the same geographical boundaries often exhibit similar buying patterns. This is in part due to similarities in demographic and psychographic

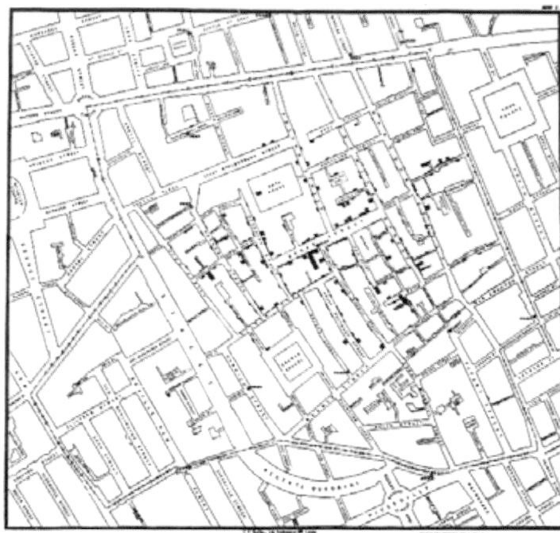

FIGURE 7.1 Original map by Dr. John Snow showing the clusters of cholera cases in the London epidemic of 1854.

characteristics of residents. This phenomena is further enforced by local weather, environment, and cultural differences. Geographic segmentation is useful if marketing or promotion medium is mass media or mass public gatherings. Spatial clustering is a common method for geographic segmentation. In this application, the locations of customer points (spatial objects) supplied by Harbin dairy in Heilongjiang, China are used. A total of 1229 customer points are collected and then a clustering method is applied to this data set [21].

Figure 7.2 shows the clustering result. There are 16 obvious clusters, while the others are isolated points. This result shows a very high correlation with the actual population of Heilongjiang. For example, the Harbin-Daqing-Suihua-Minghui-Nehe cluster is the biggest one. The next biggest ones are the Qiqihaer town cluster and the Jiamusi-Shuangyashan town cluster.

Partitioning methods are the most suitable for this application because they partition a set of spatial objects based on closeness between objects. In addition, sometimes the number of desired clusters can be predetermined according to the budget for mass marketing (e.g., television, radio, and leaflet dropping). Partitioning methods are explained in detail in Section 7.2.3.

7.2.1.3 Crime Hot-Spot Analysis

Much of crime mapping is devoted to detecting crime hot spots, i.e., geographic areas of elevated criminal activity. Hot-spot analysis helps police identify high-crime areas, types of crimes being committed, and the best way to respond. In many cases, a crime hot spot is defined as an area where crime incidents are geographically concentrated. Thus, spatial clustering can be used for crime hot-spot analysis. In this application, the locations of crime incidents in Brisbane, Australia are collected, and then a clustering method is applied to this data set [8].

FIGURE 7.2 Geographic customer segmentation for Harbin dairy in Heilongjiang, China. (From Wan, L.-H., Li, Y.-J., Liu, W.-Y., and Zhang, D.-Y., Application and study of spatial cluster and customer partitioning. In *Proc. 2005 Intl. Conf. Machine Learning and Cybernetics*, Guangzhou, China, August 2005, pp. 1701–1706.) © 2005 IEEE.

Figure 7.3 shows the clustering result. Cluster *A* is identified in the first level, directly overlaying the main urban area. This active cluster is further divided into two clusters AA and AB in the second level. Again, cluster AB is further divided into two clusters ABA and ABB in the third level. In this way, hot spots are discovered in a hierarchical manner, i.e., ranging from the largest one to smaller ones.

Hierarchical methods are the most suitable for this application because they can discover a hierarchical structure of data. One of the unique characteristics of geographical data is its generic complexity. Clusters may contain several numbers of subclusters; these subclusters may, in turn, consist of smaller subclusters. This kind of hierarchy can be easily found in the real world due to its hierarchical nature. For example, a county encloses several cities, and each city may enclose several small towns. Hierarchical methods are explained in detail in Section 7.2.4.

7.2.1.4 Land Use Detection

Spatial clustering is one of the basic methods for automatic land use detection from remote sensing data. In this application, a five-dimensional feature space is created from several satellite images of a region on the surface of the earth covering California. These images are taken from the raster data of the SEQUOIA 2000 Storage Benchmark. Through some preprocessing, a feature vector of a point is generated to have 1,024,000 intensity values each for five spectral channels—one visible, two reflected infrared, and two emitted (thermal) infrared. Here, a point represents an earth surface area of 1000 by 1000 meters. Then, a clustering method is applied to this data set.

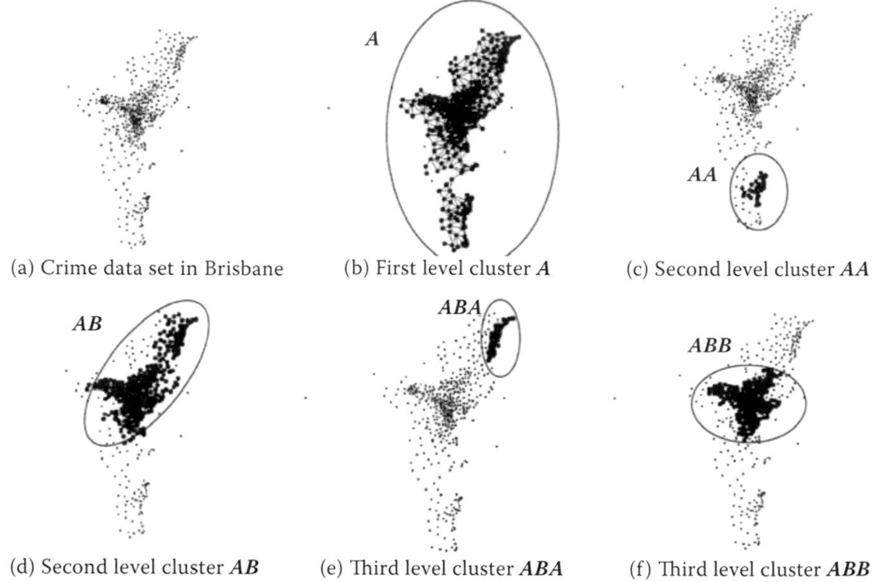

(a) Crime data set in Brisbane (b) First level cluster ***A*** (c) Second level cluster ***AA***

(d) Second level cluster ***AB*** (e) Third level cluster ***ABA*** (f) Third level cluster ***ABB***

FIGURE 7.3 Crime hot spots generated from a crime data set in Brisbane, Australia. (From Estivill-Castro, V. and Lee, I., Amoeba: Hierarchical clustering based on spatial proximity using delaunty diagram, in *Proc. 9th Intl. Symp. Spatial Data Handling (SDH '00)*, Beijing, China, August 2000, pp. 26–41.)

Figure 7.4 shows the clustering result. Nine clusters are obtained with sizes ranging from 2,016 points to 598,863 points. Each cluster is displayed using a different color. That is, a point in the image is colored according to the cluster to which the point belongs. We can easily know that there is a high degree of correspondence between Figure 7.4 and a physical map of California.

Density-based methods are the most suitable for this application because they can discover clusters of arbitrary shape. The image in Figure 7.4 is, in fact, obtained by DBSCAN [7], which is the representative density-based method. Density-based methods are explained in detail in Section 7.2.5.

7.2.2 Categorization of Major Clustering Methods

A large number of clustering methods are reported in the literature. It is difficult to provide a crisp categorization of clustering methods because these categories may overlap so that a method may have features from several categories. In general, the major clustering methods can be classified into the following categories:

Partitioning methods: Given a database of n objects or data tuples, a partitioning method constructs $k(\leq n)$ partitions of the data, where each partition represents a cluster. That is, it classifies the data into k groups, which together satisfy the following requirements: (1) each group must

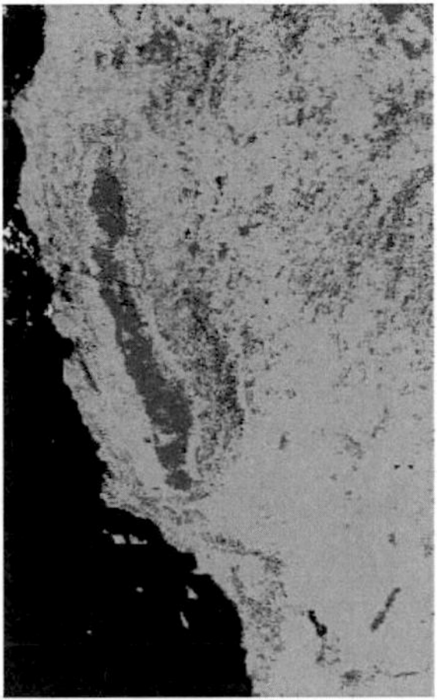

FIGURE 7.4 Visualization of the clustering result for the SEQUOIA 2000 raster data.

contain at least one object, and (2) each object must belong to exactly one group. Notice that the second requirement can be relaxed in some fuzzy partitioning techniques. Such a partitioning method creates an initial partitioning. It then uses an *iterative relocation technique* that attempts to improve the partitioning by moving objects from one group to another. Representative algorithms include k-means [16], k-medoids [14], CLARANS [18], and the EM algorithm [6]. These algorithms are studied in Section 7.2.3.

Hierarchical methods: A hierarchical method creates a hierarchical decomposition of a given set of data objects. Hierarchical methods can be classified as *agglomerative* (*bottom-up*) or *divisive* (*top-down*), based on how the hierarchical decomposition is formed. AGNES and DIANA [14] are examples of agglomerative and divisive methods, respectively. Hierarchical methods suffer from the fact that once a step (merge or split) is done, it can never be undone. Two approaches are developed to alleviate this problem. Representative algorithms include BIRCH [23] and Chameleon [13]. These algorithms are studied in Section 7.2.4.

Density-based methods: Most partitioning methods cluster objects based on the distance between objects. Such methods can find only spherical-shaped clusters and encounter difficulty in discovering clusters of arbitrary shape. Other clustering methods have been developed based on the notion of *density*. Their general idea is to continue growing a given cluster as long as the density (the number of objects or data points) in the "neighborhood" exceeds a threshold. Such a method is able to filter out noises (outliers) and discover clusters of arbitrary shape. Representative algorithms include DBSCAN [7], OPTICS [3], and DENCLUE [12]. These algorithms are studied in Section 7.2.5.

Grid-based methods: Grid-based methods quantize the object space into a finite number of cells that form a grid structure. All of the clustering operations are performed on the grid structure (i.e., on the quantized space). The main advantage of this approach is its fast processing time, which is typically independent of the number of data objects and dependent only on the number of cells in each dimension in the quantized space. Representative algorithms include STING [22], WaveCluster [20], and CLIQUE [1].

The choice of clustering algorithm depends both on the type of data available and on the particular purpose of the application. If cluster analysis is used as a descriptive or exploratory tool, it is possible to try several algorithms on the same data to see what the data may disclose.

7.2.3 Partitioning Methods

Given D, a data set of n objects, and k, the number of clusters to form, a **partitioning algorithm** organizes the objects into k partitions ($k \leq n$), where each partition represents a cluster. The clusters are formed to optimize an objective partitioning criterion, such as a dissimilarity function based on distance, so that the objects within a cluster are "similar," whereas the objects of different clusters are "dissimilar."

7.2.3.1 Centroid-Based Technique: The *k*-Means Method

The k-**means algorithm** takes the input parameter, k, and partitions a set of n objects into k clusters so that the resulting intracluster similarity is high, but the intercluster similarity is low. Cluster similarity is measured with respect to the *mean* value of the objects in a cluster, which can be viewed as the cluster's *centroid* or *center of gravity*.

How does the k-means algorithm work? The k-means algorithm proceeds as follows. First, it randomly selects k of the objects, each of which initially represents a cluster mean or center. For each of the remaining objects, an object is assigned to the cluster to which it is the most similar, based on the distance between the object and the cluster mean. It then computes the new mean for each cluster. This process iterates until the criterion function converges. Typically, the **square-error criterion** is used, defined as

$$E = \sum_{i=1}^{k} \sum_{p \in C_i} |p - m_i|^2, \tag{7.1}$$

where E is the sum of the square-error for all objects in the data set; p is the point representing a given object; and m_i is the mean of a cluster C_i (both p and m_i are multidimensional). In other words, for each object in each cluster, the distance from the object to its cluster center is squared, and the distances are summed up. This criterion tries to make the resulting k clusters as compact and as separate as possible. The k-means procedure is summarized in Figure 7.5.

Input:

- k: the number of clusters,
- D: a data set containing n objects.

Output: A set of k clusters.

Method:

(1) arbitrarily choose k objects from D as the initial cluster centers;
(2) repeat;
(3) (re)assign each object to the cluster to which the object is the most similar, based on the distance between the object and the cluster mean;
(4) update the cluster means, i.e., calculate the mean value of the objects for each cluster;
(5) until no change;

FIGURE 7.5 The k-means partitioning algorithm.

EXAMPLE 7.1

Suppose there is a set of objects shown in Figure 7.6(a). Let $k = 2$; that is, the user would like the objects to be partitioned into two clusters. In Figure 7.6(a), two objects marked by "×" are randomly selected as initial cluster centers. In Figure 7.6(b), each object is distributed to the nearest cluster. Objects in one cluster are represented by circles, and those in the other by rectangles. Next, the cluster centers are moved as depicted by arrows. This process iterates nine times, leading to Figure 7.6(c). The process of iteratively reassigning objects to clusters to improve the partitioning is referred to as *iterative relocation*. Eventually, no redistribution of the objects in any cluster occurs and so the process terminates.

The algorithm attempts to determine k partitions that minimize the square-error function. It works well when the clusters are compact clouds that are rather well separated from one another. The method is relatively scalable and efficient in processing large data sets because the computational complexity of the algorithm is $O(nkt)$, where n is the total number of objects, k is the number of clusters, and t is

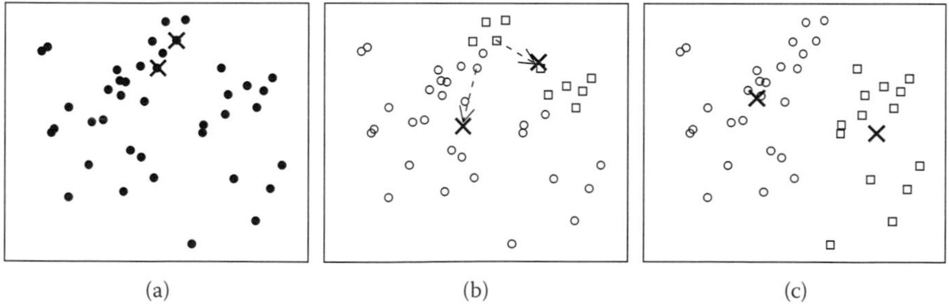

FIGURE 7.6 Clustering of a set of objects based on the k-means method. (The mean of each cluster is marked by "×".)

the number of iterations. Normally, $k \ll n$ and $t \ll n$. The method often terminates at a local optimum.

> **Algorithm: *k*-means.** The *k*-means algorithm for partitioning, where each cluster's center is represented by the mean value of the objects in the cluster.

The necessity for users to specify *k*, the number of clusters, in advance can be seen as a disadvantage. The *k*-means method is not suitable for discovering clusters with nonconvex shapes or clusters of very different size. Moreover, it is sensitive to noise and outlier data points because a small number of such data can substantially influence the mean value.

7.2.3.2 Representative Object-Based Technique: The *k*-Medoids Method

The *k*-means algorithm is sensitive to outliers because an object with an extremely large value may substantially distort the distribution of data. This effect is particularly exacerbated due to the use of the *square*-error function (Equation 7.1).

How might the algorithm be modified to diminish such sensitivity? Instead of taking the mean value of the objects in a cluster as a reference point, we pick an actual object to represent each cluster. This representative object, called a **medoid**, is meant to be the most centrally located object within the cluster. Each remaining object is clustered with the representative object to which it is the most similar. The partitioning method is then performed based on the principle of minimizing the sum of the dissimilarities between each object and its corresponding reference point. That is, an **absolute-error criterion** is used, defined as

$$E = \sum_{j=1}^{k} \sum_{p \in C_j} |p - o_j|, \tag{7.2}$$

where E is the sum of the absolute-error for all objects in the data set; p is the point representing a given object in cluster C_j; and o_j is the representative object of C_j.

The overall procedure of *k*-**medoids** clustering is as follows. Initial representative objects (or seeds) are chosen arbitrarily. The iterative process of replacing representative objects by nonrepresentative ones continues as long as the quality of the resulting clustering is improved. Here, the quality of a clustering is measured by the average dissimilarity between an object and the representative object of its cluster. When replacing a current representative object by a nonrepresentative one, the following four cases are examined as illustrated in Figure 7.7.

Suppose there are two representative objects o_i and o_j. If o_i is replaced with a nonrepresentative object o_h, for all the objects $\mathcal{I}$ that are originally in the cluster represented by o_i, we need to find the most similar representative object. In Case 1, $p \in \mathcal{I}$ moves to the cluster represented by o_j that was the second most similar one. In Case 2, $p \in \mathcal{I}$ moves to the new cluster represented by o_h, and the cluster represented by o_j is not affected. In addition, we need to examine all the objects $\mathcal{J}$ that are

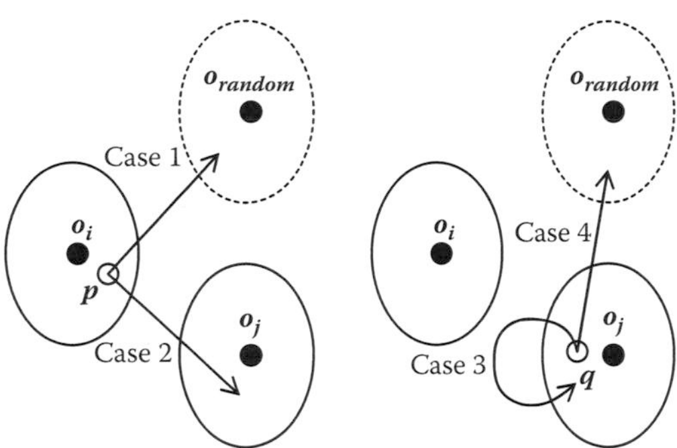

FIGURE 7.7 Four cases of the cost function for *k*-medoids clustering.

originally in the cluster represented by o_j. Due to the replacement, $q \in \mathcal{J}$ stays with o_j (Case 3) or moves to the new cluster represented by o_h (Case 4).

The *k*-medoids method computes the *difference* in the absolute-error value if a current representative object is swapped with a nonrepresentative object. The total cost of swapping is the sum of differences incurred by all nonrepresentative objects. If the total cost is negative, o_i is replaced or swapped with o_h since the actual absolute-error E would be reduced. Otherwise, the current representative object o_i is considered acceptable, and nothing is changed in the iteration.

PAM (partitioning around medoids) was one of the first *k*-medoids algorithms introduced. The PAM algorithm is summarized in Figure 7.8. Initial *k* representative objects are selected arbitrarily. It next computes the total cost TC_{ih} of a swapping for

Input:

- *k:* the number of clusters,
- *D*: a data set containing *n* objects.

Output: A set of *k* clusters.

Method:

(1) arbitrarily choose *k* objects in *D* as the initial representative objects or seeds;
(2) repeat;
(3) for each nonrepresentative object o_h do;
(4) for each representative object o_i do;
(5) calculate the total cost TC_{ih} of a swapping between o_i and o_h;
(6) find *i* and *h* where TC_{ih} is the smallest;
(7) if $TC_{ih} < 0$ then replace o_i with o_h;
(8) until $TC_{ih} \geq 0$;
(9) assign each nonrepresentative object to the cluster with the nearest representative object;

FIGURE 7.8 PAM, a *k*-medoids partitioning algorithm.

every pair of objects o_i and o_h, where o_i is a representative one, and o_h is not. It then selects the pair of o_i and o_h that achieves the minimum of TC_{ih}. If the minimum is negative, o_i is swapped with o_h, and the same procedure is repeated until no swapping occurs. The final set of representative objects are the respective medoids of the clusters. The complexity of each iteration is $O(k(n - k)^2)$. For large values of n and k, such computation becomes very costly.

Algorithm: *k*-medoids. PAM, a *k*-medoids algorithm for partitioning based on medoids or central objects.

7.2.3.3 A Model-Based Method: Expectation-Maximization (EM)

In practice, each cluster can be represented mathematically by a parametric probability distribution. The entire data is a *mixture* of these distributions, where each individual distribution is typically referred to as a *component density*. We can therefore cluster the data using a finite **mixture density model** of M probability distributions, where each distribution represents a cluster. The problem is to estimate the parameters of the probability distributions so as to best fit the data. Figure 7.9 is an example of a simple finite mixture density model. There are two clusters. Each follows a normal or Gaussian distribution with its own mean and standard deviation.

The mixture model is formalized as Equation (7.3), where the parameters are $\Theta = (\alpha_1,\ldots, \alpha_M, \theta_1,\ldots, \theta_M)$ such that $\Sigma_{i=1}^{M}\alpha_i = 1$, and each p_i is a density function parameterized by θ_i. That is, M component densities are mixed together with M mixing coefficients α_i.

$$p(x|\Theta) = \sum_{i=1}^{M} \alpha_i p_i(x|\theta_i) \tag{7.3}$$

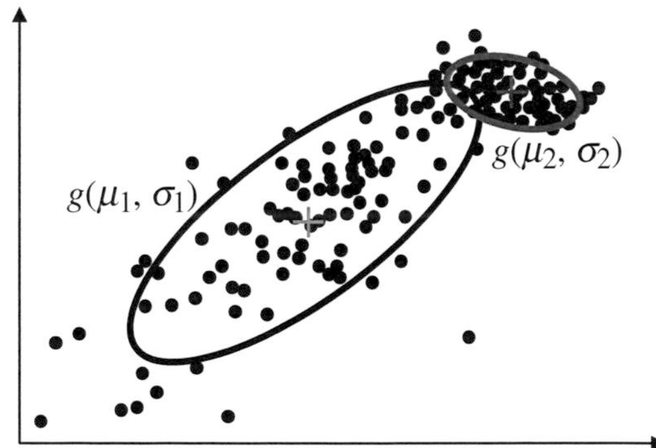

FIGURE 7.9 Each cluster can be represented by a probability distribution, centered at a mean, and with a standard deviation. Here, we have two clusters, corresponding to the Gaussian distributions $g(\mu_1, \sigma_1)$ and $g(\mu_2, \sigma_2)$.

The log-likelihood expression for this density from observed data $\mathcal{X}$ is given by Equation (7.4), if we posit the existence of unobserved data $\mathcal{Y} = \{y_i\}_{i=1}^N$ whose values inform us which component density "generated" each data item. That is, if the i-th sample was generated by the k-th mixture component, $y_i = k$ ($y_i \in \{1, \ldots, M\}$). Now, the goal is to find Θ that maximizes $\mathcal{L}$. Often, $\log(\mathcal{L}(\Theta \mid \mathcal{X}, \mathcal{Y}))$ is maximized because it is analytically easier. A variety of techniques can be used for this optimization.

$$\log(\mathcal{L}(\Theta \mid \mathcal{X},\mathcal{Y})) = \log(P(\mathcal{X},\mathcal{Y} \mid \Theta))$$

$$= \sum_{i=1}^{N} \log(P(x_i|y_i)P(y_i)) = \sum_{i=1}^{N} \log(\alpha_{yi} p_{yi}(x_i|\theta_{y_i})) \tag{7.4}$$

The **(Expectation-Maximization) EM** algorithm is one of such techniques. The EM algorithm is a general method of finding the maximum-likelihood estimate of the parameters of an underlying distribution from a given data set when the data is incomplete. It can be viewed as an extension of the k-means paradigm. Instead of assigning each object to a dedicated cluster, EM assigns each object to a cluster according to a weight representing the probability of membership. In other words, there are no strict boundaries between clusters.

EM starts with an initial estimate or "guess" of the parameters of the mixture model. It iteratively re-scores the objects against the mixture density produced by the parameters. The re-scored objects are then used to update the parameter estimates. The algorithm is described as follows [4].

1. Make an initial guess of the parameters $\Theta = (\alpha_1^g, \ldots, \alpha_M^g, \theta_1^g, \ldots, \theta_M^g)$.
2. Iteratively refine the parameters (or clusters) based on the following two steps:

   (a) Expectation Step: Given Θ^g, compute $p_j(x_i \mid \theta_j^g)$ for each i and j. Notice that α_j can be considered as prior probabilities of each mixture component, i.e., $\alpha_j = p$ (component j).

$$p(y_i|x_i,\Theta^g) = \frac{\alpha_{y_i}^g p_{y_i}\left(x_i|\theta_{y_i}^g\right)}{p(x_i|\Theta^g)} = \frac{\alpha_{y_i}^g p_{y_i}\left(x_i|\theta_{y_i}^g\right)}{\sum_{k=1}^M \alpha_k^g p_k\left(x_i|\theta_k^g\right)} \tag{7.5}$$

   In other words, this step calculates the probability of cluster membership of an object x_i, for each of the clusters. These probabilities are the "expected" cluster memberships for the object x_i.

   (b) Maximization Step: Use the probability estimates from above to re-estimate (or refine) the parameters. The estimates of the new parameters in terms of the old parameters are as follows:

$$\alpha^{new} = \frac{1}{N} \sum_{i=1}^{N} p(\mid x_i, \Theta^g) \tag{7.6}$$

$$\mu^{new} = \frac{\sum_{i=1}^{N} x_i p(\mid x_i, \Theta^g)}{\sum_{i=1}^{N} p(\mid x_i, \Theta^g)} \tag{7.7}$$

$$\sum^{new} = \frac{\sum_{i=1}^{N} p(\mid x_i, \Theta^g)(x_i - \mu^{new})(x_i - \mu^{new})^T}{\sum_{i=1}^{N} p(\mid x_i, \Theta^g)} \tag{7.8}$$

This step is the "maximization" of the likelihood of the distributions given the data.

The EM algorithm is simple and easy to implement. In practice, it converges fast, but may not reach the global optimum. Convergence is guaranteed for certain forms of optimization functions. The computational complexity is linear in d (the number of input features), n (the number of objects), and t (the number of iterations).

How is spatial information taken into account? Ambroise and Govaert [2] proposed the *neighborhood EM (NEM)* algorithm. The likelihood is penalized using a regularizing term, which takes into account the spatial information relative to the data. Let V be the neighborhood matrix, where $v_{ij} > 0$ if x_i and x_j are neighbors, and $v_{ij} = 0$ otherwise. Besides, c_{ik} denotes the probability of x_i belonging to a cluster k, i.e., $p(k \mid x_i, \Theta)$. The regularizing term is defined as follows:

$$\mathcal{G}(c) = \frac{1}{2} \sum_{k=1}^{M} \sum_{i=1}^{N} \sum_{j=1}^{N} c_{ik} c_{jk} v(i,j) \tag{7.9}$$

Then, the penalized likelihood is constructed by adding $\beta \cdot G(c)$ to the original likelihood. Here, β is a coefficient given by a user. At each iteration of the NEM algorithm, the spatial information modifies the partitioning according to the importance of the β coefficient.

7.2.3.4 Partitioning Methods in Large Databases: From *k*-Medoids to CLARANS

How efficient is the *k*-medoids algorithm on large data sets? A typical *k*-medoids algorithm like PAM works effectively for small data sets, but does not scale well for large data sets. To deal with larger data sets, a *sampling*-based method, called **CLARA** (Clustering LARge Applications), can be used.

The idea behind CLARA is as follows: Instead of finding representative objects for the entire data set, CLARA draws a sample of the data set, applies PAM on the sample, and finds the medoids of the sample. The point is that, if the sample is drawn in a sufficiently random way, the medoids of the sample would approximate the medoids of the entire data set. To come up with better approximations, CLARA draws multiple samples and gives the best clustering as the output. Here, for accuracy, the quality of a clustering is measured based on the average dissimilarity of all objects in the entire data set, not only of those objects in the samples. Experiments indicate that five samples of size $40 + 2k$ give satisfactory results. The time complexity of each iteration is $O(k(40 + k)^2 + k(n - k))$. However, CLARA cannot find the best clustering if any of the best k medoids are not selected during sampling.

How might we improve the quality and scalability of CLARA? Another algorithm called **CLARANS** (Clustering Large Applications based on RANdomized Search) was proposed, which combines the sampling technique with PAM. Unlike CLARA, CLARANS does not confine itself to any sample at any given time. Conceptually, the clustering process can be viewed as a search through a graph, where each node is a potential solution (a set of k medoids). Two nodes are *neighbors* (that is, connected by an arc in the graph) if their sets differ by only one object. Each node can be assigned a cost that is defined by the total dissimilarity between every object and the medoid of its cluster. At each step, PAM examines all neighbors of the current node in its search for a minimum cost solution. The current node is then replaced by the neighbor with the largest descent in costs. Because CLARA works on a sample of the entire data set, it examines fewer neighbors and restricts the search to subgraphs that are smaller than the original graph. Whereas CLARA draws a sample of nodes at the beginning of a search, CLARANS dynamically draws a random sample of neighbors in each step of a search. The number of neighbors to be randomly sampled is restricted by a user-specified parameter. In this way, CLARANS does not confine the search to a localized area. If a better neighbor is found (i.e., having a lower error), CLARANS moves to the neighbor's node, and the process starts again. Otherwise, the current clustering produces a local minimum, and CLARANS starts with new randomly selected nodes in search for a new local minimum. Once a user-specified number of local minima has been found, the algorithm outputs the best local minimum.

CLARANS has been experimentally shown to be much more efficient than PAM. In addition, given the same amount of time, CLARANS is able to find clusterings of better quality than CLARA.

7.2.4 Hierarchical Methods

A hierarchical clustering method works by grouping data objects into a tree of clusters. Hierarchical clustering methods can be further classified as either *agglomerative* or *divisive*, depending on whether the hierarchical decomposition is formed in a bottom-up (merging) or top-down (splitting) fashion. The quality of a pure hierarchical clustering method suffers from its inability to perform adjustment once a merge or split decision has been executed. That is, if a particular merge or split decision later turns out to have been a poor choice, the method cannot backtrack and correct it. Recent studies have emphasized the integration of hierarchical agglomeration with iterative relocation methods.

7.2.4.1 Agglomerative and Divisive Hierarchical Clustering

AGNES and DIANA are two earlier hierarchical clustering algorithms. **AGNES** (AGglomerative NESting) is an agglomerative (bottom-up) algorithm which starts by placing each object in its own cluster and then merging these atomic clusters into larger and larger clusters until all of the objects are in one cluster or until a certain termination condition is satisfied. On the other hand, **DIANA** (DIvisive ANAlysis) is a divisive (top-down) algorithm that does the reverse of AGNES by starting with all objects in one cluster. It subdivides the cluster into smaller and smaller pieces until each object forms a cluster on its own or until a certain termination condition is satisfied. In either AGNES or DIANA, one can specify a desired number of clusters as a termination condition.

EXAMPLE 7.2

Figure 7.10 shows the application of AGNES and DIANA to a data set of five objects $\{p, q, r, s, t\}$. Initially, AGNES places each object into a cluster of its own. At each stage, the algorithm joins the two clusters that are closest together (i.e., most similar). The cluster merging process repeats until all of the objects are eventually merged to form one cluster. DIANA does the reverse of AGNES.

Four widely used measures for the distance between clusters are formulated by Equation (7.10) through Equation (7.13). Here, $|p - p'|$ is the distance between two objects or points, p and p'; m_i is the mean for a cluster, C_i; and n_i is the number of objects in C_i.

$$\textbf{Minimum distance}: d_{\min}(C_i, C_j) = \min_{p \in C_i, p' \in C_j} |p - p'| \tag{7.10}$$

$$\textbf{Maximum distance}: d_{\max}(C_i, C_j) = \max_{p \in C_i, p' \in C_j} |p - p'| \tag{7.11}$$

$$\textbf{Mean distance}: d_{mean}(C_i, C_j) = |m_i - m_j| \tag{7.12}$$

$$\textbf{Average distance}: d_{avg}(C_i, C_j) = \frac{1}{n_i n_j} \sum_{p \in C_i} \sum_{p' \in C_j} |p - p'| \tag{7.13}$$

When an algorithm uses the *minimum distance*, $d_{min}(C_i, C_j)$, to measure the distance between clusters, it is called a **single-linkage algorithm**. It is also known as a nearest neighbor clustering algorithm. On the other hand, when an algorithm uses the *maximum distance*, $d_{max}(C_i, C_j)$, it is called a **complete-linkage algorithm**. It is also

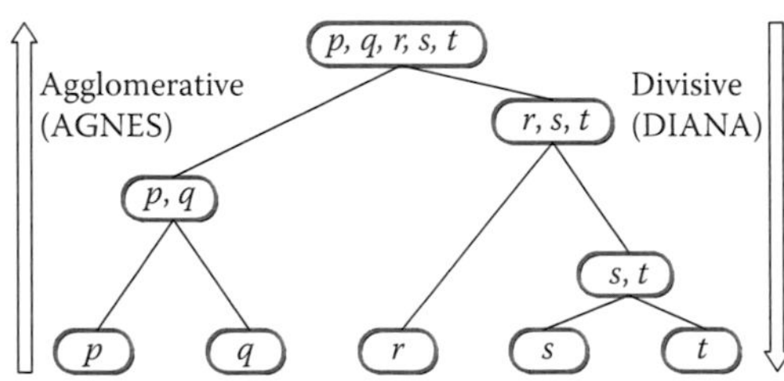

FIGURE 7.10 Agglomerative and divisive hierarchical clustering on a set of data objects $\{p, q, r, s, t\}$.

known as a farthest neighbor clustering algorithm. The above minimum and maximum measures represent two extremes in measuring the distance between clusters. They tend to be overly sensitive to outliers or noisy data. The use of *mean* or *average distance* is a compromise between the minimum and maximum distances and overcomes the outlier sensitivity problem.

What are some of the difficulties with hierarchical clustering? The hierarchical clustering method, though simple, often encounters difficulties regarding the selection of merge or split points. Such a decision is critical because once a group of objects is merged or split, the process at the next step will operate on the newly generated clusters. It will neither undo what was done previously, nor perform object swapping between clusters. Thus, merge or split decisions, if not well chosen at some step, may lead to low-quality clusters. Moreover, the method does not scale well because each decision of merge or split needs to examine and evaluate a good number of objects or clusters.

One promising direction for improving the clustering quality of hierarchical methods is to integrate hierarchical clustering with other clustering techniques, resulting in multiple-phase clustering. Two such methods are introduced in the following subsections. The first, called BIRCH, begins by partitioning objects hierarchically using tree structures, where the leaf or low-level nonleaf nodes can be viewed as "microclusters" depending on the scale of resolution. It then applies other clustering algorithms to perform macroclustering on the microclusters. The second method, called Chameleon, explores dynamic modeling in hierarchical clustering.

7.2.4.2 BIRCH: Balanced Iterative Reducing and Clustering Using Hierarchies

BIRCH is designed for clustering a large amount of numerical data by integration of hierarchical clustering (at the initial *microclustering* stage) and other clustering methods such as iterative partitioning (at the later *macroclustering* stage). It overcomes the two difficulties of agglomerative clustering methods: (1) scalability and (2) the inability to undo what was done in the previous step.

BIRCH introduces two concepts, *clustering feature* (CF) and *clustering feature tree* (*CF tree*), which are used to summarize cluster representations. These structures help the clustering method achieve good speed and scalability in large databases, and also make it effective for incremental and dynamic clustering of incoming objects.

A clustering feature is a triple summarizing the information about a cluster. Given N d-dimensional data points in a cluster, $\{\vec{X}_i\}$, where $i = 1, 2, \ldots, N$, the **CF** vector of the cluster is defined as a tuple:

$$\mathbf{CF} = (N, LS, SS) \tag{7.14}$$

where N is the number of data points in the cluster, LS is the linear sum of the N data points, i.e., $\Sigma_{i=1}^{N} X_i$, and SS is the square sum of the N data points, i.e., $\Sigma_{i=1}^{N} X_i^2$. CFs are sufficient for calculating all of the measurements that are needed for making clustering decisions in BIRCH. This avoids the necessity of storing all objects. An important property is that the CF vectors are additive. Suppose that the CF vectors of two disjoint clusters are $CF_1 = (N_1, LS_1, SS_1)$ and $CF_2 = (N_2, LS_2, SS_2)$, respectively. Then, the CF vector of the cluster formed by merging the two clusters is

$$CF_1 + CF_2 = (N_1 + N_2, LS_1 + LS_2, SS_1 + SS_2) \tag{7.15}$$

EXAMPLE 7.3

Suppose that there are three points, (2, 5), (3, 2), and (4, 3), in a cluster, C_1. The CF of C_1 is

$$CF_1 = (3, (2 + 3 + 4, 5 + 2 + 3), (2^2 + 3^2 + 4^2 +, 5^2 + 2^2 + 3^2)) = (3, (9, 10), (29, 38)).$$

Suppose that the CF of another cluster, C_2, is $CF_2 = (3, (35, 36), (417, 440))$. The CF of a new cluster, C_3, that is formed by merging C_1 and C_2, is derived by adding CF_1 and CF_2. That is,

$$CF_3 = (3 + 3, (9 + 35, 10 + 36), (29 + 417, 38 + 440)) = (6, (44, 46), (446, 478)).$$

A **CF tree** is a height-balanced tree with two parameters: *branching factor* (*B* for nonleaf nodes and *L* for leaf nodes) and *threshold T*. Each nonleaf node contains at most *B* entries of the form $[CF_i, child_i]$ $(i = 1, 2, \ldots, B)$, where $child_i$ is a pointer to its *i*-th child node, and CF_i is the sum of the CFs of their children. A leaf node contains at most *L* entries, and each entry is a CF. In addition, each leaf node has two pointers, *prev* and *next*, which are used to chain all leaf nodes together for efficient scans. All entries in a leaf node must satisfy a threshold requirement: the diameter of each leaf entry has to be less than *T*. The tree size is a function of *T*. The larger *T* is, the smaller the tree is. Figure 7.11 shows an example of a CF tree.

The CF tree is built dynamically as objects are inserted. An object is inserted to the appropriate leaf entry by recursively descending the CF tree and choosing the closest child node according to a chosen distance metric. Next, it is checked if the leaf entry can absorb the node without violating the threshold. If there is no room, the node is split. After inserting an object to a leaf, the CF information for each nonleaf entry on the path to the leaf should be updated.

BIRCH tries to produce the best clusters with the available resources. Given a limited amount of main memory, an important consideration is to minimize the time required for I/O. BIRCH applies a *multiphase* clustering technique: a single scan of the data set yields a basic good clustering, and one or more additional scans can

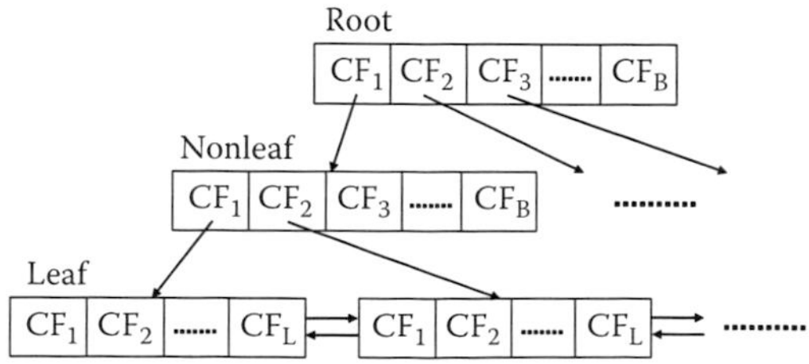

FIGURE 7.11 A CF tree structure.

(optionally) be used to further improve the quality. BIRCH performs the following four phases:

- Phase 1: Load data into memory by building a CF tree.
- Phase 2 (optional): Condense the initial CF tree into a desirable range by building a smaller CF tree.
- Phase 3: Perform global clustering.
- Phase 4 (optional): Perform cluster refining.

For Phase 1, the CF tree is built dynamically as objects are inserted. Thus, the method is incremental. The size of the CF tree can be changed by modifying the threshold. If the size of the memory that is needed for storing the CF tree is larger than the size of the main memory, a smaller CF tree is rebuilt by increasing the threshold. The rebuild process is performed by building a new tree from the leaf nodes of the old tree. Thus, the process of rebuilding the tree is done without the necessity of rereading all of the objects or points. Therefore, for building the tree, data has to be read just once. Some heuristics and methods have been introduced to deal with outliers and improve the quality of CF trees by additional scans of the data. Once the CF tree is built, an existing clustering algorithm can be used with the CF tree in Phase 3.

How effective is BIRCH? The computation complexity of the algorithm is $O(n)$, where n is the number of objects to be clustered. Experiments have shown the linear scalability of the algorithm with respect to the number of objects, and good quality of clustering of the data. However, since each node in a CF tree can hold only a limited number of entries due to its size, a CF tree node does not always correspond to what a user may consider a natural cluster. Moreover, if the clusters are not spherical in shape, BIRCH does not perform well because it uses the notion of radius or diameter to control the boundary of a cluster.

7.2.4.3 Chameleon: A Hierarchical Clustering Algorithm Using Dynamic Modeling

Chameleon is a hierarchical clustering algorithm that uses dynamic modeling to determine the similarity between two clusters. It was derived based on the observed weaknesses of the agglomerative algorithms: one set of schemes ignores the information about the aggregate interconnectivity of objects in two clusters, whereas the other set of schemes ignores the information about the closeness of two clusters as defined by the similarity of the closest objects across two clusters. Chameleon determines the pair of most similar sub-clusters by taking into account both the *interconnectivity* as well as the *closeness* of the clusters, and thus it overcomes the limitations above. It does not depend on a static, user-supplied model and can automatically adapt to the internal characteristics of the clusters being merged.

EXAMPLE 7.4

Figure 7.12 explains the weaknesses of existing agglomerative algorithms. In Figure 7.12(a), an algorithm that focuses only on the closeness of two clusters will incorrectly prefer to merge clusters C_3 and C_4 over clusters C_1 and C_2. In Figure 7.12(b),

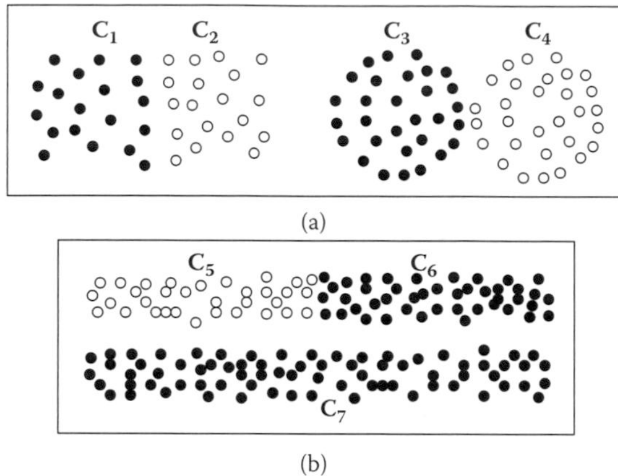

FIGURE 7.12 Examples of clusters for merging choices.

an algorithm that focuses only on the interconnectivity of two clusters will incorrectly prefer to merge a cluster C_5 with a cluster C_7 rather than with C_6.

How does Chameleon work? The main approach of Chameleon is illustrated in Figure 7.13. Chameleon models the data using a *k*-nearest neighbor graph, where each vertex of the graph represents a data object, and there exists an edge between two vertices (objects) if one object is among the *k*-most similar objects of the other. The edges are weighted to reflect the similarity between objects. Chameleon uses an algorithm that consists of two distinct phases. In the first phase, it uses a graph partitioning algorithm to partition the *k*-nearest neighbor graph into a large number of relatively small subclusters. In the second phase, it uses an agglomerative hierarchical clustering algorithm that repeatedly merges subclusters based on their similarity.

Note that the *k*-nearest neighbor graph captures the concept of neighborhood dynamically: the neighborhood radius of an object is determined by the *density* of the region in which the object resides. In a dense region, the neighborhood is defined narrowly; in a sparse region, it is defined more widely. This tends to result in more natural clusters, in comparison with density-based methods like DBSCAN

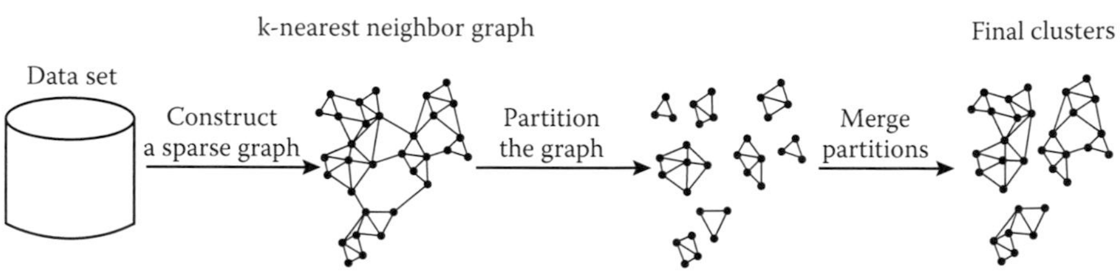

FIGURE 7.13 Chameleon: Hierarchical clustering based on *k*-nearest neighbors and dynamic modeling. (Based on Karypis, G., Han, E.-H., and Kumar, CHAMELEON: A hierarchical clustering algorithm using dynamic modeling, *Computer*, 32:68–75, 1999.) © 1999 IEEE.

(described in Section 7.2.5.1) that instead use a *global* neighborhood. Moreover, the density of the region is recorded as the weight of the edges. That is, the edges of a dense region tend to weigh more than that of a sparse region.

The graph partitioning algorithm partitions the k-nearest neighbor graph into several partitions such that the **edge cut**, i.e., the sum of the weight of the edges that straddle partitions, is minimized. Since each edge in the k-nearest neighbor graph represents the similarity among objects, a partitioning that minimizes the edge cut effectively minimizes the relationship (affinity) among objects across the resulting partitions. Edge cut is denoted $EC(C_i, C_j)$ and assesses the *absolute* interconnectivity between clusters C_i and C_j.

Chameleon determines the similarity between each pair of clusters C_i and C_j according to their *relative interconnectivity*, $RI(C_i, C_j)$, and their *relative closeness*, $RC(C_i, C_j)$:

- The **relative interconnectivity**, $RI(C_i, C_j)$, between two clusters, C_i and C_j, is defined as the absolute interconnectivity between C_i and C_j, normalized with respect to the internal interconnectivity of the two clusters, C_i and C_j. That is,

$$RI(C_i,C_j)=\frac{\left|EC_{\{C_i,C_j\}}\right|}{\frac{\left|EC_{C_i}\right|+\left|EC_{C_j}\right|}{2}}, \tag{7.16}$$

  where $EC_{\{C_i,C_j\}}$ is the edge cut as defined above for a cluster containing both C_i and C_j. Similarly, EC_{C_i} (or EC_{C_j}) is the minimum sum of the cut edges that partition C_i (or C_j) into two roughly equal parts.

- The **relative closeness**, $RC(C_i, C_j)$, between two clusters, C_i and C_j, is the absolute closeness between C_i and C_j, normalized with respect to the internal closeness of the two clusters, C_i and C_j. That is,

$$RC(C_i,C_j)=\frac{\bar{S}_{EC_{\{C_i,C_j\}}}}{\frac{|C_i|}{|C_i|+|C_j|}\bar{S}_{EC_{C_i}}+\frac{|C_j|}{|C_i|+|C_j|}\bar{S}_{EC_{C_j}}}, \tag{7.17}$$

  where $\bar{S}_{EC_{\{C_i,C_j\}}}$ is the average weight of the edges that connect vertices in C_i to vertices in C_j, and $\bar{S}_{EC_{\{C_i}}$ (or $\bar{S}_{EC_{C_j}}$) is the average weight of the edges that belong to the min-cut bisector of cluster C_i (or C_j).

Chameleon decides to merge the pair of clusters for which both $RI(C_i, C_j)$ and $RC(C_i, C_j)$ are high; i.e., it selects to merge clusters that are well interconnected as well as close together with respect to the internal interconnectivity and closeness of the clusters. In fact, it selects the pair of clusters that maximizes $RI(C_i, C_j) \times RC(C_i, C_j)^{\alpha}$, where α is a user-specified parameter. If $\alpha > 1$, Chameleon gives a higher importance to the relative closeness; otherwise, it gives a higher importance on the relative interconnectivity.

Chameleon has been shown to have greater power at discovering arbitrarily shaped clusters of high quality than several well-known algorithms such as BIRCH and the density-based algorithm DBSCAN (Section 7.2.5.1). However, the processing cost for high-dimensional data may require $O(n^2)$ time for n objects in the worst case.

7.2.5 Density-Based Methods

To discover clusters with arbitrary shape, density-based clustering methods have been developed. They typically regard clusters as dense regions of objects in the data space that are separated by regions of low density (representing noise).

7.2.5.1 DBSCAN: A Density-Based Clustering Method Based on Connected Regions with Sufficiently High Density

DBSCAN (density-based spatial clustering of applications with noise) is a density-based clustering algorithm. The algorithm grows regions with sufficiently high density into clusters and discovers clusters of arbitrary shape in spatial databases with noise. It defines a cluster as a maximal set of *density-connected* points.

The basic ideas of density-based clustering involve a number of new definitions. We intuitively present these definitions, and then follow up with an example.

- The neighborhood within a radius ε of a given object is called the **ε-neighborhood** of the object.
- An object is a **core object** if its ε-neighborhood contains at least a minimum number, *MinPts*, of objects.
- Given a set of objects, D, we say that an object p is **directly density-reachable** from object q if p is within the ε-neighborhood of q, and q is a core object.
- An object p is **density-reachable** from object q with respect to ε and *MinPts* in a set of objects, D, if there is a chain of objects $p_1, \ldots, p_n$, $p_1 = q$, and $p_n = p$ such that p_{i+1} is directly density-reachable from p_i with respect to ε and *MinPts*, for $1 \leq i \leq n$, $p_i \in D$.
- An object p is **density-connected** to object q with respect to ε and *MinPts* in a set of objects, D, if there is an object $\boldsymbol{o} \in D$ such that both p and q are density-reachable from o with respect to ε and *MinPts*.

Density reachability is the transitive closure of direct density reachability, and this relationship is asymmetric. Only core objects are mutually density reachable. Density connectivity, however, is a symmetric relation.

EXAMPLE 7.5

Consider Figure 7.14 for a given ε represented by the radius of the circles, and let $MinPts = 3$. Based on the above definitions,

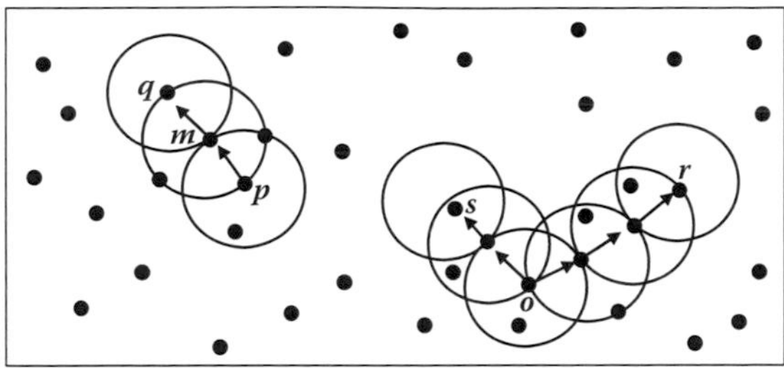

FIGURE 7.14 Density reachability and density connectivity in density-based clustering. (Based on Ester, M. et al., A density-based algorithm for discovering clusters in large spatial databases, in *Proc. 1996 Intl. Conf. Knowledge Discovery and Data Mining (KDD'96)*, Portland, OR August 1996, pp. 226–231.) Used with permission, Association for the Advancement of Artificial Intelligence. © 1996 AAAI Press.

- Of the labeled points, *m*, *p*, *o*, and *r* are core objects since each is in an ε-neighborhood containing at least three points.
- *q* is directly density-reachable from *m*. *m* is directly density-reachable from *p* and vice versa.
- *q* is (indirectly) density-reachable from *p* since *q* is directly density-reachable from *m*, and *m* is directly density-reachable from *p*. However, *p* is not density-reachable from *q* since *q* is not a core object. Similarly, *r* and *s* are density-reachable from *o*, and *o* is density-reachable from *r*.
- *o*, *r*, and *s* are all density-connected.

A **density-based cluster** is a set of density-connected objects that is maximal with respect to density-reachability. Every object not contained in any cluster is considered to be *noise*.

How does DBSCAN find clusters? DBSCAN searches for clusters by checking the ε-neighborhood of each point in the database. This process starts with an arbitrary point *p*. If the ε-neighborhood of a point *p* contains more than *MinPts*, a new cluster with *p* as a core object is created. DBSCAN then iteratively collects directly density-reachable objects from these core objects, which may involve the merge of a few density-reachable clusters. The process terminates when no new point can be added to any cluster.

If a spatial index is used, the computational complexity of DBSCAN is $O(n \log n)$, where *n* is the number of database objects. Otherwise, it is $O(n^2)$. With appropriate settings of the user-defined parameters, ε and *MinPts*, the algorithm is effective at finding arbitrary shaped clusters. However, DBSCAN still leaves the user with the responsibility of selecting parameter values that will lead to the discovery of acceptable clusters. Actually, this is a problem associated with many other clustering algorithms. Such parameter settings are usually empirically set and difficult to determine, especially for real-world, high-dimensional data sets. Most algorithms are very sensitive to such parameter values.

Many spatial databases contain extended objects such as polygons rather than points. Then, any reflexive and symmetric predicate (e.g., two polygons have a nonempty intersection) suffices to define a neighborhood. Additional measures (e.g., nonspatial attributes such as the average income of a city) can be used to define the cardinality of the neighborhood. These two generalizations lead to the algorithm GDBSCAN, which uses the two parameters *NPred* and *MinWeight*.

7.2.5.2 OPTICS: Ordering Points to Identify the Clustering Structure

An important property of many real-world data sets is that their intrinsic cluster structure cannot be characterized by *global* density parameters. Very different local densities may be needed to reveal clusters in different regions of the data space. Thus, DBSCAN may fail to find the optimal clustering when the data space has both dense and sparse regions. A possible solution is to use a density-based clustering algorithm with different parameter settings. However, there are an infinite number of possible parameter values.

To help overcome this difficulty, a cluster analysis method called **OPTICS** was proposed. Rather than produce a data set clustering explicitly, OPTICS computes an augmented *cluster ordering* for automatic and interactive cluster analysis. This ordering represents the density-based clustering structure of the data. It contains information that is equivalent to density-based clustering obtained from a wide range of parameter settings. The cluster ordering can be used to extract basic clustering information (such as cluster centers, or arbitrary-shaped clusters), as well as provide the intrinsic clustering structure.

To introduce the notion of a cluster ordering, we first make the following observation. Density-based clusters with respect to a higher density (i.e., a lower value for ε) are *completely contained* in density-connected sets obtained with respect to a lower density (i.e., a higher value for ε). Thus, in order to produce a set or ordering of density-based clusters, we can extend the DBSCAN algorithm to process a set of distance parameter values at the same time. To construct the different clusterings simultaneously, the objects should be processed in a specific order. This order selects an object that is density-reachable with respect to the lowest ε value so that clusters with higher density (lower ε) will be finished first. Based on this idea, two values need to be stored for each object—*core-distance* and *reachability-distance*:

- The **core-distance** of an object p is the smallest ε' value that makes p a core object. If p is not a core object, the core-distance of p is undefined.
- The **reachability-distance** of an object q with respect to another object p is the greater value of the core-distance of p and the Euclidean distance between p and q. If p is not a core object, the reachability-distance between p and q is undefined.

EXAMPLE 7.6

Figure 7.15 illustrates the concepts of core-distance and reachability-distance. Suppose that $\varepsilon = 6$ mm and $MinPts = 5$. The core-distance of p is the distance, ε', between p and the fourth closest data object. The reachability-distance of q_1 with

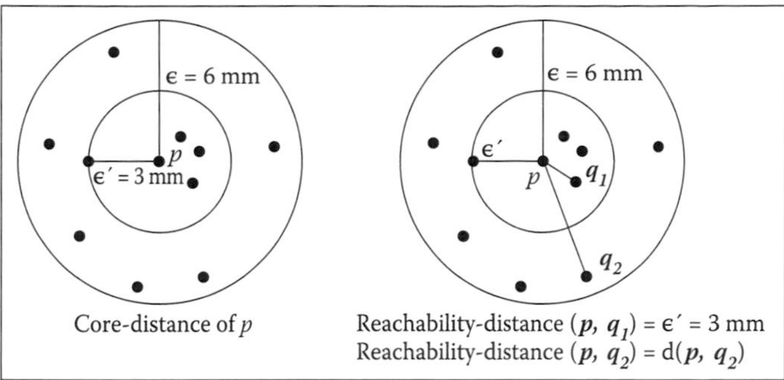

FIGURE 7.15 OPTICS terminology.

respect to p is the core-distance of p (i.e., $\varepsilon' = 3$ mm) since this is greater than the Euclidean distance from p to q_1. The reachability-distance of q_2 with respect to p is the Euclidean distance from p to q_2 since this is greater than the core-distance of p.

How are these values used? The OPTICS algorithm creates an ordering of the objects in a database, additionally storing the core-distance and a suitable reachability-distance for each object. An algorithm was proposed to extract clusters based on the ordering information produced by OPTICS. Such information is sufficient for the extraction of all density-based clusterings with respect to any distance ε' that is smaller than the distance ε used in generating the order.

The cluster ordering of a data set can be represented graphically, which helps in its understanding. For example, Figure 7.16 is the reachability plot for a simple two-dimensional data set, which presents a general overview of how the data are structured and clustered. The data objects are plotted in cluster order (horizontal axis) together with their respective reachability-distance (vertical axis). The three Gaussian "bumps" in the plot reflect three clusters in the data set. Methods have also been developed for viewing clustering structures of high-dimensional data at various levels of detail.

Due to the structural equivalence of the OPTICS algorithm to DBSCAN, the OPTICS algorithm has the same run-time complexity as that of DBSCAN, i.e., $O(n \log n)$ if a spatial index is used, where n is the number of objects.

7.2.5.3 DENCLUE: Clustering Based on Density Distribution Functions

DENCLUE (DENsity-based CLUstEring) is a clustering method based on a set of density distribution functions. The method is built on the following ideas: (1) the influence of each data point can be formally modeled using a mathematical function, called an *influence function*, which describes the impact of a data point within its neighborhood; (2) the overall density of the data space can be modeled analytically as the sum of the influence function applied to all data points; and (3) clusters can then be determined mathematically by identifying *density attractors*, where density attractors are local maxima of the overall density function.

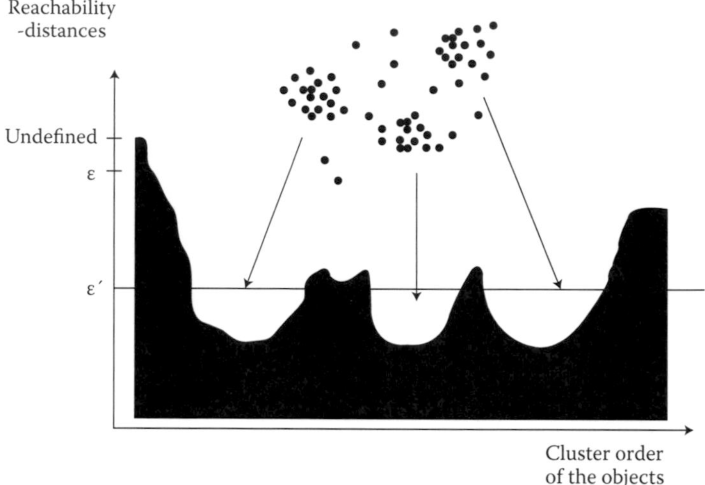

FIGURE 7.16 Cluster ordering in OPTICS. (Based on Ankerst, M. et al., OPTICS: Ordering points to identify the clustering structure, in *Proc. ACM-SIGMOD Intl. Conf. Management of Data (SIGMOD'99),* Philadelphia, PA, June 1999, pp. 49–60.) © 1999 ACM, Inc. Reprinted by permission.

Let x and y be objects or points in F^d, a d-dimensional input space. The **influence function** of data object y on x is a function, $f_B^Y : F^d \to R_0^+$, which is defined in terms of a basic influence function f_B:

$$f_B^Y(x) = f_B(x, y) \tag{7.18}$$

This reflects the impact of y on x. In principle, the influence function can be an arbitrary function that can be determined by the distance between two objects in a neighborhood. The distance function, $d(x, y)$, should be reflexive and symmetric, such as the Euclidean distance function. Two examples of an influence function are a *square wave influence function* and a *Gaussian influence function*:

$$f_{square}(x, y) = \begin{cases} 0 & \text{if } d(x, y) > \sigma \\ 1 & \text{Otherwise} \end{cases} \tag{7.19}$$

$$f_{Gauss}(x, y) = e^{-\frac{d(x,y)^2}{2\sigma^2}} \tag{7.20}$$

The **density function** at an object or point $x \in F^d$ is defined as the sum of influence functions of all data points. That is, it is the total influence on x of all of the data points. Given n data objects, $D = \{x_1, \ldots, x_n\} = \subset F^d$, the density function at x is defined as

$$f_B^D(x) = \sum_{i=1}^{n} f_B^{x_i}(x) = f_B^{x_1}(x) + f_B^{x_2}(x) + \cdots + f_B^{x_n}(x) \tag{7.21}$$

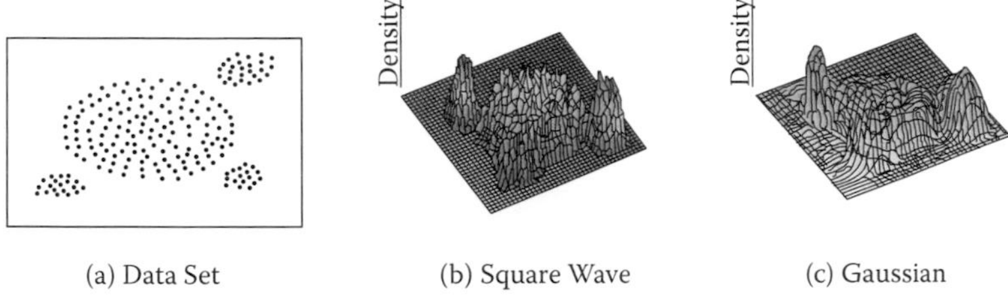

(a) Data Set (b) Square Wave (c) Gaussian

FIGURE 7.17 Possible density functions for a 2-D data set. (From Hinneburg, A. and Keim, D.A., An efficient approach to clustering in large multimedia databases with noise, in *Proc. 1998 Intl. Conf. Knowledge Discovery and Data Mining (KDD'98),* New York, August 1998, pp. 58–65.) Used with permisson, Association for the Advancement of Artificial Intelligence, © 1998 AAAI Press.

For example, the density function that results from the Gaussian influence function (7.20) is

$$f^D_{Gauss}(x) = \sum_{i=1}^{n} e^{-\frac{d(x,x_i)^2}{2\sigma^2}} \tag{7.22}$$

Figure 7.17 shows an example of a set of data points in a 2-dimensional space, together with the corresponding overall density functions for the square wave influence function and the Gaussian influence function.

From the density function, one can define the *gradient* function at a point x which is in fact a vector that indicates the strength and direction where most of x's influence comes from. The density function is also used to locate the *density attractor,* which is the local maxima of the overall density function. A point x is said to be *density attracted* to a density attractor x^* if there exists a set of points $x_0, x_1, \ldots, x_k$ such that $x_0 = x$, $x_k = x^*$ and the gradient of x_{i-1} is in the direction of x_i for $0 < i < k + 1$. For a continuous and differentiable influence function, a hill-climbing algorithm guided by the gradient can be used to determine the density attractor of a set of data points.

In general, points that are density attracted to x^* may form a cluster. Based on the above notions, both *center-defined clusters* and *arbitrary-shape clusters* can be formally defined. A **center-defined cluster** for a density attractor, x^*, is a subset of points, $C \subseteq D$, that are *density-attracted* by x^*, and where the density function at x^* is no less than a threshold, ξ. Points that are density-attracted by x^*, but for which the density function value is less than ξ, are considered outliers. That is, intuitively, points in a cluster are influenced by many points, but outliers are not. An **arbitrary-shape cluster** for a set of density attractors is a set of C's, each being density-attracted to its respective density-attractor, where (1) the density function value at each density-attractor is no less than a threshold, ξ, and (2) there exists a path, P, from each density-attractor to another, where the density function value for each point along the path is no less than ξ. Examples of center-defined and arbitrary-shape clusters are shown in Figure 7.18.

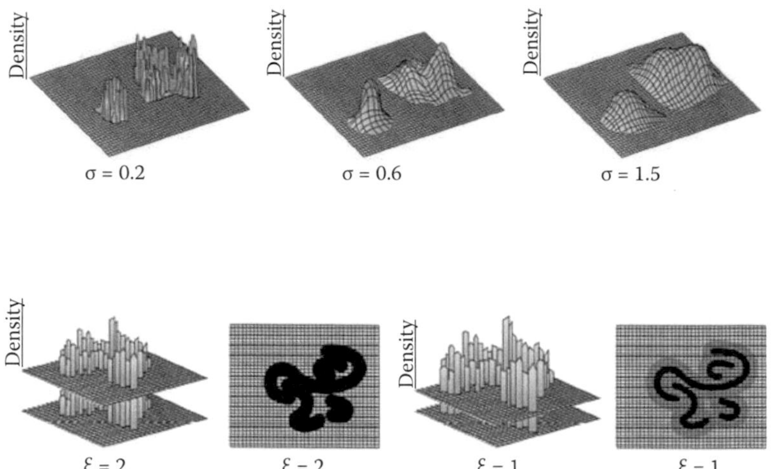

FIGURE 7.18 Examples of center-defined clusters (top row) and arbitrary-shape clusters (bottom row). (From Hinneburg, A. and Keim, D. A., An efficient approach to clustering in large multimedia databases with noise, in *Proc. 1998 Intl. Conf. Knowledge Discovery and Data Mining (KDD'98),* New York, August 1998, pp. 58–65 .) Used with permisson, Association for the Advancement of Artificial Intelligence, © 1998 AAAI Press.

7.3 CLUSTERING OF TRAJECTORY DATA

In this section, we present an overview of clustering methods developed for trajectory data. First, we introduce an application of trajectory data clustering in Section 7.3.1. Then, we explain the major clustering methods. Notice that very few clustering methods have been developed thus far. We study earlier methods that do not use trajectory partitioning in Section 7.3.2. These methods, however, have limited capability in discovering clusters of *sub*-trajectories. Very recently, the partition-and-group framework [15] has been proposed to overcome this drawback. As the final topic, in Section 7.3.3 we study the partition-and-group framework.

7.3.1 Application: Tropical Cyclone Trajectory Analysis

Tropical cyclones (TCs) are crucial dynamical ingredients of the atmospheric circulation, directly impacting local weather. A better understanding of the behavior of TCs in the context of climate variability and change could have important societal implications. Cluster analysis of TC trajectories provides a natural way to extract the common behavior of TCs.

Camargo et al. [5] and Gaffney et al. [9] performed cluster analysis of TC trajectories. They applied a probabilistic clustering technique [10] to the best track data set of the Joint Typhoon Warning Center for the period 1950 to 2002. The best track contains the TC's latitude, longitude, maximum sustained surface wind,

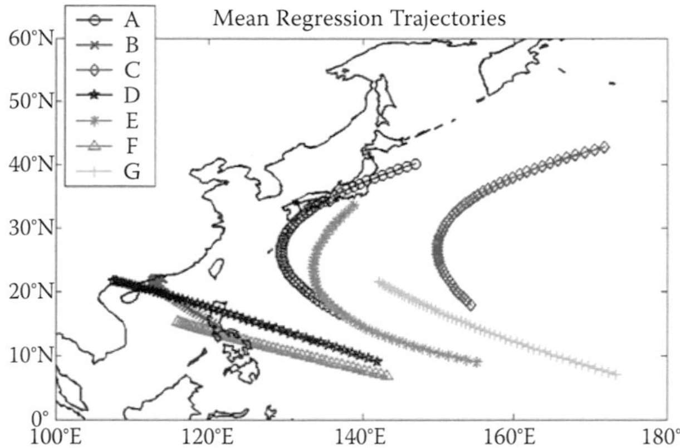

FIGURE 7.19 Mean regression trajectories relative to the initial positions. (From Camargo, S. et al., Cluster analysis of western north pacific tropical cyclone tracks, in *Technical Report*, International Research Institute for Climate and Society, Columbia University, 2005.)

and minimum sea-level pressure at 6-hour intervals. Only TCs with tropical storm intensity or higher were included, a total of 1393 TCs. The aim of this analysis is to identify different types of track, their seasonality, and their relationship to the large-scale circulation and El Niño-Southern Oscillation (ENSO).

The clustering technique consists of building a mixture of polynomial regression models (i.e., curves), which are used to fit the geographical "shape" of trajectories. The model is fit to the data by maximizing the likelihood of the parameters, given the data set. The mixture model framework allows the clustering problem to be posed in a rigorous probabilistic context. This technique will be explained in Section 7.3.2.1.

Figure 7.19 shows seven underlying regression models, relative to their starting points. Two main trajectory types are "straight-movers" and "recurvers." Two tropical cyclone clusters A and G are are shown in Figure 7.20. Thin lines are trajectories, and thick lines are mean regression trajectories. The main variables analyzed per cluster

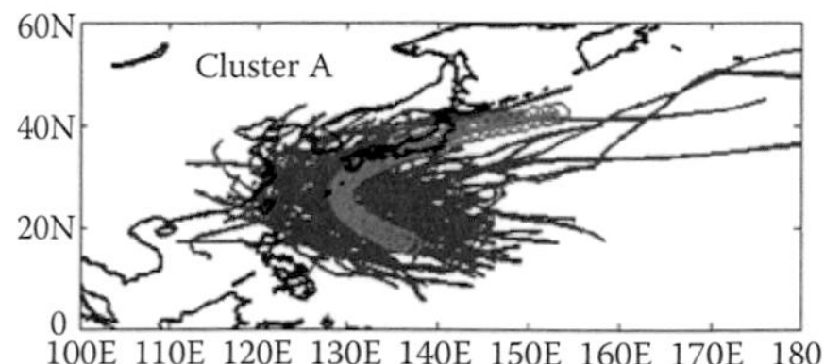

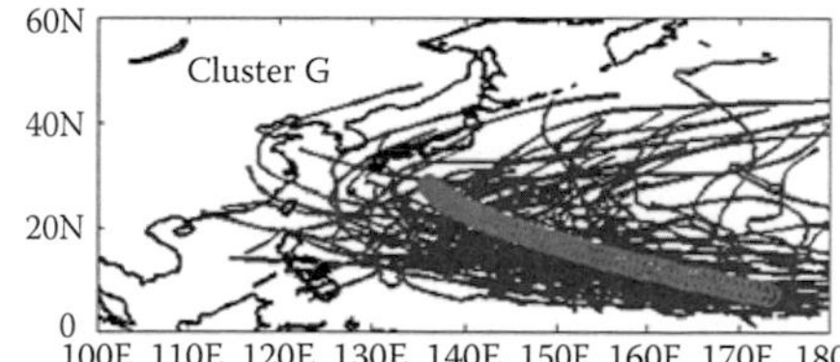

FIGURE 7.20 Two tropical cyclone clusters A and G in Figure 7.19. (From Camargo, S. et al., Cluster analysis of western north pacific tropical cyclone tracks, in *Technical Report*, International Research Institute for Climate and Society, Columbia University, 2005.)

are the number of tropical cyclones; number of tropical cyclones with tropical storm (TS), typhoon (TY—Dvorak's scale 1–2), and super-typhoon (STY—Dvorak's scale 3–5) intensities; location and distribution of first position (genesis); track types and density; ACE (accumulated cyclone energy); and lifetime.

7.3.2 Whole Trajectory Clustering

7.3.2.1 Probabilistic Method

This method is based on *probabilistic clustering*. In probabilistic clustering, it is assumed that the data are being produced in the following "generative" manner [10]:

1. An individual is drawn randomly from the population of interest.
2. The individual has been assigned to a cluster k with probability w_k, $\Sigma_{k=1}^{K} w_k = 1$. These are the "prior" weights on the K clusters.
3. Given that an individual belongs to a cluster k, there is a density function $f_k(y_j \mid \theta_k)$ which generates an observed data item y_j for the individual j.

From the generative model above, the probability density function of observed trajectories is a mixture density, Equation (7.23), where $f_k(y_j|x_j, \theta_k)$ is the density component, w_k is the weight, and θ_k is the set of parameters for the k-th component.

$$P(y_j|x_j,\theta) = \sum_{k}^{K} f_k(y_j|x_j,\theta_k)w_k \tag{7.23}$$

Due to conditional independence between trajectories, they are regarded as a random sample from a population of individuals. This property allows the full joint density to be written as Equation (7.24). The log-likelihood of the parameters θ given the data set S, Equation (7.25), is directly derived from Equation (7.24).

$$P(Y|X,\theta) = \prod_{j}^{M} \sum_{k}^{K} w_k \prod_{j}^{n_j} f_k(y_j(i) \mid x_j(i),\theta_k) \tag{7.24}$$

$$\mathcal{L}(\theta|S) = \sum_{j}^{M} \log \sum_{k}^{K} w_k \prod_{i}^{n_j} f_k(y_j(i) \mid x_j(i),\theta_k) \tag{7.25}$$

Here, θ_k and w_k can be estimated from the trajectory data using the EM algorithm. The EM algorithm allows us to estimate the hidden data so that the log-likelihood $\mathcal{L}(\theta| S)$ is guaranteed to never decrease. The *expectation* and *maximization* steps are repeated until a stopping criterion is satisfied. In practice, the procedure is stopped when the marginal change in the log-likelihood falls below a certain threshold. Finally, the estimated density components $f_k(y_j|x_j, \theta_k)$ are interpreted as clusters. Please refer to the original paper [10] for details of the EM algorithm.

7.3.2.2 Density-Based Method

This method relies on *density-based clustering*. In this method, a trajectory is represented as a sequence of the location and timestamp. Then, the distance between two trajectories τ_1 and τ_2 is defined as the average distance between objects, Equation (7.26), where $d()$ is the Euclidean distance over $\mathbb{R}^2$, T is the time interval over which the trajectories τ_1 and τ_2 exist, and $\tau_i(t)$ ($i \in \{1, 2\}$) is the position of the object of τ_i at time t. Using this distance function, the density-based clustering method *OPTICS* (explained in Section 7.2.5.2) is applied to clustering of trajectory data. This method is called **T-OPTICS**.

$$D(\tau_1, \tau_2)|_T = \frac{\int_T d(\tau_1(t), \tau_2(t))dt}{|T|} \tag{7.26}$$

Figure 7.21 shows an execution result of T-OPTICS over a synthetic data set. The data set in Figure 7.21(a) is composed of 250 trajectories organized into four natural clusters plus noise. As indicated by the resulting reachability plot in Figure 7.21(b), T-OPTICS finds the four natural clusters, which can be easily isolated by selecting a proper value for the ε parameter ($\varepsilon = 24$ in this example).

Next, the concept of **temporal focusing** is introduced. In a real environment, not all time intervals have the same importance. A meaningful example is urban traffic: in rush hours, many people move from home to work, and vice versa. Thus, it would be interesting if trajectory clustering is performed with a focus on the temporal dimension. In other words, clustering trajectories only in meaningful time intervals can produce more interesting results. An algorithm called **TF-OPTICS** is presented for this temporal focusing. TF-OPTICS aims at searching *the most meaningful time intervals*, which allows us to isolate the (density-based) clusters of higher quality.

The first issue is to define a quality function. The quality function used in this work takes account of both high-density clusters and low-density noise. To compute the quality function, the reachability plot generated by OPTICS is used. $R(D, I, \varepsilon')$

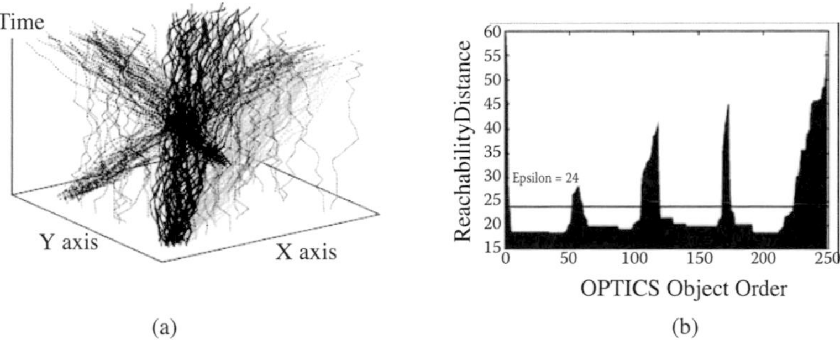

FIGURE 7.21 An execution result of T-OPTICS. (a) A synthetic data set. (b) The corresponding reachability plot. (From Nanni, M. and Pedreschi, D., Time-focused clustering of trajectories of moving objects, *Journal of Intelligent Information Systems*, 27: 267–289, 2006.) © 2006 Springer.

denotes the average reachability-distance of nonnoise objects, where D is an input data set, I is a time interval, and ε' is a density threshold parameter. Then, the quality measure Q_1 is defined as $Q_1(D, I, \varepsilon') = -R(D, I, \varepsilon')$. Dispersed clusters yield high reachability distances, and thus highly negative values of Q_1, whereas compact clusters yield values of Q_1 closer to zero. In practice, a variation of Q_1 is used to promote larger intervals as follows: $Q_2(D, I, \varepsilon') = Q_1(D, I, \varepsilon') / \log_{10}(10 + |I|)$.

A basic approach for finding the time interval that maximizes Q_2 is an exhaustive search over all possible intervals. Such an approach is obviously very expensive. A natural alternative is to find a local optimum, adopting a greedy search paradigm. The procedure is described in the following, where t_0 is the chosen temporal granularity, and I_{ALL} is the largest time interval.

1. Choose an initial random time interval $I \subseteq I_{ALL}$.
2. Let $I' = \text{argmax}_{T \in NeighI}, Q_2(D, T, \varepsilon')$, where $I = [T_s, T_e]$ and $Neigh_I = \{[T_s \pm t_0, T_e], [T_s, T_e \pm t_0]\}$.
3. If $Q_2(D, I', \varepsilon') > Q_2(D, I, \varepsilon')$, then let $I := I'$ and return to Step 2; otherwise, stop.

7.3.3 Partial Trajectory Clustering: The Partition-and-Group Framework

A key observation is that clustering trajectories *as a whole* could not detect *similar portions* of the trajectories. Even though some portions of trajectories show a common behavior, the whole trajectories may not. For example, consider the five trajectories in Figure 7.22. It is obvious that there is a common behavior, denoted by the thick arrow, in the dotted rectangle. In earlier clustering methods, however, this common behavior cannot be detected because the trajectories move to totally different directions.

The solution is to partition a trajectory into a set of line segments and then group similar line segments. This framework is called **partition-and-group** framework. The primary advantage of the partition-and-group framework is the discovery of common *sub*-trajectories from a trajectory database. As indicated by its name, trajectory clustering based on this framework consists of two phases: the **partitioning** and **grouping** phases. We give an overview of the partition-and-group framework in Section 7.3.3.1. Then, we explain the partitioning and grouping phases in Sections 7.3.3.2 and 7.3.3.3.

7.3.3.1 Overall Procedure

We first define necessary terminology. A *trajectory* is a sequence of multidimensional points. A *cluster* is a set of trajectory partitions. A *trajectory partition* is a line

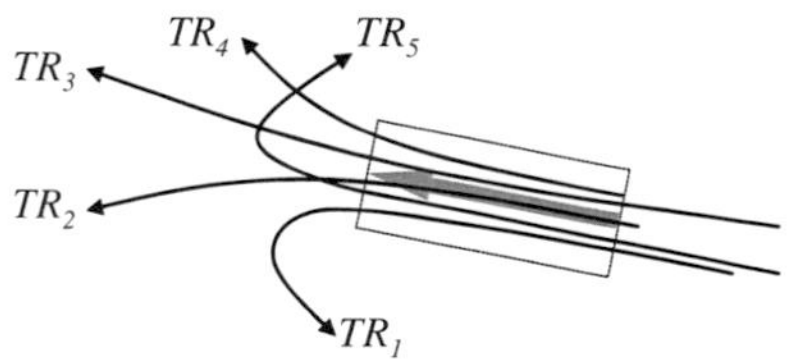

FIGURE 7.22 An example of a common sub-trajectory.

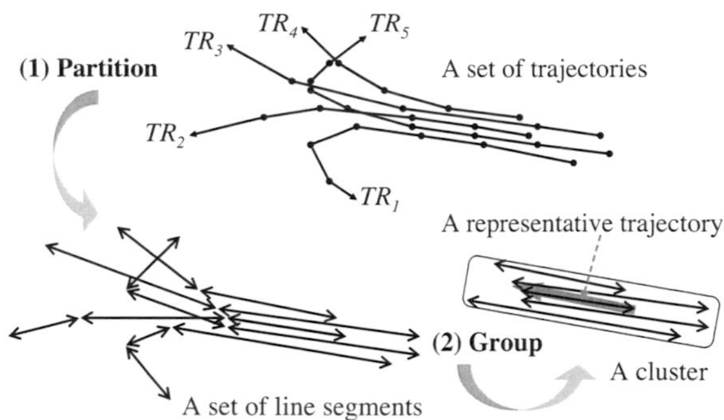

FIGURE 7.23 The overall procedure of trajectory clustering in the partition-and-group framework. (From Lee, J.-G., Han, J., and Whang, K.-Y., Trajectory clustering: A partition-and-group framework, in *Proc. 2007 ACM-SIGMOD Intl. Conf. Management of Data (SIGMOD'07)*, Beijing, China, June 2007, pp. 593–604.) © 2007 ACM, Inc. Reprinted by permission.

segment $p_i p_j$ $(i < j)$, where p_i and p_j are the points chosen from the same trajectory. A *representative trajectory* is an imaginary trajectory that indicates the major behavior of the trajectory partitions (i.e., line segments) belonging to the cluster.

Figure 7.23 shows the overall procedure of trajectory clustering in the partition-and-group framework. First, each trajectory is partitioned into a set of line segments. Second, line segments that are close to each other according to the distance measure are grouped together into a cluster. The distance measure is composed of three components: the *perpendicular* ($d_{\perp}$), *parallel* ($d_{||}$), and *angle* (d_{θ}) distances. Then, a representative trajectory is generated for each cluster.

Figure 7.24 shows the skeleton of the trajectory algorithm **TRACLUS**. Notice that TRACLUS can discover the superset of the clusters that can be discovered by earlier methods.

Algorithm: TRACLUS

Input: A set of trajectories $I = \{TR_1, \cdots, TR_n\}$.

Output: A set of clusters $O = \{ C_1, \cdots, C_m \}$ with representative trajectories.

Method:

```
    /* Partitioning Phase */
(1) for each TR ∈ I do
(2)     Partition TR into a set L of line segments and accumulate L
            into a set D;
    /* Grouping Phase */
(3) Group D into a set O of clusters;
(4) for each C ∈ O do
(5)     Generate a representative trajectory for C;
```

FIGURE 7.24 The skeleton of the trajectory clustering algorithm TRACLUS.

7.3.3.2 The Partitioning Phase

What properties should be satisfied in trajectory partitioning? The optimal partitioning of a trajectory should possess two desirable properties: *preciseness* and *conciseness*. Preciseness means that the difference between a trajectory and a set of its trajectory partitions should be as small as possible. Conciseness means that the number of trajectory partitions should be as small as possible. Preciseness and conciseness are contradictory to each other. Hence, it is required to find an optimal tradeoff between the two properties.

How is the optimal tradeoff between preciseness and conciseness found? The proposed method uses the minimum description length (MDL) principle. The MDL cost consists of two components: $L(H)$ and $L(D|H)$. Here, H means the hypothesis, and D the data. The two components are informally stated as follows [11]: "$L(H)$ is the length, in bits, of the description of the hypothesis; and $L(D|H)$ is the length, in bits, of the description of the data when encoded with the help of the hypothesis." The best hypothesis H to explain D is the one that minimizes the sum of $L(H)$ and $L(D|H)$.

The MDL principle fits the trajectory partitioning problem very well. A set of trajectory partitions corresponds to H, and a trajectory to D. Most importantly, $L(H)$ measures conciseness, and $L(D|H)$ preciseness. Thus, finding the optimal trajectory partitioning translates to finding the best hypothesis using the MDL principle.

Figure 7.25 shows the formulation of $L(H)$ and $L(D|H)$. $L(H)$ is formulated by Equation (7.27). $L(H)$ represents the sum of the length of a trajectory partition. On the other hand, $L(D|H)$ is formulated by Equation (7.28). $L(D|H)$ represents the sum of the difference between a trajectory and a trajectory partition. The sum of the perpendicular distance and the angle distance is considered to measure this difference.

$$L(H) = \sum_{j=1}^{par_i - 1} \log_2(len(p_{c_j} p_{c_{j+1}})) \tag{7.27}$$

$$L(D|H) = \sum_{j=1}^{par_i - 1} \sum_{k=c_j}^{c_{j+1}-1} \{\log_2(d_\perp(p_{i_j} p_{c_{j+1}}, p_k p_{k+1})) + \log_2(d_\theta(p_{c_j} p_{c_{j+1}}, p_k p_{k+1}))\} \tag{7.28}$$

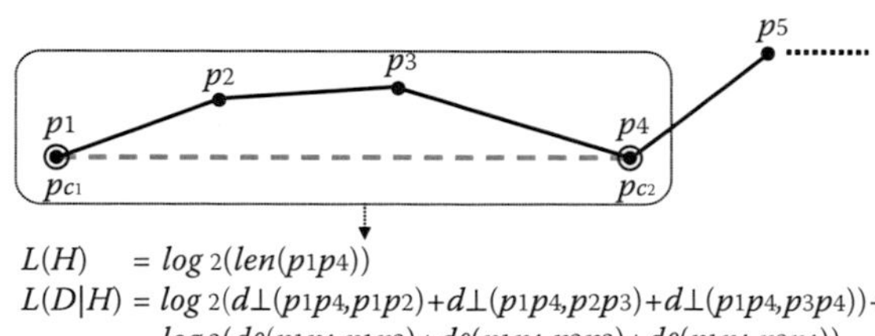

FIGURE 7.25 Formulation of the MDL cost.

The optimal trajectory partitioning is generated so as to minimize $L(H) + L(D|H)$. This is exactly the tradeoff between preciseness and conciseness. In practice, an $O(n)$ greedy algorithm is adopted for the sake of efficiency.

7.3.3.3 The Grouping Phase

Which clustering methods are the most suitable for line segment clustering? Density-based clustering methods are the most suitable for this purpose because they can *discover clusters of arbitrary shape* and can *filter out noises*. We can easily see that line-segment clusters are usually of arbitrary shape, and a trajectory database typically contains a large amount of noise (i.e., outliers).

Among density-based methods, the algorithm DBSCAN [7] is adopted. The definitions for points, originally proposed in the context of DBSCAN, are changed to those for line segments. The procedure for line-segment clustering is summarized as follows.

1. Select an unprocessed line segment L.
2. Retrieve all line segments density-reachable from L w.r.t. ε and *MinLns*. If L is a core line segment, a cluster is formed. Otherwise, L is marked as a noise.
3. Continue this process until all line segments have been processed.
4. Filter out clusters whose trajectory partitions have been extracted from too few trajectories.

How is a representative trajectory generated? Figure 7.26 illustrates the approach of generating a representative trajectory. A representative trajectory is a sequence of points obtained by a *sweep line* approach. While sweeping a vertical line across line segments in the direction of the *major axis* of a cluster, the number of the line segments hitting the sweep line is counted. If this number is equal to or greater than *MinLns*, the average coordinate of those line segments with respect to the major axis is inserted into the representative trajectory; otherwise, the current point (e.g., the 5th and 6th positions in Figure 7.26) is skipped. Besides, if a previous point is located too close (e.g., the 3rd position in Figure 7.26), then the current point is skipped to smooth the representative trajectory.

7.3.3.4 Clustering Result

Figure 7.27 shows the clustering result for the Atlantic hurricane (1950 to 2004) data set. The data set contains 570 trajectories and 17736 points. The result in Figure 7.27

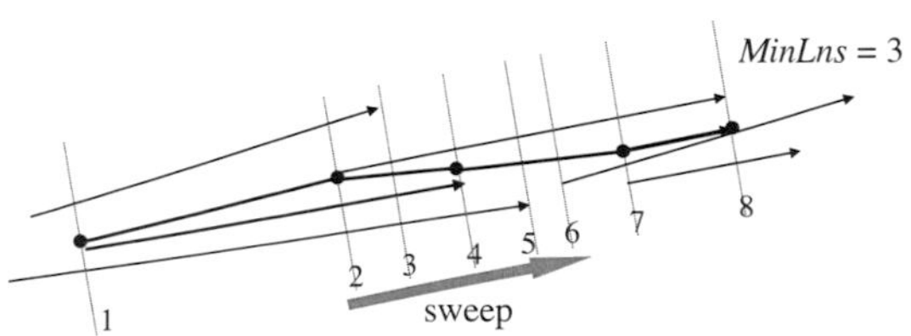

FIGURE 7.26 An example of a cluster and its representative trajectory.

FIGURE 7.27 Clustering result for the Atlantic hurricane data. (From Lee, J.-G., Han, J., and Whang, K.-Y., Trajectory clustering: A partition- and -group framework, in *Proc. 2007 ACM-SIGMOD Intl. Conf. Management of Data (SIGMOD'07)*, Beijing, China, June 2007, pp. 593–604. © 2007 ACM, Inc. Reprinted by permission.

is quite reasonable. We know that some hurricanes move along a curve, changing their direction from east-to-west to south-to-north, and then to west-to-east. On the other hand, some hurricanes move along a straight east-to-west line or a straight west-to-east line. The lower horizontal cluster represents the east-to-west movements, the upper horizontal one the west-to-east movements, and the vertical ones the south-to-north movements.

7.4 SUMMARY

We have presented an overview of clustering methods in geographic data analysis. In the first part, the major clustering methods for **point** data are presented. Partitioning methods, hierarchical methods, and density-based methods are often used for geographic data.

- A **partitioning method** first creates an initial set of k partitions, where the parameter k is the number of partitions to construct. It then uses an *iterative relocation technique* that attempts to improve the partitioning by moving objects from one group to another. Typical partitioning methods include k-means, k-medoids, CLARANS, and their improvements. Whereas k-means and k-medoids perform hard clustering, the EM algorithm performs *fuzzy* clustering: each object is assigned to *each* cluster according to a weight representing its probability membership.

- A **hierarchical method** creates a hierarchical decomposition of a given set of data objects. The methods can be classified as being either *agglomerative* (*bottom-up*) or *divisive* (*top-down*), based on how the hierarchical decomposition is formed. To compensate for the rigidity of *merge* or *split*, the quality of hierarchical agglomeration can be improved by analyzing object linkages at each hierarchical partitioning (such as in Chameleon), or by first performing *microclustering* and then operating on the microclusters with other clustering techniques, such as iterative relocation (as in BIRCH).
- A **density-based method** clusters objects based on the notion of density. It either grows clusters according to the density of neighborhood objects (such as in DBSCAN) or according to some density function (such as in DENCLUE). OPTICS is a density-based method that generates an augmented ordering of the clustering structure of the data.

In the second part, the major clustering methods for **trajectory** data are presented. They are categorized depending on whether they can discover clusters of *sub*-trajectories.

- A **probabilistic method** clusters trajectories based on a regression mixture model. The EM algorithm is employed to determine cluster memberships. A **density-based method** applies OPTICS to trajectory clustering. To achieve this, the distance function is defined between two trajectories. These methods have limited capability in discovering clusters of sub-trajectories because they cluster trajectories *as a whole*.
- The **partition-and-group** framework goes through two phases. In the *partitioning* phase, each trajectory is optimally partitioned into a set of line segments. These line segments are provided to the next phase. In the *grouping* phase, similar line segments are grouped into a cluster. Here, a variation of DBSCAN is exploited. The clustering algorithm *TRACLUS* is implemented based on this framework. TRACLUS allows us to discover clusters of sub-trajectories.

REFERENCES

[1] R. Agrawal, J. Gehrke, D. Gunopulos, and P. Raghavan. Automatic subspace clustering of high dimensional data for data mining applications. In *Proc. 1998 ACM-SIGMOD Intl. Conf. Management of Data (SIGMOD'98)*, Seattle, WA, June 1998, pp. 94–105.

[2] C. Ambroise and G. Govaert. Convergence of an EM-type algorithm for spatial clustering. *Pattern Recognition Letters*, 19:919–927, 1998.

[3] M. Ankerst, M. Breunig, H.-P. Kriegel, and J. Sander. OPTICS: Ordering points to identify the clustering structure. In *Proc. 1999 ACM-SIGMOD Intl. Conf. Management of Data (SIGMOD'99)*, Philadelphia, PA, June 1999, pp. 49–60. (http://doi.acm.org/10.1145/304182.304187).

[4] J. A. Bilmes. A gentle tutorial of the EM algorithm and its applications to parameter estimation for Gaussian mixture and hidden Markov models. In *Technical Report*, International Computer Science Institute, 1998.

[5] S. Camargo, A. Robertson, S. Gaffney, P. Smyth, and M. Ghil. Cluster analysis of western north pacific tropical cyclone tracks. In *Technical Report*, International Research Institute for Climate and Society, Columbia University, 2005.

[6] A. Dempster, N. Laird, and D. Rubin. Maximum likelihood from incomplete data via the EM algorithm. *J. Royal Statistical Society*, 39:1–38, 1977.

[7] M. Ester, H.-P. Kriegel, J. Sander, and X. Xu. A density-based algorithm for discovering clusters in large spatial databases. In *Proc. 1996 Intl. Conf. Knowledge Discovery and Data Mining (KDD'96)*, Portland, OR, August 1996, pp. 226–231.

[8] V. Estivill-Castro and I. Lee. Amoeba: Hierarchical clustering based on spatial proximity using delaunaty diagram. In *Proc. 9th Intl. Symp. Spatial Data Handling (SDH'00)*, Beijing, China, August 2000, pp. 26–41.

[9] S. Gaffney, A. Robertson, P. Smyth, S. Camargo, and M. Ghil. Probabilistic clustering of extratropical cyclones using regression mixture models. In *Technical Report*, Bren School of Information and Computer Sciences, University of California, Irvine, 2006.

[10] S. Gaffney and P. Smyth. Trajectory clustering with mixtures of regression models. In *Proc. 1999 Intl. Conf. Knowledge Discovery and Data Mining (KDD'99)*, San Diego, CA, August 1999, pp. 63–72.

[11] P. Grünwald, I. J. Myung, and M. Pitt. *Advances in Minimum Description Length: Theory and Applications*. MIT Press, 2005.

[12] A. Hinneburg and D. A. Keim. An efficient approach to clustering in large multimedia databases with noise. In *Proc. 1998 Intl. Conf. Knowledge Discovery and Data Mining (KDD'98)*, New York, August 1998 , pp. 58–65.

[13] G. Karypis, E.-H. Han, and V. Kumar. CHAMELEON: A hierarchical clustering algorithm using dynamic modeling. *Computer*, 32:68–75, 1999.

[14] L. Kaufman and P. J. Rousseeuw. *Finding Groups in Data: An Introduction to Cluster Analysis*. John Wiley & Sons, 1990.

[15] J.-G. Lee, J. Han, and K.-Y. Whang. Trajectory clustering: A partition-and-group framework. In *Proc. 2007 ACM-SIGMOD Intl. Conf. Management of Data (SIGMOD'07)*, Beijing, China, June 2007, pp. 593–604. (http://doi.acm.org/10.1145/1247480.1247546).

[16] S. P. Lloyd. Least Squares Quantization in PCM. *IEEE Trans. Information Theory*, 28:128-137, 1982, (original version: Technical Report, Bell Labs, 1957).

[17] M. Nanni and D. Pedreschi. Time-focused clustering of trajectories of moving objects. *Journal of Intelligent Information Systems*, 27:267–289, 2006.

[18] R. Ng and J. Han. Efficient and effective clustering method for spatial data mining. In *Proc. 1994 Intl. Conf. Very Large Data Bases (VLDB'94)*, Santiago, Chile, September 1994, pp. 144–155.

[19] J. Sander, M. Ester, H.-P. Kriegel, and X. Xu. Density-based clustering in spatial databases: The algorithm GDBSCAN and its applications. *Data Mining and Knowledge Discovery*, 2:169–194, 1998.

[20] G. Sheikholeslami, S. Chatterjee, and A. Zhang. WaveCluster: A multiresolution clustering approach for very large spatial databases. In *Proc. 1998 Intl. Conf. Very Large Data Bases (VLDB'98)*, New York, August 1998, pp. 428–439.

[21] L.-H. Wan, Y.-J. Li, W.-Y. Liu, and D.-Y. Zhang. Application and study of spatial cluster and customer partitioning. In *Proc. 2005 Intl. Conf. Machine Learning and Cybernetics*, Guangzhou, China, August 2005, pp. 1701–1706.

[22] W. Wang, J. Yang, and R. Muntz. STING: A statistical information grid approach to spatial data mining. In *Proc. 1997 Intl. Conf. Very Large Data Bases (VLDB'97)*, Athens, Greece, August 1997, pp. 186–195.

[23] T. Zhang, R. Ramakrishnan, and M. Livny. BIRCH: an efficient data clustering method for very large databases. In *Proc. 1996 ACM-SIGMOD Intl. Conf. Management of Data (SIGMOD'96)*, Montreal, Canada, June 1996, pp. 103–114.

8 Computing Medoids in Large Spatial Datasets

Kyriakos Mouratidis
Dimitris Papadias
Spiros Papadimitriou

CONTENTS

8.1 INTRODUCTION

In this chapter, we consider a class of queries that arise in spatial decision making and resource allocation applications. Assume that a company wants to open a number of warehouses in a city. Let P be the set of residential blocks in the city. P represents customer locations to be potentially served by the company. At the same time, P also comprises the candidate warehouse locations because the warehouses themselves must be opened in some residential blocks. In this context, an analyst may ask any of the following questions:

Q1. ***k*-Medoid query:** If the number k of warehouses is known, in which residential blocks should they be opened, so that the average distance from each location in P to its closest warehouse is minimized?

Q2. **Medoid-aggregate query:** If the average distance should be around a given value, what is the smallest number of warehouses (and their locations) that best approximates this value?
Q3. **Medoid-optimization query:** If the warehouse opening/maintenance overhead and the transportation cost per mile are given, what is the number of warehouses (and their locations) that minimizes the total cost?

The warehouse locations correspond to the *medoids*. Since the k-medoid problem (Q1) is NP-hard (Garey and Johnson, 1979), research has focused on approximate algorithms, most of which are suitable only for datasets of small and moderate sizes. On the contrary, this chapter focuses on very large databases. In addition to conventional k-medoids, we introduce and solve the alternative queries Q2 and Q3, which have practical relevance.

To formalize, given a set P of data points, we wish to find a set of medoids $R \subseteq P$, subject to certain optimization criteria. The average (*avg*) Euclidean distance $\|p - r(p)\|$ between each point $p \in P$ and its closest medoid $r(p) \in R$ is denoted by

$$C(R) = \frac{1}{|P|} \sum_{p \in P} \|p - r(p)\|.$$

Letting $|R|$ represent the cardinality of R, the *k-medoid query* can be formally stated as: "Given dataset P and integer parameter k, find $R \subseteq P$, such that $|R| = k$ and $C(R)$ is minimized." Figure 8.1 shows an example, where the dots represent points in P (e.g., residential blocks), $k = 3$ and $R = \{h, o, t\}$. The three medoids h, o, t are candidate locations for service facilities (e.g., warehouses or distribution centers), so that the average distance $C(R)$ from each block to its closest facility is minimized.

The *medoid-aggregate* (MA) query is defined as: "Given P and a value T, find $R \subseteq P$, such that $|R|$ is minimized and $C(R) \approx T$." In other words, k is not specified in advance. Instead, a target value T for the average distance is given, and we want to select a minimal set R of medoids, such that $C(R)$ best approximates T. Finally, the *medoid-optimization* (MO) query is formalized as: "Given P and a cost function f that is monotonically increasing with both the number of medoids $|R|$ and with $C(R)$,

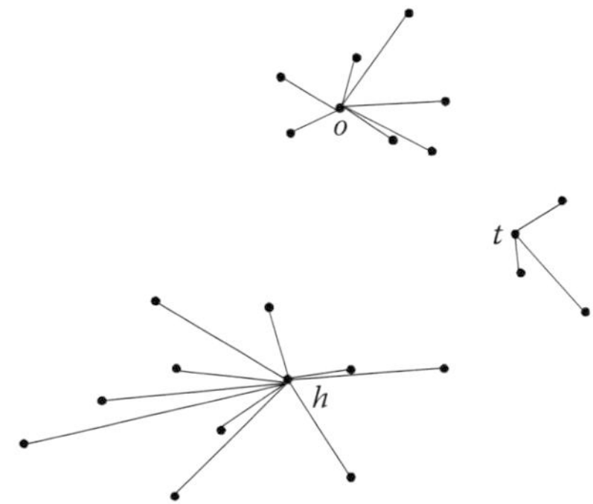

FIGURE 8.1 Example of 3-medoids.

find $R \subseteq P$ such that $f(C(R), |R|)$ is minimized." For example, in Q3 above, function f may be defined as $f(C(R), |R|) = C(R) + Cost_{pm} \times |R|$, where $Cost_{pm}$ is the opening/maintenance cost per warehouse. The goal is to achieve the best tradeoff between the number of warehouses and the average distance achieved.

Interesting variants of the above three query types arise when the quality of a medoid set is determined by the maximum distance between the input points and their closest medoid; i.e., when

$$C(R) = max_{p \in P} \|p - r(p)\| .$$

For instance, the company in our example may want to minimize the maximum distance (instead of the average one) between the residential blocks and their closest warehouse, potentially achieving a desired $C(R)$ with the minimal set of warehouses (MA), or minimizing a cost function (MO).

In this chapter, we present *Tree-based PArtition Querying* (TPAQ) (Mouratidis, Papadias, and Papadimitriou, 2008), a methodology that can efficiently process all of the previously mentioned query types. TPAQ avoids reading the entire dataset by exploiting the grouping properties of a data partition method on P. It initially traverses the index top-down, stopping at an appropriate level and placing the corresponding entries into groups according to proximity. Finally, it returns the most centrally located point within each group as the corresponding medoid. Compared to previous approaches, TPAQ achieves solutions of comparable or better quality, at a small fraction of the processing cost (seconds as opposed to hours). The rest of the chapter is organized as follows. Section 2 reviews related work. Section 3 introduces key concepts and outlines the general TPAQ framework. Section 4 considers k-medoid queries, while Section 5 and Section 6 focus on MA and MO queries, respectively. Section 7 presents experimental results and Section 8 concludes the chapter.

8.2 BACKGROUND

Although TPAQ can be used with any data partition method, we assume R*-trees (Beckmann, et al., 1990) due to their popularity. Section 8.2.1 overviews R*-trees and their application to nearest neighbor queries. Section 8.2.2 presents existing algorithms for k-medoids and related problems.

8.2.1 R-Trees and Nearest Neighbor Search

We illustrate our examples with the R*-tree of Figure 8.2 that contains the data points of Figure 8.1, assuming a capacity of four entries per node. Points that are nearby in space (e.g., a, b, c, d) are inserted into the same leaf node (N_3). Leaf nodes are recursively grouped in a bottom-up manner according to their proximity, up to the top-most level that consists of a single root. Each node is represented as a minimum bounding rectangle (MBR) enclosing all the points in its sub-tree. The nodes of an R*-tree are meant to be compact, have small margin, and achieve minimal overlap among nodes of the same level (Theodoridis, Stefanakis, and Sellis, 2000). Additionally, in practice, nodes at the same level contain a similar number of data

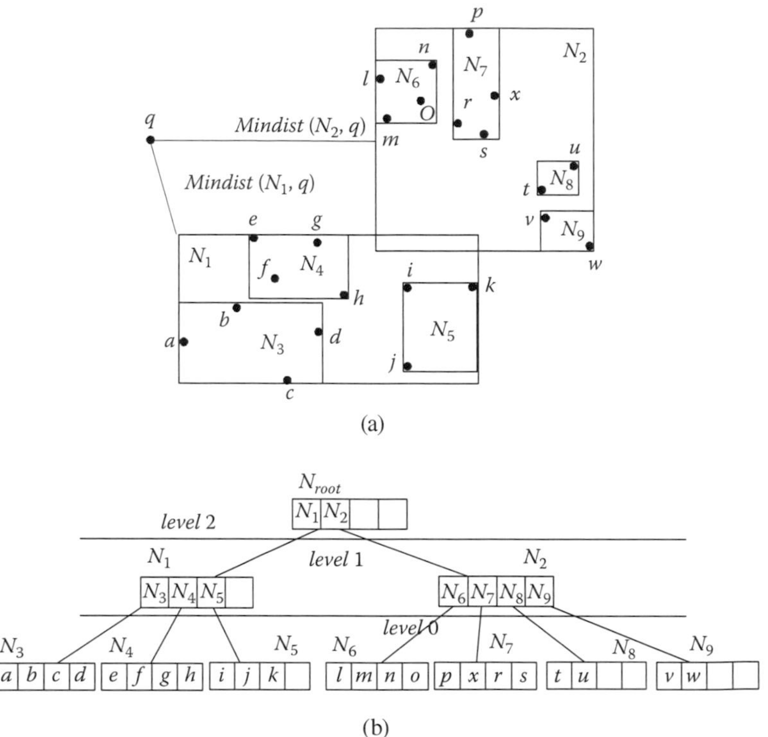

FIGURE 8.2 R-tree example.

points, due to a minimum utilization constraint (typically 40%). These properties imply that the R*-tree (or any other data partition method based on similar concepts) provides a natural way to partition P according to data proximity and group cardinality criteria. Furthermore, the R*-tree is a standard index for spatial query processing. Specialized structures may yield solutions of better quality for k-medoid problems, but would have limited applicability in existing systems, where R-trees are prevalent.

The R-tree family of indexes has been used for spatial queries such as range search, nearest neighbors, and spatial joins. A nearest neighbor (NN) query retrieves the data point that is closest to an input point, q. R-tree algorithms for processing NN queries utilize some metrics to prune the search space. The most common such metric is *mindist*(N,q), which is defined as the minimum possible distance between q and any point in the sub-tree rooted at node N. Figure 8.2 shows the *mindist* between q and nodes N_1 and N_2. The algorithm of Roussopoulos, Kelly, and Vincent (1995), shown in Figure 8.3, traverses the tree in a depth-first manner: starting from the root, it first visits the node with the minimum *mindist* (i.e., N_1 in our example). The process is repeated recursively until a leaf node (N_4) is reached, where the first potential nearest neighbor (point e) is found. Let *bestNN* be the best NN found thus far (e.g., *bestNN*=e) and *bestDist* be its distance from q (e.g., *bestDist*=$\|e-q\|$). Subsequently, the algorithm only visits entries whose minimum distance is less than *bestDist*. In

```
Algorithm NN (q,N)
If N is a leaf node
For each point p∈N
If ||p-q||<bestDist
best NN = p; bestDist =||p-q||
Else // N is an internal node
For each child N_i of N do
If mindist(q, N_i) < bestDist
NN(q, N_i)
```

FIGURE 8.3 The *NN* algorithm. (From Roussopoulos, N., Kelly, S., and Vincent, F. Nearest neighbor queries. *SIGMOD*, 1995.)

the example, N_3 and N_5 are pruned since their *mindist* from q is greater than $||e - q||$. Similarly, when backtracking to the upper level, node N_2 is also excluded and the process terminates with e as the result. The extension to k (>1) NNs is straightforward. Hjaltason and Samet (1999) propose a best-first NN algorithm that is I/O optimal (i.e., it only visits nodes that may contain NNs) and incremental (the number k of NNs does not need to be known in advance).

8.2.2 *k*-Medoids and Related Problems

A number of approximation schemes for k-medoids and related problems appear in the literature (Arora, Raghavan, and Rao, 1998). Most of them, however, are largely theoretical in nature. Kaufmann and Rousseeuw (1990) propose *partitioning around medoids* (PAM), a practical algorithm based on the hill climbing paradigm. PAM (illustrated in Figure 8.4) starts with a random set of k medoids $R_0 \subseteq P$. At each iteration i, it updates the current set R_i of medoids by exhaustively considering all *neighbor sets* R_i' that result from R_i by exchanging one of its elements with another data point. For each of these $k \cdot (|P| - k)$ alternatives, it computes the function $C(R_i')$ and chooses as R_{i+1} the one that achieves the lowest value. It stops when no further improvement is possible. Since computing $C(R_i')$ requires $O(|P|)$ distance calculations, PAM is prohibitively expensive for large $|P|$. *Clustering large applications* (CLARA) (Kaufmann and Rousseeuw, 1990) alleviates the problem by generating random samples from P and executing PAM on them. Ng and Han (1994) propose *clustering large applications based on randomized search* (CLARANS) as an extension to PAM. CLARANS draws a random sample of size *maxneighbors* from all the $k \cdot (|P| - k)$ possible neighbor sets R_i' of R_i. It performs *numlocal* restarts and selects the best local minimum as the final answer.

Although CLARANS is more scalable than PAM, it is inefficient for disk-resident datasets because each computation of $C(R_i')$ requires a scan of the entire database.

Algorithm **PAM** (P, k)

Initialize $R_0 = \{r_1, r_2, ..., r_k\}$ to a random subset of P with k elements, and set $i = 0$
Repeat
bestNeighbor $= R_i$
For each position $j = 1$ to k do
For each point $p \in P$ do
$R_i' = R_i - \{r_j\} \cup \{p\}$
If $C(R_i') < C(bestNeighbor)$
bestNeighbor $= R_i'$
$R_{i+1} = bestNeighbor; i = i + 1$
Until $R_i = R_{i-1}$ // no improvement was made
Return R

FIGURE 8.4 The *PAM* algorithm. (From Kaufman, L. and Rousseeuw, P. *Finding Groups in Data*. Wiley-Interscience, 1990.)

Assuming that P is indexed with an R-tree, Ester, Kriegel, and Xu (1995a,b) developed *focusing on representatives* (FOR). FOR takes the most centrally located point of each leaf node and forms a sample set, which is considered as representative of the entire set P. Then, it applies CLARANS on this sample to find the k medoids. FOR is more efficient than CLARANS, but it still has to read the entire dataset in order to extract the representatives. Furthermore, in very large databases, the leaf level population may still be too high for the efficient application of CLARANS (the experiments of Ester, Kriegel, and Xu use R-trees with only 50,559 points and 1,027 leaf nodes).

To the best of our knowledge, no existing method for the *max* case is suitable for disk-resident data. For in-memory processing, the *k-centers algorithm* (CTR) of Gonzales (1985) answers *max* k-medoid queries in $O(k \times |P|)$ time with an approximation factor of 2; i.e., the returned medoid set is guaranteed to achieve a maximum distance $C(R)$ that is no more than two times larger than the optimal one. The algorithm is shown in Figure 8.5. The first medoid is randomly selected from P and forms set R_1. The second medoid is the point in P that lies furthest from the point in R_1. These two medoids form R_2. In general, the i-th medoid is the one that has the maximum distance from any point in R_{i-1}. Finally, set R_k is returned as the result. The algorithm is simple and works well in practice. However, its adaptation to large datasets would be very expensive in terms of both CPU and I/O cost, since in order to find the i-th medoid it has to scan the entire dataset and compute the distance between every data point and all elements of R_{i-1}.

A problem related to k-medoids is *min-dist optimal-location* (MDOL) computation. Given a set of data points P, a set of existing facilities, and a user-specified

Algorithm **CTR** (P, k)

1. Choose a point $p \in P$ randomly, and set $R_1 = \{p\}$
2. For $i = 2$ to k do
3. Let p be the point in $P - R_{i-1}$ that is furthest from any medoid in R_{i-1}
4. $R_i = R_{i-1} \cup \{p\}$
5. Return R_k

FIGURE 8.5 The CTR algorithm for *max k*-medoids. (From Gonzalez, T. Clustering to minimize the maximum intercluster distance. *Theoretical Computer Science*, 38: 293–306, 1985.)

spatial region Q (i.e., range for a new facility), an MDOL query computes the location in Q which, if a new facility is built there, minimizes the average distance between each data point and its closest facility. The main difference with respect to k-medoids is that the output of an MDOL query is a single point (as opposed to k) that does not necessarily belong to P, but it can be anywhere in Q. Zhang et al. (2006) propose an exact method for this problem. This technique is complementary to the proposed algorithms because it can be used to increase the cardinality of an existing medoid set when there is a need for incremental processing (e.g., the company of our example may decide to open an additional warehouse in a given area).

The k-medoid problem is related to clustering. Clustering methods designed for large databases include DBSCAN (Ester et al., 1996), BIRCH (Zhang, Ramakrishnan, and Livny, 1996), CURE (Guha, Rastogi, and Shim, 1998), and OPTICS (Ankerst et al., 1999). However, the objective of clustering in general and of these techniques in particular is inherently different. Extensive work on medoids and clustering has been carried out in the areas of statistics (Hartigan, 1975; Kaufman and Rousseeuw, 1990; Hastie, Tibshirani, and Friedman, 2001), machine learning (Pelleg and Moore, 1999, 2000; Hamerly and Elkan, 2003), and data mining (Ester et al., 1996; Fayyad et al., 1996). However, the focus there is on assessing the statistical quality of a given clustering, usually based on assumptions about the data distribution (Hastie et al., 2001; Kaufman and Rousseeuw, 1990; Pelleg and Moore, 2000). Only few approaches aim at dynamically discovering the number of clusters (Pelleg and Moore, 2000; Hamerly and Elkan, 2003). Besides tackling problems of a different nature, these algorithms are computationally intensive and unsuitable for disk-resident datasets.

8.3 FRAMEWORK OVERVIEW AND BASIC DEFINITIONS

The TPAQ framework traverses the R-tree in a top-down manner, stopping at the topmost level that provides enough information for answering the given query. In the case of k-medoids, this decision depends on the number of entries at the level.

On the other hand, for MA and MO queries, the selection of the partitioning level is also based on the spatial extents and (in the *avg* case) on the expected cardinality of its entries. Next, TPAQ groups the entries of the partitioning level into *slots*. For a given k, this procedure is performed by a fast slotting algorithm. For MA and MO, multiple calls of the slotting algorithm might be required. The last step returns the NN of each slot center as the medoid of the corresponding partition. We first provide some basic definitions, which are used throughout the chapter.

Definition 1 [*Extended entry*]: An *extended entry* e consists of an R-tree entry N, augmented with information about the underlying data points, i.e., $e = \langle c, w, N \rangle$, where the *weight* w is the expected number of points in the sub-tree rooted at N. The center c is a vector of coordinates that corresponds to the *geometric centroid* of N, assuming that the points in the sub-tree of N are uniformly distributed.

Definition 2 [*Slot*]: A *slot* s consists of a set E of extended entries, along with aggregate information about them. Formally, a slot s is defined as $s = \langle c, w, E \rangle$, where w is the expected number of points represented by s,

$$w = \sum_{e \in E} e.w.$$

In the *avg* case, vector c is the weighted center of s,

$$c = \frac{1}{w} \sum_{e \in E} e.w \cdot e.c.$$

In the *max* case, vector c is the center of the *minimum enclosing circle* of all the entry centers $e.c$ in s; i.e., c is the center of the circle enclosing $e.c$ $\forall e \in E$ that has the minimum possible radius.

A fundamental operation is the insertion of an extended entry e into a slot s. The pseudo-code for this function in the *avg* case is shown in Figure 8.6. The insertion computes the new center, taking into account the relative positions and weights of the slot s and the entry e, e.g., if s and e have the same weights, the new center is at the midpoint of the line segment connecting $s.c$ and $e.c$. In the *max* case, the new slot center is computed as the center of the minimum circle enclosing $e.c$ and all the entry centers currently in s. We use the incremental algorithm of Welzl (1991), which finds the new slot center in expected constant time.

Function **InsertEntry** (extended entry e, slot s)

1. $s.c = (e.w \cdot e.c + s.w \cdot s.c)/(e.w + s.w)$
2. $s.w = e.w + s.w$
3. $s.E = s.E \cup \{e\}$

FIGURE 8.6 The *InsertEntry* function for *avg*.

In the subsequent sections, we describe the algorithmic details for each query type. For every considered medoid problem, we first present the *avg* case, followed by *max*. Note that, similar to PAM, CLARA, CLARANS, and FOR, TPAQ aims at efficient processing without theoretical guarantees on the quality of the medoid set. Meaningful quality bounds are impossible because TPAQ is based on the underlying R-trees, which are heuristic-based structures. Nevertheless, as we show in the experimental evaluation, TPAQ computes medoid sets that are better than those of the existing methods at a small fraction of the cost (usually several orders of magnitude faster). Furthermore, it is more general in terms of the problem variants it can process.

8.4 *k*-MEDOID QUERIES

Given an *avg* k-medoid query, TPAQ finds the top-most level with $k' \geq k$ entries. For example, if $k = 3$ in the tree of Figure 8.2, TPAQ descends to level 1, which contains $k' = 7$ entries, N_3 through N_9. The weights of these entries are computed as follows. Since $|P| = 23$, the weight of the root node N_{root} is $w_{root} = 23$. Assuming that the entries of N_{root} are equally distributed between the two children N_1 and N_2, $w_1 = w_2 = N/2 = 11.5$ (the true cardinalities are 11 and 12, respectively). The process is repeated for the children of N_1 ($w_3 = w_4 = w_5 = w_1/3 = 3.83$) and N_2 ($w_6 = w_7 = w_8 = w_9 = w_2/4 = 2.87$). Figure 8.7 illustrates the algorithm for computing the initial set of entries. Note that *InitEntries* assumes that k does not exceed the number of leaf nodes. This is not restrictive because the lowest level typically contains several thousand nodes (e.g., in our datasets, between 3,000 and 60,000), which is sufficient for all ranges of k that are of practical interest. If needed, larger values of k can be accommodated by conceptually splitting leaf level nodes.

```
Function InitEntries (P, k)
Load the root of the R-tree of P
Initialize list = {e}, where e = (N_root.c, |P|, N_root)
While list contains fewer than k extended entries do
Initialize an empty list next_level_entries
For each e = (c, w, N) in list do
Let num be the number of child entries in node N
For each entry N_i in node N do
w_i = w/num // the expected cardinality of N_i
Insert extended entry (N_i.c, w_i, N_i) into next_level_entries
Set list = next_level_entries
Return list
```

FIGURE 8.7 The *InitEntries* function.

The next step merges the k' initial entries in order to obtain exactly k groups. First, k out of the k' entries are selected as slot *seeds*, i.e., each of the chosen entries forms a singleton slot. Clearly, the seed locations play an important role in the quality of the final answer. The seeds should capture the distribution of points in P, i.e., dense areas should contain many seeds. Our approach for seed selection is based on *space-filling curves*, which map a multidimensional space into a linear order. Among several alternatives, Hilbert curves best preserve the locality of points (Korn, Pagel, and Faloutsos, 2001; Moon et al., 2001). Therefore, we first Hilbert-sort the k' entries and select every m-th entry as a seed, where $m = k'/k$. This procedure is fast and produces well-spaced seeds that follow the data distribution. Returning to our example, Figure 8.8a shows the level 1 MBRs (for the R-tree of Figure 8.2) and the output seeds $s_1 = N_4$, $s_2 = N_9$, and $s_3 = N_7$ according to their Hilbert order. Recall that each slot is represented by its weight (e.g., $s_1.w = w_4 = 3.83$), its center (e.g., $s_1.c$ is the centroid of N_4), and its MBR. Then, each of the remaining $(k' - k)$ entries is inserted into the k slots, based on proximity. More specifically, for each entry e, we choose the slot s whose weighted center $s.c$ is closest to the entry's center $e.c$. In the running example, assuming that N_3 is considered first, it is inserted into slot s_1 using the *InsertEntry* function of Figure 8.6. The center of s_1 is updated to the midpoint of N_3 and N_4's centers, as illustrated in Figure 8.8b. TPAQ proceeds in this manner, until the final slots and weighted centers are computed as shown in Figure 8.8c.

After grouping all entries into exactly k slots, we find one medoid per slot by performing an NN query. The query point is the slot's weighted center $s.c$, and the search space is the set of entries $s.e$. Since all the levels of the R-tree down to the partition level have already been loaded in memory, the NN queries incur very few node accesses and negligible CPU cost. Observe that an actual medoid (i.e., a point in P that minimizes the average distance) is more likely to be closer to $s.c$ than simply to the center of the MBR of s. The intuition is that $s.c$ captures information about the point distribution within s. The NN queries on these points return the final medoids $R = \{h, o, t\}$.

Figure 8.9 shows the complete TPAQ k-medoid computation algorithm. The problem of seeding the slot table is similar to that encountered in spatial hash joins, where the number of buckets is bounded by the available main memory (Lo and Ravishankar, 1995, 1998; Mamoulis and Papadias, 2003). However, our ultimate goals are different. First, in the case of hash joins, the table capacity is an upper bound. Reaching it is desirable in order to exploit available memory as much as possible, but falling slightly short is not a problem. In contrast, we want *exactly* k slots. Second, in our case, slots should minimize the average distance $C(R)$ on one dataset, whereas slot selection in spatial joins attempts to minimize the number of intersection tests that must be performed between points that belong to different datasets.

TPAQ follows similar steps for the *max* case. The function *InitEntries* proceeds as before, but without computing the expected cardinality for entries and slots; in the *max* version of the problem, we use only the geometric centroids of the R-tree entries. Let E be the set of entries in the partitioning level. We apply the CTR algorithm (described in Section 8.2.2) to select k slot seeds among the entry centers $e.c$ in E. Then, we insert the remaining entries in E one by one into the slot with the

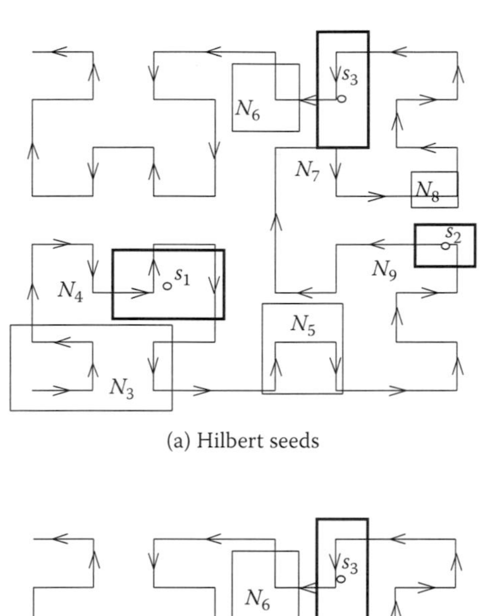

(a) Hilbert seeds

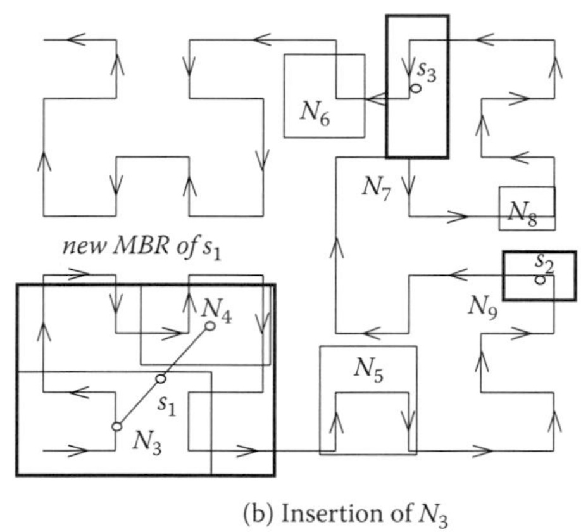

(b) Insertion of N_3

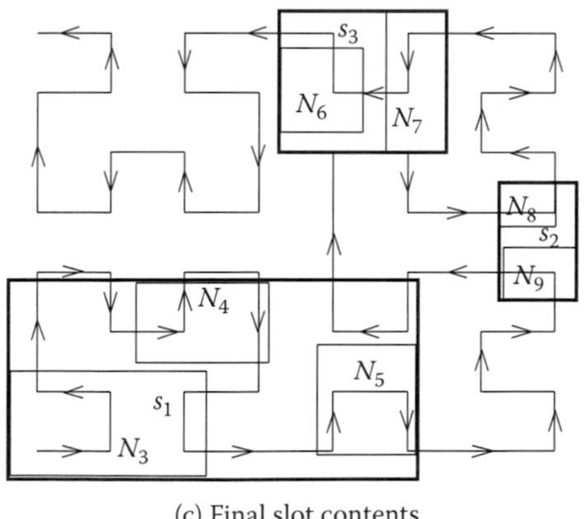

(c) Final slot contents

FIGURE 8.8 Insertion of entries into slots.

```
Algorithm TPAQ (P, k)
Initialize a set S = Ø, and an empty list
Set E = the set of entries returned by InitEntries (P, k)
Hilbert-sort the centers of the entries in E and store them in a sorted list sorted_list
For i = 1 to k do // compute the slot seeds
Form a slot containing the (i·|E|/k)-th entry of sorted_list and insert it into S
For each entry e in E (apart from the ones selected as seeds) do
Find the slot s in S with the minimum distance ||e.c – s.c||
InsertEntry (e, s)
For each s∈S do
Perform a NN search at s.c on the points under s.E
Append the retrieved point to list
Return list
```

FIGURE 8.9 The *TPAQ* algorithm.

closest center. Finally, we perform an NN search at the center of each slot to retrieve the actual corresponding medoid. Recall that the center of each slot is the center of the minimum circle enclosing its entries' centers. Returning to our running example, if a 3-medoid query is given in the tree of Figure 8.2, level 1 is chosen as the partitioning level. Among the entries of level 1, assume that CTR returns the centers of N_4, N_6, and N_9 as the seeds. The insertion of the remaining entries into the created slots (s_1, s_2, and s_3) results in the partitioning shown in Figure 8.10. The three circles

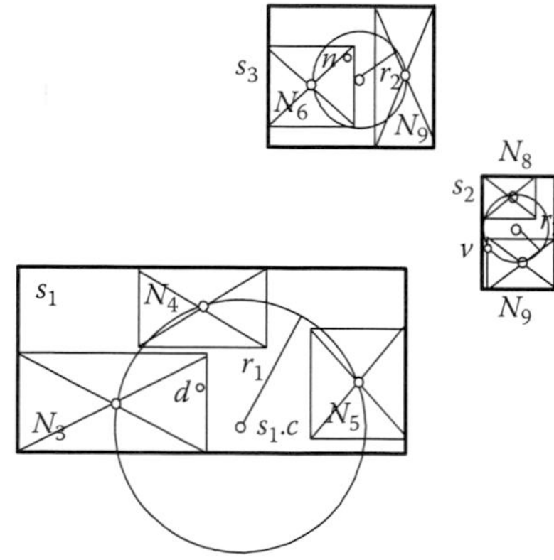

FIGURE 8.10 3-medoids in the *max* case.

correspond to the minimum circles enclosing the centers of nodes in each slot. The final step of the TPAQ algorithm retrieves the NNs of $s_1.c$, $s_2.c$, and $s_3.c$, which are points d, v, and n, respectively. The returned medoid set is $R = \{d, v, n\}$.

8.5 MEDOID-AGGREGATE QUERIES

A medoid-aggregate (MA) query specifies the desired distance T (between points and medoids), and asks for the minimal medoid set R that achieves $C(R) = T$. The proposed algorithm, TPAQ-MA, is based on the fact that as the number of medoids $|R|$ increases, the corresponding $C(R)$ decreases, in both the *avg* and the *max* case. TPAQ-MA first descends the R-tree of P down to an appropriate partitioning level. Next, it estimates the value of $|R|$ that achieves the average distance $C(R)$ closest to T and returns the corresponding medoid set R. Consider first the *avg* case. The initial step of TPAQ-MA is to determine the partitioning level. The algorithm selects for partitioning the top-most level whose *minimum possible distance* (MPD) does not exceed T. The MPD of a level is the smallest $C(R)$ that can be achieved if partitioning takes place in this level. According to the methodology of Section 8.4, MPD is equal to the $C(R)$ resulting if we extract one medoid from each entry in the level. Since computing the exact $C(R)$ requires scanning the entire dataset P, we use an estimate of $C(R)$ as the MPD. In particular, for each entry e of the level, we assume that the underlying points are distributed uniformly* in its MBR, and that the corresponding medoid is at $e.c$. The average distance $\bar{C}(e)$between $e.c$ and the points in e is given by the following lemma.

Lemma 8.1: If the points in e are uniformly distributed in its MBR, then their average distance from $e.c$ is

$$\bar{C}(e) = \frac{1}{3}\left(\frac{D}{2} + \frac{B^2}{8A}\ln\left(\frac{D+A}{D-A}\right) + \frac{A^2}{8B}\ln\left(\frac{D+B}{D-B}\right)\right),$$

where A and B are the side lengths of the MBR of e, and D is its diagonal length.

Proof: If we translate the MBR of e so that its center $e.c$ falls at the origin (0,0), $\bar{C}(e)$ is the average distance of points $(x,y) \in [-A/2, A/2] \times [-B/2, B/2]$ from (0,0). Hence,

$$\bar{C}(e) = \frac{1}{AB}\int_{-A/2}^{A/2}\int_{-B/2}^{B/2}\sqrt{x^2+y^2}\ dxdy,$$

which evaluates to the quantity of Lemma 8.1.

The MPD of each level is estimated by averaging $\bar{C}(e)$ over all $e \in E$, where E is the set of entries at the level:

$$\text{MPD} = \frac{1}{|P|}\sum_{e \in E} e.w \cdot \bar{C}(e).$$

* This is a reasonable assumption for low-dimensional R-trees (Theodoridis et al., 2000).

TPAQ-MA applies the *InitEntries* function to select the top-most level that has MPD $\leq T$. The pseudo-code of *InitEntries* is the same as shown in Figure 8.7, after replacing the while-condition of line 3 with the expression: "the estimated MPD is more than T." Returning to our running example, the root node N_{root} of the R-tree of P has MPD=$\bar{C}(N_{root})$, which is higher than T. Therefore, *InitEntries* proceeds with level 2 (containing entries N_1 and N_2), whose MPD is also higher than T. Next, it loads the level 1 nodes and computes the MPD over entries N_3 to N_9. The MPD is less than T, and level 1 is selected for partitioning. *InitEntries* returns a list containing seven extended entries corresponding to N_3 up to N_9.

The next step of TPAQ-MA is to determine the number of medoids that best approximate value T. If E is the set of entries in the partitioning level, the candidate values for $|R|$ range between 1 and $|E|$. TPAQ-MA assumes that $C(R)$ decreases as $|R|$ increases, and performs binary search in order to find the value of $|R|$ that yields the average distance closest to T. This procedure considers $O(\log|E|)$ different values for $|R|$, and creates slots for each of them as discussed in Section 8.4. Since the exact evaluation of $C(R)$ for every examined $|R|$ would be very expensive, we produce an estimate $\bar{C}(S)$ of $C(R)$ for the corresponding set of slots S. Particularly, we assume that the medoid of each slot s is located at $s.c$, and that the average distance from the points in every entry $e \in s$ is equal to distance $\|e.c - s.c\|$. Hence, the estimated value for $C(R)$ is given by the formula

$$\bar{C}(S) = \frac{1}{|P|}\sum_{s\in S}\sum_{e\in s} e.w\cdot \|e.c - s.c\|,$$

where S is the set of slots produced by partitioning the entries in E into $|R|$ groups. Note that we could use a more accurate estimator assuming uniformity within each entry $e \in s$, similar to Lemma 8.1. However, the derived expression would be more complex and more expensive to evaluate, because now we need the average distance from $s.c$ (as opposed to the center $e.c$ of the entry's MBR). The TPAQ-MA algorithm is shown in Figure 8.11.

In the example of Figure 8.2, the partitioning level contains entries $E = \{N_3, N_4, N_5, N_6, N_7, N_8, N_9\}$. The binary search considers values of $|R|$ between 1 and 7. Starting with $|R| = (1 + 7)/2 = 4$, the algorithm creates S with four slots, as shown in Figure 8.12. It computes $\bar{C}(S)$, which is lower than T. It recursively continues the search for $|R| \in [1,4]$ in the same way, and decides that $|R| = 4$ yields a value of $\bar{C}(S)$ that best approximates T. Finally, similar to TPAQ, TPAQ-MA performs an NN search at the center $s.c$ of the slots corresponding to $|R| = 4$, and returns the retrieved points (f, k, t, and o) as the result.

Consider now the *max* version of the MA problem. *InitEntries* chooses for partitioning the top-most level with MPD less than or equal to T. The MPD of a level is an estimated upper bound for the maximum distance $C(R)$, assuming that we return a medoid at the center of each of the level's entries. Given an R-tree entry e and assuming that we can find a medoid at $e.c$ (i.e., the crossing point of its MBR diagonals), then the maximum possible distance of any point in e from the medoid is half the MBR diagonal length. Therefore, the MPD of a level is computed as the half of the maximum entry diagonal in the level. In other words, $\bar{C}(e) = D/2$ (where D is the diagonal of e), and MPD $= max_{e\in E}\bar{C}(e)$ (where E is the set of entries in the given level).

```
Algorithm TPAQ-MA (P, T)
Initialize an empty list
Set E = set of the entries at the topmost level with MPD≤T
low = 1; high = |E|
While low ≤ high do
mid = (low + high)/2
Group the entries in E into mid slots
S = the set of created slots
If C̄(S) < T, set high = mid
Else, set low = mid
For each s∈S do
Perform a NN search at s.c on the points under s.E
Append the retrieved point to list
Return list
```

FIGURE 8.11 The *TPAQ-MA* algorithm.

Similar to the *avg* case, in order to determine the number of medoids that best approximate the target distance *T*, we perform a binary search. If *E* is the set of entries in the partitioning level, then the candidate |*R*| values range between 1 and |*E*|. For each considered |*R*|, we use the *max* slotting algorithm (described in Section 8.4). Let *S* be the set of slots for a value of |*R*|. To estimate the achieved *C*(*R*) [i.e., to

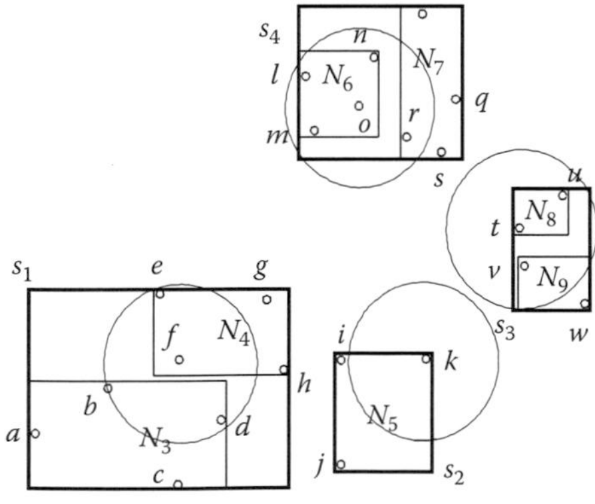

FIGURE 8.12 Entries and final slots.

compute $\bar{C}(S)$], we assume that the maximum distance within each slot *s* is equal to the radius of the minimum circle enclosing the entry centers in *s*. For example, if level 1 is selected for partitioning and $|R| = 3$, the slotting produces the grouping shown in Figure 8.10. $C(R)$ is estimated as the maximum radius of the three circles, that is, $\bar{C}(S) = max\{r_1, r_2, r_3\} = r_1$. Formally, if *MincircRadius*(*s*) is the radius of the smallest circle enclosing *e.c* $\forall e \in s$, then $\bar{C}(S) = max_{s \in S}\, MincircRadius(s)$. When the binary search terminates, we retrieve the medoids corresponding to the best value of $|R|$. The algorithm of Figure 8.11 directly applies to *max* MA queries, by using the *max* versions of MPD and $\bar{C}(S)$, and by implementing line 6 with the *max* slotting algorithm.

8.6 MEDOID-OPTIMIZATION QUERIES

In real-world scenarios, opening a facility has some cost. Thus, users may wish to find a good tradeoff between overall cost and coverage (i.e., the average or maximum distance between clients and their closest facilities). If the relative importance of these conflicting factors is given by a user-specified cost function $f(C(R), |R|)$, the aim of an MO query is to find the medoid set *R* that minimizes *f*. The TPAQ methodology applies to this problem, provided that *f* is increasing on both $C(R)$ and $|R|$. Consider the example of Figure 8.1 in the *avg* case, and let $f(C(R), |R|)$ be $C(R) + Cost_{pm} \times |R|$, where $Cost_{pm}$ is the cost per medoid. Assume that we know *a priori* all the optimal *i*-medoid sets R^i and the corresponding $C(R^i)$, for $i = 1,...,23$. If the plot of $f(C(R^i), |R^i|)$ vs. $|R^i|$ is shown in Figure 8.13, then the optimal $|R|$ is 3 and the result of the query is $\{h, o, t\}$ (as in Figure 8.1). TPAQ-MO is based on the observation that $f(C(R^i), |R^i|)$ has a single minimum. Hence, it applies a gradient descent technique to decide the partitioning level and the optimal number of medoids $|R|$.

In both the *avg* and *max* cases, TPAQ-MO initially descends the R-tree of *P* and for each candidate level, it computes its *cost*. We define the cost of a level as the value $f(\text{MPD}, |E|)$, where *E* is the set of its entries. TPAQ-MO selects for partitioning the top-most level whose cost is greater than the cost of the previous one (i.e., at the first

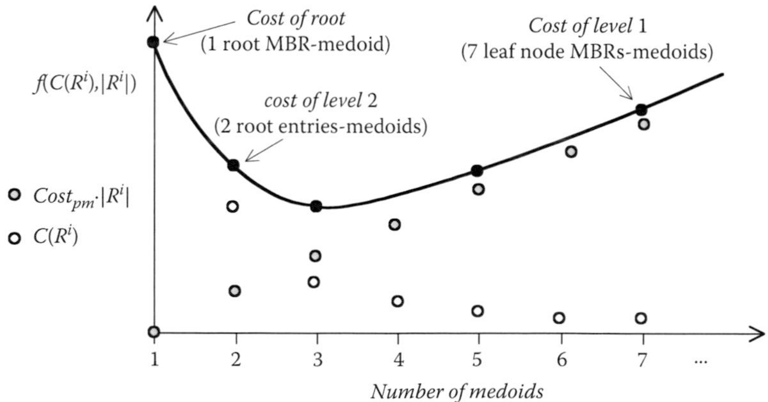

FIGURE 8.13 $f(C(R^i), |R^i|)$ versus number of medoids.

detected increase in the curve of Figure 8.13). If the MPD estimations are accurate, then the medoid set that minimizes f has size $|R|$ between 1 and $|E|$ (the number of entries at the partitioning level). The traversal of the R-tree down to the appropriate level is performed by the *InitEntries* function of Figure 8.7 by modifying the while-condition in line 3 to "the cost of the current level is less than the cost of the previous one." In Figure 8.2, *InitEntries* compares the costs of the root entry (1 medoid) and level 2 (two medoids — one for each root entry). Since the cost of level 2 is less than that of the root, it proceeds with level 1, whose cost is larger than level 2. Thus, level 1 is selected for partitioning and *InitEntries* returns the set of extended entries from N_3 to N_9.

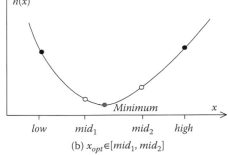

Given the set of entries E at the partitioning level, the next step of TPAQ-MO is to compute the optimal value for $|R|$, which lies between 1 and $|E|$. To perform this task, TPAQ-MO uses a gradient descent method which considers $O(\log_{3/2}|E|)$ different values for $|R|$. Consider the example of Figure 8.14, where we want to find the value $x_{opt} \in [low, high]$ that minimizes a given function $h(x)$. We split the search interval into three equal sub-intervals, defined by $mid_1 = (2 \cdot low + high)/3$ and $mid_2 = (low + 2 \cdot high)/3$. Next, we compute $h(mid_1)$ and $h(mid_2)$. Assuming that $h(mid_1) < h(mid_2)$, we distinguish two cases; either $x_{opt} \in [low, mid_1]$ (as shown in Figure 8.14a), or $x_{opt} \in [mid_1, mid_2]$ (Figure 8.14b). In other words, the search interval is restricted to $[low, mid_2]$. Symmetrically, if $h(mid_1) > h(mid_2)$, then the search interval becomes $[mid_1, high]$. Otherwise, if $h(mid_1) = h(mid_2)$, the search is restricted to interval $[mid_1, mid_2]$.

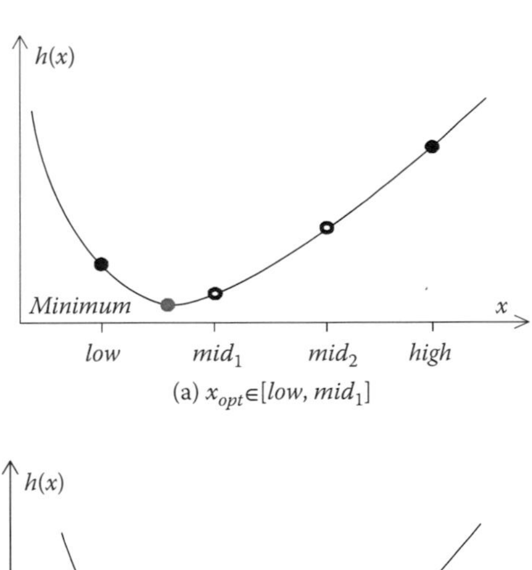

FIGURE 8.14 Computing the minimum of a function h.

The x_{opt} can be found by recursively applying the same procedure to the new search interval. If x_{opt} is an integer, then the search terminates in $O(\log_{3/2}(high\text{-}low))$ steps.

We use the above technique to determine the optimal value of $|R|$, starting with $low = 1$ and $high = |E|$. For each considered $|R|$, we compute the set of slots S in the way presented in Section 8.4, and estimate the corresponding $C(R)$ as the quantity $\bar{C}(S)$ discussed in Section 8.5. The gradient descent method returns the value of $|R|$ that minimizes $f(\bar{C}(S), |R|)$. Finally, the result of TPAQ-MO is the set of points retrieved by an NN search at the center of each slot $s \in S$ of the corresponding partitioning. TPAQ-MO is illustrated in Figure 8.15. The algorithm works for both *avg* and *max* MO queries, by using the corresponding MPD and $\bar{C}(S)$ functions, and the appropriate slotting strategies. In our running example, for the *avg* case, level 1 is the

Algorithm **TPAQ-MO** (P, f)

Initialize an empty *list*

Set E = set of the entries at the topmost level with cost greater than that of the previous level

$low = 1$; $high = |E|$

While $low + 2 < high$ do

$mid_1 = (2 \cdot low + high)/3$; $mid_2 = (low + 2 \cdot high)/3$

Group the entries in E into mid_1 slots

S_1 = the set of created slots

Group the entries in E into mid_2 slots

S_2 = the set of created slots

If $f(\bar{C}(S_1), mid_1) < f(\bar{C}(S_2), mid_2)$

Set $high = mid_2$ and $S = S_1$

Else, if $f(\bar{C}(S_1), mid_1) > f(\bar{C}(S_2), mid_2)$

Set $low = mid_1$ and $S = S_2$

Else, if $f(\bar{C}(S_1), mid_1) = f(\bar{C}(S_2), mid_2)$

Set $low = mid_1$, $high = mid_2$ and $S = S_1$

For each $s \in S$ do

Perform a NN search at $s.c$ on the points under $s.E$

Append the retrieved point to *list*

Return *list*

FIGURE 8.15 The *TPAQ-MO* algorithm.

partitioning level and $|R| = 3$ is selected as the best medoid set size. The slots and the returned medoids (i.e., h, o, and t) are the same as in Figure 8.8.

8.7 EXPERIMENTAL EVALUATION

In this section we evaluate the performance of the proposed methods for k-medoid, medoid-aggregate, and medoid-optimization queries. For each of these three problems, we first present our experimental results for *avg*, and then for *max*, using both synthetic and real datasets. The synthetic ones (SKW) follow a Zipf distribution with parameter $\alpha = 0.8$, and have cardinality 256K, 512K, 1M, 2M and 4M points (with 1M being the default). The real dataset (LA) contains 1,314,620 points (available at www.rtreeportal.org). All datasets are normalized to cover the same space with extent $10^4 \times 10^4$ and indexed by an R*-tree (Berchtold, Keim, and Kriegel, 1996) with a 2Kbyte page size. For the experiments, we use a 3GHz Pentium CPU.

8.7.1 *k*-MEDOID QUERIES

First, we focus on k-medoid queries and compare TPAQ against FOR, which as discussed in Section 2.2, is the only other method that utilizes R-trees. For TPAQ, we use the depth-first algorithm of Roussopoulos et al. (1995) to retrieve the nearest neighbor of each computed slot center. In the case of FOR, we have to set the parameters *numlocal* (number of restarts) and *maxneighbors* (sample size of the possible neighbor sets) of the CLARANS component. Ester et al. (1995a) suggest setting $numlocal = 2$ and $maxneighbors = k \times (M - k)/800$, where M is the number of leaf nodes in the R-tree of P. With these parameters, FOR terminates in several hours for most experiments. Therefore, we set $maxneighbors = k \times (M - k)/(8000 \times \log M)$ and keep $numlocal = 2$. These values speed up FOR considerably, while the deterioration of the resulting solutions is small (with respect to the suggested values of *numlocal* and *maxneighbors*). Regarding the *max* case, there is currently no other algorithm for disk-resident data. For the sake of comparison, however, we adapted FOR to *max* k-medoid queries by defining $C(R)$ to be the maximum distance between data points and medoids; that is, the CLARANS component of FOR exchanges the current medoid set R_i with a neighbor one R_i', only if the maximum distance achieved by R_i' is smaller than that of R_i. All FOR results presented in this section are average values over 10 runs of the algorithm. This is necessary because the performance of FOR depends on the random choices of CLARANS. The algorithms are compared for different data cardinality $|P|$ and number of medoids k; for k, the tested values are from 1 to 512, and its default is 32. In each experiment we fix either parameter (i.e., $|P|$ or k) to its default value and vary the other one.

We first measure the effect of $|P|$ in the *avg* case. Figure 8.16a shows the CPU time of TPAQ and FOR for SKW, when $k = 32$ and $|P|$ ranges between 256K and 4M. TPAQ is 2 to 4 orders of magnitude faster than FOR. Even for $|P| = $ 4M points, our method terminates in less than 0.04 sec (while FOR needs more than 3 min). Figure 8.16b shows the I/O cost (number of node accesses) for the same experiment. FOR is approximately 2 to 3 orders of magnitude more expensive than TPAQ because it reads the entire dataset once. Both the CPU and the I/O costs of TPAQ are relatively stable and small because partitioning takes place at a high tree level. The cost

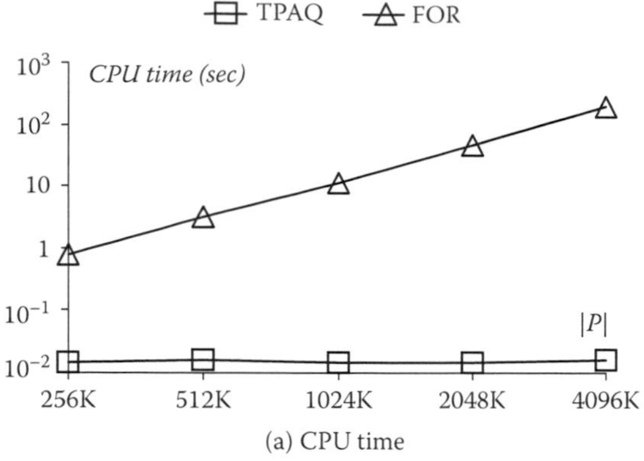

(a) CPU time

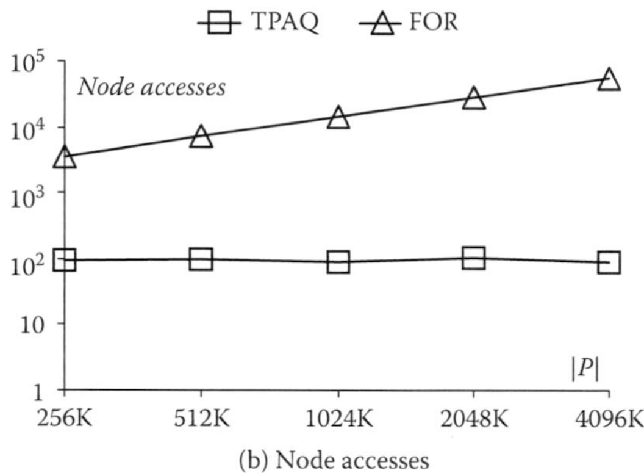

(b) Node accesses

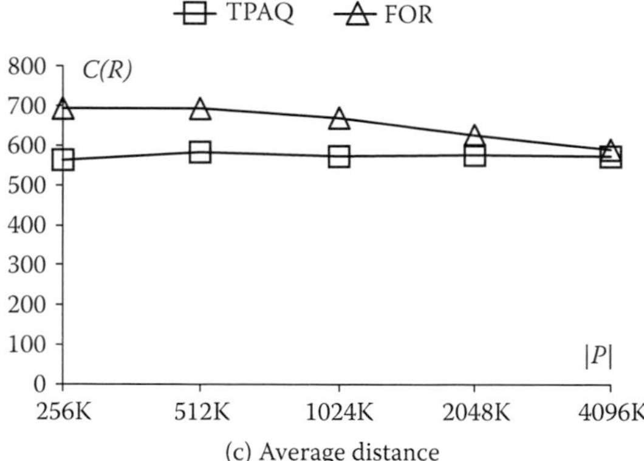

(c) Average distance

FIGURE 8.16 Performance versus $|P|$ (SKW, *avg*).

improvements of TPAQ come with no compromise in answer quality. Figure 8.16c shows the average distance $C(R)$ achieved by the two algorithms. TPAQ outperforms FOR in all cases. An interesting observation is that the average distance for FOR drops when the cardinality of the dataset $|P|$ increases. This happens because a higher $|P|$ implies more possible "paths" to a local minimum. To summarize, the results of Figure 8.16 verify that TPAQ scales gracefully with the dataset cardinality and incurs much lower cost than FOR, without sacrificing medoid quality.

The next set of experiments studies the performance of TPAQ and FOR in the *avg* case, when k varies between 1 and 512, using an SKW dataset of cardinality $|P| = 1M$. Figure 8.17a compares the CPU time of the methods. In all cases, TPAQ is three orders of magnitude faster than FOR. It is worth mentioning that for $k = 512$ our method terminates in 2.5 sec, while FOR requires approximately 1 hour and 20 min. For $k = 512$, both the partitioning into slots of TPAQ and the CLARANS component of FOR are applied on an input of size 14,184; the input of the TPAQ partitioning algorithm consists of the extended entries at the leaf level, while the input of CLARANS is the set of actual representatives retrieved in each leaf node. The large difference in CPU time verifies the efficiency of our partitioning algorithm.

Figure 8.17b shows the effect of k on the I/O cost. The node accesses of FOR are constant and equal to the total number of nodes in the R-tree of P (i.e., 14,391). On the other hand, TPAQ accesses more nodes as k increases. This happens because (1) it needs to descend more R-tree levels in order to find one with a sufficient number (i.e., k) of entries, and (2) it performs more NN queries (i.e., k) at the final step. However, TPAQ is always more efficient than FOR; in the worst case, TPAQ reads all R-tree nodes up to level 1 (this is the situation for $k = 512$), while FOR reads the entire dataset P for any value of k. Figure 8.17c compares the accuracy of the methods. TPAQ achieves lower $C(R)$ for all values of k. In order to confirm the generality of our observations, Figure 8.18 repeats the above experiment for the real dataset LA. TPAQ outperforms FOR by orders of magnitude in terms of both CPU time (Figure 8.18a) and number of node accesses (Figure 8.18b). Regarding the average distance $C(R)$, the methods achieve similar results (Figure 8.18c), with TPAQ being the winner.

Next, we focus on *max* k-medoid queries. We perform the same experiments as in the *avg* case, with identical test ranges and default values for $|P|$ and k. Figure 8.19 compares TPAQ and FOR on 32-medoid queries over SKW datasets of varying cardinality. As in Figure 8.16, our method significantly outperforms FOR in terms of both CPU and I/O cost because FOR reads the entire input dataset and its CLARANS component is much more expensive than our *max* slotting algorithm. TPAQ is also considerably better on the quality of the retrieved medoids (Figure 8.19c). This is expected because FOR is originally designed for the *avg* k-medoid problem. FOR converges to poor local minima when CLARANS considers swapping a current medoid with another representative because it selects the latter randomly among the set of representatives. Since the representatives follow the data distribution, the choices of CLARANS are biased toward dense areas of the workspace. Even though this behavior is desirable in *avg* k-medoid queries, it is clearly unsuitable for the *max* case because even a single point in a sparse area can lead to a large $C(R)$.

Figure 8.20 and Figure 8.21 examine the effect of k on TPAQ and FOR over the SKW and LA datasets. The CPU cost of both methods increases with k. Larger values

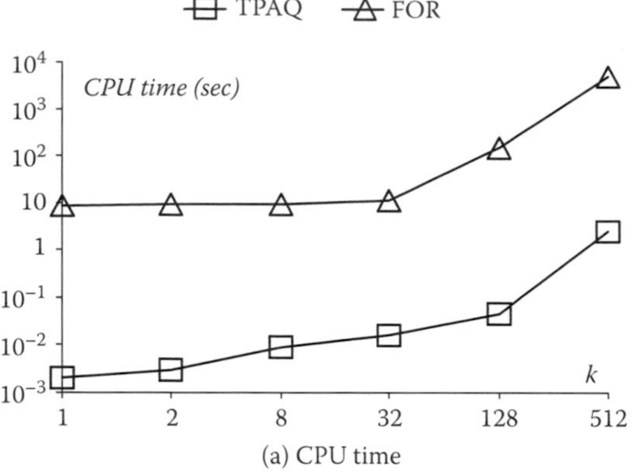

(a) CPU time

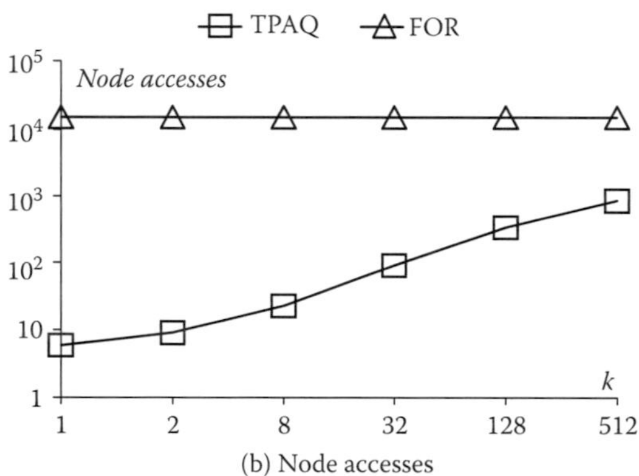

(b) Node accesses

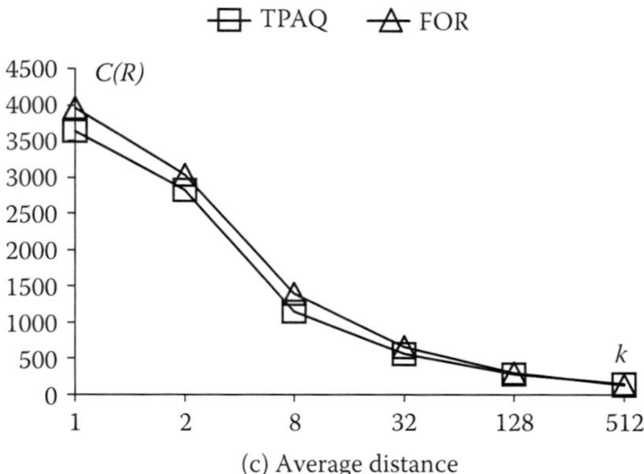

(c) Average distance

FIGURE 8.17 Performance versus *k* (SKW, *avg*).

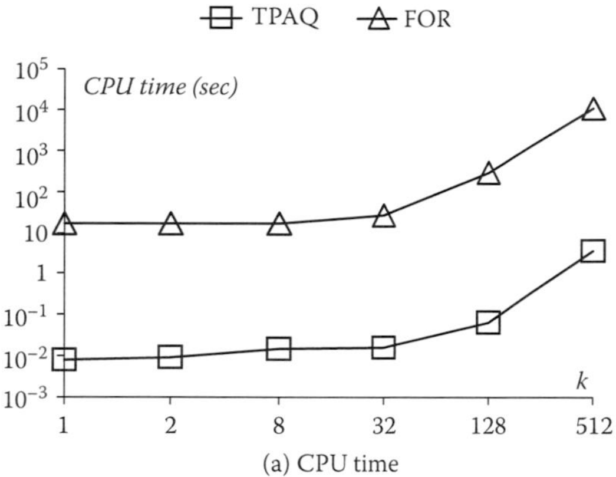

(a) CPU time

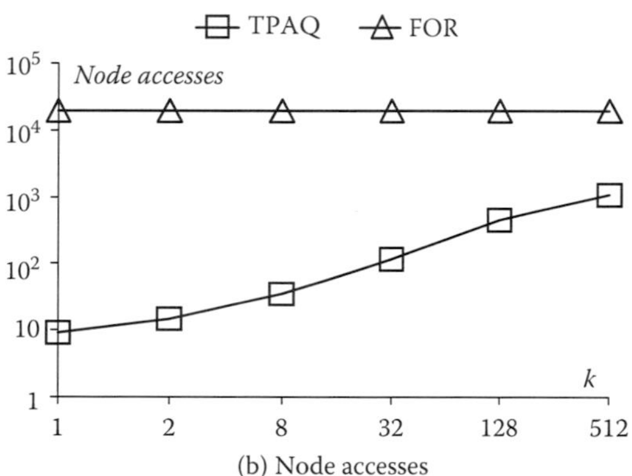

(b) Node accesses

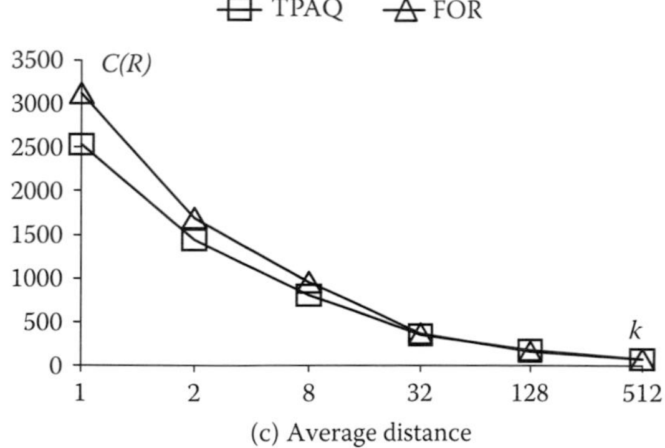

(c) Average distance

FIGURE 8.18 Performance versus k (LA, *avg*).

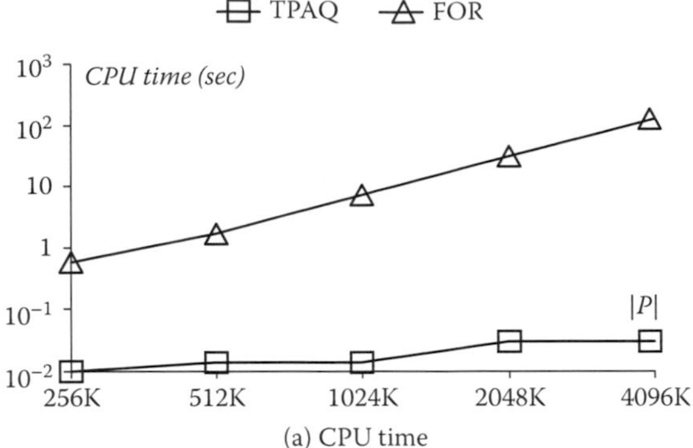

(a) CPU time

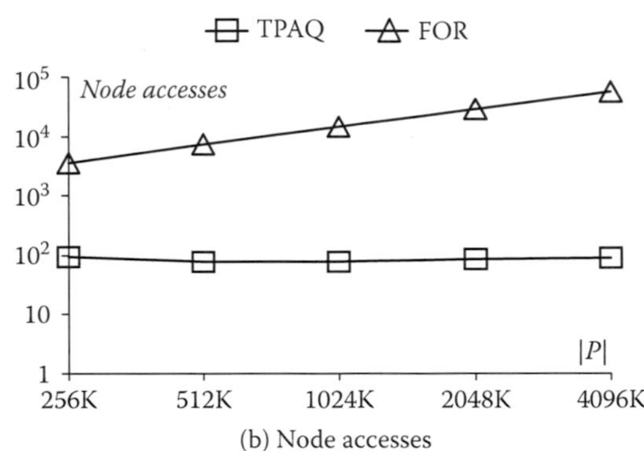

(b) Node accesses

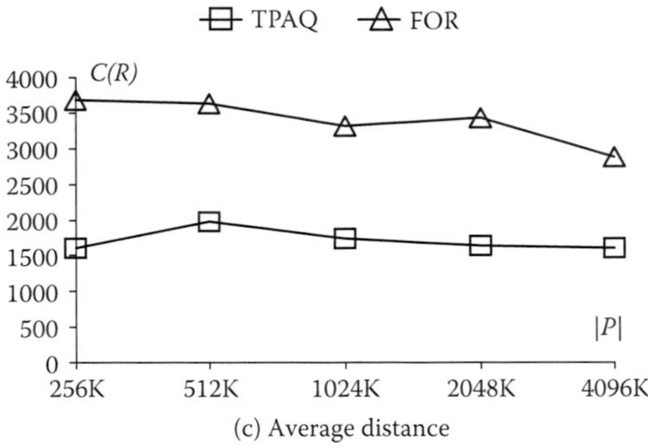

(c) Average distance

FIGURE 8.19 Performance versus |P| (SKW, *max*).

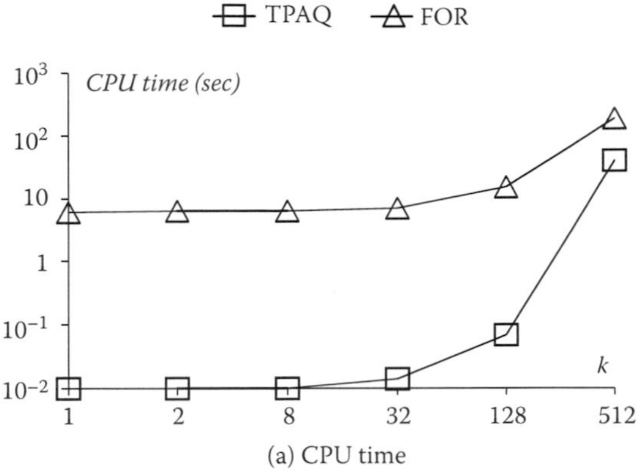

(a) CPU time

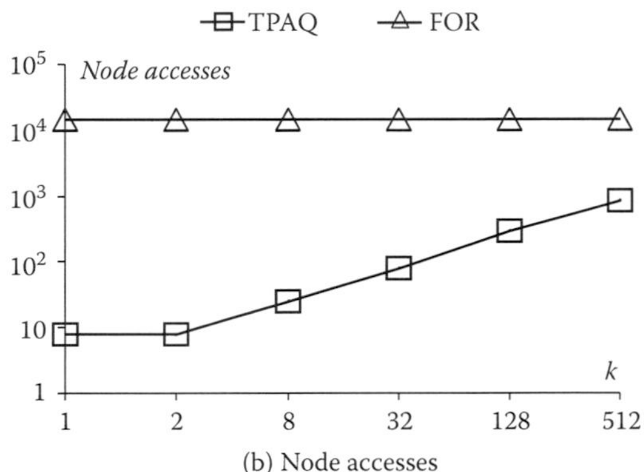

(b) Node accesses

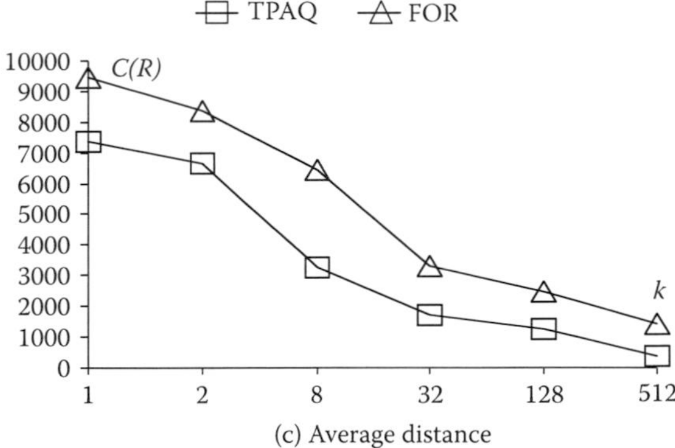

(c) Average distance

FIGURE 8.20 Performance versus *k* (SKW, *max*).

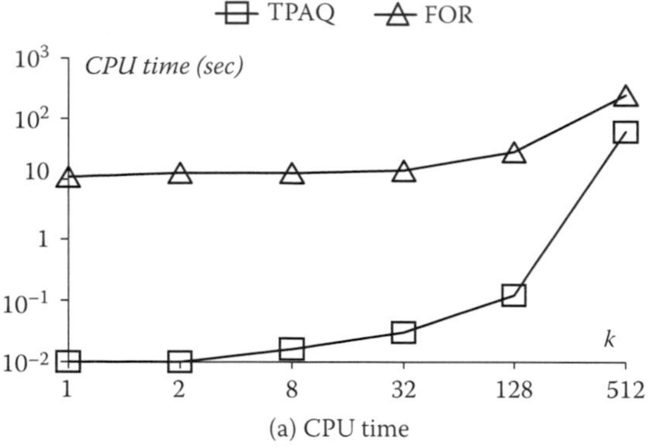

(a) CPU time

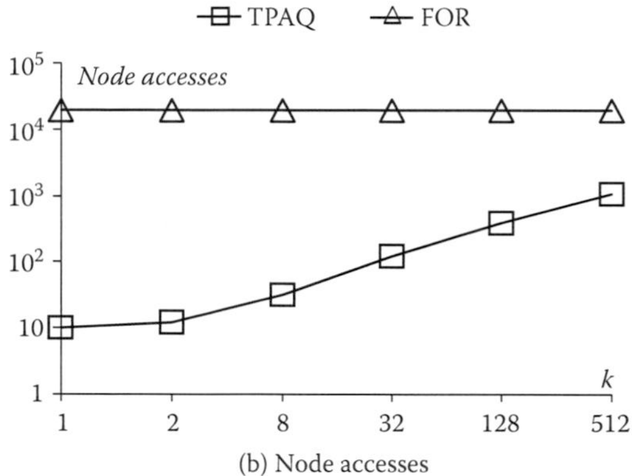

(b) Node accesses

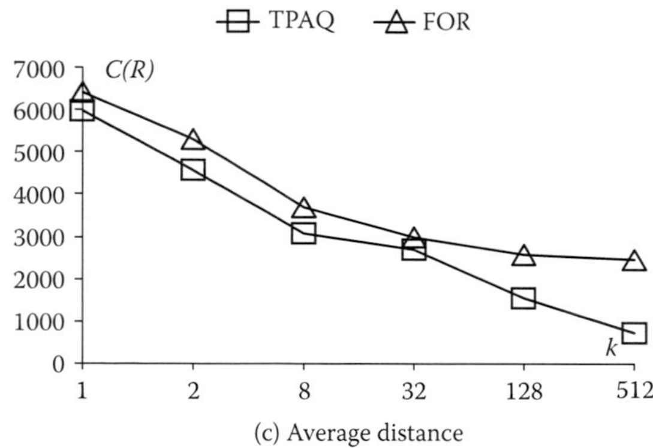

(c) Average distance

FIGURE 8.21 Performance versus k (LA, *max*).

of k incur higher I/O costs for TPAQ for the reasons explained in the context of Figure 8.17b. FOR performs a constant number of node accesses because it always reads the entire dataset. Regarding the quality of the returned medoid sets, our algorithm achieves much lower maximum distance $C(R)$.

8.7.2 Medoid-Aggregate Queries

In this section we study the performance of TPAQ-MA, starting with the *avg* case. We use datasets SKW (with 1M points) and LA, and vary T from 100 to 1500 (recall that our datasets cover a space with extent $10^4 \times 10^4$). Since there is no existing algorithm for processing such queries on large indexed datasets, we compare TPAQ-MA against an exhaustive algorithm (EXH) that works as follows. Let E be the set of entries at the partitioning level of TPAQ-MA. EXH computes and evaluates all the medoid sets for $|R| = 1$ up to $|R| = |E|$, by performing partitioning of E into slots with the technique presented in Section 4. EXH returns the medoid set that yields the closest average distance to T. Note that EXH is prohibitively expensive in practice because, for each examined value of $|R|$, it scans the entire dataset P in order to exactly evaluate $C(R)$. Therefore, we exclude EXH from the CPU and I/O cost charts.

Figure 8.22a shows the $C(R)$ for TPAQ-MA versus T on SKW. Clearly, the average distance returned by TPAQ-MA approximates the desired distance (dotted line)

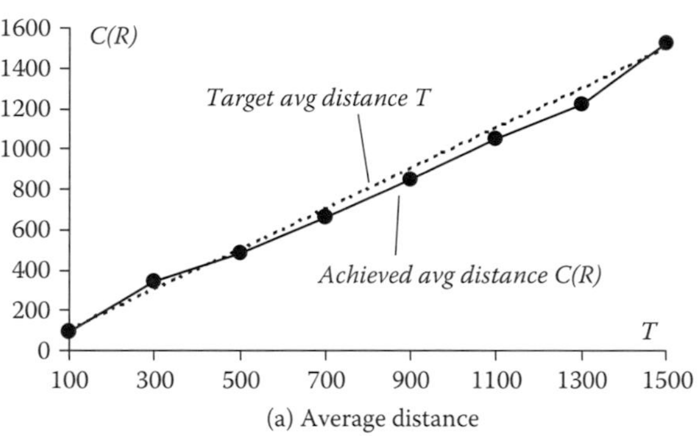

(a) Average distance

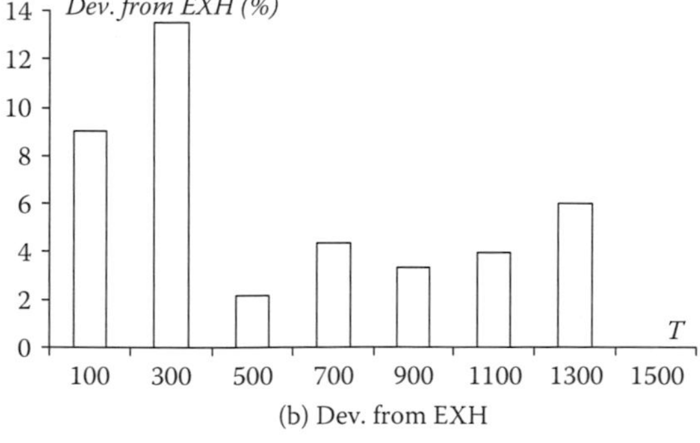

(b) Dev. from EXH

FIGURE 8.22 Performance versus T (SKW, *avg*).

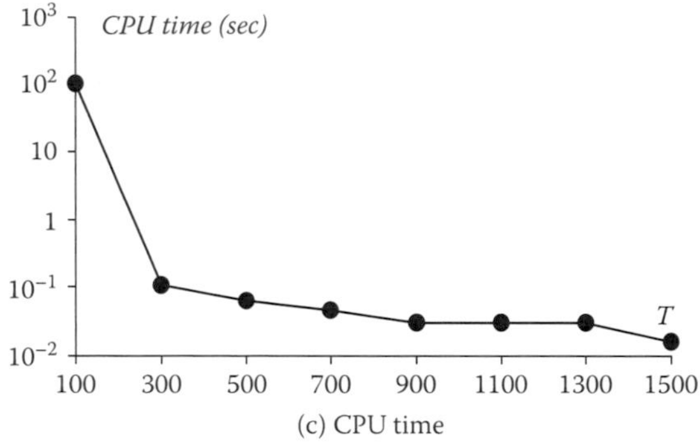

(c) CPU time

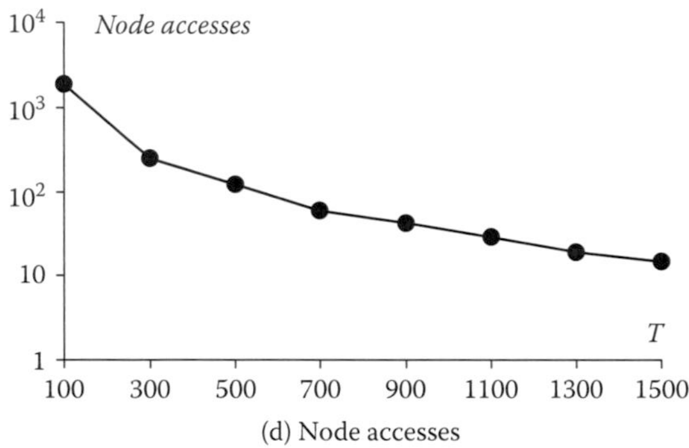

(d) Node accesses

FIGURE 8.22 (Continued).

very well. Figure 8.22b plots the deviation percentage between the average distances achieved by TPAQ-MA and EXH. The deviation is below 9% in all cases, except for $T = 300$ where it is equal to 13.4%. Interestingly, for $T = 1500$, TPAQ-MA returns exactly the same result as EXH with $|R| = 5$. Figure 8.22c and Figure 8.22d illustrate the CPU time and the node accesses of our method, respectively. For $T = 100$, both costs are relatively high (100.8 sec and 1839 node accesses) compared to larger values of T. The reason is that when $T = 100$, partitioning takes place at level 1 (i.e., the leaf level, which contains 14,184 entries) and returns $|R| = 1272$ medoids, incurring many computations and I/O operations. In all the other cases, partitioning takes place at level 2 (containing 203 entries), and TPAQ-MA runs in less than 0.11 sec and reads fewer than 251 pages.

Figure 8.23 repeats the above experiment for the LA dataset. Figure 8.23a and Figure 8.23b compare the average distance achieved by TPAQ-MA with the input value T and the result of EXH, respectively. The deviation from EXH is always smaller than 8.6%, while for $T = 1500$ the answer of TPAQ-MA is the same as EXH. Concerning the efficiency of TPAQ-MA, we observe that the algorithm has, in general, very low CPU and I/O cost. The highest cost is again in the case of $T = 100$ for the reasons explained in the context of Figure 8.22; TPAQ-MA partitions

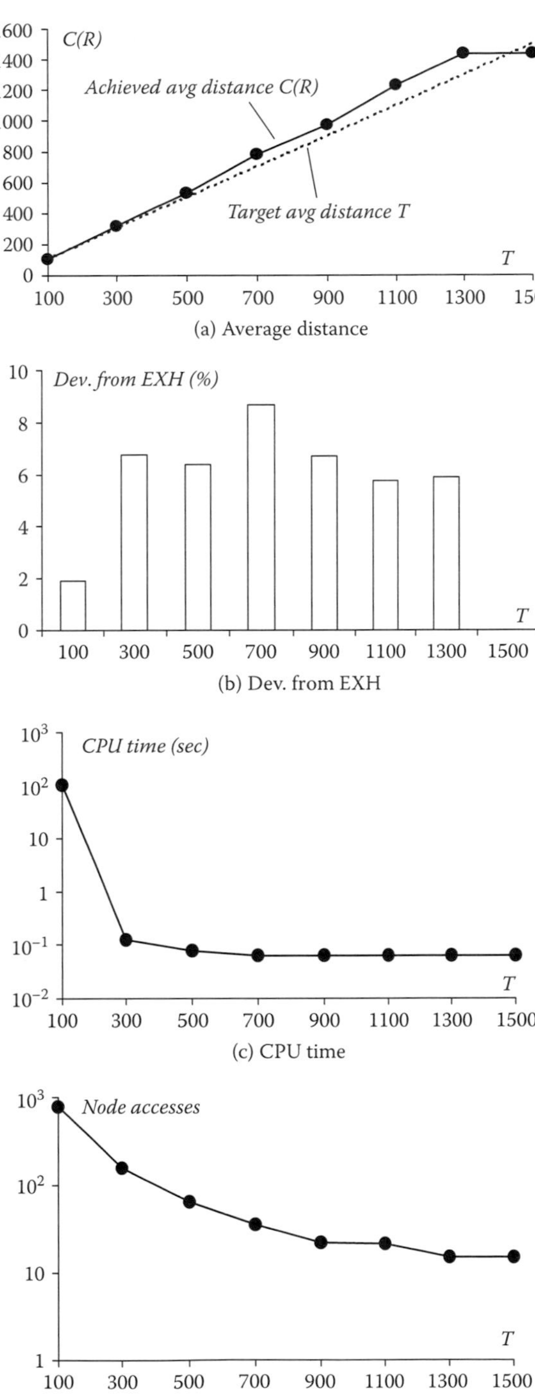

FIGURE 8.23 Performance versus T (LA, *avg*).

19,186 entries into slots and extracts $|R| = 296$ medoids, taking in total 105.6 sec and performing 781 node accesses.

In Figure 8.24 and Figure 8.25 we examine the performance of TPAQ-MA in the *max* case, using datasets SKW and LA. We compare again with the EXH algorithm. It is implemented as explained in the beginning of the subsection, the difference being that now it uses the *max* k-medoid TPAQ algorithm. For *max*, the range of T is from 500 to 1500. We do not use the same range as in the previous two experiments (i.e., 100 to 1500) because for T<500 the number of required medoids becomes very high and EXH requires several hours to terminate. As shown in Figure 8.24a and Figure 8.25a, the maximum distance of TPAQ-MA is close to the desired value T. In general, the deviation from EXH (illustrated in Figure 8.24b and Figure 8.25b) is low, and in the worst case it reaches 6.1% for SKW and 11.6% for LA. The algorithm terminates in less than 21 sec in all cases, and incurs a small number of node accesses.

8.7.3 Medoid-Optimization Queries

Finally, we experiment on the performance of TPAQ-MO, using datasets SKW (with 1M points) and LA. We process medoid-optimization queries with $f(C(R), |R|) = C(R) + Cost_{pm} \times |R|$, where $Cost_{pm}$ is the cost per medoid and ranges between 1 and 256. TPAQ-MO is again compared with an exhaustive algorithm (EXH), which in the MO case (1) computes all the medoid sets with $|R|$ from 1 to $|E|$, by performing partitioning into slots in the same level as TPAQ-MO, (2) calculates the (average or maximum) distance $C(R)$ achieved for each considered set, and (3) returns the one that minimizes function f.

First, we experiment on *avg* MO queries using the SKW dataset. Figure 8.26a plots the deviation percentage (between the values of f achieved by TPAQ-MO and EXH) as a function of the cost $Cost_{pm}$ per medoid. The deviation does not exceed 1.8% in any case. Interestingly, TPAQ-MO returns exactly the same medoid sets as EXH for many values of $Cost_{pm}$, verifying the effectiveness of the gradient descent technique and the accuracy of the estimators described in Section 6. Figure 8.26b and Figure 8.26c show the CPU and I/O costs of the algorithm. In both charts, the cost of TPAQ-MO is much higher when $Cost_{pm} \leq 8$. In these cases, the CPU time is between 147 and 157 sec and the number of node accesses ranges between 251 and 430. The returned medoid sets have size $|R|$ between 33 and 174. On the other hand, when $Cost_{pm} > 8$ the CPU time is less than 0.1 sec and the incurred node accesses are fewer than 60. The answer contains from 3 to 24 medoids. This large difference is explained by the fact that when $Cost_{pm} \leq 8$ partitioning takes place in level 1 (with 14,184 entries), while for $Cost_{pm} > 8$ the partitioning level is level 2 (with 203 entries).

In Figure 8.27 we repeat the above experiment for the LA dataset. The performance of TPAQ-MO is very similar to the SKW case. The deviation of TPAQ-MO from EXH is 0.07% and 1.82% for $Cost_{pm}$ equal to 4 and 8, respectively. For all the other values of $Cost_{pm}$, our algorithm retrieves the same medoid set as EXH. The cost of TPAQ-MO is plotted in Figure 8.27b and Figure 8.27c. There is a large difference in both the CPU time and the node accesses for $Cost_{pm} \leq 4$ and $Cost_{pm} > 4$. The reason for this behavior is the same as in Figure 8.26.

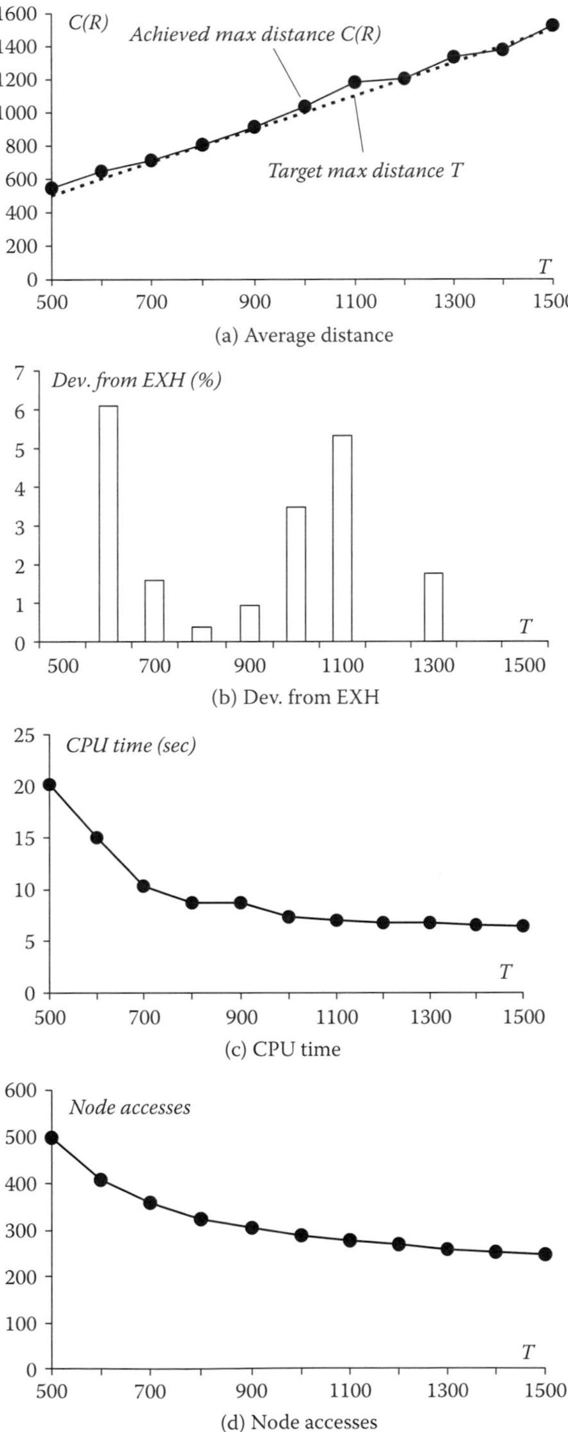

FIGURE 8.24 Performance versus T (SKW, *max*).

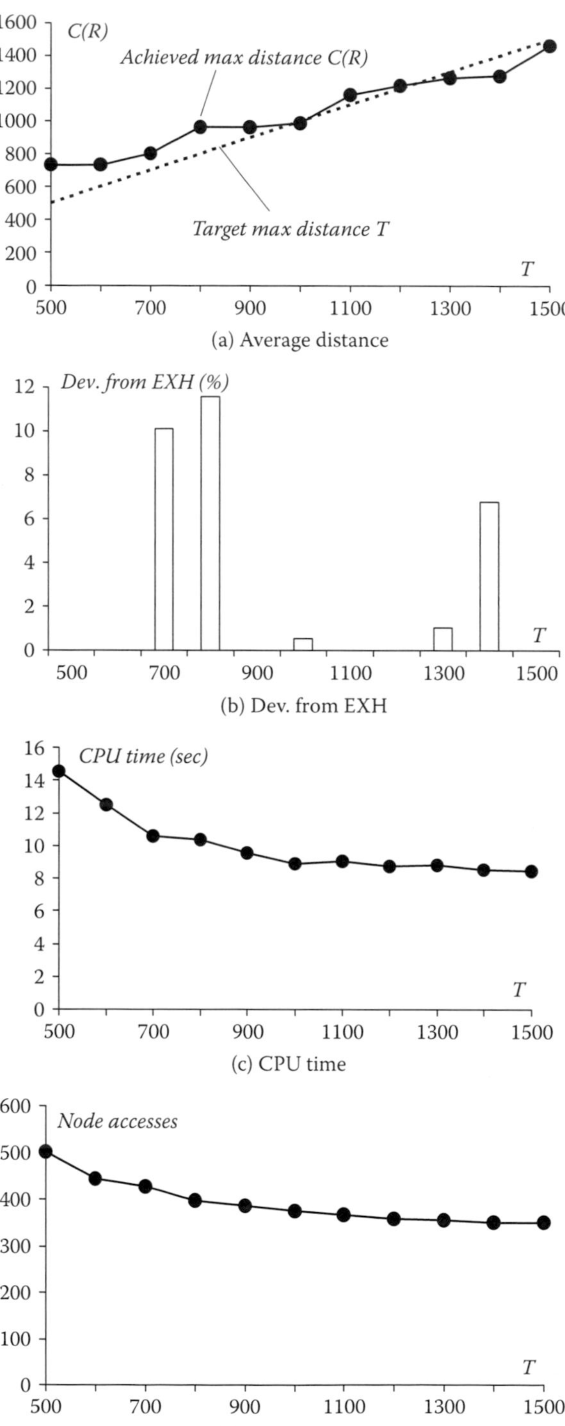

FIGURE 8.25 Performance versus T (LA, *max*).

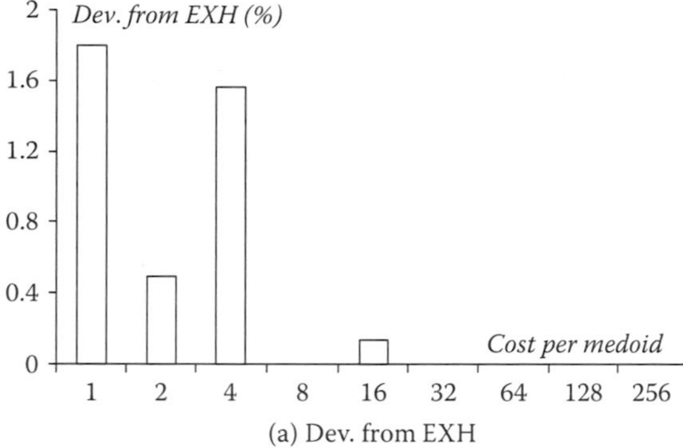

(a) Dev. from EXH

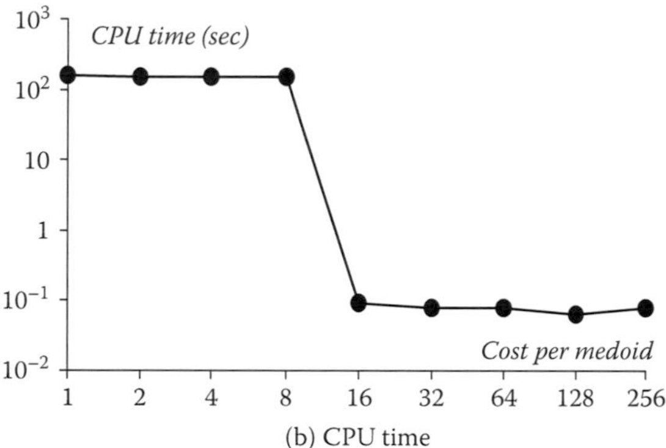

(b) CPU time

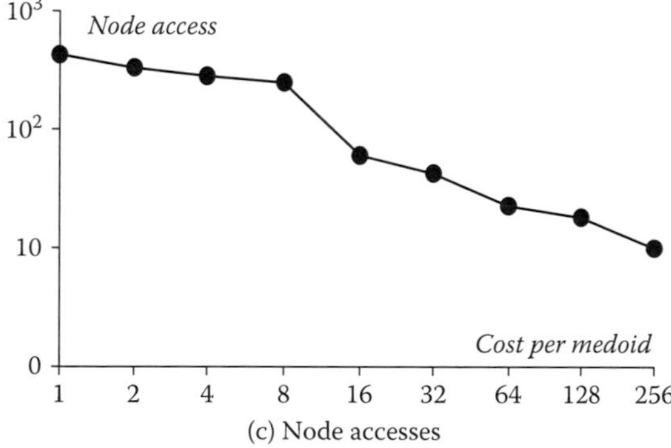

(c) Node accesses

FIGURE 8.26 Performance versus $Cost_{pm}$ (SKW, *avg*).

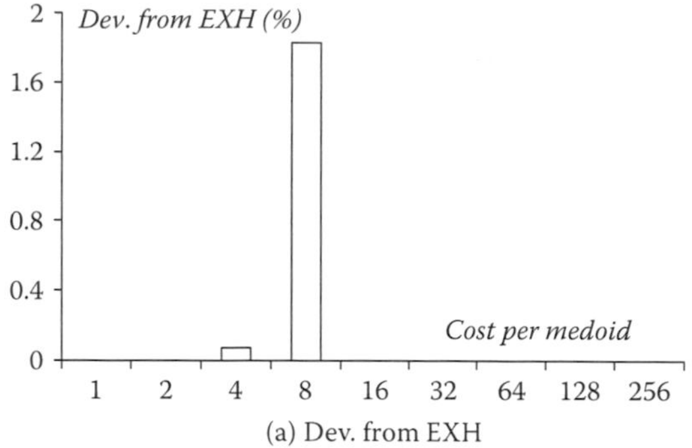

(a) Dev. from EXH

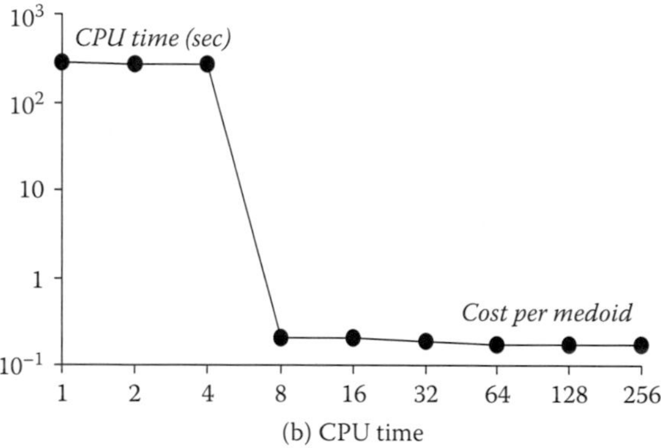

(b) CPU time

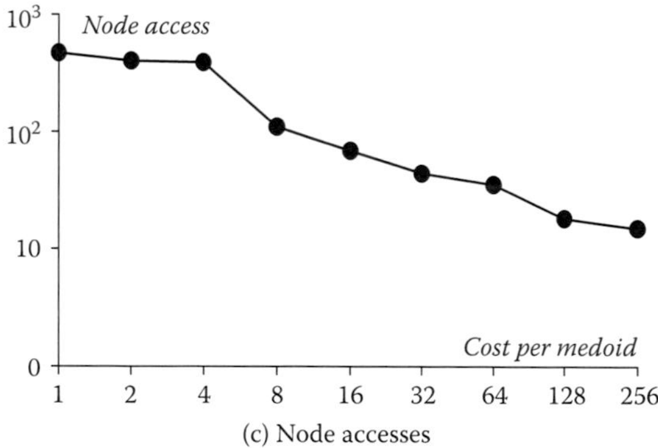

(c) Node accesses

FIGURE 8.27 Performance versus $Cost_{pm}$ (LA, *avg*).

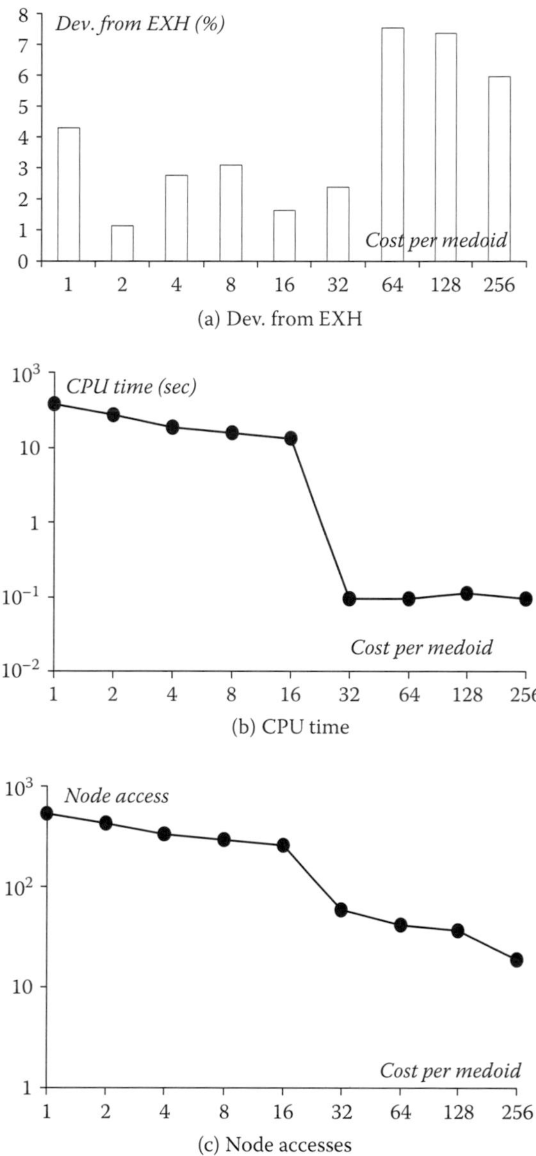

(a) Dev. from EXH

(b) CPU time

(c) Node accesses

FIGURE 8.28 Performance versus $Cost_{pm}$ (SKW, *max*).

In the last two experiments we focus on *max* MO queries. Figure 8.28 and Figure 8.29 illustrate the performance of TPAQ-MO when $Cost_{pm}$ varies between 1 and 256, using datasets SKW and LA, respectively. The deviation from EXH is usually small. For SKW, the maximum deviation is 7.5%. For LA, the deviation is in general higher; on the average it is around 10% with maximum value 22.3% (for $Cost_{pm} = 8$). TPAQ-MO performs worse for LA because it contains large empty

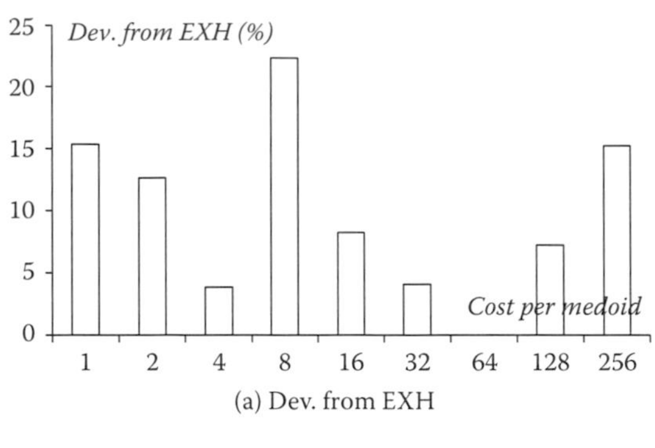

(a) Dev. from EXH

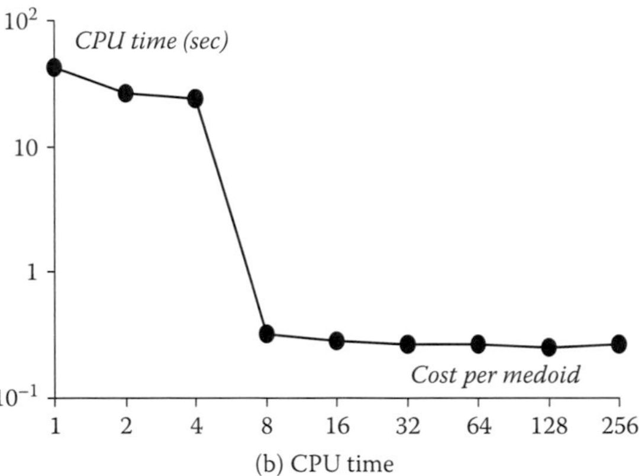

(b) CPU time

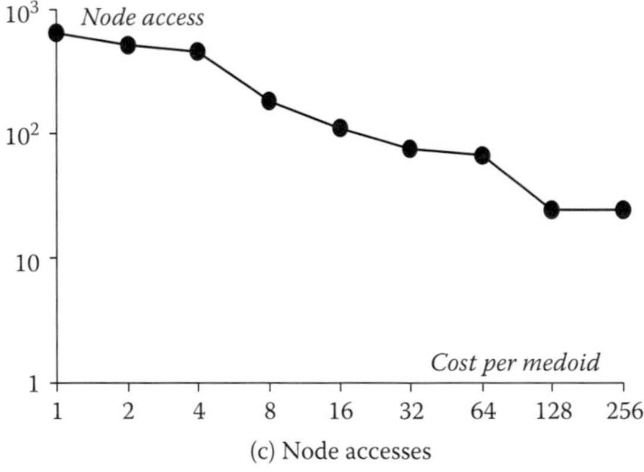

(c) Node accesses

FIGURE 8.29 Performance versus $Cost_{pm}$ (LA, *max*).

areas. On the other hand, SKW (even though it is very skewed) covers the whole workspace. Concerning the CPU time of TPAQ-MO, it does not exceed 43 sec in any case. As in Figure 8.26 and Figure 8.27, both the I/O and the CPU costs drop when partitioning takes place at a higher level. For SKW (for LA), the partitioning level is level 1 for $Cost_{pm} \leq 16$ (for $Cost_{pm} \leq 4$), while for higher $Cost_{pm}$ it is level 2.

8.8 CONCLUSION

This chapter studies k-medoids and related problems in large spatial databases. In particular, we consider k-medoid, MA, and MO queries. We present TPAQ, a framework that efficiently processes all three query types, and is applicable to both their *avg* and *max* versions. TPAQ provides high-quality answers almost instantaneously, by exploiting the data partitioning properties of a spatial access method on the input dataset. TPAQ is a three-step methodology that works as follows. Initially, it descends the index, and stops at the topmost level that provides sufficient information about the underlying data distribution. Next, it partitions the entries of the selected level into a number of slots. Finally, it performs a NN query to retrieve one medoid for each slot. Extensive experiments with synthetic and real datasets demonstrate that (1) TPAQ outperforms the state-of-the-art method for k-medoid queries by orders of magnitude, while achieving results of better or comparable quality, and (2) TPAQ is also very efficient and effective in processing MA and MO queries. TPAQ relies on spatial indexing, which is known to suffer from the dimensionality curse (Korn, Pagel, and Faloutsos, 2001). A challenging direction for future work is to extend it to high-dimensional spaces, using appropriate data partition indexes (Berchtold et al., 1996).

REFERENCES

Ankerst, M., Breunig, M., Kriegel, H.P., and Sander, J. OPTICS: Ordering points to identify the clustering structure. *SIGMOD*, 1999.

Arora, S., Raghavan, P., and Rao, S. Approximation schemes for Euclidean k-medians and related problems. *STOC*, 1998.

Beckmann, N., Kriegel, H.P., Schneider, R., and Seeger, B. The R*-tree: An efficient and robust access method for points and rectangles. *SIGMOD*, 1990.

Berchtold, S., Keim, D., and Kriegel, H. The X-tree: An index structure for high-dimensional data. *VLDB*, 1996.

Ester, M., Kriegel, H.P., Sander, J., and Xu, X. A density-based algorithm for discovering clusters in large spatial databases with noise. *KDD*, 1996.

Ester, M., Kriegel, H.P., and Xu, X. A database interface for clustering in large spatial databases. *KDD*, 1995a.

Ester, M., Kriegel, H.P., and Xu, X. Knowledge discovery in large spatial databases: focusing techniques for efficient class identification. *SSD*, 1995b.

Fayyad, U., Piatetsky-Shapiro, G., Smyth, P., and Uthurusamy, R. *Advances in Knowledge Discovery and Data Mining*. AAAI/MIT Press, 1996.

Garey, M. and Johnson, D. *Computers and Intractability: A Guide to the Theory of NP-Completeness*. W.H. Freeman, 1979.

Gonzalez, T. Clustering to minimize the maximum intercluster distance. *Theoretical Computer Science*, 38: 293–306, 1985.

Guha, S., Rastogi, R., and Shim, K. CURE: An efficient clustering algorithm for large databases. *SIGMOD*, 1998.

Hamerly, G. and Elkan, C. Learning the k in k-means. *NIPS*, 2003.

Hartigan, J.A. *Clustering Algorithms*. Wiley, 1975.

Hastie, T., Tibshirani, R., and Friedman, J. *The Elements of Statistical Learning*. Springer-Verlag, 2001.

Hjaltason, G. and Samet, H. Distance browsing in spatial databases. *ACM TODS*, 24(2): 265–318, 1999.

Kamel, I. and Faloutsos, C. On packing r-trees. *CIKM*, 1993.

Kaufman, L. and Rousseeuw, P. *Finding Groups in Data*. Wiley-Interscience, 1990.

Korn, F., Pagel, B.U., and Faloutsos, C. On the 'dimensionality curse' and the 'self-similarity blessing'. *TKDE*, 13(1): 96–111, 2001.

Lo, M.L. and Ravishankar, C.V. Generating seeded trees from data sets. *SSD*, 1995.

Lo, M.L. and Ravishankar, C.V. The design and implementation of seeded trees: An efficient method for spatial joins. *TKDE*, 10(1): 136–151, 1998.

Mamoulis, N. and Papadias, D. Slot index spatial join. *TKDE*, 15(1): 211–231, 2003.

Moon, B., Jagadish, H.V., Faloutsos, C., and Saltz, J.H. Analysis of the clustering properties of the hilbert space-filling curve. *TKDE*, 13(1): 124–141, 2001.

Mouratidis, K., Papadias, D., and Papadimitriou S. Tree-based partition querying: a methodology for computing medoids in large spatial datasets. *VLDB Journal*, 17(4):923–945, 2008.

Ng, R. and Han, J. Efficient and effective clustering methods for spatial data mining. *VLDB*, 1994.

Pelleg, D. and Moore, A.W. Accelerating exact k-means algorithms with geometric reasoning. *KDD*, 1999.

Pelleg, D. and Moore, A.W. X-means: Extending k-means with efficient estimation of the number of clusters. *ICML*, 2000.

Roussopoulos, N., Kelly, S., and Vincent, F. Nearest neighbor queries. *SIGMOD*, 1995.

Theodoridis, Y., Stefanakis, E., and Sellis, T. Efficient cost models for spatial queries using r-trees. *TKDE*, 12(1): 19-32, 2000.

Welzl, E. Smallest enclosing disks (balls and ellipsoids). *New Results and New Trends in Computer Science*, 555: 359–370, 1991.

Zhang, D., Du, Y., Xia, T., and Tao, Y. Progressive computation of the min-dist optimal-location query. *VLDB*, 2006.

Zhang, T., Ramakrishnan, R., and Livny, M. BIRCH: An efficient data clustering method for very large databases. *SIGMOD*, 1996.

9 Looking for a Relationship? Try GWR*

A. Stewart Fotheringham

Martin Charlton

Urška Demšar

CONTENTS

9.1 INTRODUCTION

It is often desirable in an analysis to examine the relationship between two or more variables. By relationship, we mean the manner in which one variable changes given change in another variable, *ceteris paribus.* An increase in the value of one variable might be associated with an increase in another; conversely, an increase in one variable might be associated with a decrease in another. It is very tempting when faced

* Research presented in this chapter was funded by a Strategic Research Cluster grant (07/SRC/I1168) by Science Foundation Ireland under the National Development Plan. The authors gratefully acknowledge this support.

with a large number of variables, perhaps for thousands of observations, to reach for the principal components option in a statistical package and let the computer report the relationships it has found. Such a course of action would be abrogating one's responsibility as an analyst to a computer and, more importantly, the resulting analysis would be partial, at best.

We shall assume that, long before you start using any of the regression techniques that appear in this chapter, you have carried out a substantial univariate analysis. What are the properties of each variable's distribution? What shape are the distributions — symmetrical or skew? Are there any outliers? If so, where are they? Some simple descriptive statistics and graphical devices such as maps, histograms, and boxplots will go a long way toward helping you understand your data although they will probably be limited in their ability to convey information about relationships, particularly in a multivariate context.

Thinking about relationships between variables involves complexity. We shall assume that there is one variable of interest and some suggestions that the changes in its values are associated in some systematic way with changes in one or more other variables. The former we shall refer to as the "dependent" variable and the latter as the "independent" variable(s). Dependent variables are sometimes also known as response variables or regressands; independent variables are sometimes known as explanatory or predictor variables, or regressors. A convenient way of summarizing the relationship between a dependent variable and a set of independent variables is via regression.

A common technique for fitting regression models is known as ordinary least squares (OLS). The assumptions behind OLS can be violated when modeling with spatial data; the potential nonindependence of the observations is a particular concern. Spatial dependency can exist within the individual variables — they exhibit spatial autocorrelation; it can also exist in the residuals from the model itself. Spatial lag and spatial error models have been developed to deal with these situations (Anselin, 1988). Alternative models have also been proposed to deal with situations where the model structure is not spatially stationary — early attempts include parameter expansion (Casetti, 1972) and spatial adaptive filtering (Forster and Gorr, 1986), and more recently locally weighted regression (McMillen, 1996). Geographically weighted regression (GWR) (Brunsdon, Fotheringham, and Charlton, 1996; Fotheringham, Brunsdon, and Charlton, 2002) is perhaps the most well known of the methods that produce locally varying parameter estimates.

GWR is a useful exploratory technique. A "classic" regression implies that the model structure is spatially constant across the study area. With a large and complex dataset, this may not be the case. GWR allows the analyst to model the spatial nonstationarity, and then seek evidence for whether what has been found is systematic or not. In doing so, the analyst has the opportunity of asking further questions about the structures in the data. In this chapter, we use some techniques from the field of geovisual analytics (GA) to examine some of the interactions in the GWR parameter surfaces and highlight some local areas of interest. As well as spatially varying parameters, the outputs from GWR include local estimates of parameter standard error, goodness of fit, and influence. GWR can also be used to validate a model that has been fitted to a subsample of the data on another subsample of the data.

The rest of this chapter examines GWR as a data-mining tool. In Section 2, we present a brief outline of linear regression. In Section 3, we then consider briefly some of the problems that face the analyst when using OLS regression with spatial data and some possible solutions. Section 4 presents GWR in some detail. Section 5 introduces GA techniques and shows how they might be used with GWR. In Section 6, we present an example of the use of GA and GWR in a case study of voter turnout in Dublin in the 2004 general election. Finally, we draw some conclusions about the interplay of GA and GWR in data mining.

9.2 LINEAR REGRESSION

It is worth rehearsing briefly an outline of linear regression before we consider GWR. Regression is a technique that allows us to model the relationship between a dependent variable and a set of independent variables. A typical *linear* model will take the following form:

$$y_i = \beta_0 + \beta_1 x_{i1} + \beta_2 x_{i2} + \cdots + \beta_n x_{in} + \varepsilon_i \tag{9.1}$$

where for the *i*th observation y_i is the dependent variable, $\boldsymbol{x}_i$ is the vector of independent variables, β is the vector of unknown parameters, and ε_i is a residual. The residuals should be independent, they should have a mean of zero, they should be normally distributed, and they should have constant variance. The parameters can be estimated from the sample data using the estimator,

$$\hat{\beta} = (X^T X)^{-1} X^T y \tag{9.2}$$

This approach, OLS, yields unbiased estimates of the parameters, if the conditions for the residuals are met. The parameter estimates are those for which the quantity $\Sigma_{i=1}^{n}(y_i - \hat{y}_i)^2$ is minimized (this is sometimes written as $\varepsilon^T\varepsilon$), where $\hat{y}_i$ is the predicted value of y_i using the estimated parameters in the model.

If the predictions of the model were identical to the sample y values, a plot of y against $\hat{y}$ would be a straight line. This is rarely the case; a measure of the goodness of fit is the R^2 value: $1 - \Sigma_{i=1}^{n}(y_i - \hat{y}_i)^2 / \Sigma_{i=1}^{n}(y_i - \bar{y})^2$. This is 1 less the ratio between the error sum of squares and the total sum of squares. If the predicted values are close to the sample values, then residuals will be small and the numerator in the expression for R^2 will be close to zero, with the result that R^2 is close to 1. If the fit is poor, the residuals will be large, the ratio approaches 1, and R^2 approaches 0. The R^2 statistic is commonly interpreted as the proportion of variance in the independent variable explained ("accounted for") by the model.

The parameter estimates may also be tested against a particular value. The OLS methodology also yields a standard error for the parameter estimates. The parameter estimate divided by its standard error gives a T statistic for the null hypothesis that the parameter is zero. If the parameter estimate for a particular variable is insignificantly different from zero, then the variation in that variable does not contribute to the variation in the y values — that is, it has no effect. Parameter estimates for which the null hypothesis of no relationship can be rejected are said to be *significant* — we should note that this does not mean *important*, merely that the unknown parameter is unlikely to be zero.

9.3 REGRESSION WITH SPATIAL DATA

The consequences of ignoring any undesirable characteristics in the parameters are unfortunate. Nonindependence can lead to biased parameter estimates — they may be too high or too low. Nonnormality and nonuniform variance (heteroscedasticity) can produce underestimates of the standard errors; the misleading T statistics may indicate spuriously significant variables.

Unfortunately, spatial data may exhibit some particularly awkward characteristics — samples that are spatially proximate are likely to be more similar than those that are distant. This phenomenon was neatly aphorized by Tobler (1970), who observed that "everything is related to everything else, but near things are more related than distant things."

If spatial dependence is one problem, then another associated problem is spatial heterogeneity. Spatial heterogeneity exists when the structure of some spatial process is not uniform over a study area — in other words, the parameters of the spatial process are not spatially uniform. This might seem a rather strange thing to assert, but consider orographic precipitation. Rainfall generally increases with elevation. However, the west sides of mountain ranges are frequently subjected to wetter conditions than the east sides because of the upward movement of the air masses. A unit increase in elevation on the west side of a range yields greater rainfall than a corresponding unit increase on the east side. A model that attempted to predict rainfall from elevation would show a distinct spatial pattern in the residuals, with positive ones on the west side (where the model will under predict) and negative ones on the east side (where the model will over predict). What would be desirable for such a model would be one in which the parameters might be allowed to vary spatially to reflect the different processes being modeled.

There have been numerous attempts to produce models with spatially varying parameters. The study area may be divided into spatial subsets and the same model fitted to data for those subsets. Dummy variables may be used — one for each region in the study area; these may be combined with suitable interaction terms. Perhaps most notable among these is Casetti's expansion method (Casetti, 1972), in which the parameter estimates are conditioned on the coordinates of the observations. The analyst must specify the nature of the functional form for the coordinate expansions. As Jones (1984) illustrates, this can quickly lead to a rather complex model for an apparently simple process. If z and w are the dependent and independent variables measured at locations with coordinates x and y in the study area, and we assume a quadratic expansion, then the nonspatial model,

$$z = a + bw, \tag{9.3}$$

metamorphoses into

$$z = a + (b_0 + b_1 x + b_2 y + b_3 x^2 + b_4 y^2 + b_5 xy)w \tag{9.4}$$

where a, $b_0 \ldots b_5$ are the parameters to be estimated. Jones (1984) shows how a three-variable model was initially estimated with a cubic expansion, leading to 30 initial parameters estimates of which 4 were significant in the final model.

9.4 GEOGRAPHICALLY WEIGHTED REGRESSION

An alternative approach is offered by GWR (Brunsdon et al. 1996; Fotheringham et al., 1998, 2002). We assume that spatial coordinates are available for the sample observations, that the ith observation is given by the vector **i**. We shall refer to the locations where data are sampled as sample points, and those where parameters are to be estimated as regression points. We will also term a model fitted using OLS as *global*, and one fitted using GWR OLS as *local*.

Initially we shall consider the case where the sample points and regression points are coincident. Fitting a GWR model using OLS involves estimating the location specific parameters $\beta(\mathbf{i})$, where there will be one set of parameter estimates for each regression point.

$$y_i = \beta_{i0}(\mathbf{i}) + \beta_{i1}(\mathbf{i})x_{i1} + \beta_{i2}(\mathbf{i})x_{i2} \ldots + \beta_{in}(\mathbf{i})x_{in} + \varepsilon_i \tag{9.5}$$

The estimator is a weighted OLS estimator:

$$\hat{\beta}(\mathbf{i}) = (X^T W(\mathbf{i}) X)^{-1} X^T W(\mathbf{i}) y \tag{9.6}$$

The geographical weighting for the ith observation is given by a kernel—for example, a Gaussian-like kernel. This is a square matrix whose leading diagonal contains the weights for the observations j relative to location **i**, the current regression point. The weights are obtained thus,

$$w_j(\mathbf{i}) = e^{-\frac{1}{2\pi}\left(d_j(\mathbf{i})/h\right)^2} \tag{9.7}$$

where $d_j(\mathbf{i})$ is the distance from observation j to observation i and h is a smoothing parameter known as the bandwidth. There are other kernels that can be used with GWR.

Parameter estimates are obtained by solving the estimator for each regression point. Standard errors are a little more involved and the interested reader is directed to Fotheringham et al. (2002) for more detail. The same dependent and independent variables are used each time; only the weights change. As the regression points and sample points are identical, a prediction of y may be made from the observation specific parameter estimates and a residual computed.

9.4.1 How Well Does the Model Fit?

We have to consider two aspects here — first the complexity of the model and second how well the model replicates the sampled y values. In regression, there exists a matrix, the *hat matrix*, which maps the predicted ys onto the sampled ys, thus

$$\hat{y} = \mathbf{S}y \tag{9.8}$$

The trace of this matrix gives us the number of independent parameter estimates in the model — for global OLS this is equivalent to the number of independent variables plus one. For a local model, it yields the *effective* number of parameters. Each row $\mathbf{r}_i$ is

$$\mathbf{r}_i = X_i(X^T W(\mathbf{i})X)^{-1} X^T W(\mathbf{i}) \tag{9.9}$$

If $v_1 = \text{tr}(\mathbf{S})$ and $v_2 = \text{tr}(\mathbf{S}^T\mathbf{S})$, then the effective number of parameters in the model is $2v_1 - v_2$; as v_1 and v_2 are often very similar in value, tr(**S**) is a reasonable approximation to the effective number of parameters.

It should be clear that if there are n observations in the dataset, there will be n sets of parameter estimates, the hat matrix will have n rows, and the estimator will have been evaluated n times. However, we do not run out of degrees of freedom because we are re-using the sample data during each estimation — the contribution of an individual observation i is approximately the value found in S_{ii}. The effective number of parameters will always be larger than or equal to that for an equivalent global model.

Now we can turn to measuring how well the model fits. In a global model, the R^2 can often be increased by the addition of extra variables. It is usual to report the R^2 as computed above, and an adjusted R^2 that incorporates a correction for the number of variables in the model, relative to the number of observations used to fit it:

$$\bar{R}^2 = 1 - R^2 \frac{n-1}{n-p-1} \tag{9.10}$$

However, this can no longer be interpreted as a percentage in the way that the R^2 may be.

An alternative measure of goodness of fit is provided by a version of the information criterion originally proposed by Akaike (1973) modified by Hurvich, Simonoff, and Tsai (1998):

$$AICc = 2n\log_e(\hat{\sigma}) + n\log_e(2\pi) + n\left(\frac{n + tr(\mathbf{S})}{n - 2 - tr(\mathbf{S})}\right) \tag{9.11}$$

where n is the number of observations, $\hat{\sigma}$ is the estimated standard deviation of the error term, and $tr(S)$ is the trace of the hat matrix. This combines the badness of fit measure with a penalty for model complexity. We refer to this measure as the *corrected Akaike information criterion* (AICc). It has the advantage that for the same y variable different models with very different structures may be compared; for example, a global model and a local model fitted to the same data.

When two or more models are being compared for goodness of fit using the AICc, the model with the lowest AICc is taken as the "best" model — it is closer to the unknown "true" model. Two models are held to be similar if the difference in the AICcs is three or less (Burnham and Anderson, 2002). The AICc is a relative measure, not an absolute measure, and is proportional, *inter alia*, to the sample size n. Therefore, to compare the fit of the global and local models, all that is necessary is to compare with the AICc values — if the global model's value is lower by three or more, then there is reasonable evidence that the local model is a better fit to the data, given the different model structures.

9.4.2 What Is the Best Bandwidth?

In the discussion of the kernel above, we introduced the concept of the bandwidth and described it as a smoothing parameter. From the equation, we can see that a very large bandwidth (relative to the size of the study area) will result in weights that are almost 1 — a global model. Very small bandwidths will give undue prominence to the values at or very close to the regression point — the resulting parameter estimates will not be a smooth surface. There will be one bandwidth at which the predicted and observed values will be closest; this is the "optimal" bandwidth. There are various automatic techniques for determining the minimum of a function; in our case, a function that fits a GWR to a set of data with a given bandwidth and that returns a goodness of fit measure should be used.

As well as the AICc, an alternative is the cross-validation score $\Sigma_{i=1}^{n}(y_i - \hat{y}_{\neq i}(h))^2$; each fitted value is computed with the weight for the observation at the regression point set to zero. When h is very small, any "wrap-around" is avoided as the predicted value at **i** includes only nearby observations but not the observation itself.

The AICc has some slight advantages over the cross-validation score in that it can be used to compare the local and global models, and can also be used with other model forms, such as Poisson or logistic.

9.5 GEOVISUAL ANALYTICS AND GWR

Spatial variability of the GWR parameters is usually examined by producing a choropleth map of each separate parameter surface and the statistical summaries for parameter estimates (local t-values, standard residuals, local R^2). These univariate maps are then scrutinized for patterns that give an idea about the spatial behavior of parameter values (Fotheringham et al., 2002). To identify multivariate spatial and nonspatial patterns, relationships, and other structures in the GWR result space that might provide crucial information about the causes of nonstationarity of spatial processes that the GWR models, the GWR results space might be treated as a highly dimensional spatial dataset and visually explored using a GA environment. GA is a subdiscipline of visual analytics (NVAC, 2005), an emergent research area in information visualization. It has evolved from geovisualization (MacEachren and Kraak, 2001), provides theory, methods, and tools for the visual exploration of geospatial data and includes pattern discovery, knowledge construction, and analytical reasoning.

Visual exploration of a spatial dataset in a GA environment is performed as a perceptual-cognitive process of alternatively interpreting and interacting with or manipulating multiple georeferenced visual displays. This means that the user is virtually looking and interactively searching for relationships and patterns in the data. Linked displays can include geographic visualizations, such as maps or cartograms, as well as any other multivariate visualizations or even constructs containing several visualizations, such as bivariate matrices or similar multidisplays. All displays are normally interactively connected by the concept of brushing and linking, which means that data elements that are interactively selected in one display are simultaneously highlighted or selected everywhere, which facilitates visual pattern recognition across multiple displays (Dykes, MacEachren, and Kraak, 2005).

In order to examine the GWR results space, the parameter estimates can be visualized using different techniques for different purposes. Correlations between pairs of parameter estimates can be examined in a matrix of bivariate visualizations, such as scatterplots or spacefills. A combination of a map with multivariate visualizations helps with the analysis of multivariate spatial patterns. A computational data mining method, such as for example a self-organizing map, can be linked with a map for a cluster analysis of the parameter-space and to see if eventual nonspatial clusters have a particular spatial distribution that is related to the underlying geographical processes. With such tasks in mind, visual exploration of the GWR results space can provide new insights into the result space of a statistical method that would otherwise remain unnoticed. The approach facilitates interpretation of the GWR results and supports analytical reasoning about the underlying spatial processes (Demšar, Fotheringham, and Charlton, 2008). How looking for patterns in the GWR space actually works is presented in the next section.

9.6 AN EXAMPLE — VOTER TURNOUT IN DUBLIN

To demonstrate both GWR and GA in tandem, we work through an example that is concerned with modeling variation in voter turnout in Greater Dublin for the general election held in Ireland in 2002. The turnout for an election is the proportion of the electorate that casts a vote. It is desirable that the turnout should be as high as possible; low turnouts are an indication of the electorate's withdrawal from the political process and are a cause for concern. There is some evidence that there is a socioeconomic dimension to turnout variation, and this forms the underlying rationale for our model. There is extensive analysis of Irish electoral participation (for example, Kavanagh, 2004, 2005; Kavanagh, Mills, and Sinnott, 2004).

9.6.1 The Data

For confidentiality reasons, the data on voter turnout have been aggregated to electoral divisions (ED). There are 322 EDs in the four counties that make up Greater Dublin: Fingal, Dublin City, South Dublin, and Dun Laoghaire-Rathdown. The area had a population of 1,122,821 in 2002, representing about 29% of the population of the Republic of Ireland. Figure 9.1 shows the four counties together with the areas mentioned in this chapter.

The turnout data are taken from the marked voting registers, which are used at each polling station. Every elector who casts a vote has a mark placed against his or her name in the register. From this, the totals for each polling station are available. For the study, counts for the individual polling stations inside each ED have been aggregated to provide an ED level count. The ED is the finest spatial unit for which census data are released, so this aggregation is appropriate.

The variation in turnout is shown in Figure 9.2. The highest turnouts are in the middle class parts of Dublin — Clontarf, Druncondra, Clonskeagh, and Dundrum — as well as in the more rural parts of Fingal. The lower turnouts in are in the more deprived areas — inner city Dublin, Ballymun, North Clondalkin, and West Tallaght.

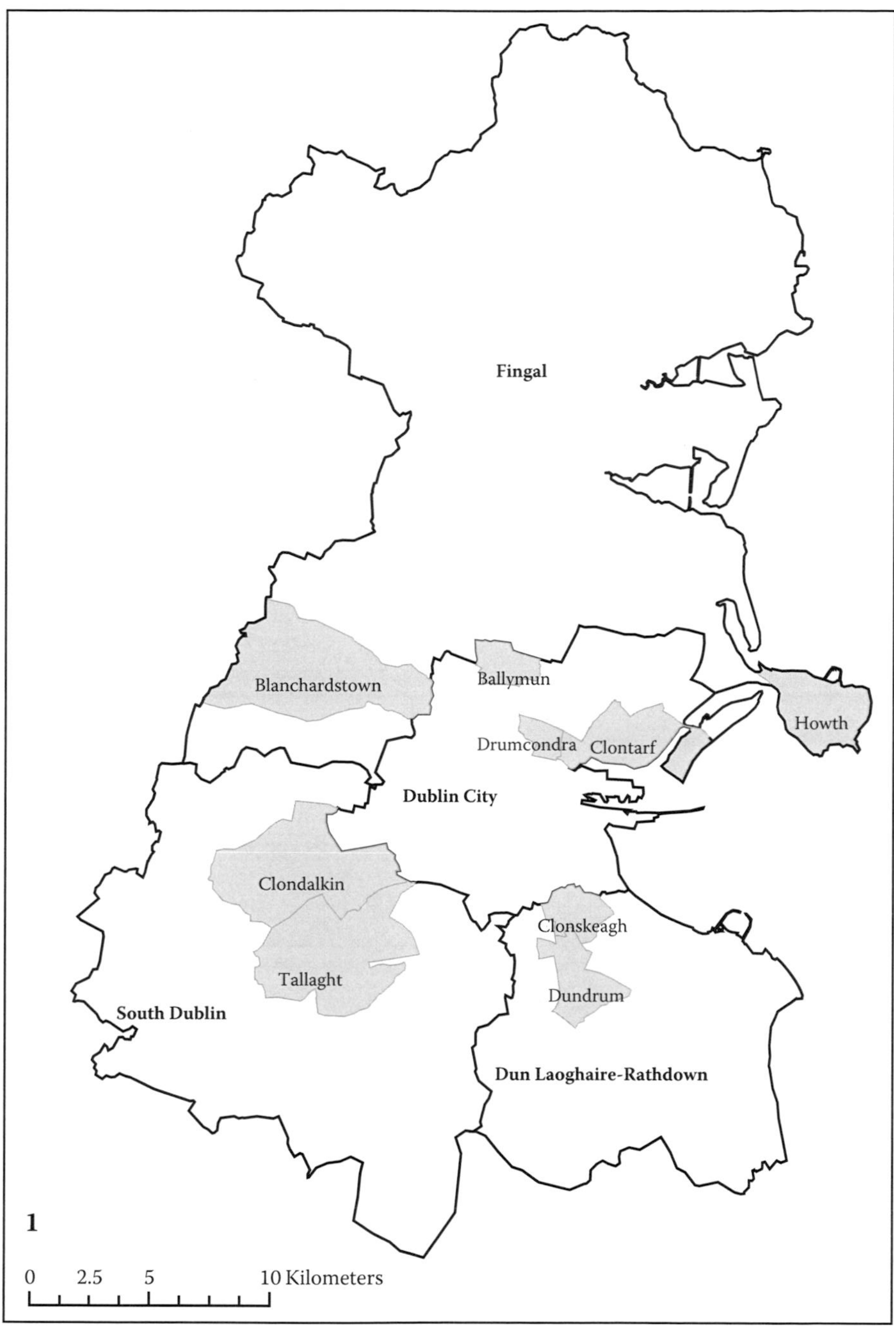

FIGURE 9.1 Index map showing counties and other locations mentioned in the text.

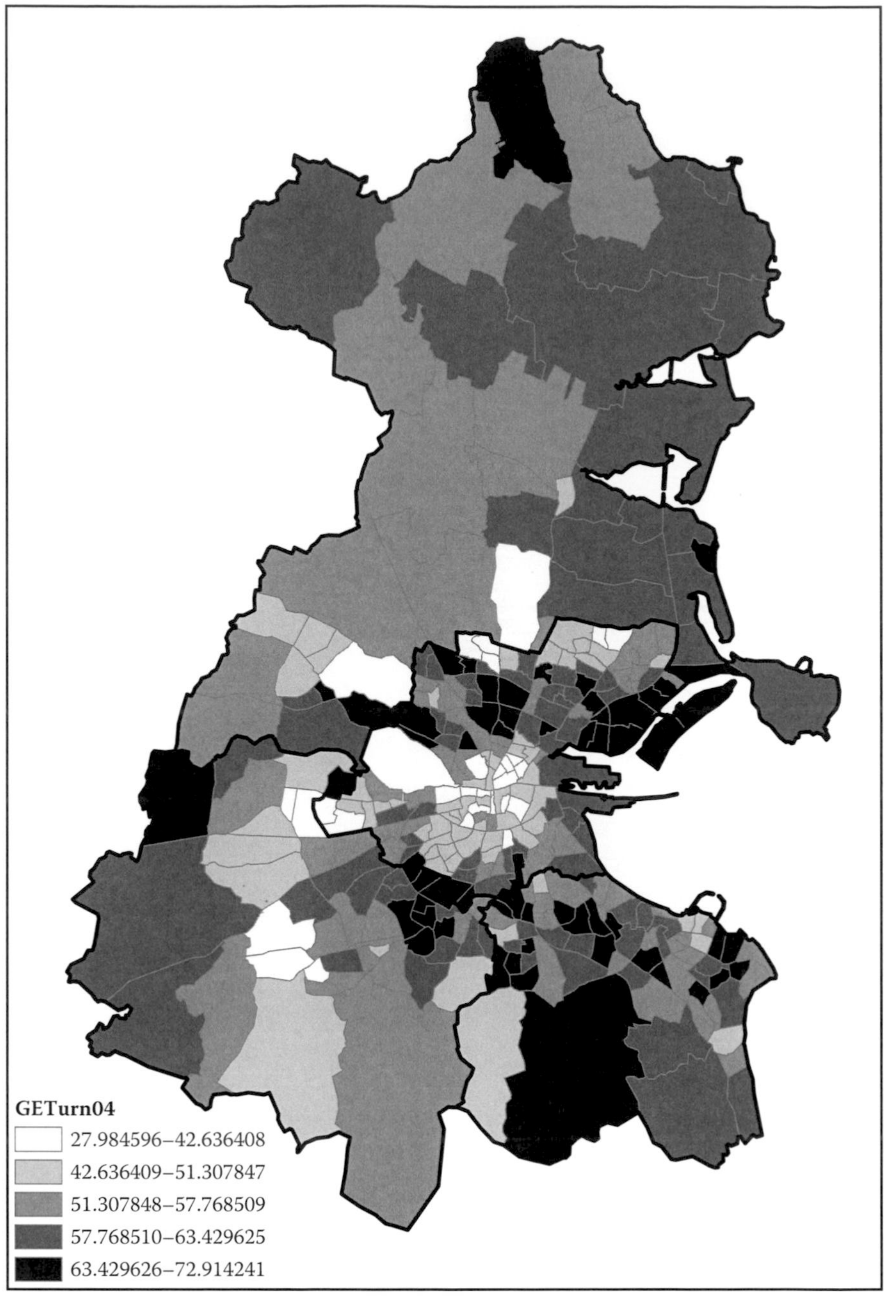

FIGURE 9.2 Percentage turnout by electoral division in Greater Dublin.

Turnout variation may be related to a number of social and economic factors. Directly measuring disenchantment with the political process is not easy, so we are forced toward proxies from the census. Eight regressors have been chosen:

1. Proportion of persons whose address was different 12 months previously
2. Proportion of households in property rented from the council
3. Proportion of persons in social class 1
4. Proportion of the labor force that is unemployed
5. Proportion of residents aged 15+ with no formal education, or education to primary or lower secondary level
6. Proportion of the electorate aged 18–24
7. Proportion of the electorate aged 25–44
8. Proportion of the electorate aged 45–64

The data were extracted from the 2002 Small Area Population Statistics made available by the Central Statistics Office, Ireland. The turnout counts were normalized by the count of the electorate (persons aged 16 or over) in each ED.

GWR requires location coordinates for each observation. The EDs in Dublin have simple shapes so it was appropriate to take the geometric centroid as the location of both the sample points and the regression points.

9.6.2 General Patterns in the Data

An initial correlation analysis reveals moderate levels of correlation between the dependent and independent variables. Two variables have significant positive correlations (Social Class 1 0.35, Electorate 45–64 0.42). The rest of the independent variables have significant negative correlations (Different Address –0.31, Local Authority Renters –0.64, Unemployment –0.63, Educational Disadvantage –0.32, Electorate 18–24 –0.39, Electorate 25–44, –0.47). Largely these are in agreement with some of the theoretical views of the drivers of voter turnout — the affluent are more likely to participate in the voting process than the more disadvantaged. There are some relationships, at a global level, between the independent variables. Unemployment and Local Authority Renters are strongly positively correlated (0.85), and there are a few pairs of moderate negative correlations: Educational Disadvantage with Social Class 1 (–0.87), Different Address with Electorate 45-64 (–0.71) and Electorate 25–44 with Electorate 45–64.

These are to some extent expected. Residents in local authority property are more likely to be in low earning jobs or unemployed; members of higher social classes are less likely to have low educational attainment; the older electorate tend not to be recent migrants; and stages in the family life cycle might well be responsible for those in older age groups not to be mixed with those in younger age groups. One might transform the variables to principal components before entering them into the model, but the interpretation of the regression coefficients becomes a problem.

Two interesting variables are the percentage of residents in Social Class 1 and the percentage of the electorate with educational disadvantage. Figure 9.3 shows the variation in the percent of residents in Social Class 1: this group is concentrated in the southern parts of Dublin City and the northern parts of Dun Laoghaire-Rathdown.

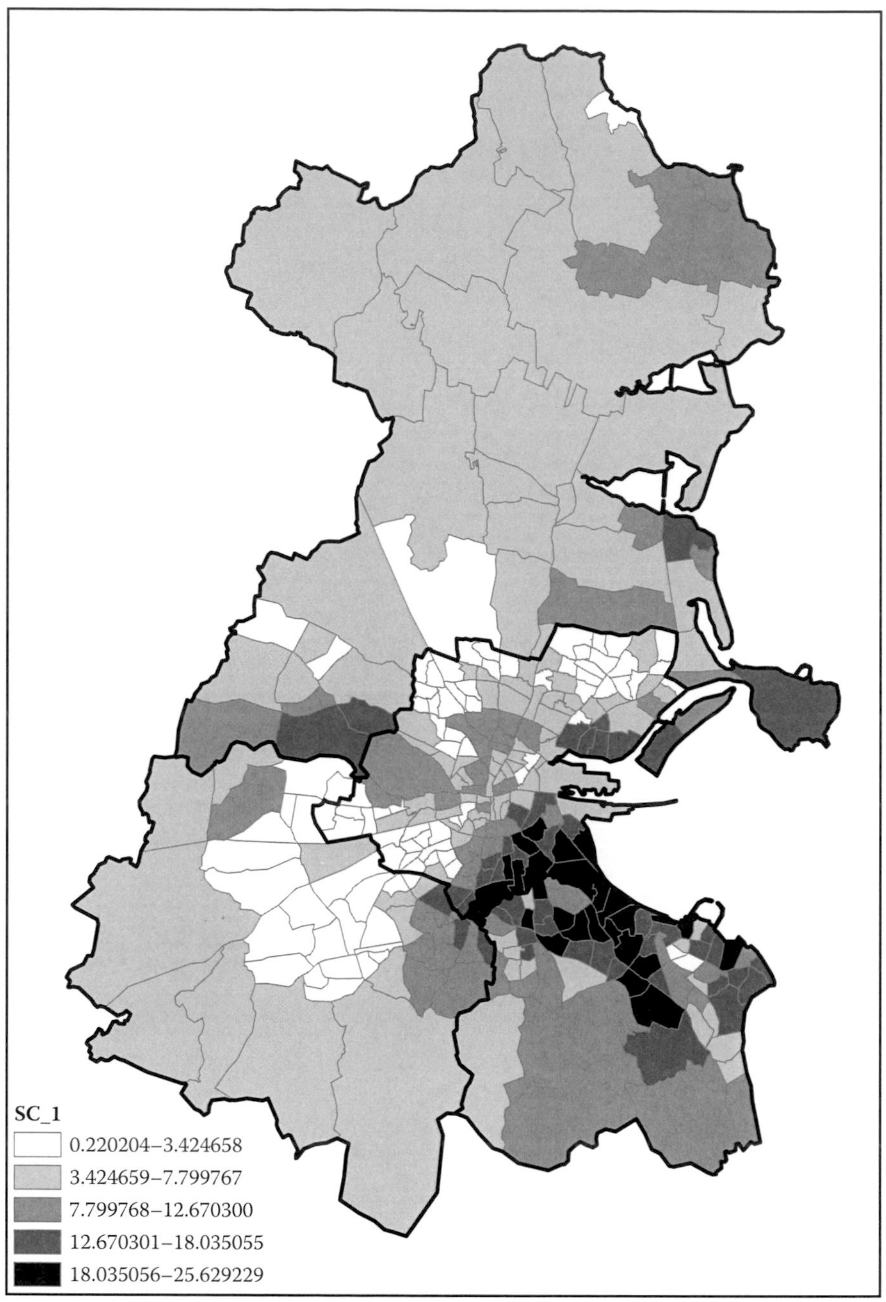

FIGURE 9.3 Variation in the percentage of residents in Social Class 1.

Figure 9.4 shows that within Dublin City there are pockets of educational disadvantage, mostly in peripheral EDs in the northern and west parts of the city. Parts of Blanchardstown in Fingal, and Clondalkin in South Dublin are also notable as areas of deprivation.

Some initial visual exploration of the relationships in the data is desirable, and the parallel coordinates plot is an ideal tool for this. In Figure 9.5, the eight independent variables are presented on the first eight parallel ordinates. The turnout is shown on the rightmost parallel ordinate. Each polygonal line in the plot represents a single ED. Although the data are in percentage form, there is quite a wide variety in the mean values and variances, so to aid clarity, the values have been rescaled so that the maximum and minimum of each variable corresponds to the top and bottom of its ordinate. The lines have been colored by their turnout quintile with red lines indicating a high turnout, and green lines indicating a low turnout. Low turnouts would appear to be associated with high values of Different Address, Local Authority Renters, Unemployment, Educational Disadvantage, and Electorate 18–24; it also appears to be associated with low values of Electorate 45–64. However, while the plot hints at some relationships, it does not give an idea of how these relationships vary across Dublin itself.

We begin a more formal identification of the relationships by fitting a global model; that is, a multiple linear regression using OLS. This provides a baseline against which to compare the GWR model.

9.6.3 Global Model

The global model has an adjusted R^2 of 0.61 ($R^2 = 0.62$). The AICc is 2012.20. The parameter estimates are in the following table (significant variables are shown in normal type):

Parameter	**Estimate**	**T**
Intercept	86.19	14.90
Different Address	–0.56	–5.73
Local Authority Renters	–0.19	–4.26
Social Class 1	0.18	1.55
Unemployment	–0.86	–3.61
Educational Disadvantage	–0.03	–0.60
Electorate 18–24	–0.13	–1.77
Electorate 25–44	–0.27	–4.97
Electorate 45–64	–0.21	–2.44

Variables that do not contribute significantly to explain the variation in the model because their parameters are not significantly different from zero are Social Class 1, Educational Disadvantage, and Electorate Aged 18–24. The other variables all have a negative sign; this is not unexpected in the case of migrants, council renters, and the unemployed. The interpretation of the negative signs on the electorate variables suggests some interesting interactions with the other variables, and perhaps some colinearity.

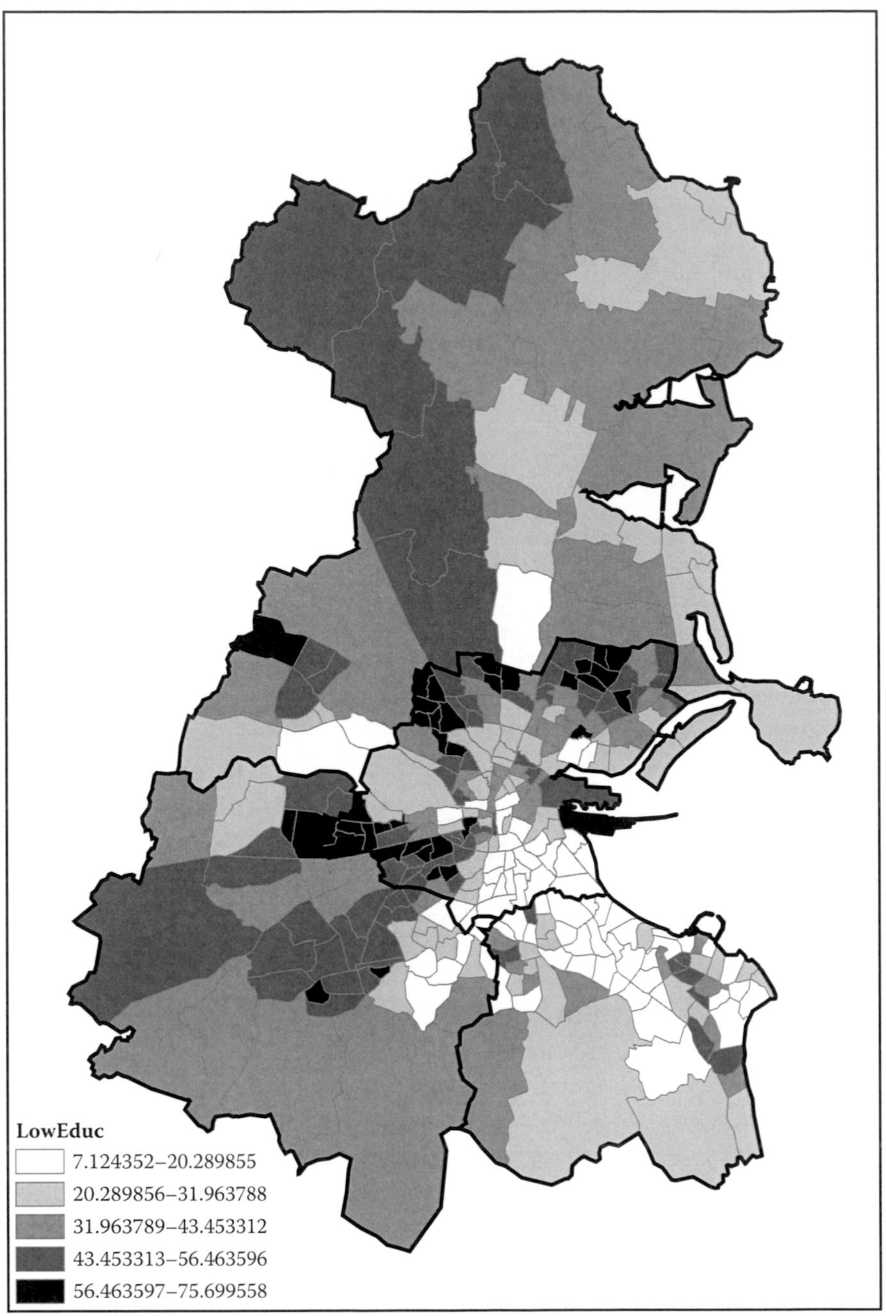

FIGURE 9.4 Variation in percentage of residents with an educational disadvantage.

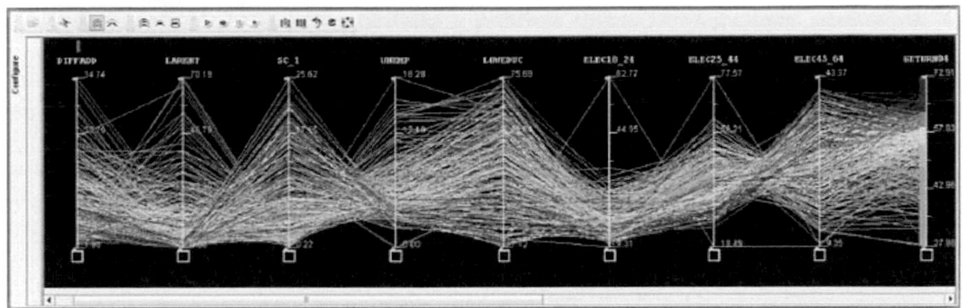

FIGURE 9.5 Rescaled parallel coordinates plot showing relationship between variables in the regression colored by quintile. Red: high turnout; yellow: middle turnout; green: low turnout. See color insert after page 148.

9.6.4 Local Model

The local model was fitted with OLS GWR. An adaptive kernel was used — in the less urbanized parts of the county, the EDs are larger, so we need a larger kernel for these areas. We elected to find the bandwidth that minimized the AICc. The local sample size for the kernel was 61 — each kernel covers approximately 20% of the sample points.

The initial results suggest that the local model is a "better" model, in that it provides a closer fit to the original data. The residual sum of squares (RSS) is 3302 compared with the 9149 in the global model. The adjusted R^2 is 0.80, which is further evidence of more accurate predictions, and the AICc was reduced from 2012.20 to 1972.84 — strong evidence that we are closer to the "true" model for the turnout variable. The number of degrees of freedom has increased from 9 to 103.19.

9.6.5 Results

While the AICc change suggests that the GWR model is an improvement over the global model, the next question to ask is whether there is any evidence for spatial variability in the parameter estimates. A Monte Carlo test is used to determine whether the observed variance in the parameter estimates could have arisen by chance — the test is based on Hope (1968). A separate test is carried out for each of the regressors, with the observed variance compared with the results from 99 simulations. The *p* values from the tests are shown in the table below with the global model's t value for comparison.

Variable	p	Global t
Different Address	0.41	–5.73
Local Authority Renters	0.65	–4.26
Social Class 1	0.00	1.55
Unemployment	0.48	–3.61
Educational Disadvantage	0.01	–0.60
Electorate 18–24	0.28	–1.77
Electorate 25–44	0.00	–4.97
Electorate 45–64	0.02	–2.44

Four sets of local parameter estimates exhibit significant spatial variability. These are those associated with Social Class 1, Educational Disadvantage, Electorate 25–44 and Electorate 45–64. The remaining local parameter estimates do not exhibit any significant spatial variation and hence the global estimates are appropriate. It is interesting that two of the parameters, those associated with Social Class 1 and Educational Disadvantage, that are not significant globally exhibit significant local variation suggesting that quite different relationships exist over the study area.

9.6.6 Local Parameter Estimate Maps

Many of the applications of GWR that have appeared show maps of the local parameter variation. This is a useful first step toward interpreting the results. This has always been the intention of the developers of GWR and is strongly evident from their work (e.g., Brunsdon et al., 1996; Fotheringham et al., 1998, 2002). However, the processes are sometimes a little difficult to unravel in the local maps and require the comparison of multiple maps.

The parameter surface for Social Class 1 shown in Figure 9.6 has both negative and positive values — the positive values are intuitive, we might expect relatively affluent and educated individuals to wish to take part in the democratic process. However, the negative values are perplexing — why should an increase in the proportion of the residents in Social Class 1 lead to a decrease in voter turnout?

The parameter surface for Educational Disadvantage shown in Figure 9.7 has some counter-intuitive elements as well. We might expect there to be a negative relationship — low educational attainment implies poorer awareness of political processes and their effects, and perhaps contributes toward an unwillingness to take part in those processes. The positive parts of this surface are in areas with low proportions of residents with educational disadvantage. However, not everyone with a poor educational attainment refuses to take part in the electoral processes, and this is suggested by the positive values in the inner city, and parts of rural Fingal.

The t surface map in Figure 9.8 provides the clue to the interpretation of the Social Class parameter estimates. In examining parameter variation we need to take into account the variability of the parameter estimates themselves, which variability is measured by the standard error of the individual parameter estimates? The construction and visualization of a 95% confidence interval would be rather complex, but if the interval included locally zero, we would have some evidence that the variable in that location was contributing to the variability of the dependent variable. This would appear to be the case in areas where the proportion in Social Class 1 is high or moderately high. The t surface for the negative areas would suggest that the local estimates are not dissimilar to zero — in other words, the variation in Social Class 1 has no influence in these areas.

Figure 9.9 shows the general distribution of the t surface for the Educational Disadvantage parameter estimate. Again, in examining parameter variation we need to be aware of the variability of the estimates themselves. An interesting issue is raised here. If we regard the surfaces as 322 significance tests, then some adjustment

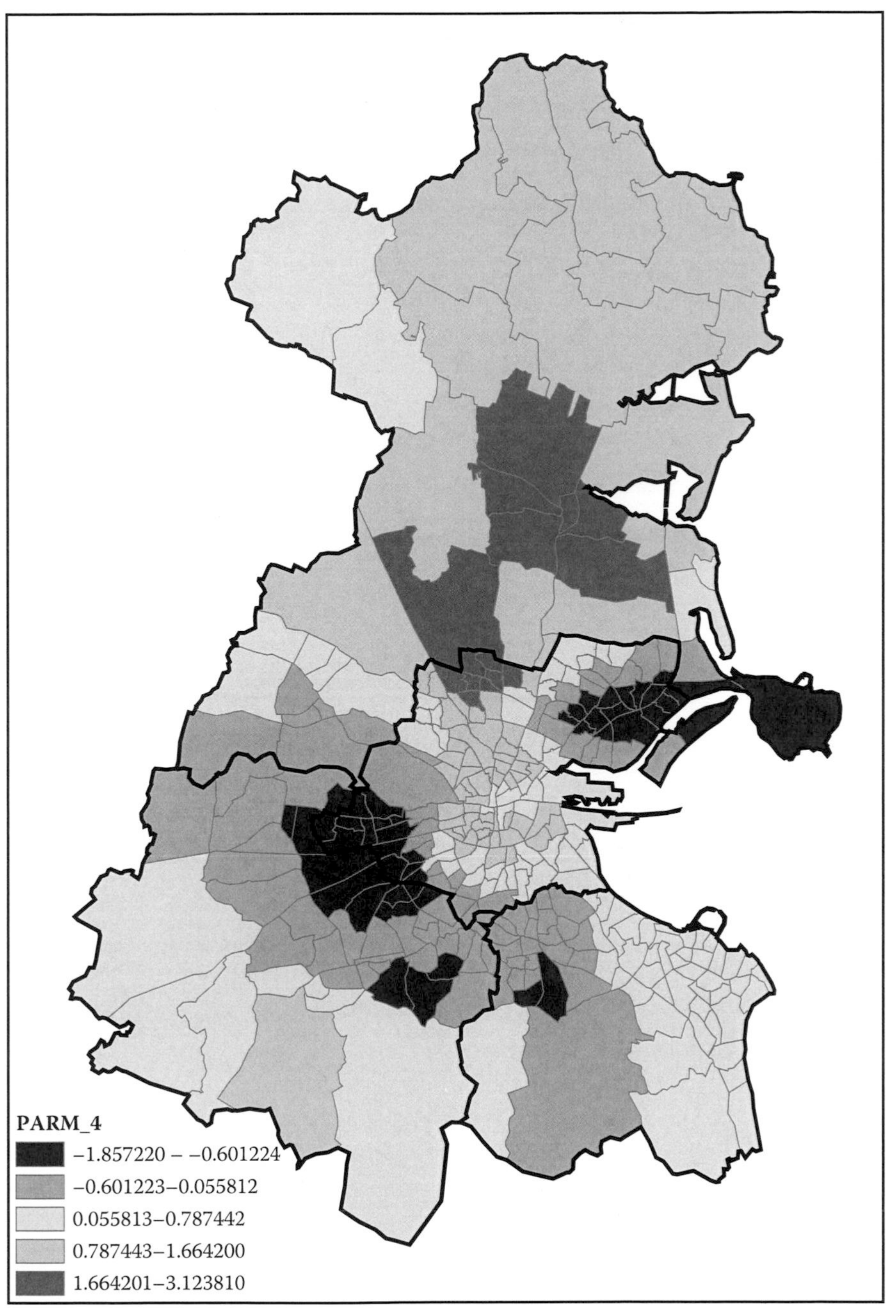

FIGURE 9.6 GWR parameter estimates for Social Class 1.

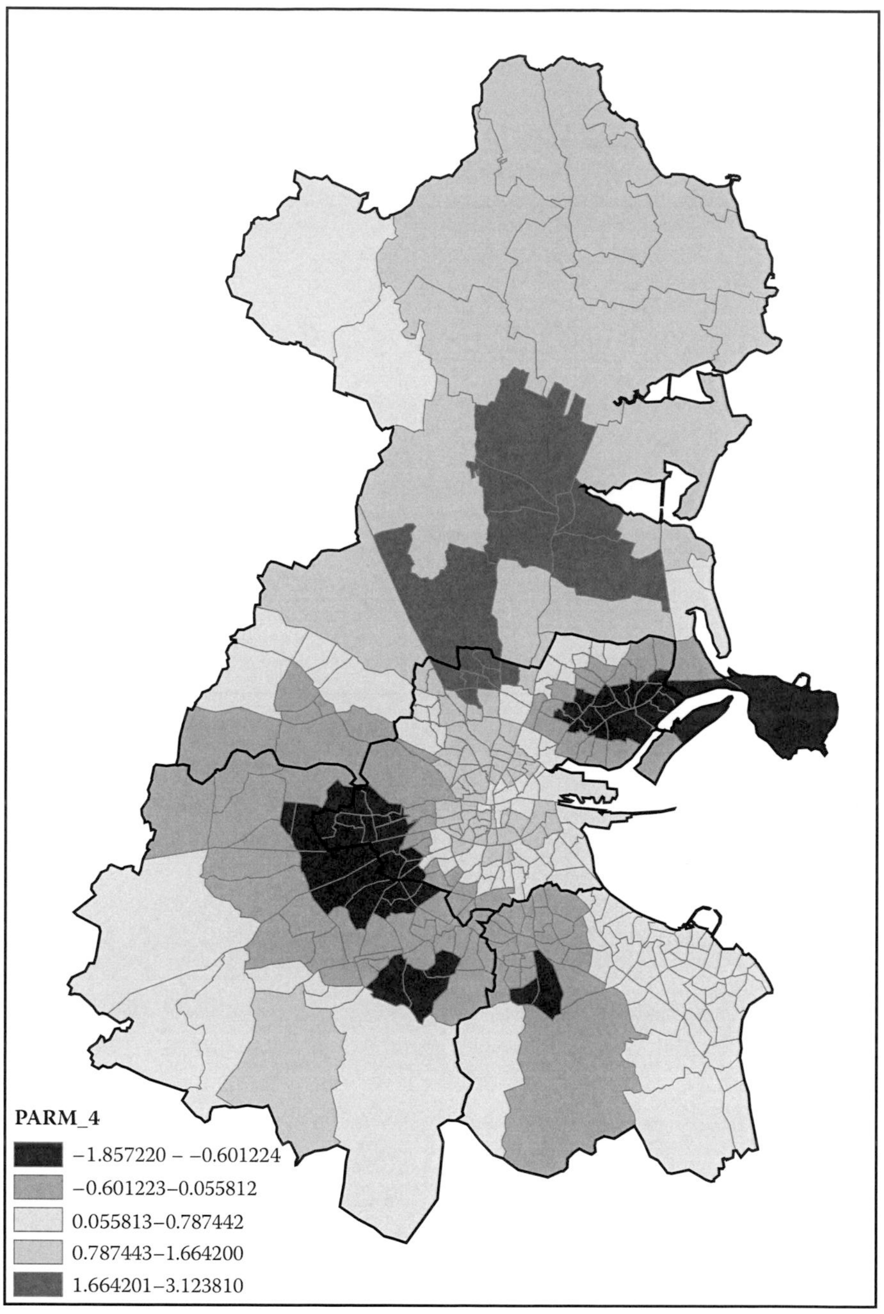

FIGURE 9.7 GWR parameter estimates for Educational Disadvantage.

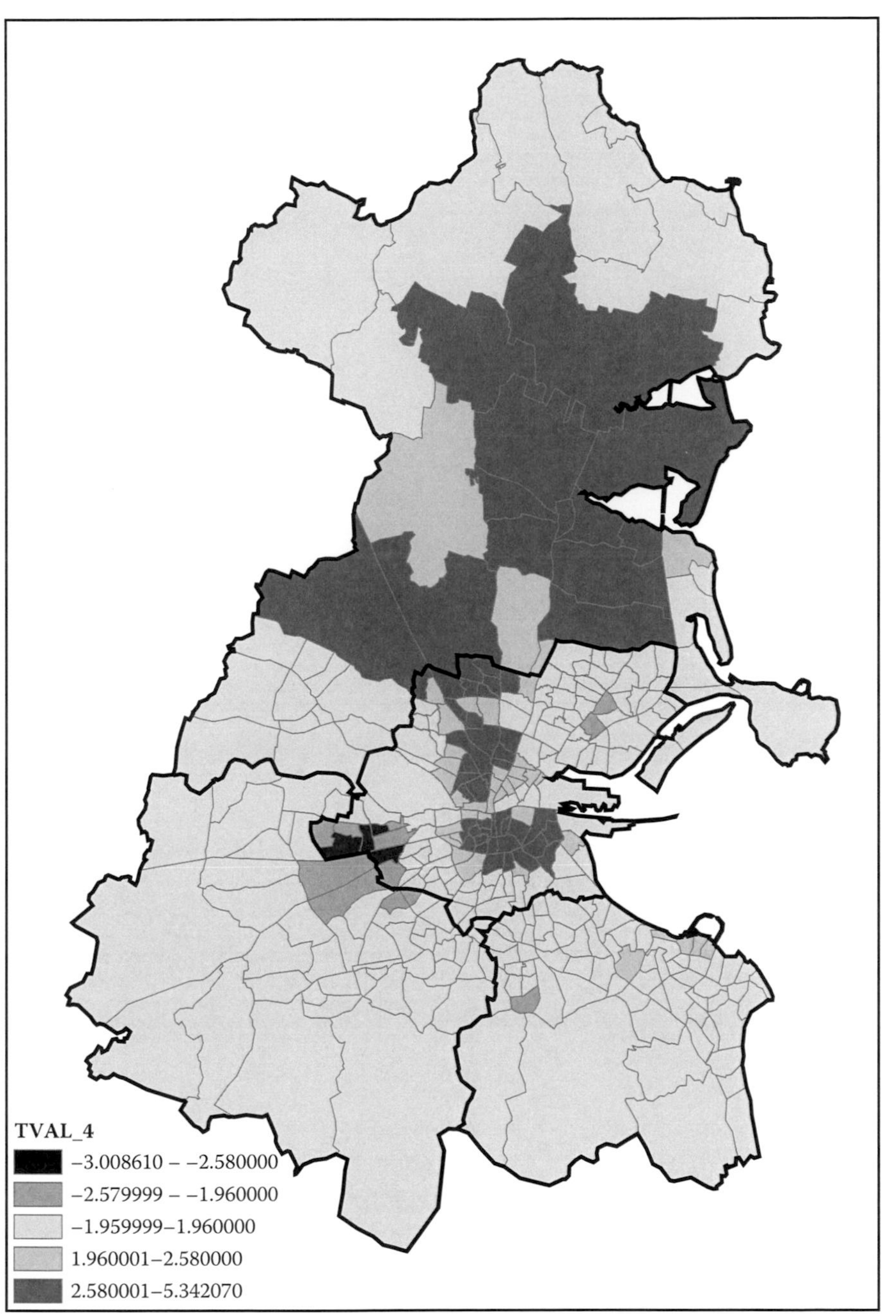

FIGURE 9.8 Local t statistics for Social Class 1.

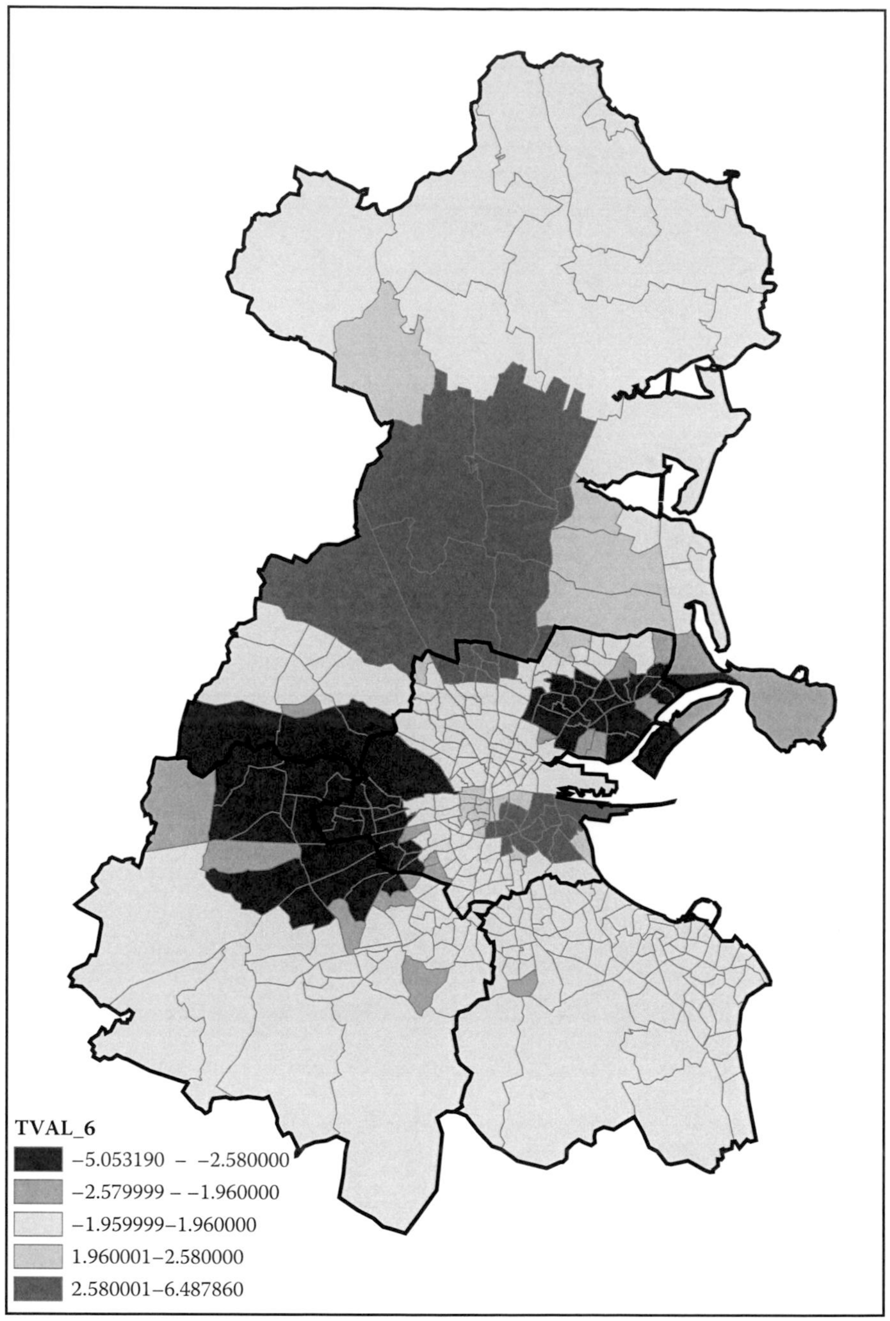

FIGURE 9.9 Local t statistics for Educational Disadvantage.

of the percentage points of the test statistic might be appropriate. Bonferroni adjustment may be too crude. Further research is needed in this area of GWR.

However, what the examination of the parameter surfaces for two variables out of the eight regressors shows is the difficulty of addressing one variable at a time, when the results of multivariate analysis where the outputs are as complex as the inputs or more so demand multivariate methods of exploration. This is where GA has a useful role to play.

9.6.7 Visualizing the Results — Further Insights

The GWR result space was explored using a geovisual exploratory environment based on GeoVISTA Studio, which is a collection of various geographic and other visualizations and computational data mining methods (Gahegan et al., 2002). The environment consisted of three interactively linked visualizations: a map, a parallel coordinates plot (PCP) and a visualization of a computational data mining method — the self-organizing map (SOM).

The SOM is an unsupervised neural network, which projects the multidimensional data onto a two-dimensional lattice of cells while preserving the topology and the probability density of the input data space. This means that similar input data elements are mapped to neighboring cells and the similarity patterns that exist in the higher dimensional space correspond to patterns in the SOM lattice (Kohonen, 1997, Silipo, 2003), which is easy to visualize because of its two-dimensionality (Vesanto, 1999). The SOM visualization in GeoVISTA Studio, which was used in this example, is a hexagonal U-matrix, where a grey shade is assigned to each hexagonal lattice cell according to the cell's distance from its immediate neighbors. Light areas in the lattice indicate areas with similar cells and represent clusters of similar elements. Dark areas indicate borders between clusters. The distribution of data in the SOM lattice is represented by the size of the circles that are projected in a regular pattern over the grey hexagonal cells. The color of the circles helps to transfer the information about clusters to other visualizations through visual brushing. It is defined by draping a smooth color map over the circles in the SOM lattice. Through visual brushing the hue of each circle is then inherited by graphic entities belonging to the same data elements in other visualizations (Vesanto, 1999, Takatsuka, 2001), in our case in the map and in the PCP.

Figure 9.10 shows two visualizations of the subspace of the GWR result space, consisting of parameter estimates for all eight GWR variables. Other attributes, such as t-values, s-values, etc. could also be visualized, but to keep this example simple, we limited the subspace to parameter estimates only. Figure 9.10a shows the SOM clustering based on these eight parameter estimates. The EDs in the map in Figure 9.10b inherited their colors from their respective location in the SOM, as explained above. There are several light areas in the SOM that define clusters: the green cluster in the top-left corner, the yellow cluster in the center on the top, the orange-red cluster in the top-right corner, the violet cluster in the bottom-right corner, the light blue cluster in the bottom-left corner, and the turquoise cluster on the left. It should be pointed out that these are nonspatial clusters, meaning that they are based only on attribute

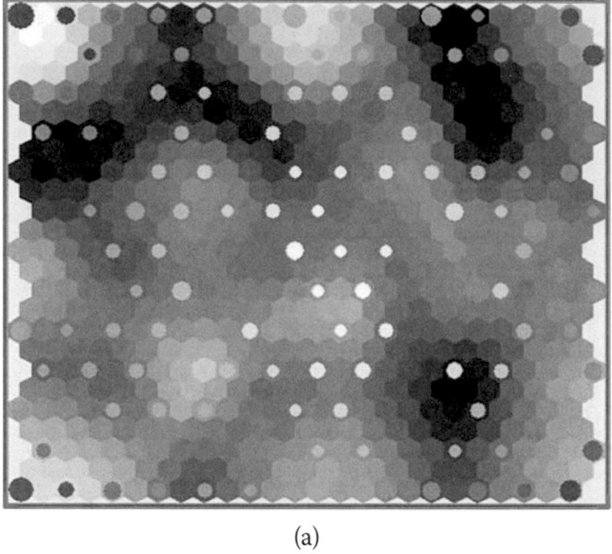

(a)

(b)

FIGURE 9.10 Visualizing the structure of the GWR space with (a) a self-organizing map (SOM) and (b) a map. See color insert after page 148.

data (i.e., the parameter estimates) and not on geographical location. Yet once the clusters are transferred to the map in Figure 9.10b with the help of visual brushing using color, it becomes apparent that similarly colored EDs (which are similar to each other in the attribute space and are located in neighboring cells in the SOM) are mostly also adjacent to each other in the geographical space.

In the following, we look at two selections of clusters that identify different areas in Dublin in the third visualization, the PCP, to attempt to explain the differences in processes that are at work in each of the selected areas. In a PCP, each axis represents one variable of the input data space, in our case one parameter estimate. Axes are linearly scaled from the minimum to the maximum value of each dimension. Each data element is displayed as a polygonal line intersecting each of the axes at the point that corresponds to the respective attribute value for this data element (Inselberg, 2002). The colors of the polygonal lines in the GeoVISTA PCP used in our example are again inherited from the SOM as described above. Additionally, the GeoVISTA PCP has boxplots assigned to each axis, for a better impression of how the values at each axis are distributed.

We have used the PCP in conjunction with selection of clusters in the SOM in order to try to describe the characteristics of each cluster. This can be achieved by looking at the trajectory that the lines belonging to each cluster form in the PCP. Figure 9.11 shows a selection of two clusters that identify the central area of Dublin: the yellow and the violet cluster. The map of the EDs belonging to these two clusters is shown in Figure 9.11a. Figure 9.11b shows the PCP of the same selection. In this PCP, there are clearly two different trajectories: a group of yellow polygonal lines for the areas in the northern part of the city center and a group of violet polygonal lines for the southern areas (and a few spatial outliers located northwest from the center). The trajectories are most clearly separated at parm_5 axis, which is the parameter estimate for the percentage of unemployment and at parm_8 axis, which is the parameter estimate for population of 25–44 years of age. At these two axes, the yellow and violet lines lay on different sides of the quartile box. Looking at the values of the parameter estimates on the respective two axes, we can conclude that percentage of unemployment (parm_5) has a strongly positive influence on voter turnout in yellow areas (north city center) and a strongly negative influence in violet areas (south city center). Conversely, the percentage of population of age 25–44 (parm_8) has a strongly negative influence on voter turnout in the north and a strongly positive influence in the south. At other parameter estimates, both the yellow and the violet group of lines intersect the respective axis inside or near the quartile box of the boxplot and mostly on the same side of the average.

Another interesting pattern can be seen in the selection of red and blue clusters in the SOM. The map of this selection is shown in Figure 9.12a and the PCP is shown in Figure 9.12b. The EDs in the red cluster are located in the Ballymun area, which is known as one of the problematic areas in Dublin. Blue EDs are located along the coast north of Dublin city center and on the Howth peninsula — these are the most affluent areas in Dublin. The GWR results space clearly captures these differences: the red and blue clusters are located very far from each other in the SOM in Figure 9.12a and the PCP in Figure 9.12b shows different red and blue trajectories. In this PCP the biggest difference in parameter estimates between these two trajectories is at parm_4 (percentage of social class 1) and at parm_6 axis (low education). Social class 1 (parm_4) has a strongly negative influence on voter turnout in Howth and surrounding blue areas, but a strongly positive influence in the red Ballymun and surroundings. The same is true for parm_6, i.e., percentage of low education. The red and blue trajectories differ in

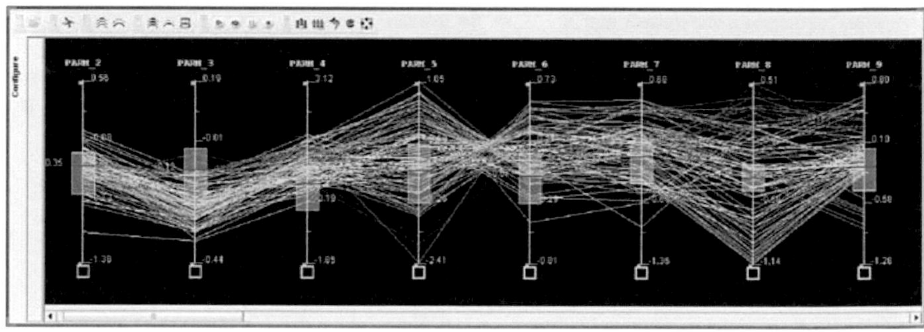

FIGURE 9.11 (a) The map and (b) the parallel coordinates plot (PCP) showing the selection of the two clusters from the SOM (violet and yellow, respectively) that represent two areas in the center of Dublin. River Liffey (running through the center in the east-west direction) divides the center in the north yellow area and south violet areas where the social processes behind voting mechanisms have different characteristics. This can be clearly seen by comparing the violet and yellow trajectories in the PCP. See color insert after page 148.

most of the other parameter estimates as well, but interestingly not at parm_5, which represents the influence of percentage of unemployment. Both in the affluent blue Howth and in poorer red Ballymun this variable has a strongly negative influence on voter turnout.

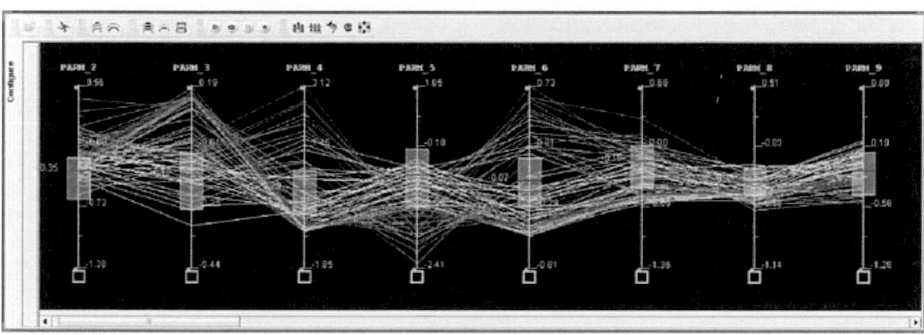

FIGURE 9.12 (a) The map and (b) the PCP showing the selection of the two clusters in SOM (red and blue) that represent two areas north of city center. Blue EDs are located around Howth and are the most affluent areas in Dublin. Red EDs, on the other hand, represent one of the more problematic areas in Dublin (Ballymun). The two different trajectories in the PCP again capture the differences between areas with different social situations. See color insert after page 148.

9.7 CONCLUSIONS

The inputs to any modeling exercise require thorough exploration — not just singly but where necessary in combination. This helps to establish what is unusual in a multivariate sense over and above what is unusual when the variables of interest are taken singly.

The outputs from a technique such as GWR are often voluminous and the analyst needs to make sense of them. Clearly one can map the surfaces for parameter estimates, and standard errors for individual variables. One can map residuals, leverage statistics, and local measures of goodness of fit. The generating process for the dependent variable is multivariate, the modeling is multivariate, and the results are multivariate.

This chapter has demonstrated that the methods and techniques of GA allow us to understand the structure in the outputs from GWR. When faced with a number of univariate displays, the analyst may be misled or slowed in his or her ability to comprehend the structures and interactions in the outputs.

We are familiar with the concept of data reduction — reducing the complexity of the data to something manageable. We often use boxplots, scatterplots, and parallel coordinate plots to help us understand and explore the data space. This chapter has shown how we can also use the techniques of GA to explore the result space from a modeling procedure such as GWR.

In a "classic" regression, we assume spatial stationarity. If this is not the case, any evidence will be apparent in the residuals — structure in the residuals is a problem. The assumption that a single global model is sufficient is not always warranted — with GWR the assumption is that the structures vary locally — there are suitable tests to determine whether individual variable parameter surfaces are spatially stationary. GWR helps to reveals patterns that are local and, coupled with multivariate visualization, we can examine how the interactions between the variables vary locally. These are local exploratory approaches designed to help reveal patterns in the datasets and suggest further directions for analysis. Not all spatial processes are heterogeneous. Nakaya et al. (2005) have developed a semi-parametric model that they use to model variation in premature mortality in Tokyetropolitan Area. In this model form, some variables are assumed to have stationary parameters, while others have spatially varying parameters. This opens an interesting problem in model selection where variables to be entered into a model can be tested with either fixed or varying parameters to assess their contribution to the improvement in model fit.

The techniques reveal the importance of looking at results from techniques such as GWR in a multivariate context rather than a univariate context. Rather than looking at maps of one local parameter estimate distributed across the study region, for example, and looking for areas of stability or change in that one relationship, we have shown how we can now investigate areas where sets of parameter estimates are relatively stable or where differences between sets of parameter estimates are most noticeable. In this way, we are exploring ever deeper into the measurement and understanding of spatial relationships and, ultimately, spatial processes.

REFERENCES

Akaike, H., 1973, Information theory as an extension of the maximum likelihood principle. In B.N. Petrov and F. Csaksi, Eds., *2nd International Symposium on Information Theory*. Akademiai Kiado, Budapest, Hungary, pp. 267–281.

Anselin, L., 1988, Spatial Econometrics: Methods and Models, Springer, Berlin.

Brunsdon, C., Fotheringham, A.S., and Charlton, M.E., 1996, Geographically weighted regression: A method for exploring spatial nonstationarity, *Geographical Analysis*, 28(4), 281–298

Burnham, K.P. and Anderson, D.R., 2002, *Model Selection and Multimodel Inference: A Practical Information Theoretic Approach*, 2nd ed., Springer, Berlin.

Casetti, E., 1972, Generating models by the expansion method: Applications to geographic research. *Geographical Analysis*, 4, 81–91.

Demšar, U., Fotheringham, A.S., and Charlton, M., 2008, Combining geovisual analytics with spatial statistics: The example of geographically weighted regression. Under review for a special issue of *The Cartographic Journal*.

Dykes, J.A., MacEachren, A.M., and Kraak, M.-J. (Eds.), 2005, *Exploring Geovisualization*. Elsevier, Amsterdam.

Foster, S.A. and Gorr, W.L., 1986, An adaptive filter for estimating spatially varying parameters: Application to modeling police hours spent in response to calls for service. *Management Science* 32, 878–889.

Fotheringham, A.S., Brunsdon, C., and Charlton, M.E., 1998, Geographically weighted regression: A natural evolution of the expansion method for spatial data analysis, *Environment and Planning A*, 30(11), 1905–1927.

Fotheringham, A.S., Brunsdon, C., and Charlton, M., 2002, *Geographically Weighted Regression: The Analysis of Spatially Varying Relationships*. Wiley, Chichester.

Gahegan, M., Takatsuka, M., Wheeler, M., and Hardisty, F., 2002, Introducing Geo-VISTA Studio: an integrated suite of visualization and computational methods for exploration and knowledge construction in geography, *Computers, Environment and Urban Systems*, 26, 267–292.

Hope, A.C.A., 1968, A simplified Monte Carlo significance test procedure, *Journal of the Royal Statistical Society, Series B*, 30(3), 582–598.

Hurvich, C.F., Simonoff, J.S., and Tsai, C.-L., 1998, Smoothing parameter selection in non-parametric regression using an improved Akaike information criterion, *Journal of the Royal Statistical Society series B*, 60, 271–293.

Inselberg, A., 2002, Visualization and data mining of high-dimensional data, *Chemometrics and Intelligent Laboratory Systems*, 60, 147–159.

Jones III, J.P., 1984, A spatially varying parameter model of AFDC participation: Empirical analysis using the expansion method, *Professional Geographer*, 36(4), 455–461.

Kavanagh, A., 2004, The 2004 local elections in the Republic of Ireland, *Irish Political Studies*, 19(2), 64–84.

Kavanagh, A., 2005, The 2005 Meath and Kildare North by-elections, *Irish Political Studies*, 20(2), 201–211.

Kavanagh, A., Mills, G., and Sinnott, R., 2004, The geography of Irish voter turnout: A case study of the 2002 general election, *Irish Geography*, 37(2), 177–186.

Kohonen, T., 1997, *Self-Organizing Maps*, 2nd ed., Springer Verlag, Berlin.

MacEachren, A.M. and Kraak, M.-J., 2001, Research challenges in geovisualization, *Cartography and Geographic Information Science*, 28(1), 3–12.

McMillen, D.P., 1996, One hundred fifty years of land values in Chicago: a nonparametric approach, *Journal of Urban Economics*, 40, 100–124.

Nakaya, T., Fotheringham, A.S., Brunsdon, C., and Charlton, M.E., 2005, Geographically weighted Poisson regression for disease association mapping, *Statistics in Medicine*, 24(17), 2695–2717.

National Visualization and Analytics Center (NVAC), 2005, Illuminating the path: Creating the R&D agenda for visual analytics. Available at: http://nvac.pnl.gov/agenda.stm.

Silipo, R., 2003, Neural networks, in: Berthold, M. and Hand, D.J. (Eds.), *Intelligent Data Analysis*, 2nd ed., Springer Verlag, Berlin, 269–320.

Takatsuka, M., 2001, An application of the self-organizing map and interactive 3-D visualization to geospatial data, in: *Proceedings of the Sixth International Conference on Geocomputation*, Brisbane, Australia.

Tobler, W., 1970, A computer model simulation of urban growth in the Detroit region, *Economic Geography*, 46(2), 234–240.

Vesanto, J., 1999, SOM-based data visualization methods, *Intelligent Data Analysis*, 3, 111–126.

10 Leveraging the Power of Spatial Data Mining to Enhance the Applicability of GIS Technology

Donato Malerba

Antonietta Lanza

Annalisa Appice

CONTENTS

10.1 INTRODUCTION

In a large number of application domains (e.g., traffic and fleet management, environmental and ecological modeling), collected data are measurements of one or more attributes of objects that occupy specific locations with respect to the Earth's surface. Collected geographic objects are characterized by a geometry (e.g., point, line, or polygon) which is formulated by means of a reference system and stored under a geographic database management system (GDBMS). The geometry implicitly defines both spatial properties, such as orientation, and spatial relationships of a different nature, such as topological (e.g., intersects), distance, or direction (e.g., north of) relations.

A GIS is the software system that provides the infrastructure for editing, storing, analyzing, and displaying geographic objects, as well as related data on geoscientific, economic, and environmental situations [11]. Popular GISs (e.g., ArcView, MapInfo, and Open GIS) have been designed as a toolbox that allows planners to explore geographic data by means of geo-processing functions, such as zooming, overlaying, connectivity measurements, or thematic map coloring. Consequently, these GISs are provided with functionalities that make the geographic visualization of individual variables effective, but overlook complex multivariate dependencies. Traditional GIS technology does not address the requirement of complex geographic libraries which search for relevant information, without any *a priori* knowledge of data set organization and content. In any case, GIS vendors and researchers now recognize this limitation and have begun to address it by adding spatial data interpretation capabilities to the systems.

A first attempt to integrate a GIS with a knowledge-base and some reasoning capabilities is reported in [43]. Nevertheless, this system has a limited range of applicability for a variety of reasons. First, providing the GIS with operational definitions of some geographic concepts (e.g., morphological environments) is not a trivial task. Generally only declarative and abstract definitions, which are difficult to compile into database queries, are available. Second, the operational definitions of some geographic objects are strongly dependent on the data model adopted for the GIS. Finding relationships between density of vegetation and climate is easier with a raster data model, while determining the usual orientation of some morphological elements is simpler in a topological data model [15]. Third, different applications of a GIS will require the recognition of different geographic elements in a map. Providing the system in advance with all the knowledge required for its various application domains is simply illusory, especially in the case of wide-ranging projects like those set up by governmental agencies.

The solution to these difficulties can be found in spatial data mining [22], which investigates how interesting, but not explicitly available, knowledge (or pattern) can be extracted from spatial data. This knowledge may include classification rules, which describe the partition of the database into a given set of classes [22], clusters of spatial objects [19, 42], patterns describing spatial trends, that is, regular changes of one or more nonspatial attributes when moving away from a given start object [26], and subgroup patterns, which identify subgroups of spatial objects with an unusual, an unexpected, or a deviating distribution of a target variable [21].

Following the mainstream of research in spatial data mining, there have been several atttempts to enhance the applicability of GIS technology by leveraging the power of spatial data mining [6, 16, 18, 32, 34]. In all these cases, the GIS users are not interested in processing the geometry of geographic objects collected in spatial database, but in working at higher conceptual levels, where human-interpretable properties and relationships between geographic objects are expressed.[1] To bridge the gap between geometrical representation and conceptual representation of geographic objects, GISs are provided with facilities to compute the properties and relationships (features), which are implicit in the geometry of data. In most cases, these features are then stored as columns of a single double entry data table (or relational table), such that a classical data mining algorithm can be applied to transformed data within the GIS platform. Unfortunately, the representation in a single double entry data table offers inadequate solutions with respect to spatial data analysis requirements. Indeed, information on the original heterogeneous structure of geographic data is partially lost: for each unit of analysis, a single row is constructed by considering the geographic objects which are spatially related to the unit of analysis. Properties of objects of the same type are aggregated (e.g., by sum or mode) to be represented in a single value.

In this chapter, we present a prototype of GIS, called INGENS 2.0, that differs from most existing GISs in the fact that the data mining engine works in a first-order logic, thus providing functionalities to navigate relational structures of geographic data and generate potentially new forms of evidence. Originally built around the idea of applying the classification patterns induced from georeferenced data to the task of topographic map interpretation [31], INGENS 2.0 now extends its predecessor INGENS [32] by combining several technologies, such as spatial Data Base Management System (DBMS), spatial data mining, and GIS within an open extensible Web-based architecture. Vectorized topographic maps are now stored in a spatial database [40], where mechanisms for accessing, filtering, and indexing spatial data are available free of charge for the GIS requests. Data mining facilities include the possibility of discovering operational definitions of geographic objects (e.g., fluvial landscape) not directly stored in the GIS database, as well as regularities in the spatial arrangement of geographic objects stored in the GIS database. The former are discovered in the form of classification rules, while the latter are discovered in the form of association rules. The operational definitions can then be used for predictive purpose, that is, to query a new map and recognize instances of geographic objects not directly modeled in the map itself. Efficient procedures are implemented to model spatial features not explicitly encoded in the spatial database. Such features are associated with clear semantics and represented in a first-order logic formalism. In addition, INGENS 2.0 integrates a spatial data mining query language, called SDMOQL [28], which interfaces users with the whole system and hides the different technologies. The entire spatial data mining process is condensed in a query written in SDMOQL and run on the server side. The query is graphically composed by means of a wizard

[1] A typical example is represented by the possible relations between two roads, which either cross each other, or run parallel, or can be confluent, independently of the fact that they are geometrically represented as lines or regions in a map.

on the client side. The GUI (graphical user interface) is a Web-based application that is designed to support several categories of users (administrators, map managers, data miners, and casual users) and allows them to acquire, update, or navigate vectorized maps stored in the spatial database, formulate SDMOQL queries, explore data mining results, and so on. Logging data and the history of users are maintained in the database.

The chapter is organized as follows. In the next section, we discuss issues and challenges of leveraging the power of spatial data mining to enhance the applicability of GIS technology. We present the architecture and data model of INGENS 2.0 in Section 10.3 and the spatial data mining process in Section 10.4. The syntax of SDMOQL is described in Section 10.5. An application of INGENS 2.0 is reported and discussed in Section 10.6. Finally, Section 10.7 gives conclusions and presents ideas for further work.

10.2 SPATIAL DATA MINING AND GIS

Empowering a GIS with spatial data mining facilities presents some difficulties, since the design of a spatial data mining module depends on several aspects. The first aspect is the representation of spatial objects. In the literature, there are two types of data representations for the spatial data, that is, tessellation and vector [39]. They differ in storage, precision, and complexity of the spatial relation computation. The second aspect is the implicit definition of spatial relationships among objects. The three main types of spatial relationships are topological, distance, and directional relationships, for which several models have been proposed for the definition of their semantics (e.g., "9-intersection model" [14]). The third aspect is the heterogeneity of spatial objects. Spatial patterns often involve different types of objects (e.g., roads or rivers), which are described by completely different sets of features. The fourth aspect is the interaction between spatially close objects, which introduces different forms of spatial autocorrelation: spatial error (correlations across space in the error term), and spatial lag (the dependent variable in space i is affected by the independent variables in space i, as well as those, dependent or independent, in space j).

Classical data mining algorithms, such as those implemented in Weka [45], offer inadequate solutions with respect to these aspects. In fact, they work under the single table assumption [46], that is, units of analysis are represented as rows of a classical double-entry table (or database relation), where columns correspond to elementary (nominal, ordinal, or numeric) single-valued attributes. In any case, this representation neither deals with geographic data characterized by geometry, nor handles observations belonging to separate relations, nor naturally represents spatial relationships, nor takes them into account when mining patterns. Differently, geographic (or spatial) data are naturally modeled as a set of relations $R_1,...,R_n$, such that each R_i has a number of elementary attributes and possibly a geometry attribute (in which case a relation is a layer). In this perspective, a multirelational data mining approach seems the most suitable for spatial data mining tasks, since multirelational data mining tools can be applied directly to data distributed on several relations and since they discover relational patterns [13].

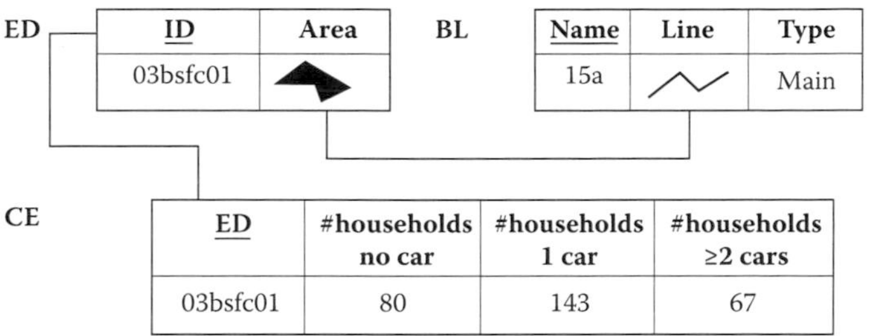

FIGURE 10.1 Representation of geographic data on the social effects of public transportation in a British city.

Example. To investigate the social effects of public transportation in a British city, a geographic data set composed of three relations is considered (see Figure 10.1). The first relation, ED, contains information on enumeration districts, which are the smallest areal units for which census data are published in the U.K. In particular, ED has two attributes, the identifier of an enumeration district and a geometry attribute (a closed polyline), which describes the area covered by the enumeration district. The second relation, BL, describes all the bus lines which cross the city. In this case, relevant attributes are the name of a bus line, the geometry attribute (a line) represents the route of a bus and the type of bus line (classified as main or secondary). The third relation, CE, contains some census data relevant for the problem, namely, the number of households with 0, 1, or "more than 1" cars. This relation also includes the identifier of the enumeration district, which is a foreign key for the table ED. A unit of analysis corresponds to an enumeration district (the target object), which is described in terms of the number of cars per household and crossing bus lines (bus lines are the task-relevant objects). The relationship between reference objects and task-relevant objects is established by means of a spatial join, which computes the intersection between the two layers ED and BL.

Although several spatial data mining methods have already been designed by resorting to the multirelational approach [4, 7, 21, 29, 30], most GISs which integrate data mining facilities [6, 16, 18] continue to frame the requests made by the spatial dimension within the classical data mining solution. Spatial properties and relationships of geographic objects are computed and stored as columns of a classical double-entry table, such that a classical data mining algorithm can be applied to the transformed data table.

At present, only two of the GISs reported in the literature integrate spatial data mining algorithms designed according to the multirelational approach. They are SPIN! [34] and INGENS [32]. SPIN! is the spatial data mining platform developed within the European Union (EU) research project of the same name. SPIN! assumes an object-relational data representation and offers facilities for multirelational sub-group discovery and multirelational association rule discovery. Subgroup discovery [21] is approached by taking advantage of a tight integration of the data mining algorithm with the database environment. Spatial relationships and attributes are then dynamically derived by exploiting spatial DBMS extension facilities (e.g., packages, cartridges, or extenders) and used to guide the subgroup discovery. Association rule discovery [4] works in first-order logic and is only loosely integrated with a spatial database by means of some middle layer module that extracts spatial attributes and relationships independently of the mining step and represents these features in a first-order logic formalism. INGENS is our first attempt to empower a GIS with inductive learning capabilities. Indeed, it integrates the inductive learning system, ATRE, which can induce first-order logic descriptions of some concepts from a set of training examples. INGENS assumes an object-oriented representation of data organized in topographic maps. The geographic data collection is organized according to an object-oriented data model and is stored in the object store object-oriented DBMS. Since object store does not provide automatic facilities for storing, indexing, and retrieving geographic objects, these facilities are completely managed by the GIS. In addition, INGENS integrates a Web-based GUI, where the user is simply asked to provide a set of (counter-) examples of geographic concepts of interest and a number of parameters that define the classification task more precisely. First-order descriptions learned by ATRE are only visualized in a textual format. The data mining process is condensed in a query written in SDMOQL [28], but the textual composition of the query is completely managed by the user.

10.3 INGENS 2.0 ARCHITECTURE AND SPATIAL DATA MODEL

The architecture of INGENS 2.0 is illustrated in Figure 10.2. It is designed as an open, highly extensible, Web-based architecture, where spatial data mining services are integrated within a GIS environment. The GIS functionalities are distributed among the following software components:

- a Web-based *GUI* for supporting users in all activities, that is, user log-in and log-out, acquisition and editing of a topographic map, visualization and exploration of a topographic map, execution of a data mining request formulated by means of a spatial data mining query;

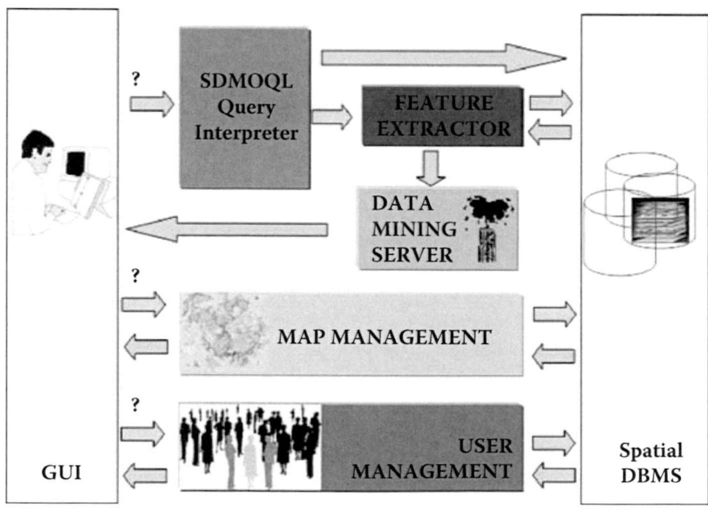

FIGURE 10.2 INGENS 2.0 software architecture.

- the *User Management* module for managing the access to the GIS (user creation, authentication, and history) for the different categories of users;
- the *Map Management* module for managing requests of map creation, acquisition, update, delete, visualization, and exploration;
- the *Query Interpreter* module for running user-composed SDMOQL queries and performing a spatial data mining task of classification or association rule discovery;
- the *Feature Extractor* module for automatically generating conceptual descriptions (in first-order logic) of geographic objects, by making explicit (spatial) properties and relationships, which are implicit in the spatial dimension of data;
- the *Data Mining Server* for running data mining algorithms;
- the *Spatial Database* for storing both map data and information on the user history (logging user identifier and password, privileges, and spatial data mining queries executed in the past).

The GUI can be accessed by four categories of users, namely, the GIS administrators, the map maintenance users, the data miners, and the casual end users. User profiles (e.g., authentication information, list of privileges) are stored in the database. The profile lists the topographic maps and the GIS functionalities to be accessed by the user. The administrator is the only user authorized to create, delete, or modify profiles of all other users of the GIS. The map maintenance user is in charge of upgrading the map repository stored in the spatial database by creating, updating, or deleting a map. The data miner can ask the GIS to discover either the operational definition of a geographic object or a spatial arrangement of geographic objects that are frequent on the topographic map under analysis. Finally, the casual end user is provided with geo-processing functionalities to navigate the topographic map, visualize geographic objects, belonging to one or more map layers (roads, parcels, and so on), and perform zooming operations.

The user management module is in charge of the activities of creating, modifying, or deleting a user profile. Users are authorized to use only the GIS functionalities that match the privileges provided in their profiles.

The map management module executes the requests of the map maintenance users. This component interfaces with the spatial database in order to create or drop an instance of a topographic map, as well as retrieve and display geographic objects belonging to one or more layers of a map.

The query interpreter runs the SDMOQL queries composed by data miners. A query refers to one of the topographic maps accessible to the data miner and specifies the set of objects relevant to the task at hand, the kind of knowledge to be discovered (classification or association rules), the set of descriptors to be extracted from the map, the set of descriptors to be used for pattern description and optionally the background knowledge to be used in the discovery process, the geographic hierarchies, and the interestingness measures for pattern evaluation. The query interpreter's responsibility is to ask the feature extractor to generate conceptual descriptions of the geographic objects extracted from the spatial database and then to invoke the inference engine of the data mining server. The conceptual descriptions are conjunctive formulae in a first-order logic language, involving both spatial and nonspatial descriptors specified in the query. SDMOQL queries are maintained in the user workspace and can be reconsidered in a new data mining process. Due to the complexity of the SDMOQL syntax, a user-friendly wizard is designed on the GUI side to graphically support data miners in formulating SDMOQL queries.

The data mining server provides a suite of data mining systems that can be run concurrently by multiple users to discover previously unknown, useful patterns in geographic data. Currently, the data mining server provides data miners with two systems, ATRE [27] and SPADA [24]. ATRE is an inductive learning system that generates models of geographic objects from a set of training examples and counter-examples. SPADA is a spatial data mining system to discover multilevel spatial association rules, that is, association rules involving spatial objects at different granularity levels. In both cases, discovered patterns are returned to the GUI to be visualized and interpreted by data miners.

The spatial database (SDB) can run on a separate computational unit, where topographic maps are stored according to an object-relational data model. The object-relational DBMS used to store data is a commercial one (Oracle 10*g*) that includes spatial cartridges and extenders, so that full use is made of a well-developed, technologically mature spatial DBMS. Moreover, the object-relational technology facilitates the extension of the DBMS to accommodate management of geographic objects.

At a conceptual level, the geographic information is modeled according to an object-based approach [41], which sees a topographic map as a surface littered with distinct, identifiable, and relevant objects that can be punctual, linear, or surfacic. Interactions between geographic objects are then described by means of topological, directional, and distance-based operators. In addition, geographic objects are organized in a three-level hierarchy expressing the semantics of geographic objects independently of their physical representation (see Figure 10.3). The entity object is a total generalization of eight distinct entities, namely, hydrography, orography, land administration, vegetation, administrative (or political) boundary, ground transportation network, construction, and built-up area. Each of these is in turn a generalization, for example, administrative boundary generalizes the entity's city, province, county, or state.

At a logical level, geographic information is represented according to a hybrid model, which combines both a tessellation and a vector model [39]. The tessellation model partitions the space into a number of cells, each of which is associated with a value of a given attribute. No variation is assumed within a cell and values correspond to some aggregate function (e.g., average) computed on the original values in the cell. A grid of square cells is a special tessellation model called raster. In the vector model the geometry is represented by a vector of coordinates, which define points, lines, or polygons. Both data structures are used to represent geographic information in INGENS 2.0. The partitioning of a map into a grid of square cells simplifies the localization and indexing process. For each cell, the raster image in GIF format is stored, together with its coordinates and component geographic objects. These are represented by a vector of coordinates stored in the field *Geometry* of the database relation PHYSICAL_OBJECT (see Figure 10.4), while their semantics are

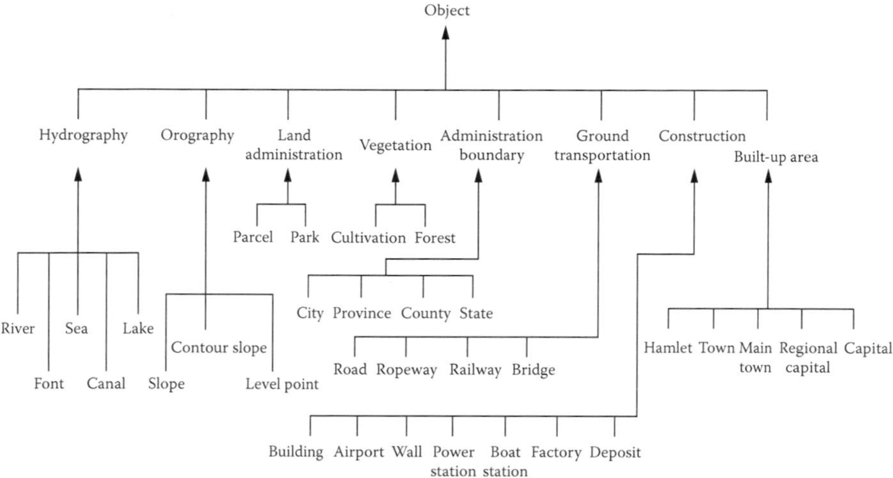

FIGURE 10.3 Hierarchical representation of geographic objects at different levels of granularity.

defined in the field *LogicalObject* of the database relation LOGICAL_OBJECT. A foreign key constraint relates each tuple of PHYSICAL_OBJECT to one tuple of LOGICAL_OBJECT. Type inheritance is exploited to represent the conceptual hierarchy in Figure 10.3 at the logical level. Indeed, the type of the attribute *LogicalObject* (LOGICAL_OBJECT_TY) has eight subtypes, namely, HYDROGRAPHY_TY, OROGRAPHY_TY, LAND_ADMINISTRATION_TY, VEGETATION_TY, ADMINISTRATIVE_BOUNDARY_TY, GROUND_TRANSPORTATION_TY,

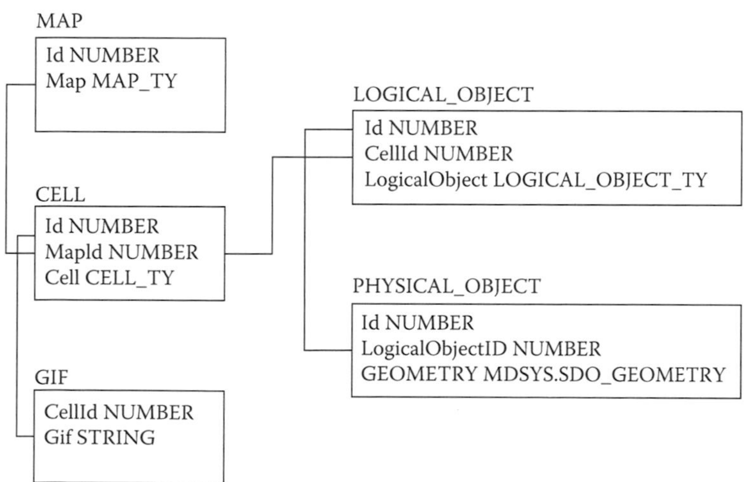

FIGURE 10.4 Spatial data schema.

CONSTRUCTION_TY, and BUILDUP_AREA_TY. Each of these is in turn a generalization of new types according to the conceptual hierarchy.

Spatial and nonspatial features can be extracted from geographic objects stored in the SDB. Feature extraction requires complex data transformation processes to make spatial properties and relationships explicit. This task is performed by the feature extractor module, which makes possible a loose coupling between data mining services and the SDB. The feature extractor module is implemented as an Oracle package of PL/SQL functions to be used in the spatial SQL queries.

10.4 SPATIAL DATA MINING PROCESS IN INGENS 2.0

In INGENS 2.0 the spatial data mining process is activated and controlled by means of a query expressed in SDMOQL (see Figure 10.5). Initially, the query is syntactically and semantically validated. Then the feature extractor generates the conceptual representation of the geographic objects selected by the query. This representation, which is in a first-order logic language, is input to multirelational data mining systems, which return spatial classification rules or association rules. Finally, the results of the mining process are presented to the user.

10.4.1 Conceptual Description Generation

A set of descriptors used in INGENS 2.0 is reported in Table 10.1. They are either spatial or nonspatial. According to their nature, spatial descriptors can be classified as follows:

1. Geometrical, if they depend on the computation of some metric/distance. Their domain is typically numeric, for example, "extension."
2. Topological, if they are invariant under the topological transformations (translation, rotation, and scaling). The type of their domain is nominal, for example, "region_to_region" and "point_to_region."

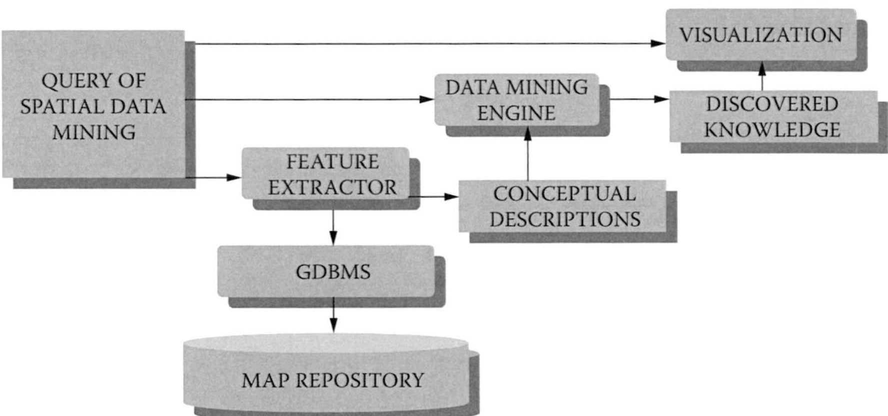

FIGURE 10.5 Spatial data mining process in INGENS 2.0.

TABLE 10.1
Set of Descriptors Extracted by the Feature Extractor

Feature	Meaning	Value
contain(C,L)	Cell C contains a logical object L	{true, false}
part_of(L,F)	Logical object L is composed of physical object F	{true, false}
type_of(L)	Type of L	33 nominal values (e.g., river, road, ...)
color(L)	Color of L	{blue, brown, black}
area(F)	Area of F	[0..MAX_AREA]
extension(F)	Extension of F	[0..MAX_EXT]
geographic_direction(F)	Geographic direction of F	{north-east, north-west, east, north}
line_shape(F)	Shape of the linear object F	{straight, curvilinear, cuspidal}
altitude(F)	Altitude of F	[0.. MAX_ALT]
line_to_line(F1,F2)	Spatial relation between lines F1 and F2	{almost parallel, almost perpendicular}
distance(F1,F2)	Distance between lines F1 and F2	[0..MAX_DIST]
region_to_region(F1,F2)	Spatial relation between regions F1 and F2	{disjoint, contains, inside, equal, meet, covers, covered by, over lap}
line_to_region(F1,F2)	Spatial relation between a line F1 and a region F2	{along edge, intersect}
point_to_region(F1, F2)	Spatial relation between a point F1 and a region F2	{inside, outside, on boundary, vertex (i.e., F1 is a vertex of F2)}

3. Directional, if they concern orientation. The type of their domain can be either numerical or nominal, for example, "geographic_direction."
4. Locational, if they concern the location of objects. Locations are represented by numeric values that express coordinates. There are no examples of locational descriptors in Table 10.1.

Some spatial descriptors are hybrid, in the sense that they merge properties of two or more of the above categories. For instance, the descriptor "line_to_line" that expresses conditions of parallelism and perpendicularity is both topological (it is invariant with respect to translation, rotation, and scaling) and geometrical (it is based on the angle of incidence).

In INGENS 2.0, geographic objects can also be described by two nonspatial descriptors, namely "type_of" and "color." The former describes the type of a geographic object, according to the layer (street, parcel, river, and so on) it belongs to, while the latter describes the color (blue, black, or brown) used in the visualization of a geographic object. The descriptor "part_of" describes the structure of complex geographic objects, i.e., a geographic object can be formed by physical component objects, represented by separate geometries.

There is no common mechanism to express the semantics of such different features. The semantics of topological relationships are based on the 9-intersection model [14], while the semantics of other features are based on mathematical methods of 2D-graphics [37] as described in [23].

Example (Geographic Direction). Let o be a geographic object associated with a line, that is,

$$o : \{P_1 = (x_1, y_1), \ldots, P_n = (x_n, y_n)\}.$$

If α is the angle defined by the straight line L connecting P_1 and P_n, that is,

$$\alpha = arctg \frac{x_n - x_1}{y_n - y_1},$$

then the geographic direction of o is computed as follows:

$$\begin{array}{rl} \text{north} & \text{if } \alpha > \left(\frac{\pi}{2} - \frac{\pi}{8}\right) \vee \alpha \leq -\left(\frac{\pi}{2} - \frac{\pi}{8}\right) \\ \text{northeast} & \text{if } \alpha \leq \left(\frac{\pi}{2} - \frac{\pi}{8}\right) \wedge \alpha > -\frac{\pi}{8} \\ \text{east} & \text{if } \alpha \leq \frac{\pi}{8} \wedge \alpha > -\frac{\pi}{8} \\ \text{northwest} & \text{if } \alpha \leq -\frac{\pi}{8} \wedge \alpha > -\left(\frac{\pi}{2} - \frac{\pi}{8}\right). \end{array}$$

This feature is computed only for geographic objects physically represented as lines.

10.4.2 Classification Rule Discovery

Classification of geographic objects is a fundamental task in spatial data mining and GIS, where training data consist of multiple target geographic objects (reference objects), possibly spatially related with other nontarget geographic objects (task-relevant objects). The goal is to learn the concept associated with each class on the basis of the interaction of two or more spatially referenced objects or space-dependent attributes [22].

While a lot of research has been conducted on classification, only a few works deal with geographic classification. GISs empowered with classification facilities are reported in [6, 18]. These systems allow the learning of a classifier from data stored in a classical double-entry table (single-table assumption [46]). This is a severe restriction in GIS applications, where different geographical objects have different features (properties and relationships), which are properly modeled by as many data relations as the number of object types. To map the natural multirelational form of geographic data into a single double-entry data table, GISs must integrate a transformation module that is in charge of computing the spatial features of geographic objects (e.g., a street crosses a river) and store them as columns of the double-entry table. This table can then be input to a wide range of robust and well-known classification methods which operate on a single table. This transformation (known as propositionalization) presents some drawbacks. In fact, the full equivalence between the original and the transformed training sets is possible only in special cases. However, even when possible, the output table size is unacceptable in practice [10] and some form of feature selection is required. Therefore, the transformed problem is different from the original one, for pragmatic reasons [7].

On the other hand, INGENS 2.0 overcomes the limitations of single table assumption by integrating a classification system, named ATRE [27], which resorts to a multirelational data mining approach [13] to classify geographic objects. Indeed, a multirelational approach to data mining (or MRDM) looks for patterns that involve multiple relations of a relational data representation. Thus, data taken as input by these approaches typically consist of several relations and not just a single one, as is the case in most existing data mining approaches. Patterns found by these approaches are called relational and are typically stated in a more expressive language than patterns defined in a single data table. Typically, subsets of first-order logic, which is also called predicate calculus or relational logic, are used to express relational patterns. In this way, the expressive power of predicate logic is exploited to represent both spatial relationships and background knowledge, thus providing functionalities to navigate relational structures of geographic data and generate potentially new forms of evidence, not readily available in flattened single double-entry data table representation.

The problem solved by ATRE is formalized as follows:

Given

- a set of concepts $C_1, C_2, \ldots, C_r$ to be learned;
- a set of units of analysis (or observations) O described in a language $\mathcal{L}_O$;
- a background knowledge BK described in a language $\mathcal{L}_{BK}$;
- a language of hypotheses $\mathcal{L}_H$ that defines the space of hypotheses S_H;
- a user's preference criterion PC.

Find a logical theory $T \in S_H$, defining the concepts $C_1, C_2, \ldots, C_r$, such that T is complete and consistent with respect to the set of observations and satisfies the preference criterion PC.

The logical theory T is a set of first-order definite clauses [25], such as:

cell(X1)=fluvial_landscape ←
contain(X1,X2)=true, type_of(X2)=river, part_of(X2,X3)=true,
line_to_line(X4,X3)=almost_parallel, part_of(X5,X4), type_of(X5)=street

This clause can be interpreted easily as follows: If a cell X1 contains a river X2 with X2 represented by the line X3 and X3 almost parallel to the line X4 that represents a street X5, then the cell X1 can be classified as a "fluvial landscape." This clause contains an operational definition of the fluvial landscape morphology. This definition can be used to recognize the unknown morphology for the cells of a new topographic map.

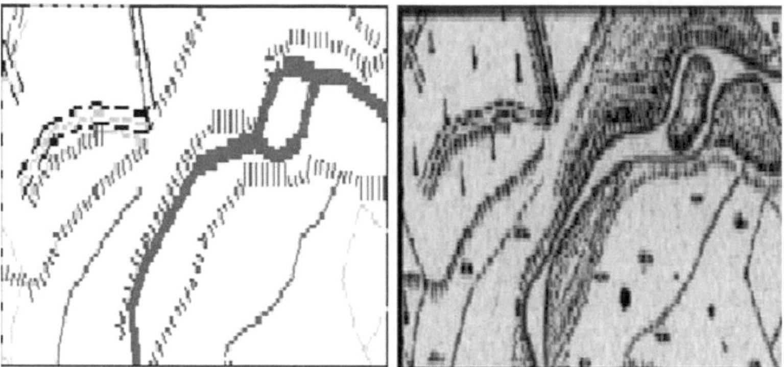

FIGURE 10.6 Raster and vector representation (left) and symbolic description of a cell (right). The cell is an example of a territory where there is a fluvial landscape. The cell is extracted from a topographic chart (Canosa di Puglia 176 IV SW—Series M891) produced by the Italian Geographic Military Institute (IGMI) at scale 1:25,000 and stored in INGENS 2.0.

The units of analysis are represented by means of a ground clause[2] called objects. For example, if the units of analysis are the cells (reference objects) of a topographic map, then the body of an object describes the spatial arrangement of the geographic objects (task-relevant objects) within the cell, while the head may describe the landscape morphology (class) associated with the cell. The literal in the head of the clause is an example (either positive or negative) of the concepts $C_1, C_2, \ldots, C_r$.

An instance of an object is reported in Figure 10.6, where the constant c8 denotes the whole cell, while the remaining constants (e.g., rv1_8, pc473_0, x20_8,…) denote the logical (river, street, parcel) or geometrical (line, point or polygon) component of the geographic objects in the cell. The descriptor cell(X) in the head denotes the known value of the morphology of the territory covered by the cell.

The background knowledge *BK* can be defined in the form of first-order definite clauses, which allow the definition of new descriptors not explicitly encoded in a conceptual description of objects. An example of a clause that is part of a *BK* is the following:

parcel_to_parcel(A,B)=C ←type_of(A)=parcel,
    type_of(B)=parcel, part_of(A,D)=true,
    part_of(B,E)=true, region_to_region(D,E)=C

This clause allows the relationship *C* between two regions *D* and *E* to be automatically renamed as "parcel_to_parcel," when *D* and *E* are parts of two parcels *A* and *B*.

The completeness property of the output theory *T* holds when *T* explains all observations in *O* of the *r* concepts C_i, while the consistency property holds when *T*

[2] A ground clause contains no variables.

explains no counter-example in O of any concept C_i. The satisfaction of these properties guarantees the correctness of the induced theory with respect to O, but not necessarily with respect to new unseen observations. The selection of the clause in T is made on the grounds of an inductive bias [35], expressed in the form of preference criterion (*PC*). For example, clauses that explain a high number of positive examples and a low number of negative examples can be preferred to others.

At the high-level, the learning strategy implemented in ATRE is sequential covering (or separate-and-conquer) [35], that is, one clause is learned (conquer stage), covered examples are removed (separate stage), and the process is iterated on the remaining examples. The conquer stage of this algorithm aims to generate a clause that covers a specific positive example, called *seed*. The most important novelty of the learning strategy implemented in ATRE is embedded in the design of the conquer stage. Indeed, the separate-and-conquer strategy is traditionally adopted by single concept learning systems that generate clauses with the same literal in the head at each step. In ATRE, clauses generated at each step may have different literals in their heads. In addition, the body of the clause generated at the i-th step may include all literals corresponding to those target concepts $C_1, C_2, \ldots, C_r$ for which at least a clause has been added to the partially learned theory in previous steps. In this way, dependencies between target concepts can be automatically discovered. An example of a logical theory, where the dependency between concepts "downtown" and "residential" is handled, is reported in the following:

class(X)=downtown ←
  on_the_sea(X)=true, business_activity(X)=high.

class(X)=residential ←
  contain(X,Y)=true, type_of(Y)=kindergarten, shopping_activity(X)=high.

class(X)=residential ←
  close to(X,Y)=true, *class(Y)=downtown,* business_activity(X)=low.

The order in which clauses of distinct target concepts have to be generated is not known in advance. This means that it is necessary to generate clauses with different literals in the head and then to pick one of them at the end of each step of the separate-and-conquer strategy. Since the generation of a clause depends on the chosen seed, several seeds have to be chosen, such that at least one seed per incomplete concept definition is kept. Therefore, the search space is actually a forest of as many search-trees (called specialization hierarchies) as the number of chosen seeds. A directed arc from a node (clause) C to a node C' exists if C' is obtained from C by adding a literal (single refinement step).

The forest can be processed in parallel by as many concurrent tasks as the number of search-trees (hence, the name of separate-and-parallel-conquer for this search strategy). Each task traverses the specialization hierarchy top-down (or general-to-specific), but synchronizes traversal with the other tasks at each level. Initially, some clauses at depth one in the forest are examined concurrently. Each task is actually free to adopt its own search strategy, and to decide which clauses are worth testing. If none of the tested clauses is consistent, clauses at depth two are considered. The search proceeds toward deeper and deeper levels of the specialization hierarchies until at least a user-defined number of consistent clauses is found. Task synchronization is performed after all "relevant" clauses at the same depth have been examined. A supervisor task decides whether the search should carry on or not, on the basis of the results returned by the concurrent tasks. When the search is stopped, the supervisor selects the "best" consistent clause according to the user's preference criterion. This separate-and-parallel-conquer search strategy provides us with a solution to the problem of interleaving the induction process for distinct concept definitions. It has the advantage that simpler consistent clauses are found first, independently of the predicates to be learned. Moreover, the synchronization allows tasks to save much computational effort when the distribution of consistent clauses in the levels of the different search-trees is uneven. A more detailed description of the search strategy implemented in ATRE and its optimization through caching techniques is reported in [5, 27].

10.4.3 Association Rule Discovery

Association rules are a class of regularities introduced by Agrawal and Srikant [1], which can be expressed by an implication of the form,

$$A \Rightarrow C\ (s,\ c),$$

where A(antecedent) and C(consequent) are sets of atoms, called *items*, with $A \cap B = \phi$. s is called support and estimates the probability $p(A \cup C)$, while c is called confidence and estimates the probability $p(C|A)$. A pattern $P\ (s\%)$ is *frequent* if $s \geq minsup$. An association rule $A \Rightarrow C\ (s\%,\ c\%)$ is *strong* if the pattern $A \cup C\ (s\%)$ is frequent and $c \geq minconf$. We call an association rule $A \Rightarrow C$ spatial, if $A \cup C$ is a spatial pattern, that is, it expresses a spatial relationship among spatial objects.

The problem of mining spatial association rules was originally tackled by Koperski [22], who implemented the module geo-associator of the spatial data mining system GeoMiner [18]. Similar to the classification task, the method implemented in geo-associator suffers from the limitations due to adapting the restrictive single-table data representation to the case geographic data. Weka-GPDM [6] is a further example of a GIS that includes facilities to discover spatial association rules. Once again, spatial features are extracted in a preprocessing step and stored as features of a single double-entry data table. Association rules are discovered in another step by applying the conventional association rule discovery algorithm included in Weka [45] to the single double-entry data table.

Similar to the classification case, INGENS 2.0 overcomes limitations of single table assumption by integrating an association rule discovery system, named SPADA [24], which exploits the expressive power of a predicate logic to deal with spatial relationships in the original relational form. In addition, SPADA automatically supports a multiplelevel analysis of geographic data. Indeed, geographic objects are organized in hierarchies of classes. By descending or ascending through a hierarchy, it is possible to view the same geographic object at different levels of abstraction (or granularity). Confident patterns are more likely to be discovered at low granularity levels. On the other hand, large support is more likely to exist at higher granularity levels. In general, the discovery of *multilevel* patterns (e.g., the most supported and confident) can be performed by forcing users to repeat independent experiments on different representations. In this way, results obtained for high granularity levels are not used at low granularity levels (or vice versa). Conversely, SPADA is able to explore altogether the search space at different granularity levels, such that patterns obtained for high granularity levels are used to control search at low granularity levels.

The problem solved by SPADA is formalized as follows:

Given

- a set S of reference objects, which is the main subject of the analysis,
- some sets R_k, $1 \leq k \leq m$ of task-relevant objects,
- a background knowledge BK including spatial hierarchies H_k on objects in R_k,
- M granularity levels in the descriptions (1 is the highest, while M is the lowest),
- a set of granularity assignments ψ_k, which associate each object in H_k with a granularity level to deal with several hierarchies at once,
- a couple of thresholds *minsup*[l] and *minconf*[l] for each granularity level l,
- a language bias LB which constrains the search space.

Find strong spatial association rules for each granularity level.

The reference objects are the main subject of the description, while task-relevant objects are geographic objects that are relevant for the task at hand and are spatially related to the reference objects. For example, the cells may be the reference objects of our analysis, while the geographic objects within the cells are the task-relevant objects. In this case, properties and relationships of task relevant objects within each cell are computed by the feature extractor and stored as ground atoms, e.g., the spatial perpendicularity between the geographic objects $g1$ and $g2$ is represented by the ground atom *almost_perpendicular*($g1$, $g2$). If g is a task-relevant object of the set R_k, then *is_a*(g, n_j) establishes the association between a geographic object g and corresponding objects at the level j ($j = 1, \ldots, M$) of the hierarchy H_k. Finally, for each cell c, the ground atom *cell*(c) identifies the unique reference object in the units of analysis.

The task of spatial association rule discovery performed by SPADA is split into two sub-tasks: find frequent spatial patterns and generate highly confident spatial association rules. The discovery of frequent patterns is performed according to the levelwise method described in [33], that is, a breadth-first search in the lattice of patterns spanned by a generality order between patterns. In SPADA the generality order is based on θ substitution [38]. The pattern space is searched one level at a time, starting from the most general patterns and iterating between candidate generation and evaluation phases. Once large patterns have been generated, it is possible to generate strong spatial association rules. For each pattern P, SPADA generates antecedents suitable for rules being derived from P. The consequent, corresponding to an antecedent, is simply obtained as the complement of atoms in P and not in the antecedent. Rule constraints are used to specify literals which should occur in the antecedent or consequent of discovered rules. In a more recent release of SPADA (3.1) [3], new pattern (rule) constraints have been introduced in order to specify exactly both the minimum and maximum number of occurrences for a literal in a pattern (antecedent or consequent of a rule). An additional rule constraint has been introduced to eventually specify the maximum number of literals to be included in the consequent of a rule. In this way, we are able to constrain the consequent of a rule requiring the presence of only the literal representing the class label and obtain useful patterns for classification purposes. Finally, the generation of patterns also takes into account a *BK* expressed in the form of first-order definite clauses. In this way, it is possible to simulate inferential mechanisms defined within a spatial reasoning theory. Moreover, by specifying both a *BK* and some suitable pattern constraints, it is possible to change the representation language used for spatial patterns, making it more abstract (human-comprehensible) and less tied to the physical representation of geographic objects.

An example of a spatial pattern discovered by SPADA is the following:

cell(A), contain(A,B), contain(A,C), is a(B,object),
is_a(C,object), extension(C,[100..200.5]) (40%),

which expresses a spatial containment relation between a cell *A* and some geographic objects *B* and *C*, where *C* is represented by a line with an extension between 100 and 200.5 m. This pattern occurs in 40% of the cells. The following spatial association rule,

cell(A), contain(A,B), contain(A,C), is_a(B,object),
is_a(C,object) ⇒ extension(C,[100..200.5]) (40%, 60%),

states that "in 60% of the cells (A), containing two geographic objects B and C, C is a line whose extension is between 100 and 200.5." Since SPADA, like many other association rule mining algorithms, cannot process numerical data properly, these are discretized in equal-width intervals which are treated as ground terms.

By taking into account hierarchies on task-relevant objects, we obtain descriptions at different granularity levels. For instance, by considering a portion of the logical hierarchy on geographic objects, in which both hydrography and administrative boundary are considered, specialization of objects is as follows:

hydrography ↘
object
administrative boundary ↗

A finer-grained spatial association rule can be the following:

cell(A), contain(A,B), contain(A,C),
is_a(B,administrativeBoundary), is_a(C,hydrography)
⇒ extension(C,[100..200.5]) (35%, 70%),

which provides better insight into the nature of the geographic objects *B* and *C*.

10.5 SDMOQL

The syntax of SDMOQL is designed according to a set of data mining primitives designed to facilitate efficient, fruitful spatial data mining in INGENS 2.0.

Seven primitives have been considered as guidelines for the design of SDMOQL. They are

1. the set of geographic objects relevant to a data mining task,
2. the kind of knowledge to be discovered,
3. the set of descriptors to be extracted from a digital map (primitive descriptors),
4. the set of descriptors to be used for pattern description (pattern descriptors),
5. the background knowledge to be used in the discovery process,
6. the concept hierarchies,
7. the interestingness measures and thresholds for pattern evaluation.

These primitives correspond directly to as many nonterminal symbols of the definition of an SDMOQL statement, according to an extended Backus Normal Form (BNF) grammar. Indeed, the SDMOQL top-level syntax is the following:

<SDMOQL> ::= <SDMOQLStatement>;
{<SDMOQLStatement>;}

<SDMOQLStatement> ::= <SDMStatement>
|<BackgroundKnowledge>
|<Hierarchy>

<SDMStatement> ::= <ObjectQuery>
mine <KindOfPattern>
analyze <PrimitiveDescriptors>
with descriptors <PatternDescriptors>
[<BackgroundKnowledge>]
{<Hierarchy>}
[**with** <InterestingnessMeasures>],

where "[]" represents 0 or one occurrence and "{}" represents 0 or more occurrences, and words in bold type represent keywords. In Sections 10.5.1 to 10.5.5 the detailed syntax for each data mining primitive is both formally specified and explained through various examples of possible mining problems.

10.5.1 Data Specification

The first step in defining a spatial data mining task is the specification of the geographic objects on which mining is to be performed. Geographic objects are selected by means of a query with a **SELECT-FROM-WHERE** structure, that is,

<Object_Query> ::= <Query_Statement>
{**UNION** <Query_Statement>}

<Query_Statement> ::=
SELECT <Object> {, <Object>}
FROM <Class> {, <Class>}
[**WHERE** <Conditions>]

The **SELECT** clause should return a cell or objects of a layer (hydrography, orography, and so on), or logical objects of a specific type (river, street, and so on). Hence, the selected geographic objects must belong to the same symbolic level, namely, cell, layer, or logic object. More formally the **FROM** clause can contain either a group of cells, a set of layers, or a set of logic objects, but never a mixture of them. Whenever the generation of the descriptions of objects belonging to different symbolic levels is necessary, the user can obtain it by means of the **UNION** operator. The following are examples of valid data queries:

Example 1 (Cell-level query). The user selects cell 26 from the topographic map of Canosa (Apulia) and the feature extractor generates the description of all the geographic objects in this cell.

SELECT x
FROM x in Cell
WHERE x->num_cell = 26 AND x->part map->map_name = "Canosa"

Example 2 (Layer-level query). The user selects the orography layer from the topographic map of Canosa and the construction layer from any map. The feature extractor generates the description of the objects in these layers for all cells of the map of Canosa.

SELECT x, y
FROM x in Orography, y in Construction
WHERE x->part_map->map_name = "Canosa"

Example 3 (Logical object-level query). The user selects the objects of the logic type river, from cell 26 of the topographic map of Canosa. The feature extractor generates the description of the rivers in this cell.

SELECT x
FROM x in River
WHERE x->part_map->map_name = "Canosa"
AND x->log_incell->num_cell = 26

10.5.2 The Kind of Knowledge to be Mined

The kind of knowledge to be discovered determines the data mining task in hand. Currently, SDMOQL supports the generation of either classification rules or association rules. The former are used for a predictive task, while the latter are used for a descriptive task. The top-level syntax is defined as follows:

<KindOfPattern> ::= <ClassificationRules>|<AssociationRules>

<ClassificationRules> ::= **classification as** <PatternName>
for <ClassificationConcept>
{, <ClassificationConcept>}

<AssociationRules> ::= **association as** <PatternName>
key is <Descriptor>

The <PatternName> denotes the name to be associated to the set of (classification or association) patterns to be discovered in the data mining task formulated within the SDMOQL statement. In a classification task, the user may be interested in inducing a set of classification rules for a subset of the classes (or concepts) to which training examples belong. In this case, the subset of interest for the user is specified in the <ClassificationConcept> list.

As pointed out, spatial association rules define spatial patterns involving both reference objects and task-relevant objects [4]. For instance, a user may be interested in describing a given area by finding associations between large towns (reference objects) and geographic objects in the road network, hydrography, and administrative boundary layers (task-relevant objects). The atom denoting the reference objects is called the key atom. The predicate name of the key atom is specified in the **key is** clause.

10.5.3 Specification of Primitive and Pattern Descriptors

The **analyze** clause specifies which descriptors, among those automatically generated by the feature extractor, can be used to describe the geographic objects extracted by means of the first primitive. The syntax of the **analyze** clause is the following:

analyze <PrimitiveDescriptors>,

where:

<PrimitiveDescriptors> ::= <Descriptor>{, <Descriptor>}
parameters <ParameterSpecs>{, <ParameterSpecs>}

<Descriptor> ::= <Predicate>/<Arity>
<ParameterSpecs> ::= <ParameterName> **threshold** <Integer>.

The specification of a set of parameters is required by the feature extractor to automatically generate some primitive descriptors. The language used to describe generated patterns is specified by means of the following clause: with descriptors <PatternDescriptors> where:

<PatternDescriptors> ::= <DescriptorSpecification>{; <DescriptorSpecification>}
<DescriptorSpecification> ::= <Descriptor> [**cost** <Integer>] | <Descriptor>
[**with** <TermsSpec>]
<TermsSpec> ::= <TermSpec>{, <TermSpec>}
<TermSpec> ::= <ConstantType> | <VariableType>
<ConstantType> ::= **constant** [<Value>]
<VariableType> ::= **variable mode** <VariableMode> role <VariableRole>
<VariableMode> ::= **old** | **new** | **diff**
<VariableRole> ::= **ro** | **tro**

The specification of descriptors to be used in the high-level conceptual descriptions can be of two types: either the name of the descriptor and its relative cost, or the name of the descriptor and the full specification of its arguments. The former is appropriate for classification.

The (classification or association) rules are expressed by means of descriptors specified in the **with descriptors** list. They are specified by Prolog programs on the basis of descriptors generated by the feature extractor. For instance, the descriptor "font_to_parcel/2" has two arguments which denote two logical objects, a font and a parcel. The topological relation between the two logical objects is defined by means of the clause

```
font_to_parcel(Font,Parcel) = TopographicRelation :-
        type_of(Font) = font, part_of(Font,Point) = true,
        type_of(Parcel) = parcel, part_of(Parcel,Region) = true,
        point_to_region(Point,Region) = TopographicRelation.
```

In association rule mining tasks, the specification of pattern descriptors corresponds to the specification of a collection of atoms: "predicateName(t_1, …, t_n)," where the name of the predicate corresponds to a <Descriptor>, while <TermSpec> describes each term *ti*, which can be either a constant or a variable. When the term is a variable, the mode and role clauses indicate, respectively, the type of variable to add to the atom and its role in a unification process. Three different modes are possible: **old** when the introduced variable can be unified with an existing variable in the pattern, **new** when it is not already present in the pattern, or **diff** when it is a new variable but its values must be different from the values of a similar variable in the same pattern. Furthermore, the variable can fill the role of reference object (**ro**) or task-relevant object (**tro**) in a discovered pattern during the unification process. The **is key** clause specifies the atom that has the key role during the discovery process. The first term of the key object must be a variable with mode **new** and role **ro**. The following is an example of specification of pattern descriptors defined by an SDMOQL statement:

with descriptors
    contain/2 **with variable mode old role ro,**
        variable mode new role tro;
    type_of/2 **with variable mode old role tro,**
        constant;
    fluvial_landscape/1 **with is key with variable mode new role ro;**

This specification helps to select only association rules where the descriptors fluvial_landscape/1, contain/2, and type_of/2 occur. The argument of "cell" is a **new** variable that plays the role of **ro**. The argument of the predicate "fluvial landscape" is always a new variable that plays the role of **ro**. The predicate "contain" links the **ro** with other geographic objects contained in the "fluvial_landscape." Finally, the first argument of the predicate "type_of" is always an **old** variable, denoting a geographic object that plays the role of **tro**, whereas the second argument is a constant value that denotes the type of object (e.g., river, street, parcel). The following association rule,

fluvial_landscape(X), contain(X,Y), type_of(Y,river), X≠Y ⇒
  contain(X,Z), type_of(Z,font), X≠Z, Y≠X

satisfies the constraints of the specification and expresses the co-presence of both a river and a font in a cell classified as a fluvial landscape.

10.5.4 Syntax for Background Knowledge and Concept Hierarchy Specification

Many data mining algorithms use background knowledge or concept hierarchies to discover interesting patterns. Background knowledge is provided by a domain expert on the domain to be discovered. This can be useful in the discovery process.
The SDMOQL syntax for background knowledge specification is the following:

<BackgroundKnowledge> ::= [<NewKnowledge>] {<UseKnowledge>}
<NewKnowledge> ::= **define knowledge** <Clause> {; <Clause>}
<UseKnowledge> ::= **use background knowledge of users** <User> {, <User>}
    on <Descriptor> {, <Descriptor>}

In INGENS 2.0, the user can define a background knowledge expressed as a set of definite clauses; alternatively, the user can specify a set of rules explicitly stored in a deductive database and possibly discovered in a previous step. An example of a background knowledge definition is reported in the following:

Example (Definition of close_to).
close_to(X,Y)=true :_region_to_region(X,Y)=meet.
close_to(X,Y)=true :_close_to(Y,X)=true.

while an example of the use of this background knowledge is reported in the following:

Example (Import of close_to).
use background knowledge of users UserName1 **on** close_to/2.

Concept hierarchies allow knowledge mining at multiple abstraction levels [17]. In SDMOQL, a specific syntax is defined for the hierarchy:

<Hierarchy> ::= [<NewHierarchy>] [<UseHierarchy>]
<NewHierarchy> ::= **define hierarchy** <Schema_Hierarchy> |
define hierarchy for <SetGroupingHierarchy>
<UseHierarchy> ::= **use hierarchy** <NameHierarchy> **of user** <User>.

The following example shows how to define some hierarchies in SDMOQL:

Example (Logical hierarchy on geographic objects).
define hierarchy LogicalObject **as**
    level1: {Hydrography,Orography, ...} < level0: Object;
    level2: {River,Lake,See,Font,Canal...} <level1:Hydrography;
    level1: {Slope,Contour slope, Level Point ...} < level0: Orography;
    ...

In INGENS 2.0, this hierarchy is automatically extracted from the GIS data model and used to discover multilevel spatial association rules.

10.5.5 Syntax for Interestingness Measure Specification

The user can control the data mining process by specifying interestingness measures for data patterns and their corresponding thresholds. The SDMOQL syntax is the following:

<InterestingnessMeasures> ::= [<Criteria>] [<Settings>

<Criteria> ::= **criteria**
(intermediate | final)(minimize | maximize) <Parameter>
with tolerance <Value> **{,(intermediate | final)**
(minimize | maximize) <Parameter> **with tolerance** <Value>}

<Settings> ::= <Parameter> := <StringValue>

Interestingness measures may include: threshold values, weights, search biases in the hypotheses space, and algorithm-specific parameters. In particular the user can bias the search in the hypotheses space by a number of preference criteria, such as the maximization of the number of covered examples or the minimization of the number of variables in the body of a learned clause. The user can also set thresholds such as confidence, support, or number of learned concepts. Finally, the user can set the value of a generic input parameter of a data mining algorithm.

10.6 MINING SPATIAL PATTERNS: A CASE STUDY

To show the potential of the integration of spatial data mining tools with GIS technology, we extend and elaborate on the case study on topographic map interpretation reported in [31]. The goal is to characterize and recognize some morphologies, which are not explicitly represented in the GIS data model.

The area considered in this application covers 90 km^2 in the surrounding area of the Ofanto River of Apulia, Italy (see Figure 10.7). The map of this area, stored in INGENS 2.0, is produced at a scale of 1:25000 by the Italian Military Geographic Institute (IGMI). The map is segmented into 90 square observation units of 1 km^2. A map maintenance user has created the vectorized map and stored it in the SDB, according to the data model reported above.

The geomorphology considered in the following sections is the fluvial landscape, which is characterized by the presence of waterways, fluvial islands, and embankments. The classification rule provides an operational definition which can be used to retrieve this geomorphology in other similar topographic maps, while spatial association rules can be used to describe the area and support the implementation of an environmental policy.

10.6.1 Mining Classification Rules

The data miner user graphically composes an SDMOQL query to mine the concept of a fluvial landscape, by using, as training data, all the cells of the map. The query interpreter analyzes the SDMOQL query and verifies its syntactic and semantic correctness. The feature extractor generates a symbolic description for each cell by computing descriptors listed in the **analyze** clause. In this study, all descriptors in

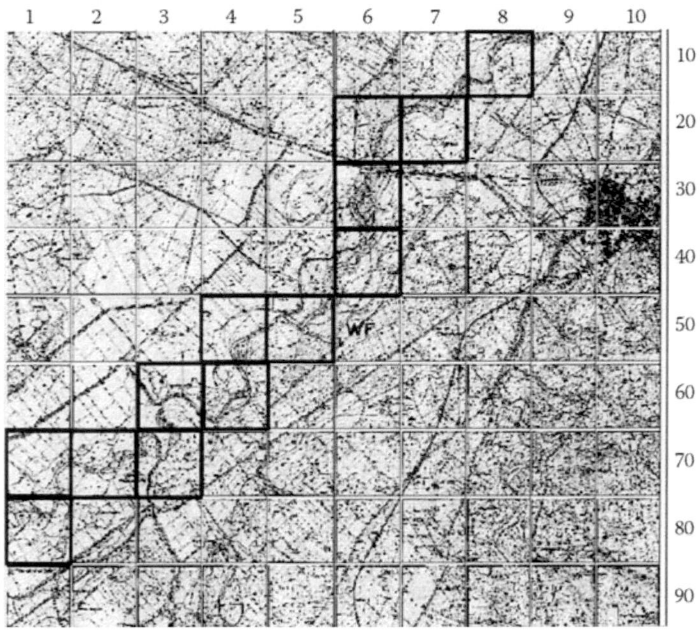

FIGURE 10.7 Surroundings of the Ofanto River. The boundary of fluvial landscape cells is blue.

Table 10.1 are extracted. The data miner then associates the conceptual description of each cell with a concept (fluvial landscape or others), thus completely defining the training data. Association is made by binding variable terms of one of the concepts to be discovered to the constants that represent the cells. This binding function is supported by the GUI of the system (see Figure 10.8).

The classification rules induced by the learning system ATRE are reported as follows:

R1: class(X1)=fluvial_landscape ←type_of(X1)=cell,
contain(X1,X2)=true, color(X2)=blue,
type_of(X2)=river, part_of(X2,X3)=true,
extension(X3)∈[653.495..1642.184],
line_to_line(X4,X3)=almost_perpendicular,
extension(X4)∈[325.576..1652.736].

R2: class(X1)=fluvial_landscape ←type_of(X1)=cell,
contain(X1,X2)=true, type_of(X2)=province,
part_of(X2,X3)=true,
line_to_line(X4,X3)=almost_parallel,
part_of(X5,X4)=true, type of(X5)=contour_slope.

R1 covers 10 examples, while R2 covers 5 examples, two of which are different from those covered by R1.

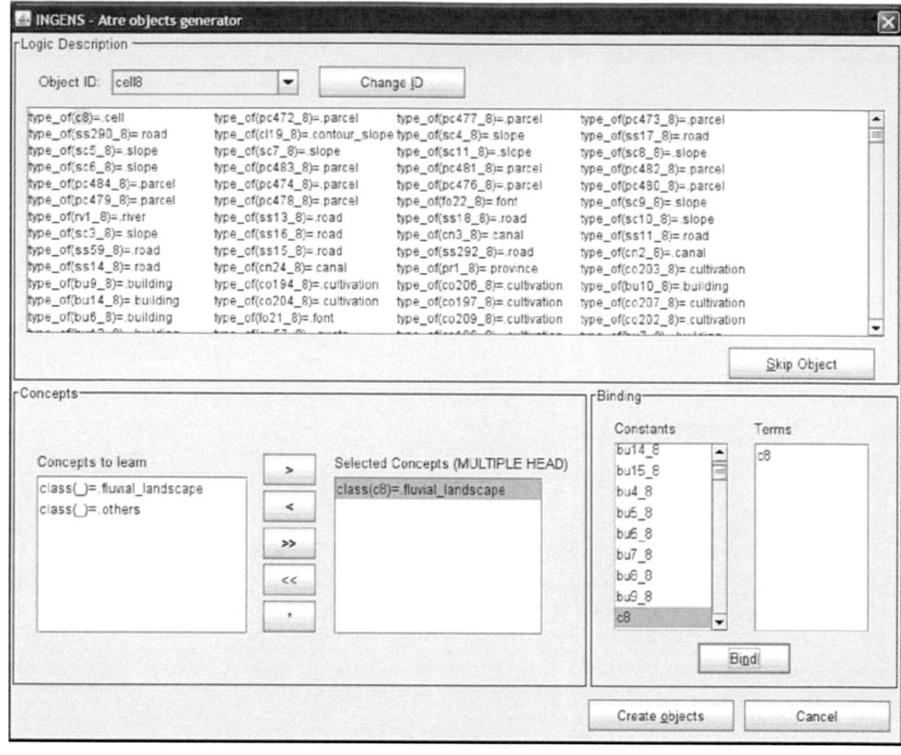

FIGURE 10.8 Associating a cell with a concept in INGENS 2.0.

According to R1, a cell is an instance of fluvial landscape if it contains geographic objects in blue classified as river, which is represented as a line (X3) with an extension between 653.495 and 1642.184 m. This line is almost perpendicular to another line (X4) with an extension between 325.576 and 1652.736 m. Unfortunately, the logical type of X4 is not specified by the rule. This is because the representation of a cell is related to the physical objects that it contains. To move from a physical to a logical level in the conceptual descriptions of the cells, some new descriptors are defined as background knowledge (see Figure 10.9). For example, the following <BackgroundKnowledge> statement

parcel_to_parcel(A,B)=C ←type_of(A)=parcel,
 type_of(B)=parcel, part_of(A,D)=true,
 part_of(B,E)=true, region_to_region(D,E)=C

describes the topological relation between the regions that physically represent the "parcels" here referred to as the variables A and B, respectively. This BK statement can be stored in the GIS repository and re-used in a new data mining task. By defining other similar descriptors and then constraining the search space only to

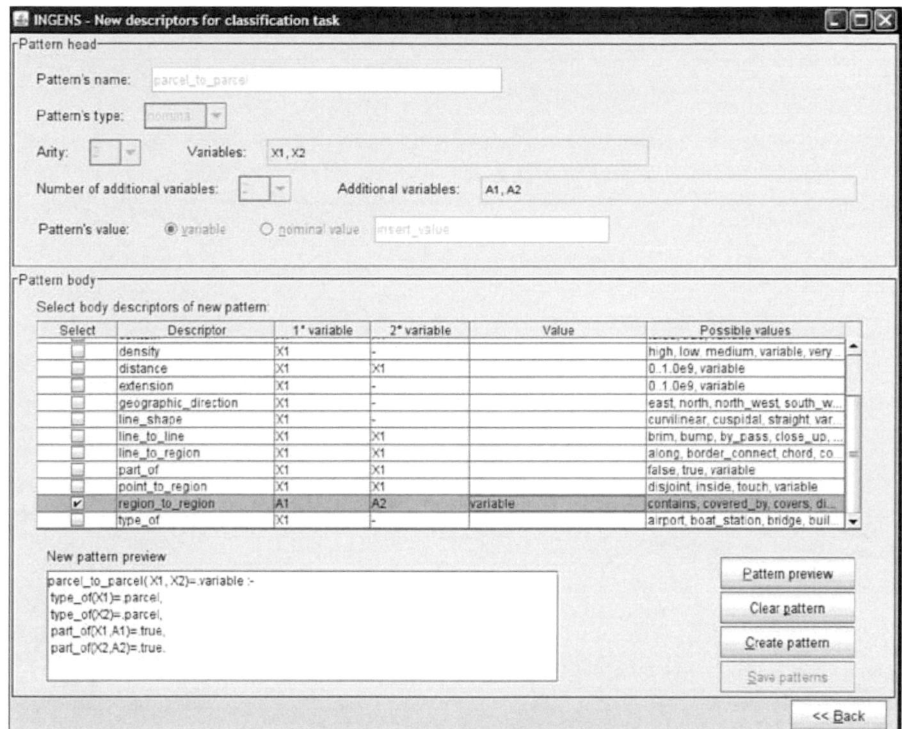

FIGURE 10.9 Specifying a new pattern descriptor in INGENS 2.0.

the definite clauses including these new descriptors, it is possible to discover a more abstract, human-interpretable operational definition of a fluvial landscape:

R3: class(X1)=fluvial_landscape ←
    contain(X1,X2)=true,
    river_extension(X2)∈[653.495..1642.184],
    river direction(X2)=north east.
R4: class(X1)=fluvial_landscape ←
    contain(X1,X2)=true,
    road_to_province(X2,X3)= almost_perpendicular,
    road_to_river(X2,X4)= almost_perpendicular,
    river_extension(X4) in [653.495..1642.184].

Rule R3 covers eight examples, while R4 covers five examples, four of which are different from those covered by R1. Both rules capture the presence of a river as a characterizing geographic object. In addition, rule R4 describes the spatial arrangement of other logical objects (road and administrative boundary) in the surroundings. The presence of an administrative boundary in this rule is not surprising because the River

Ofanto partially overlaps the boundary between the provinces of Bari and Foggia in Apulia.

A different analysis is done by randomly selecting only four positive examples (8, 16, 17, 53) and nine negative examples (5, 11, 15, 27, 29, 34, 84, 88, 89) of the fluvial landscape concept and using only this training data to discover an operational definition of a fluvial landscape. By ignoring the BK, the following rule is discovered:

R5: class(X1)=fluvial_landscape ←type_of(X1)=cell,
contain(X1,X2)=true, type_of(X2)=river,
part_of(X2,X3)=true,
line_to_line(X4,X3)=almost_perpendicular,
part_of(X5,X4)=true, type_of(X5)=road,

while considering new descriptors defined in the BK, the following rule is discovered:

R6: class(X1)=fluvial_landscape ←
contain(X1,X2)=true,
road_to_river(X2,X3)= almost_perpendicular,
river_extension(X3) in [141.623..1642.184].

Discovered rules are used to query the entire map and recognize fluvial landscape cells. Several statistics are collected in Table 10.2. "TP" is the number of true positives (correctly classified cells). "FP" is the number of false positives. "FN" is the number of false negatives. "Prec" is the precision of the concept (Prec = TP/(TP + FP)). "Recall" is the recall of concepts (Recall = TP/(TP + FN)).

10.6.2 Association Rules

A purely descriptive analysis of the fluvial landscape is performed when the data miner extracts the frequent spatial association rules which compactly describe the morphology of the fluvial landscape cells in the topographic map. Similar to the classification case, INGENS 2.0 GUI offers facilities to graphically compose the SDMOQL query. In addition, INGENS 2.0 allows users to visualize the portion of the logical hierarchy matching at least one of the geographic objects extracted within the <ObjectQuery> statement (see Figure 10.10) and to translate it in a

TABLE 10.2
Classification of the Surroundings of the Ofanto River Map (90 cells)

Rule	Time (sec)	TP	FP	FN	Prec	Recall
R5	832	12	5	1	0.706	0.923
R6	68	12	4	1	0.750	0.923

Note: The experiments are performed on Intel Pentium 4 -2.00 GHz CPU RAM 532Kb running Windows Professional 2000.

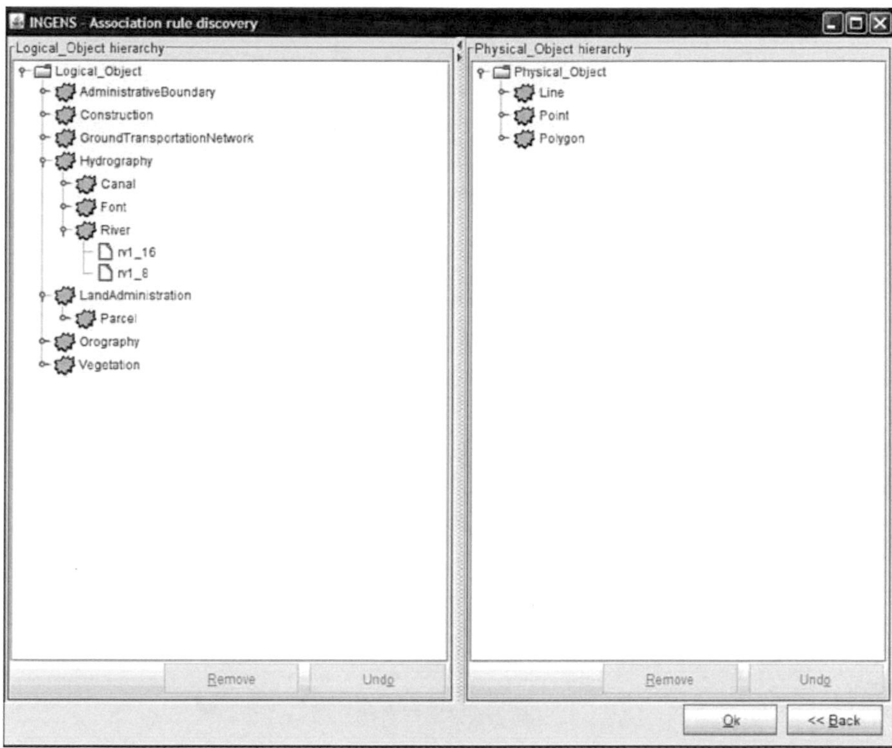

FIGURE 10.10 A portion of logical hierarchy that is automatically derived from a database. The hierarchy is visualized in the GUI of INGENS 2.0.

<NewHierarchy> statement to be added to the user-composed SDMOQL query. The logical hierarchy is then exploited to discover association rules at multiple levels of granularity without forcing data miners to repeat independent experiments on different representations. Once again, the BK is defined to move from a physical description to a logical description of the reference objects.

SPADA is run by setting *min_sup* = 0.9 and *min_conf* = 0.9 for each granularity level, and the maximum pattern length is set to eight.

Despite the above constraints, SPADA generates 25830 confident rules from a set of 15048 candidate patterns, in 1819 sec. Confident rules and frequent patterns are visualized to data miners in separate views: one view for each hierarchy level and pattern length.

An association rule discovered by SPADA at the second level of granularity is the following:

fluvial_landscape(A) ⇒
  contain(A,B), is_a(B,administration_boundary),
  almost_perpendicular(B,C), C\=B ,is_a(C,hydrography)
(92.3%, 92.3%)

At a granularity level 3, SPADA specializes the task-relevant objects B and C by generating the following rule, which preserves both support and confidence values:

fluvial_landscape(A) ⇒
contain(A,B), is a(B,province),
almost_perpendicular(B,C), C\=B, is_a(C,river)
(92.3%, 92.3%)

The rule states that A is an instance (a cell) of a fluvial landscape, then A is crossed by a province boundary B that is almost perpendicular to a river C. Once again, the frequent pattern underlying this rule suggests a correlation between a fluvial landscape and a province boundary.

10.7 CONCLUDING REMARKS AND DIRECTIONS FOR FURTHER RESEARCH

Empowering a GIS with spatial data mining capabilities is not a trivial task. First, the geometrical representation and relative positioning of geographic objects implicitly define spatial properties and relationships, whose efficient computation requires an integration of the data mining system with the GDBMS. Second, the interactions between spatially close objects introduce different forms of autocorrelation, whose effect should be considered to improve predictive accuracy of induced models and patterns. Third, the units of analysis are typically composed of several geographic objects with different properties, and their structure cannot be easily accommodated by classical double entry tabular data. In INGENS 2.0, these challenges have been dealt with by integrating (multi)relational data mining systems, which are able to navigate the relational structure of data and to generate relational patterns expressed in first-order logic or expressively equivalent formalisms. In particular, INGENS 2.0 integrates the MRDM systems ATRE and SPADA, which discover spatial classification rules and association rules, respectively. Different technologies, such as spatial database, data mining, and GIS, are hidden from users by means of a spatial data mining query language, SDMOQL, that permits condensing a data mining task in a query. Some constraints on the query language are identified by the particular mining task.

Although resorting to MRDM enables the INGENS 2.0 users to perform a sophisticated topographic map process, there are still several challenges that must be overcome and issues that must be resolved before the relational approach can effectively enhance GIS applicability.

First, several MRDM methods exploit knowledge on the data model (e.g., foreign keys), which is obtained free of charge from the database schema, in order to guide the search process. However, this approach does not fit spatial databases well, because the database navigation is also based on the spatial relationships which are not explicitly modeled in the schema. To solve this problem, a feature extraction module is implemented in INGENS 2.0 to precompute spatial properties and relationships which are converted into Prolog facts used by ATRE and SPADA. The pre-computation is justified by the fact that geographic maps are rarely updated. However, the number of spatial relationships between two layers can be very large and many of them might be unnecessarily extracted. The alternative is to dynamically perform spatial joins only for the part of the hypothesis space that is really explored during the search by a data mining algorithm. This approach has been implemented in two MRDM systems, namely SubgroupMiner for subgroup mining [21] and Mrs-SMOTI for regression analysis [30]. Both systems realize a tight integration with a spatial DBMS (namely, Oracle Spatial), but have been applied to datasets where few spatial relationships are actually computed. Hence, scalability remains a problem when many spatial predicates have to be computed.

Second, the presence of autocorrelation in spatial phenomena strongly motivates an MRDM approach to spatial data mining. In any case, it also introduces additional challenges. In particular, it has been proven that the combined effect of autocorrelation and concentrated linkage (i.e., high concentration of objects linked to a common neighbor) can bias feature selection in relational classification [20]. In fact, the distribution of scores for features formed from related objects with concentrated linkage presents a surprisingly large variance when the class attribute has a high autocorrelation. This large variance causes feature selection algorithms to be biased in favor of these features, even when they are not related to the class attribute, that is, they are randomly generated. Most MRDM algorithms, such as ATRE, do not account for this bias. A solution to be investigated in INGENS 2.0 is the generation of pseudo samples from the relational data by retaining the linkage present in the original sample and the autocorrelation among the class labels, and, at the same time, by destroying the correlation between the original attributes and the class labels [36].

Third, an inductive learning algorithm designed for the predictive tasks typically requires large sets of labeled data. However, a common situation in geographic data mining is that many unlabeled geographic objects (e.g., map cells) are available and manual annotation is fairly expensive. Inductive learning algorithms would actually use only the few labeled examples to build a prediction model, thus discarding a large amount of information potentially conveyed by the unlabeled instances. The idea of transductive inference (or transduction) [44] is to analyze both the labeled (training) data and the unlabeled (working) data to build a classifier and classify (only) the unlabeled data as accurately as possible. Transduction is based on a (semisupervised) smoothness assumption, according to which if two points in a high-density region are close, then the corresponding outputs should also be so [9]. In spatial domains, where closeness of points corresponds to some spatial distance measure, this assumption is implied by (positive) spatial autocorrelation. Therefore, the transductive setting seems especially suitable for classification and regression in GIS, and more in general, for those relational learning problems characterized by autocorrelation on the dependent variables. Only recently, a work on the transductive relational learning has been reported in the literature [8], and some preliminary results on spatial classification tasks show the effectiveness of the transductive approach [2]. No results are available on another class of predictive tasks, namely spatial regression.

Fourth, a large amount of knowledge is available in the case of geographic knowledge discovery, where relationships among geographic objects express natural geographic dependencies (e.g., a port is adjacent to a water body). These dependencies are expressed in nonnovel or uninteresting patterns but with a very high level of support and confidence. If this geographic knowledge were used to constrain the search for new patterns, the scalability of the spatial data mining algorithms would greatly increase. Actually, these dependencies are represented either in geographic database schema, through one-to-one and one-to-many cardinality constraints, or in geographic ontologies. Therefore, their usage can be done at no additional cost in MRDM perspective, thus moving a step forward toward knowledge-rich data mining [12]. In INGENS 2.0, SPADA uses knowledge to constrain the search space for spatial association rules. In any case, the use of background knowledge can be investigated in several data mining tasks.

A final consideration on spatial reasoning can be made on spatial data mining methods in general. Spatial reasoning is the process by which information about objects in space and their relationships is gathered through measurement, observation, or inference, and is used to reach valid conclusions regarding the objects' relationships. For instance, in spatial reasoning, the accessibility of a site A from a site B can be recursively defined on the basis of the spatial relationships of adjacency or contiguity. Principles of spatial reasoning have been proposed for both quantitative and qualitative approaches to spatial knowledge representation. Embedding spatial reasoning in spatial data mining is crucial to make the right inferences, either when patterns are generated or when patterns are evaluated. Surprisingly, there are few examples of data mining systems that support some form of spatial reasoning. In INGENS 2.0, SPADA supports a limited form of spatial inference if rules of spatial reasoning are encoded in the background knowledge. However, although a general-purpose theorem prover for predicate logic can be used for spatial reasoning (as in SPADA), constraints that characterize spatial problem solving have to be explicitly formulated in order to make the semantics consistent with the target domain space. Therefore, embedding specialized spatial inference engines in the GIS seems to be the most promising, but still unexplored, solution.

REFERENCES

[1] R. Agrawal and R. Srikant. Fast algorithms for mining association rules in large databases. In J. B. Bocca, M. Jarke, and C. Zaniolo, editors, *Very Large Databases, VLDB 1994*, pp. 487–499. Morgan Kaufmann, San Francisco, CA, 1994.

[2] A. Appice, N. Barile, M. Ceci, D. Malerba, and R.P. Singh. Mining geospatial data in a transductive setting. In A. Zanasi, C.A. Brebbia, and N.F.F. Ebecken, editors, *Data Mining VIII,* pp. 141–150. WIT Press, Southampton, UK, 2007.

[3] A. Appice, M. Berardi, M. Ceci, and D. Malerba. Mining and filtering multilevel spatial association rules with ares. In M.S. Hacid, Z.W. Ras, and S. Tsumoto, editors, *15th International Symposium on Methodologies for Intelligent Systems, ISMIS 2005,* volume 3488 of *LNCS*, p. 342353. Springer-Verlag, Berlin, 2005.

[4] A. Appice, M. Ceci, A. Lanza, F.A. Lisi, and D. Malerba. Discovery of spatial association rules in georeferenced census data: A relational mining approach. *Intelligent Data Analysis*, 7(6):541–566, 2003.

[5] M. Berardi, A. Varlaro, and D. Malerba. On the effect of caching in recursive theory learning. In R. Camacho, R.D. King, and A. Srinivasan, editors, *14th International Conference on Inductive Logic Programming, ILP 2004,* volume 3194 of *Lecture Notes in Computer Science,* pp. 44–62. Springer, Berlin, 2004.

[6] V. Bogorny, A.T. Palma, P. Engel, and L.O. Alvares. Weka-gdpm: Integrating classical data mining toolkit to geographic information systems. In *SBBD Workshop on Data Mining Algorithms and Aplications, WAAMD 2006*, pp. 9–16, 2006.

[7] M. Ceci and A. Appice. Spatial associative classification: propositional vs structural approach. *Journal of Intelligent Information Systems*, 27(3):191–213, 2006.

[8] M. Ceci, A. Appice, N. Barile, and D. Malerba. Transductive learning from relational data. In P. Perner, editor, *5th International Conference on Machine Learning and Data Mining in Pattern Recognition, MLDM 2007*, volume 4571 of *Lecture Notes in Computer Science*, pp. 324–338. Springer, Berlin, 2007.

[9] O. Chapelle, B. Schölkopf, and A. Zien. *Semisupervised Learning*. MIT Press, Cambridge, MA, 2006.

[10] L. De Raedt. Attribute-value learning versus inductive logic programming: The missing links (extended abstract). In D. Page, editor, *Proceedings of the 8th International Workshop on Inductive Logic Programming, ILP 1998 (extended abstract),* volume 1446 of *Lecture Notes in Computer Science*, pp. 1–8, London, UK, 1998. Springer-Verlag.

[11] P. Densham. Spatial decision support systems. *Geographical Information Systems: Principles and Applications*, pp. 403–412, 1991.

[12] P. Domingos. Toward knowledge-rich data mining. *Data Mining Knowledge Discovery*, 15(1):21–28, 2007.

[13] S. Džeroski and N. Lavrač. *Relational Data Mining*. Springer-Verlag, Berlin, 2001.

[14] M. J. Egenhofer. Reasoning about binary topological relations. In *Proceedings of the 2nd Symposium on Large Spatial Databases, VLDB 1991*, pp. 143–160, Zurich, Switzerland, 1991.

[15] A.U. Frank. Spatial concepts, geometric data models, and geometric data structures. *Computers and Geosciences*, 18(4):409–417, 1992.

[16] S. Haehnel, J. Hauf, and T. Kudrass. Design of a data mining framework to mine generalized association rules in a web-based GIS. In S.F. Crone, S. Lessmann, and R. Stahlbock, editors, *Proceedings of the 2006 International Conference on Data Mining, DMIN2006*, pp. 114–117. CSREA Press, 2006.

[17] J. Han, Y. Fu, W. Wang, K. Koperski, and O.R. Zaiane. DMQL: a data mining query language for relational databases. In *Proceedings of the Workshop on Research Issues on Data Mining and Knowledge Discovery*, pp. 27–34, Montreal, Quebec, 1996.

[18] J. Han, K. Koperski, and N. Stefanovic. *GeoMiner: a system prototype for spatial data mining*, in Proc. ACM SIGMOD Int. Conf. on Management of Data, pp. 553–556, May 1997, Tucson, Arizona, USA, ACM Press 1997.

[19] A.K.H. Tung, J. Han, and M. Kamber. Spatial clustering methods in data mining. *Geographic Data Mining and Knowledge Discovery*, pp. 188–217. Taylor & Francis, London, 2001.

[20] D. Jensen, J. Neville, and B. Gallagher. Why collective inference improves relational classification. In W. Kim, R. Kohavi, J. Gehrke, and W. DuMouchel, editors, *10th ACM SIGKDD International Conference on Knowledge Discovery and Data Mining, KDD 2004,* pp. 593–598. ACM, 2004.

[21] W. Klosgen and M. May. Spatial subgroup mining integrated in an object-relational spatial database. In T. Elomaa, H. Mannila, and H. Toivonen, editors, *European Conference on Principles and Practice of Knowledge Discovery in Databases, PKDD 2002*, volume 2431 of LNAI, pp. 275–286. Springer-Verlag, Berlin, 2002.

[22] K. Koperski. *Progressive Refinement Approach to Spatial Data Mining.* PhD thesis, Computing Science, Simon Fraser University, British Columbia, Canada, 1999.

[23] A. Lanza, D. Malerba, F.A. Lisi, A. Appice, and M. Ceci. Generating logic descriptions for the automated interpretation of topographic maps. In D. Blostein and Y.B. Kwon, editors, *Graphics Recognition: Algorithms and Applications*, volume 2390 of *Lecture Notes in Computer Science*, pp. 200–210. Springer-Verlag, Berlin, 2002.

[24] F.A. Lisi and D. Malerba. Inducing multilevel association rules from multiple relations. *Machine Learning*, 55:175–210, 2004.

[25] L. Lloyd. *Foundations of Logic Programming,* 2nd ed. Springer-Verlag, Berlin, 1987.

[26] H.P. Kriegel, J. Sander, M. Ester, S. Gundlach. Database primitives for spatial data mining. In *Proceedings of the International Conference on Database in Office, Engineering and Science, BTW 1999,* Freiburg, Germany, 1999.

[27] D. Malerba. Learning recursive theories in the normal ilp setting. *Fundamenta Informaticae,* 57(1):39–77, 2003.

[28] D. Malerba, A. Appice, and M. Ceci. A data mining query language for knowledge discovery in a geographical information system, *Database Support for Data Mining Applications*, pp. 95–116. Number 2682 in Lecture Notes in Computer Science. Springer-Verlag, Berlin, 2003.

[29] D. Malerba, A. Appice, A. Varlaro, and A. Lanza. Spatial clustering of structured objects. In S. Kramer and B. Pfahringer, editors, *ILP,* volume 3625 of *Lecture Notes in Computer Science*, pp. 227–245. Springer, Berlin, 2005.

[30] D. Malerba, M. Ceci, and A. Appice. Mining model trees from spatial data. In A. Jorge, L. Torgo, P. Brazdil, R. Camacho, and J. Gama, editors, *Conference on Principles and Practice of Knowledge Discovery in Databases*, volume 3721 of *Lecture Notes in Computer Science,* pp. 169–180. Springer, Berlin, 2005.

[31] D. Malerba, F. Esposito, A. Lanza, and F. A. Lisi. Machine learning for information extraction from topographic maps, *Geographic Data Mining and Knowledge Discovery*, pp. 291–314. Taylor & Francis, London, 2001.

[32] D. Malerba, F. Esposito, A. Lanza, F.A. Lisi, and A. Appice. Empowering a GIS with inductive learning capabilities: The case of INGENS. *Journal of Computers, Environment and Urban Systems,* 27:265–281, 2003.

[33] H. Mannila and H. Toivonen. Levelwise search and borders of theories in knowledge discovery. *Data Mining Knowledge Discovery*, 1(3):241–258, 1997.

[34] M. May and A.A. Savinov. Spin!-an enterprise architecture for spatial data mining. In V. Palade, R.J. Howlett, and L.C. Jain, editors, *KES*, volume 2773 of *Lecture Notes in Computer Science*, pp. 510–517. Springer, Berlin, 2003.

[35] T. Mitchell. *Machine Learning.* McGraw-Hill, New York, 1997.

[36] J. Neville and D. Jensen. Collective classification with relational dependency networks, 2003.

[37] T. Pavlidis. *Algorithms for Graphics and Image Processing*. Springer, Berlin, 1982.

[38] G.D. Plotkin. A note on inductive generalization. 5:153–163, 1970.

[39] D. Thompson and R. Laurini. Fundamentals of spatial information systems. *APIC Series*, 37, 1992.

[40] P. Rigaux, M. Scholl, and A. Voisard. *Spatial Databases with Application to GIS.* Morgan Kaufmann, San Francisco, CA, 2002.

[41] S. Chawla and S. Shekhar. *Spatial Databases: A Tour.* Prentice Hall, Upper Saddle River, NJ, 2003.

[42] J. Sander, M. Ester, H.P. Kriegel, and X. Xu. Density-based clustering in spatial databases: The algorithm gdbscan and its applications. *Data Mining and Knowledge Discovery*, 2(2):169–194, 1998.

[43] M. Sudhakar, A. Pankaj, T. Smith, and P. Donna. KBGIS-II: A knowledge-based geographic information system. *International Journal of Geographic Information Systems*, 1(2):149–172, 1997.

[44] V. Vapnik. *Statistical Learning Theory*. Wiley, New York, 1998.

[45] I. Witten and E. Frank. *Data Mining: Practical Machine Learning Tools and Techniques with Java Implementations*. Morgan Kaufmann, San Francisco, CA, 2000.

[46] S. Wrobel. Inductive logic programming for knowledge discovery in databases, *Relational Data Mining*, pp. 74–101. LNAI. Springer-Verlag, Berlin, 2001.

11 Visual Exploration and Explanation in Geography Analysis with Light

Mark Gahegan

CONTENTS

11.1 INTRODUCTION

Sir Francis Bacon, in Book II of his *Novum Organum* (*The New Organon*; 1620) states "Truth will sooner come out from error than from confusion." This famous epithet describes the idea that we understand the world by imposing conceptual structure upon the confusion of data we receive. Our mistakes in doing so eventually lead us to a deeper understanding. Staying with the confusion will not, by itself, bring about new insights. Methods used at the heart of knowledge discovery and data mining in essence operationalize Bacon's insight; they impose structure on collections of data, and attempt to reduce confusion by progressively modifying this structure. Structures that help simplify the data, or that describe patterns that are often repeated are likely to be favored, whereas structures that add to the confusion will be rejected. The search for useful structure typically terminates when the gains accrued from the next iteration of structure changes imposed become smaller than a given threshold.

The conceptual structures we may wish to impose on data can take many forms, but some of the most common are (1) rules, (2) categories, and (3) hypotheses or explanations. For example, "*professors are absent-minded*" is a rule; "the *absent-minded are people with the property that they forget the details of ordinary life*" might be the underlying category; and "*the mind of a professor is too focused on higher questions to remember where he put his car keys*" is a possible hypothesis or explanation to connect rule and category together in a chain of inference.

It is our ability as humans to quickly impose — and withdraw — conceptual structures that makes us so adept at discovery. However, to be successful, visualization to support knowledge discovery must achieve two consecutive goals:

1. Present data to humans, via the visual senses, that would not ordinarily be available or directly observable. This is not straightforward because the human visual system is fickle — very powerful but easily misled.
2. Provide the means to experiment with imposing different kinds of conceptual structure on the visual display, via the mechanisms that control which data are visualized and how the data attributes are mapped into visual variables (such as color, position, size, shape, and movement).

This chapter presents an alternative form of exploratory analysis and data mining, based around the use of visualization, as opposed to computational or statistical approaches. Rather than attempt to process the data in a language that a machine can understand, the visualization approach attempts to portray the data in a form that a human can understand — perhaps free from many of the representational constraints that GIS and geographic models impose. The main aim of this chapter is to show how various forms of geographic visualization can also support exploratory data analysis (sometimes called data mining or knowledge discovery). When geovisualization techniques fulfill their potential, they are not simply display technologies by which users gain a familiarity with new datasets or look for trends and outliers; they are instead environments by which new geographical concepts and processes might be uncovered or defined (providing the foundations for later data analysis) and new geographical questions formulated (hypothesis generation). Several techniques are described for visually led exploratory analysis tasks and the tasks they facilitate are characterized within the context of the cognitive reasoning that underpins the scientific discovery process. A brief discussion of the transformation from visual form to constructed knowledge is then given, following from which conclusions and challenges are presented. The technologies and supporting science are drawn from information visualization, data mining, geography, human perception and cognition, machine learning, and data modeling.

Prior to this though, two issues that are more pressing are briefly discussed by way of justification: (1) Why do we need visually led approaches to exploration and discovery? (2) Are such methods really any different from their computational and statistical counterparts?

11.1.1 Why Is There a Need for Visually Led Exploration?

To begin, it is worth examining why there is a need for discovery, and what might be gained by using these technologies. At one level, the answer is that datasets may contain important trends and relationships of which we are ignorant, so discovery offers the possibility of finding new knowledge about the world and the people in it. At a deeper level, some of the datasets and problems confronting geographers are so complex that we are not currently equipped to *discover* or *learn* all the various patterns and knowledge artifacts they contain. In this case, knowledge discovery offers mechanisms for searching and learning that can turn intractable problems into solvable forms (although as yet not perfectly so). Exploratory analysis addresses data that are not yet fully understood, classified, summarized, or otherwise formed into high-level (semantically abstract) structures. Indeed, the purpose is often to uncover structures or patterns in the data in the first instance so that they might later be formed into useful concepts and relations that later play a role in spatial analysis. A simplified overview of the process of discovery-based science, divided into a number of stages, is shown in Figure 11.1 (the details surrounding this figure are given in later sections of the chapter).

The need for these exploratory approaches has arisen in part as a consequence of the massive digital datasets now being gathered for a wide variety of applications such as marketing, sales, telecommunications, medicine, finance, and geography (Koperski, Han, and Adhikary, 1999; Roddick and Spiliopoulou, 1999; Pal and Mitra, 2004), where significant trends must first be uncovered before they can be put to use. Such datasets are considered too large for textual browsing and may contain structure (patterns and relationships) that is unexpected or hitherto unrecognized. Geographic datasets may contain huge numbers of such patterns and relationships (Bryson et al., 1999); for example, impacts of global climate change, census demographics, eco-region analysis, and epidemiology. Some of these patterns are subtle while others are more obvious, but they must be recognized before they can be used (Ester et al., 1996; 1998). Many will be entirely meaningless for a given analysis goal and unless removed will distract from, or even hide, those that might be useful. Patterns may also comprise many more data dimensions than can be displayed simultaneously using conventional map-based methods, or indeed than can be analyzed with current parametric statistical techniques (Landgrebe, 1999).

There are three immediate reasons why visualization might be useful in tackling such large datasets. First, a scene or immersive virtual reality can make a large volume of data more accessible to a human observer than can tables of figures.* Second, the process of rendering provides a number of data transformations (described later) that

* A picture really is worth a thousand words; perhaps even a million (see the description of pixel-based methods).

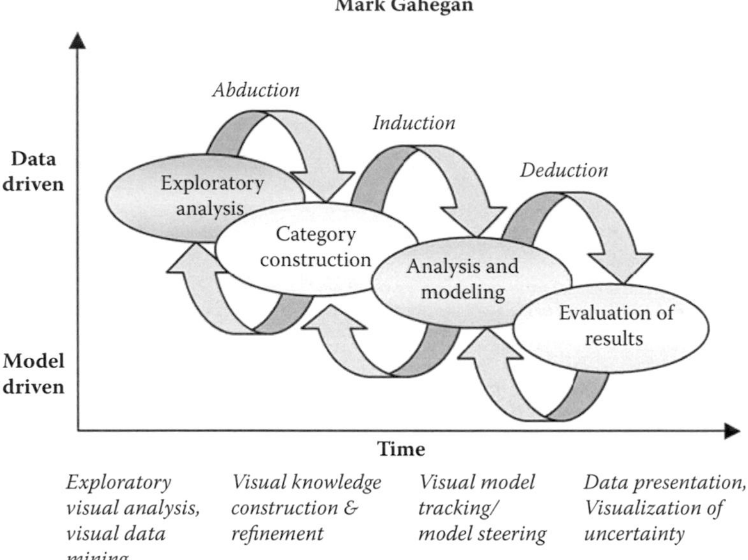

FIGURE 11.1 An overview of discovery activities as they might fit together when exploring an unfamiliar problem, positing an initial hypothesis, defining suitable conceptual structures by which to test it, performing analysis using these structures, and then evaluating the results. At any stage, if the process fails, the analyst must backtrack and begin again. The x axis also suggests some of the roles in this process that visualization techniques might fulfill.

act as querying and focusing operators, helping promote the search for, and discovery of, specific patterns. A third, deeper reason (discussed in more detail near the end of this chapter) is that the inclusion of the human expert within this activity may lead to greater insight being brought to bear. Because of this reliance on the individual, what is seen and what is inferred will be determined in part by an observer's visual system, experience, culture, and education that together form an unspecified bias (e.g. MacEachren, 1995; Mark et al., 1999; Slocum et al., 2001). This is both problematic and advantageous. Problematic because we cannot guarantee a visual stimulus will actually be observed as we might plan, and any insights gained are not represented explicitly. Advantageous because the human observer can employ much richer forms of reasoning and experience in the task of interpreting what they see. In addition, for recognition of pattern and structure, the human visual system has yet to be surpassed.

In addition, it is worth remembering that knowledge construction, for example defining categories, instances and relationships, is a fundamental activity in geography. Most, arguably all, of the concepts and relationships we use to describe and model the earth and its social systems we construct ourselves — sometimes individually, sometimes via agreements and conventions. We need help in discovering, describing, agreeing, and sharing these social constructs. Yet most existing analysis tools and systems for geographic data do not cater well to discovery and knowledge construction, being "locked in" to the cartographic paradigm. For many geographic applications, this paradigm relates best to the final stages of analysis, it presupposes that data have already been collected and processed, with the underlying categories and relationships extracted and labeled. Therefore, although the cartographic paradigm is a perfectly valid setting, it is not the *only* setting within which geographic information must be examined, just as GIS is not the only system that we can use.

11.1.2 Is a Visualization System That Different from a Database or a GIS?

Since many of the operations that visualization systems support involve selections and projections on data, it may be tempting to think that visualization systems are rather like databases, where the query language consists of similar logical operators to those we would find in, say, SQL, but expressed via a visual interface. There is some truth in this (for example, see Stolte, 2003). However, it is not the full story. The power of visualization comes from the fact that it provides a more direct connection to the conceptual structures in our minds and, largely, circumnavigates the logical organization of data in a database.* For example, it is quite a simple matter in a relational database to add in a new tuple (row) that might represent a recently uncovered new rule or fact, but harder to add or retract a new category, as this would involve at the very least the addition of new attributes (columns), possibly even the creation of new tables and reapportioning of the tuples between them. Finally, a new theory is likely to add or modify the connections between relations, hence modify the primary and foreign key structures that govern the way tables are designed and joined. These last two kinds of updates are difficult and time-consuming to achieve in current databases, often prohibitively. Therefore, while humans can suggest and retract conceptual structures with ease, databases struggle to keep up when their schemas are in constant flux!

11.1.3 Research to Date

Good, early accounts of exploratory analysis in a geographic setting are provided by Cleveland and McGill (1988), MacDougall (1992), Cook et al. (1995) and Wise, Haining, and Signoretta (1998). By now, visual approaches to data exploration are well-established, stemming from the pioneering work of Tukey (1977), Chernoff (1978), Asimov (1985), Tufte (1990) and Haslett et al. (1991), who all championed the idea of "pictures describing data," and Bertin (1985), Mackinlay (1986), and Treisman (1986) who, along with many others, studied the perception of visual variables (such as *shape*, *position*, *hue,* and *saturation*) and how they can be used in combination when building a map or a graphical display. The term exploratory visual analysis (EVA) has been coined to describe exploratory methods that are conducted primarily in the visual domain. EVA has been applied to a wide range of

* To what degree discovery is aided or hindered by the way data are initially organized is an interesting and underexplored question.

geographic settings in the work of Tang (1992), Dykes (1997), Gahegan (1998), and others (e.g., see the edited volumes by MacEachren and Kraak, 1997; and Kraak and MacEachren, 1999). Examples of specific visualization techniques used in the search for structure are shown in Figure 11.3 through Figure 11.9.

In more recent times, there has been a move toward more rigorous treatment of the process of geovisualization as it relates to exploration and discovery (Fayyad and Grinstein, 2001; Andrienko, Andrienko, and Gatalsky, 2003; Dykes, MacEachren, and Kraak, 2005), and to the activities involved in knowledge construction (Shrager and Langley, 1990; MacEachren, Gahegan, and Pike, 2004; Andrienko and Andrienko, 2005; Gahegan, 2005). A useful cross-section of the breadth of the field of visualization for data mining and knowledge discovery can be found in the 31 contributed chapters of Fayyad, Grinstein, and Wierse (2002).

Largely the aims of data mining and knowledge discovery are very similar to those of exploratory analysis: to find useful and valid structure in large volumes of data, and to provide some means of explaining it. Fayyad, Piatetsky-Shapiro, and Smyth (1996) describe the data mining and knowledge construction process as comprising five stages: (1) data selection, (2) pre-processing, (3) transformation, (4) data mining, and (5) interpretation/evaluation. These stages progressively refine a large dataset to the point where it makes sense to propose new concepts and relationships, and their instances (Piatetsky-Shapiro, Fayyad, and Smith, 1996). The stages and aims are in close agreement with approaches to exploratory visualization, where data is also selected, transformed, and interpreted in an effort to uncover structure and meaning (as shown in Figure 11.1), although in the case of visualization, the analytical workflow is less well recognized and understood.

In the past dozen or so years, many new visualization tools have been proposed to aid in knowledge discovery activities (e.g., Lee and Ong, 1996; Keim and Kriegel, 1996) under the heading of "Visual Data Mining" (VDM), and research has more recently progressed toward visually supported methods of knowledge discovery (e.g. MacEachren et al., 1999, Gahegan et al., 2001, de Oliveira and Levkowitz, 2003). Since data mining is typically performed on very large databases, the volume of data to be represented brings challenges of its own (Keim et al., 2004; Compieta et al., 2007). Any useful visualization method must be able to project key relationships or clusters, or to summarize the data by aggregation since it is not always possible to render each item individually. In addition, geographical datasets may also contain attribute spaces with high dimensionality, for example the U.S. census, which contains over 100 attributes for each region in its longer form. Visualization can help because it offers a variety of geometric and graphic devices to present more dimensions to the user than the typical two or three provided by today's GIS (e.g., MacEachren et al., 2003; Johansson, Treloar, and Jern, 2004). Even so, it is often not viable or sensible to depict all dimensions and all data instances; they must be either sampled or summarized. Otherwise the resulting scene will be too dense with visual information to be comprehendible.

To date, much of the research stemming from the GIScience community tends to concentrate on providing high levels of interactivity, interlinked and dynamic tools to encourage an exploration of the data, and draw heavily from the literature on perception and cognition (e.g., Slocum et al., 2001). Methods are often human-led or highly interactive, but without rigid control over the exact search strategies and exploratory methods used. By contrast, approaches from the database and computer science communities generally use very specific algorithms so that uncovered objects or patterns will have a pre-defined visual appearance. Such methods rely on statistical theory, pattern recognition, and machine learning, and thus are more structured and rigorous, but less flexible and perhaps less geographically intuitive. However, any current distinction is unlikely to last long as researchers work to bring these two approaches into some kind of harmony (Ribarsky, Katz, and Holland, 1999; Gahegan and Brodaric, 2002; Kovalerchuk and Schwing, 2004; MacEachren et al., 2004). Recently, the field of visual analytics (http://nvac.pnl.gov/) has arisen to cover the union of data exploration, computational discovery methods, knowledge, and workflow management (Thomas and Cook, 2005) The field is an effort to forge much stronger links between these themes, since their disconnection acts as a barrier to current investigation (Keim et al., 2006).

Having now addressed these preliminary matters, we can turn our attention to the cognitive processes and computational technologies that underpin visualization. In order to understand how best to support the act of visual discovery, we must appreciate two things:

1. What it is that humans do when they discover — how the various thought processes interconnect to support lines of investigation, and how it is that we recognize the good (useful) conceptual structures from the bad.
2. The visualization approaches and techniques available and the degree to which they can be used to support the human process of discovery.

These two topics, taken in reverse order, form the major themes of the following two sections.

11.2 TECHNIQUES AND APPROACHES FOR EXPLORATORY VISUALIZATION

11.2.1 Visually Encoding Data

There is a lot more to exploratory visualization than simply selecting among the many available techniques for creating a visual display from some dataset. The primary goal for successful data exploration is that of searching for, and finding, some unknown signal or pattern. Gaining insight and constructing knowledge demand careful consideration of how the data is to be displayed or *visually encoded* (Ware, 2000). In order to be observed, concepts of interest in the data require a *visual stimulus* that allows the observer to perceive patterns that are relevant and avoid those that are not (Grinstein and Levkowitz, 1995). So, the task of trying to locate occurrences of a certain target pattern in the data (comprising values from n data dimensions) can be expressed in the visual domain as follows:

> Display the relevant values from each supporting data dimension in such a way that their conjunction defines an observable and unique visual stimulus.

Of course, the complication here is that such a strategy presupposes that we know what we are looking for, but since this is discovery, we typically will not know beforehand what the target pattern actually is! Thus, visualization tools need to support the rapid reconfiguration of the visual display, to allow the user to search through different mappings between the information and its visualization, for the emergence of potentially useful patterns. This search is typically ad-hoc, but governed by the user's expertise and intuition — thus, subjective. More structured methods do exist though, such as the grand tour and projection pursuit strategies (see projection techniques described later).

In the following section, a number of different information visualization techniques are described (maps, scatterplots, iconographs, etc.) that might lead to the production of a useful visual stimulus, that is, one that relates to some concept of interest. First, it is important to understand the basic functionality that visualization systems can provide and upon which these different techniques are based. In other words, what are the available building blocks with which visualizations of data can be constructed? In cartography, the static visual (retinal) variables (Bertin, 1985) are often considered separately from dynamic variables because temporal behavior is a relatively new concept (and is still largely absent from many GIS). Taking this approach, visualization functionality might be divided into the following four groups:

1. **The appearance of data (visual encoding).** This is the realm of traditional cartography, and covers how data values are transformed into visual properties such as color, shape and size of symbols, contouring, and so forth. Individual data items can be displayed using the different visual variables that a visualization environment supports (e.g., shape, size, color, texture, location and so forth. Generally, visualization systems provide a greater diversity of visual variables and methods than can be found in conventional GIS.
2. **The temporal behavior.** The usefulness of time as a variable in cartographic presentation has been long understood, (e.g., DiBiase et al., 1992) but difficult to realize due to the increased computational demands and runtime complexities that animation brings. However, there are a growing number of systems of temporal behavior added via scripts that are associated with basic objects (e.g., Flash, VRML, Java3D). Viz5D (www.ssec.wisc.edu/~billh/vis5d.html) uses a data model specifically designed for geoscientific applications that supports the animation of temporal image sequences.
3. **Properties of the entire scene.** Certain functionality applies to the whole of the scene, affecting the rendering of all graphical objects within the display. The most common examples are the viewpoint (i.e., the position of the observer relative to the scene), the resolution (or granularity), and the various forms of lighting that are employed to illuminate the contained objects from various positions. The power of scene-level properties is illustrated in the pair of displays shown in Figure 11.2, where simply adjusting the angle of illumination over a surface (here showing magnetic anomalies measured over a landscape) uncovers structures that could be easily overlooked or simply not observed if using a fixed color assignment.

(a) (b)

FIGURE 11.2 Examples of the effect of moving a single light source on the perception of structure in a scene. In the left images, the light source is in the northwest; for the right image, the light is in the northeast. Notice that these pairs differ markedly in the trends that are immediately discernible. The underlying data for all images is an artificial surface constructed from a gridded *magnetics* coverage. The images use grayscale light and shadows to show structure.

4. **Interactions supported within the visualization system.** In many current systems, a high level of user interaction is supported within individual displays, and between displays, such that interaction with data in one display causes the system to perform some related behavior in a different display. The improvise visualization environment (Weaver, 2004) shows this functionality in its richest form, where any kind of event in one display can trigger any kind of corresponding event in another. For example, selecting a set of points in a map might cause a scatterplot to zoom to the extent of these same points.

For future visualization environments, these groups may represent rather false divides; the temporal properties are becoming an integral part of the behavior of any object (or group of objects) in the scene and the more global types of functionality can in fact be applied to arbitrary groups of objects as well as to the entire scene. The only really useful distinction seems to be that some properties can be used directly to *construct a stimulus* (the visual and temporal variables associated with individual scene objects and their groupings) and others can be used to *draw attention to a stimulus* (the scene lighting and viewpoint, the interactions with the data).

Visual encoding strategies need to consider the following properties that together determine what might be observed in the resulting scene:

1. The choice of data to display.
2. The assignment of the data to the visual, temporal, or scene properties that the chosen visualization techniques and environment provide.
3. The mappings used to transform data values to these visual, temporal, or scene variables.
4. The nonlinear perceptual hardware in the human visual system and its inherent biases and inadequacies.

5. The difficulty in differentiating patterns because they may visually interfere with each other.
6. The very large number of irrelevant patterns that a complex geographic dataset may contain.
7. The varying biases, assumptions, and expertise that different analysts might bring.

Gahegan (1999) provides further discussion of some of these issues and a more formal treatment of the visual assignment process by which data are transformed into graphics.

11.2.2 Visualization Techniques for Data Exploration

Here, the word "techniques" is employed to refer to the different kinds of visual displays used, so that the wide selection currently on offer might be categorized and some generalizations made. Techniques for exploratory visualization are many and varied, making the choice of a suitable technique problematic (Robertson, 1997). Hinneburg, Keim, and Wawryniuk (1999) provide a useful categorization under the heading of "visual data mining," defining four distinct categories: (1) geometric projection techniques, (2) iconographic techniques, (3) pixel techniques, and (4) hierarchical techniques. Many of these outdate the term visual data mining by quite a number of years, but it is perhaps a useful banner under which some approaches might be subsumed. These four categories are augmented here to cover the range of exploratory visual techniques common in geovisualization, resulting in six distinct visual styles.

1. Map-based techniques: Maps are considered as a separate case from ordinary graphs, largely because they require sophisticated graphical primitives and they do not use a Cartesian space (or rather they should not, given that some visualization systems still offer only unprojected maps). The map itself can, and often does, act as an exploratory tool. Commonly supported functionality allows the user to switch the underlying data attributes being mapped, for example, by changing from one attribute to another (Dykes, 1997), or to alter the mapping to the visual variables used, for example, by modifying the assignment of values to colors. This technique works especially well with map legends, as shown by Peterson (1999) and Andrienko and Andrienko (1999). Not surprisingly, maps often form the center of geovisualization environments. The choropleth map in Figure 11.3 shows rates for breast cancer and cervical cancer for all counties in the conterminous United States. The map uses a bivariate color scheme (blue for breast cancer, red for cervical cancer), so that the color used for any county is determined by both cancer rates acting together. The user can choose which data to map, which color combinations to use, how many classes to divide the data into, and how those classes are determined.
2. Chart-based techniques: Charts and other more formal graphing metaphors have proven highly successful in conveying multivariate data, with

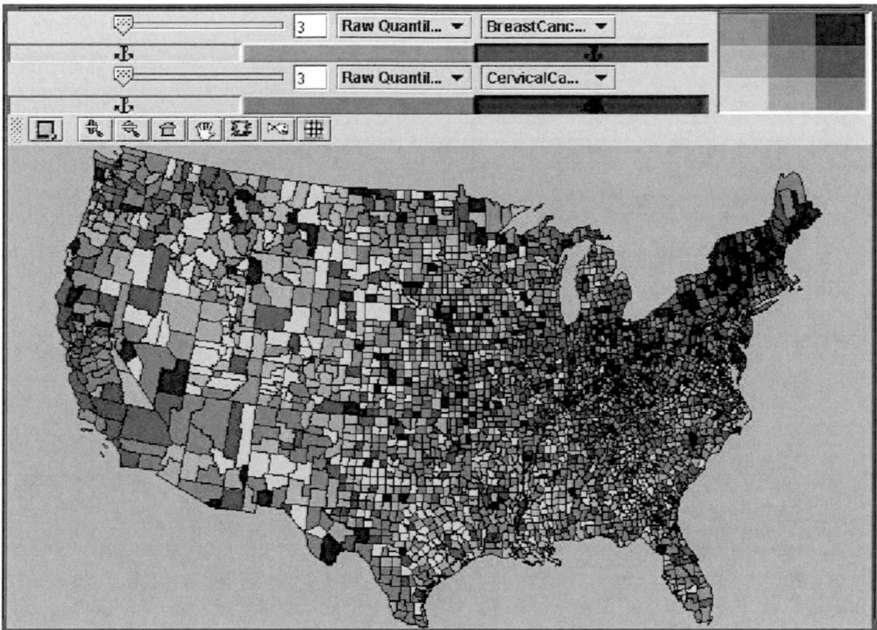

FIGURE 11.3 An example of a map-based visualization technique. A bivariate choropleth map shows cancer incidence by county for the conterminous United States. Redness increases with the rate for cervical cancer, blueness with the rate for breast cancer. Thus, counties colored light gray have low rates for both cancers, dark purple indicates places with high rates for both. See text for further details. See color insert after page 148.

the advantage that they foster perception of "quantity" due to their numeric axes. One of the most established forms, the scatterplot, uses a simple 2D or 3D graph with dots or some other symbols to mark the position of individual data items (Cleveland and McGill, 1988). Where the number of variables under investigation exceeds the capacity of the scatterplot, linking and brushing methods have been added to effectively re-join data that have been disaggregated into many graphs (e.g. Buja et al., 1991). This leads to tools like the scatterplot matrix, and many other matrix-based (multiform) displays that offer a mix of techniques (MacEachren et al., 2003), an example of which is shown in Figure 11.4. Another popular graph-based method is the parallel coordinate plot (Inselberg, 1997; Edsall, Harrower, and Mennis, 1999), which uses a number of parallel axes through which a trace of each data item can be made.

3. Projection techniques: Transformations such as principal component analysis, multidimensional scaling, and the self-organizing map (SOM) are used to project out of the data some subset of the structure — often the dominant statistical trends (e.g., Haslett et al., 1991). The idea is often to take a high-dimensional dataset and, via projection, reduce it to a simpler form with (typically) two dimensions, which is often displayed as a

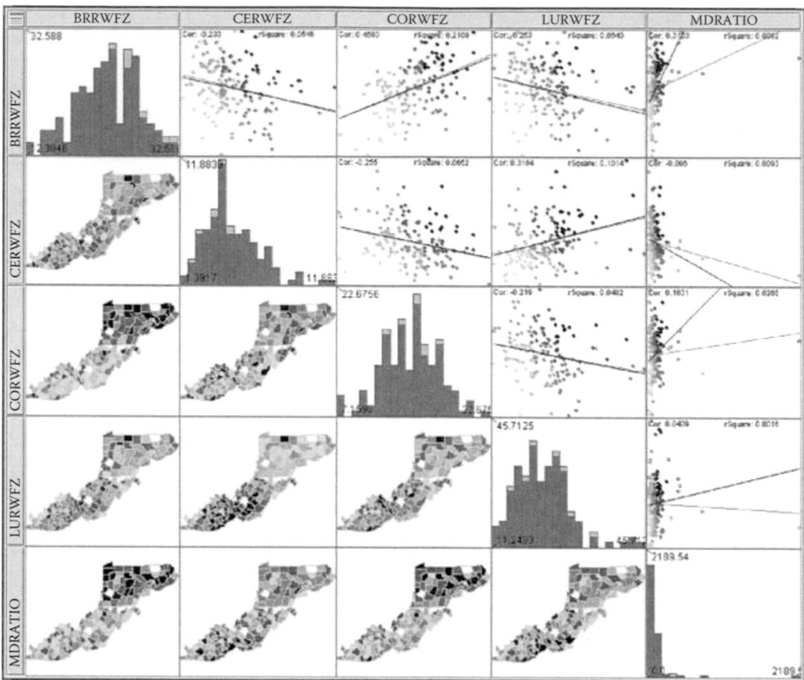

FIGURE 11.4 A matrix of maps and scatterplots, relating incidence rates for different kinds of cancer (first four rows and columns) to the ratio of doctors per 1000 population, for the Appalachian region of the United States (MD ratio, the right hand column and bottom row). This shows every pair of bivariate displays as both a scatterplot (above and right) and a map (below and left). The on-diagonal elements of the matrix simply show the distribution of values for each variable in isolation. A bivariate color scheme is used throughout. See color insert after page 148.

surface. On such a surface, similar items should appear close together, and dissimilar items far apart — although it is not possible to preserve all the topology of the original dataset. In Figure 11.5, a SOM has been used to create a surface from a complex environmental dataset (various spectral bands, geology, elevation, surface water accumulation, and so forth). The larger round symbols represent known field sites for a specific type of tree; notice that they fit quite neatly into the bottom left corner of the SOM surface (the white numbers represent other vegetation classes). One might (correctly) conclude that this tree class is characterized well in the data and is separable from the other classes.

4. Space-filling or pixel-based techniques: Efforts to perform data mining in a visual setting have produced some interesting and imaginative ways to take advantage of one of the most abundant system resources, that of the pixel (Keim, 1996). A typical display device now provides 1 to 2 million pixels, each one treated as a separate cartographic symbol, with a unique

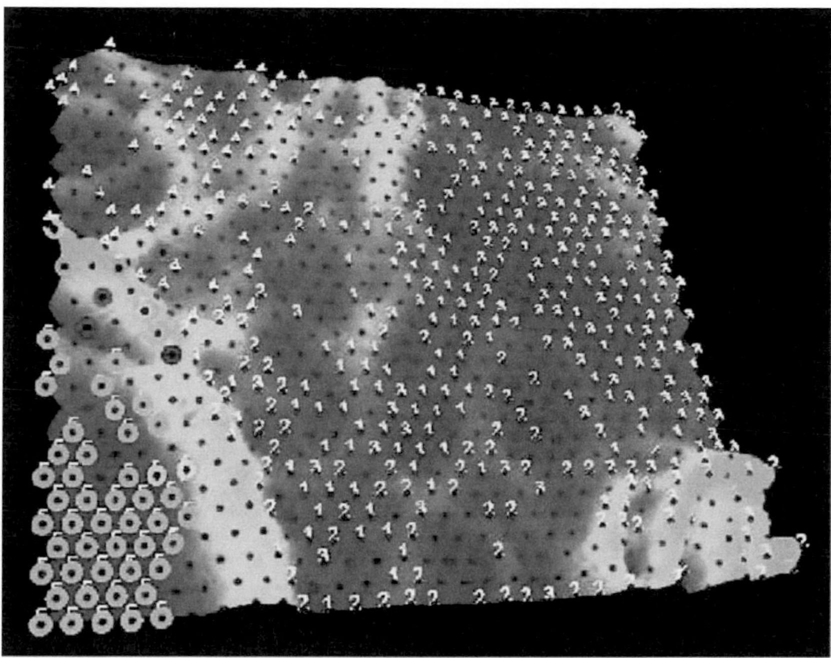

FIGURE 11.5 An example of a compositional landscape, created using a self-organizing map (SOM) that projects a high-dimensional dataset into a form that can be visualized in two or three dimensions — shown here as a surface. See text for a full explanation of how to read this figure. See color insert after page 148.

and directly observable location and color.* Data values for one attribute are mapped to pixel color via some classification method. Values for a different attribute are then used to order the pixels into the display using some kind of tiling method (for example a spiral or Peano curve). (If the two variables were perfectly correlated, then the colors would change in step with the ordering.) Figure 11.6 shows a pixel-based rendering of disease incidence data, with a scatterplot showing the same two variables. The screen can be divided into separate regions if several attributes are to be visualized concurrently, resulting in several orderings being shown. Keim and Kriegel (1994, 1996) describe sorting schemes on the data values and 2D orderings on the display space to localize groups of similar values in the scene. Jerding and Stasko (1998) instead describe a "mural" technique where data from very dense information spaces are merged together and mapped to the pixel properties of color and position — so a single pixel might be "aliased" to a number of data values. Such techniques present a useful overview of *all* the data. In the case of the mural technique, all

* While color may be available in even greater numbers (approximately 16 million are again typical), it can be difficult for humans to differentiate; our poor ability to quantify color values effectively reduces this dynamic range by several orders of magnitude.

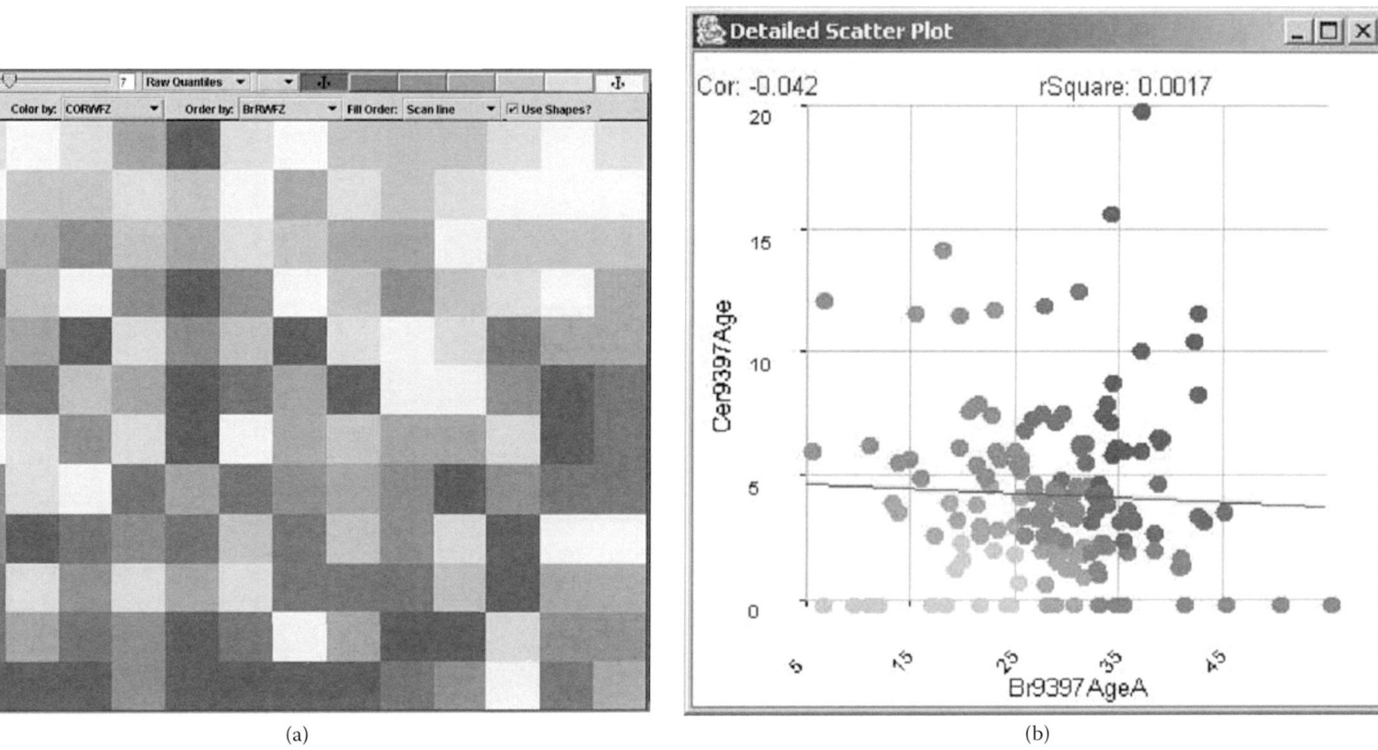

(a)

(b)

FIGURE 11.6 A space filling or pixel visual visualization, which orders data in one data dimension but colors it using another (left) is contrasted with a scatterplot showing the same data (right). The speckled nature of the plot on the left, and the cloud of points in the scatterplot show a weak but definite correlation between the two variables. See text for further details.

values in the data contribute to the scene, even if they cannot be rendered directly.

5. Iconographic/compositional techniques: By choosing a symbol more complex than a pixel, we gain access to a greater number of visual variables, allowing more data dimensions to be displayed simultaneously. Perhaps the most famous examples are the "faces" proposed by Chernoff (1973) and used so emotively with election data by Dorling (1994). Pickett and Grinstein (1988) propose a simple stick figure, where color, number of limbs, limb length, limb orientation, and figure position can be used to encode a different data value. Many other types of icon have been proposed as having desirable perceptual properties, including ribbons and arrows. The aim of iconographic displays is to promote perception of the "whole" while still allowing some differentiation of individual variables. Icons have, of course, been used a great deal in traditional cartography; for example, proportional circles used to convey the sizes of cities. The example in Figure 11.7 shows a complex visualization of a toxic plume and many

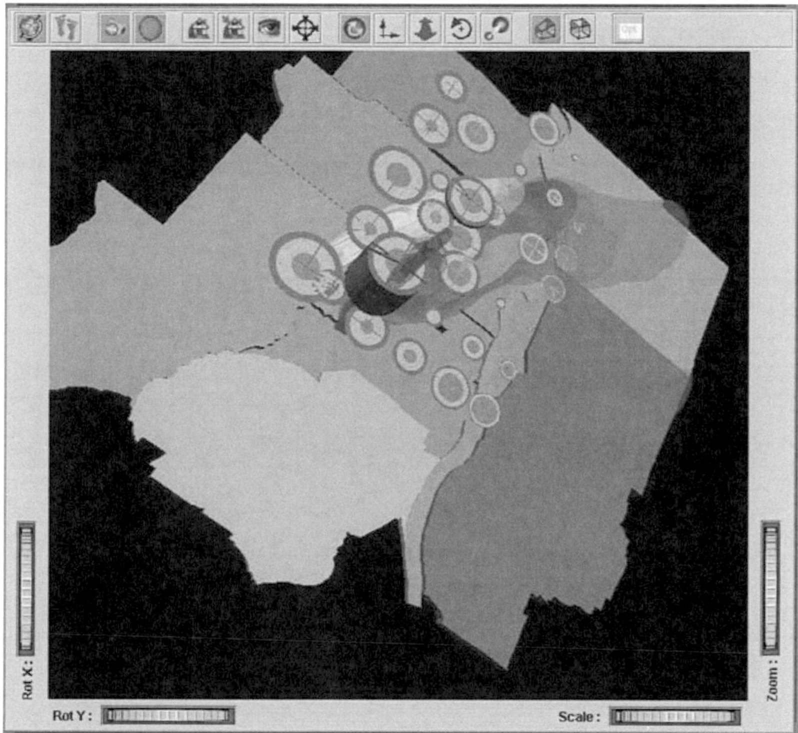

FIGURE 11.7 Toxic plumes (shown at two time intervals as yellow and red clouds) are draped over a region colored according to population. The complex target-shaped glyphs or symbols encode additional variables relating to expected dosage, uncertainty, and time to impact. Thus, the display contains a great deal of data, but all contained within a single scene, and integrated via its geography and careful mixing of graphic devices.

"target-like" circular symbols that summarize several aspects of how the plume might affect the resident population (including expected dosage, time to impact, and population total). Similar to iconographic techniques, compositional techniques typically combine a number of distinct data values into a single integrated whole — typically a surface or landscape. Gahegan (1998) and Treinish (1999) use compositional approaches on environmental data, with the advantage that several variables can be displayed together within a single scene, enabling their interactions to be studied.

6. Hierarchical and network (graph) techniques: In these displays, items are strictly organized according to a specific data structure, such as a tree or network (Robertson, 1991), with progressive levels refining the display into subspaces. In the life and earth sciences, taxonomies are often presented as a hierarchy (Tufte, 1990). Another common example of a hierarchy is the scenegraph,* which is often visualized to explore the structure and design of the visualization itself. Hierarchical techniques require knowledge of the parent–sibling relationships between data items, encoded in the metadata or as aggregation (part-of) relationships between objects. Network (or graph) techniques require similar connections to be pre-defined. Network visualization increasingly finds application in displaying ontologies, concept maps, citation links, and social networks (e.g., Mutton, and Golbeck, 2003). Such displays may well be useful tools by which to represent the knowledge discovered during exploration, as discussed later. Figure 11.8 shows an example of the *ConceptVista* concept-mapping tool (www.geovista.psu.edu/conceptvista) used to display a concept map of the GIS&T Body of Knowledge, recently developed to describe the knowledge and skills that comprise an education in the GIS field (http://www.ucgis.org/prior ities/education/modelcurriculaproject.asp).

All techniques are limited in the number of data items that can be displayed. If the screen becomes crowded, the density of information can become a barrier to comprehension. Similarly, all techniques are limited in the number of attributes that can be directly mapped. Even the most complex icons cannot hope to portray highly multivariate data. Most techniques try to present data in an integrated fashion, where related attributes are all presented together. Some approaches, such as compositional and iconographic methods, go to extreme measures to ensure visual integration. Other techniques require multiple windows (e.g., pixel-based, graph-based) and use interactive linking and brushing to connect data back together. Many of the reported techniques do not directly address any of the geographical properties of data; some are designed to uncover only one- or two- dimensional structures in isolation (e.g., pixel techniques), whereas geography usually demands a minimum of three dimensions for analysis: x, y and at least one attribute. Without this minimal spatial context, trends in attribute data can be very difficult to interpret in a geographical sense. In addition, of course, techniques tend to be combined

* The scenegraph is the supporting internal data structure in many visualization environments, describing the objects in a scene and how they are to be rendered.

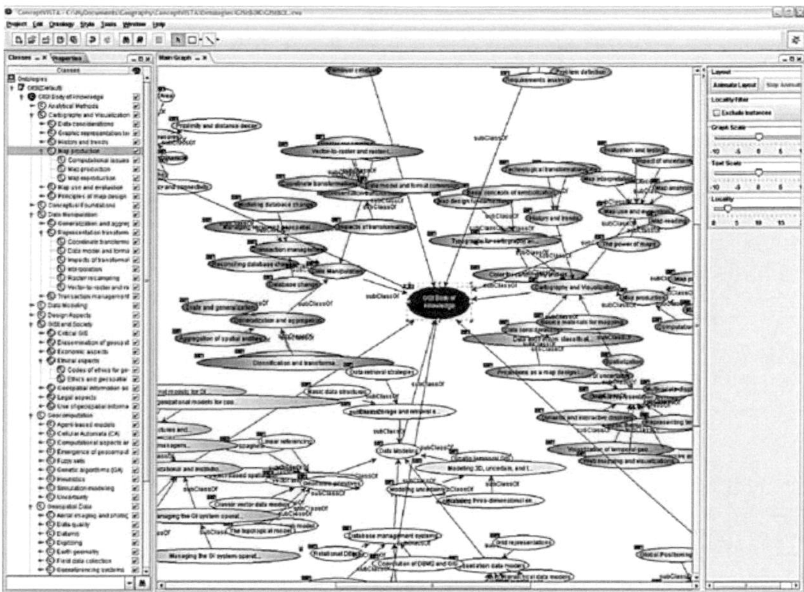

FIGURE 11.8 A graph-based visual display, showing concepts and relations contained within the GIS&T Body of Knowledge for GIS education. A hierarchical view of the concepts is shown in the left panel. Colors are used to group the major thematic areas.

together in all kinds of interesting ways and it is becoming common for systems to offer a wide palette with highly customizable interfaces for creating bespoke displays that suit specific tasks. Figure 11.9 shows a screenshot from a combined *Improvise* and *ConceptVista* application built for exploring and tracking vector-borne diseases such as West Nile Virus. It contains a map, a concept graph, several charts, a hierarchy of concepts (right panel), and even a Web browser for textual and pictorial information.

Given the previous descriptions of techniques and the process of visual encoding, it follows that a visual stimulus might take many forms, for example: (1) As a single striking feature, strongly differentiated from others due to a large difference in some visual property such as color or size; iconographic techniques often produce this kind of a stimulus. (2) As a cluster of similar objects forming a perceptual grouping, a structure that is perceived together; maps, space filling, projection, and chart-based techniques facilitate these kinds of patterns. (3) As a localized trend or anomaly, for example, a dip in a surface; compositional techniques and charts are examples. (4) As a result of some form of dynamic behavior, for example objects that all originate from, or converge on, the same location.

Table 11.1 summarizes this discussion, showing techniques against the characteristics that affect the exploratory process, that is, the visual stimulus provided, the types of data exploration supported, and the main instigator. There are bound to be a number of techniques that do not neatly fit these categories, but instead combine

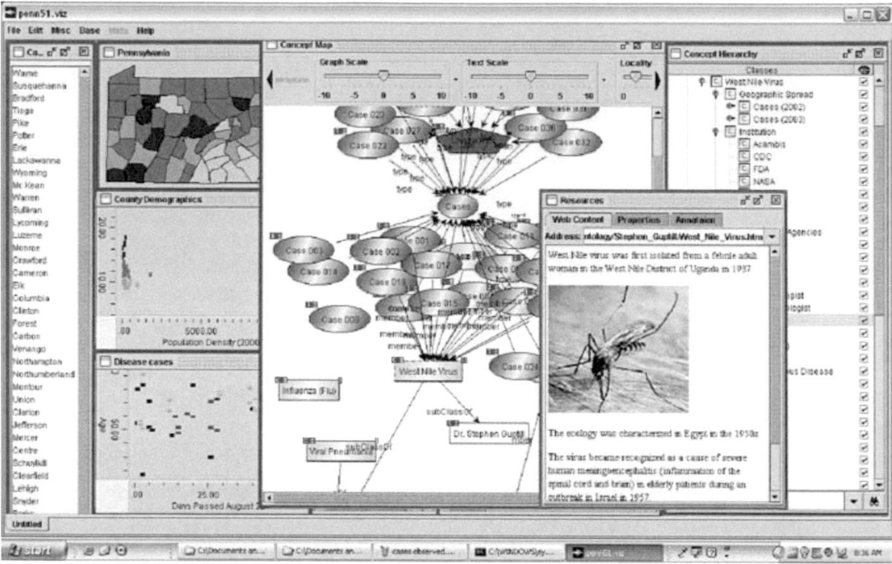

FIGURE 11.9 A screenshot of the *Improvise* visualization system, used to investigate symptoms and spread of vector-borne diseases. The display shows a mixture of many visualization techniques, to emphasize the point that techniques can be easily combined, to provide alternative views onto the same data in a coordinated manner. See text for further details. See color insert after page 148.

facets of several techniques to provide alternative views on the data. For example, a scatterplot can use icons rather than dots to increase the number of data dimensions that can be displayed concurrently, fusing ideas from chart-based and iconographic or compositional methods.

Notice that each of these techniques relies heavily on visual variables, as opposed to temporal behavior, to produce the stimulus. Future techniques may concentrate more on the untapped temporal properties that new visualization environments can now provide.

TABLE 11.1
Summary of the Characteristics of Exploratory Visualization Techniques

Method	Display Created by	Usual Exploratory Mode	Usual Visual Stimuli
Map-based	User	Interactive	Patterns in geographic space
Chart-based	User	Interactive	Clusters of points, outliers
Projection-based	Algorithm	Static, animated tour	Clusters of points, outliers
Iconographic/compositional	User	Static	Patterns in icons and surfaces
Pixel-based	Algorithm	Interactive	Clusters in dense pixel arrays
Hierarchical	Metadata	Interactive	Patterns in nodes and links

11.3 HOW DO HUMANS DISCOVER NEW KNOWLEDGE?

The application of the scientific method (e.g., Leedy, 1993) encompasses a number of stages, from data exploration and hypothesis generation to analysis and final evaluation of results, as Figure 11.1 shows. To move us through these various stages, three distinct modes of inference have been recognized for over a century (*deduction*, *induction*, and *abduction*), thanks in large part to the pioneering work of Peirce (1891), whose metaphysical writings separated them out and showed the quite different roles they play in science and in everyday life. We need to understand these modes: what they use as inputs, what they produce, to what extent they can be validated or challenged, how best to support them visually, and how to verify their effectiveness. Baker (1999) gives a useful description of how modes determine what can be known in geoscientific analyses. Within the artificial intelligence community, specific tools have been constructed to operate within these three modes (Sowa, 1999; Luger and Stubblefield, 1998). A description of each inference mode is given next. Following from this, some ideas are presented about how they might be better supported within visual displays.

11.3.1 DEDUCTION

Deductive logic applies a set of pre-defined rules that are assumed to be true to all specific instances or individual cases. The rules dictate both the questions that can be asked (hypotheses) and the answers that will be provided (interpretation). Deduction is often seen as the most reliable and rigorous mode of scientific reasoning because it relies on provable logical expressions, often in the form of syllogisms such as

Major premise: All professors are absent-minded.
Minor premise: X is a professor.
Conclusion: X is absent-minded.

Such logic fits well with most established statistical and computational approaches to analysis. A deductive expression can be directly mapped to programming and query languages such as Java or SQL so they are easy to operationalize in visual displays — they represent predefined knowledge about possible patterns of interest that we may choose to examine. However, a "closed world" assumption must be made; deduction does not produce new categories or relationships, so in the case of discovery we must suppose that the targets of interest are indeed all discoverable by the pre-defined set of expressions. We must further assume that the targets can be adequately described from existing knowledge and that there is consequently no need to learn from further examples or adapt our logic to new data.

In some of the visualization techniques described previously (e.g., projection and pixel-based techniques) the hypotheses presented to the user are pre-defined in the behavior of the system and the user plays an essentially static role (Table 11.1), so the reasoning has a strongly deductive component. If a particular pattern or shape emerges via such a transformation, then we can treat that transformation as a description that causes a pattern to emerge, and investigate it further. We then turn to induction to learn the patterns and rules underpinning such transformations.

11.3.2 Induction

Induction relaxes the "closed world" assumption, replacing it with a mechanism to learn from a (usually small) set of examples, and then generalizes their characteristics to a larger population. The inductive process is split into three parts: (1) a user detects the training examples, (2) the machine or the user builds a general visual model of the category, and (3) a (possibly different) user identifies further examples from this generalized form. To support induction it is necessary to present training examples, so that a model can be constructed to represent their generalized characteristics and then used to extract or locate additional examples. Induction forms the primary reasoning mechanism in many supervised machine-learning techniques, including neural networks and decision trees (Mitchell, 1997). Compared to deduction, this mode is more flexible because it does not rely only on pre-defined knowledge but also *learns* from specific cases. However, it introduces uncertainty because learning by generalization requires a form of bias to be introduced into the data (Gahegan, 2003). In addition, since the hypothesis is learned rather than pre-defined, it cannot be tested in the same objective manner using logical proofs. Instead, the validity of induction must be established in less deterministic ways, such as by the ability to correctly identify or classify a certain percentage of a sample.

However, induction is very natural for humans (a point that is taken up later) and our internal understanding of categories, their members, and how to differentiate them is exceptionally well developed. Visually based knowledge construction tools often rely on the analyst's inductive learning and generalization capabilities to characterize the kinds of patterns and clusters of interest. Therefore, in theory at least, it should be possible for the analyst to locate training examples for the system to "learn" from (for example, a category or a pattern of interest). The resulting generalized class description could enable an automated search for further patterns of the same type. An alternative would be to keep the analysis within the visual domain, where the training examples provided by the human expert could be used to produce a kind of generic visual representation. The resulting general form would then be suitable for use in a legend or key, and would support other users in finding further examples visually.

11.3.3 Abduction

This final mode of inference aims to connect patterns within the data with explanations (or hypotheses) by which the patterns might have come to be. It is thus the most open of the three modes because both target and hypothesis are undefined at the outset, and the explanation may require additional concepts and relationships to be defined or re-purposed. Therefore, rather than reasoning from known theory to a specific case (deduction), abduction reasons from cases to theory. To be successful, abduction requires the following:

1. The ability to posit new fragments of theory.
2. A massive set of knowledge to draw on, everything from common sense to domain expertise, by which explanations can be constructed.

3. A means of searching for connections between patterns and possible explanations, through this vast collection of knowledge (Sowa and Majumdar, 2003).
4. Complex problem-solving strategies such as using analogy, approximation, best guess, and so on.

It is no wonder then that computational abduction has been largely unsuccessful to date in any general sense (Psillos, 2000). But as with induction, humans are in fact well suited to this task, being biologically predisposed to store massive amounts of knowledge, to search efficiently through it using complex strategies, and thus to explain their observations. The trick is to find ways to present data to the analyst that allows this machinery (wetware) to be put to effective use.

In *computational* data mining, we might generally say that explanations can be constructed if we allow these explanations to take very abstract (statistical or mathematical) forms. For example, the findings of data mining are usually expressed as some kind of classification scheme or set of association rules (e.g., Han, 1999; Agrawal, Imielinski, and Swami, 1993) by which useful structure can be imposed on the data. These explanations are weak in the sense that they are not expressed in (or mapped to) the language of the domain of investigation — in this case geography. They can only describe the cluster in a mathematical sense; the expert must still translate this low-level explanation into something meaningful to humans. In most techniques for visual exploration, such as maps, scatterplots, and compositional displays, an abductive task is performed collaboratively between the observer and the creator of the visualization, since patterns are observed because of the way the display is constructed, and the way the observer perceives and comprehends it. The simultaneous task of hypothesis generation is also similarly split; the mappings used to create the scene themselves provide some form of loose hypothesis, and an observer may generate more specific internal theories to explain the observed structure.

11.3.4 Forming Chains of Inference

All three modes of inference are useful. Deduction allows us to apply rules and to test them, induction lets us construct more generic categories from a few cases, and abduction allows us to put forward explanations that connect categories and rules together. However, by itself, this understanding is not sufficient to support knowledge discovery. We also need to know how different modes of reasoning connect together to form "inferential chains" — lines of inference that connect causes and effects, hypotheses and concepts together into theories by which we can understand and model the world. When human experts show their flair for discovery, they are demonstrating their ability to move rapidly between different modes of inference, to connect their findings together into inferential chains, and to drop lines of investigation that appear to be flawed. From this point of view, it is not so much the ability to support each kind of inference in isolation that is needed, but a deeper understanding of how they connect together, and the development of visual tools that aid the user in establishing and maintaining logical connections among ideas. This is a different way of thinking about visualization, which is to some extent captured in the recent focus on visual analytics described earlier. The work of Yang et al. (2007) shows how visualization can be used to marshal supporting and refuting evidence (in their case via snapshots of visual displays called "nuggets").

Earlier attempts to understand the interplays in the discovery process typically place deduction, induction, and abduction into some kind of free-form cycle (Gahegan and Brodaric, 2002; Sowa, 1999). In the knowledge discovery from databases (KDD) field, we see similar linear arrangements in the work of Fayadd et al. (1996). But how do these activities really connect together? What follows is a deeper examination of the connections between them, with an emphasis on their interactions. Figure 11.10 shows such a free-form cycle of inference, with the more typical activities of science (analysis, verification, communication) in the center, surrounded by the conceptual structures (e.g., rules, categories, and explanations) that drive them.

The first thing to notice about the diagram is that a circle unites the various modes of inference: deduction, induction, and abduction connect together continuously, and an analyst may choose to move backward or forward from one to another at will. Ideally, any kind of system built to support discovery should not place any restrictions on such conceptual refocusing, but they often do (Gahegan, 2005). Note also in the diagram that each inferential mode is depicted close to a specific kind of conceptual structure: deduction uses rules, induction creates categories, and abduction

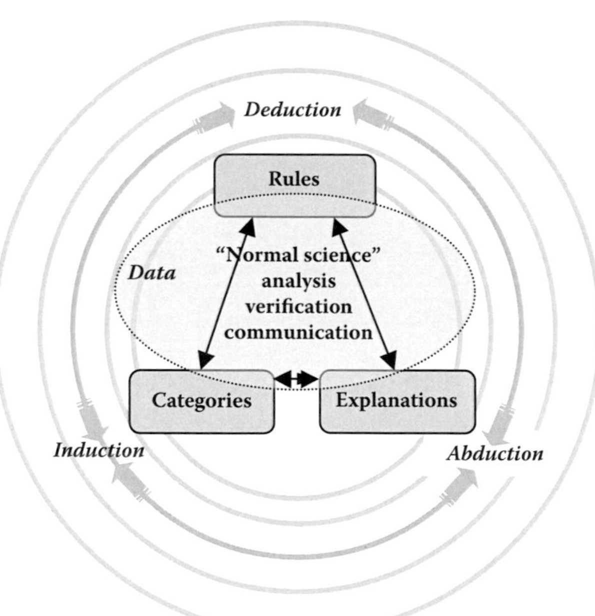

FIGURE 11.10 The "normal science" activities of analysis, verification, and communication are shown at the center of concentric wheels of inference. The three inferential modes connect together, via the conceptual structures that they create and employ. See text for full details.

links categories (and rules) together via relationships. This is not meant to imply that these conceptual structures belong only in one place, but that there is a dominant pattern to the way they are used. Additionally, many more conceptual structures than are shown here play a role in the process of knowledge discovery [e.g., a law, a methodology, a hypothesis; see Langley (2000) for a thorough account] but for brevity only three are shown here. Second, note that all of the artifacts produced (rules, categories, explanations) are typically anchored somewhere in the data; their creation depends on the act of discovery, their persistence depends on not uncovering counter-examples. All are subject to verification, although never more so than when they are first proposed. Third, a number of other faint circles are shown around the cycle to emphasize the point that inferences occur at many levels of abstraction — sometimes concurrently. In other words, explanations, categories, and rules are conceptual structures that connect together recursively, themselves building in turn on lower level rules, categories, and explanations. So, in theory at least, some higher level notions used in geography could be anchored via progressively more abstract forms, until they reach the more basic structures used in, say, physics (Wilson, 1998). However, in Popper's (1959: p. 111) words, it is far more typical for our conceptual structures to be anchored much more loosely, if at all:

> The empirical basis of objective science has thus nothing "absolute" about it. Science does not rest upon rock-bottom. The bold structure of its theories rises, as it were, above a swamp. It is like a building erected on piles. The piles are driven down from above into the swamp, but not down to any natural or "given" base; and when we cease our attempts to drive our piles into a deeper layer, it is not because we have reached firm ground. We simply stop when we are satisfied that they are firm enough to carry the structure, at least for the time being.

Finally, central to each activity is the act of *verification*, since the results from each stage must be tested before they can be safely woven in. Verification may fail either within the current inferential task or when the results are carried over for integration into some larger inferential chain or conceptual structure. When failure occurs, some part of this structure must be retracted; the conceptual structures used in science fail periodically, so we rebuild around them. These failures often become known via new evidence that contravenes existing rules, categories, or theories. As Kuhn (1962) would have it: "Discovery commences with the awareness of anomaly, i.e., with the recognition that nature has somehow violated the paradigm-induced expectations that govern normal science." To return to the previous example, if one encounters a professor who is not absent-minded, then the rule given above fails and must be replaced or further qualified. Perhaps a new category is needed for less befuddled scholars and additional rules to tell them apart. (Then again, perhaps not.)

A more detailed version of the inference diagram is shown in Figure 11.11, concentrating only on those activities that lead to conceptual modification, that is, where discovery or learning occurs outside the bounds of "normal science."

The more typical science activities of applying established rules, categories, and theories to data are not shown. Three examples of Kuhn's notion of a "paradigm induced anomaly" are included: dashed arrows (a), (d), and (e), where the analyst discovers an instance in the data that contravenes an existing category (a) or explanation

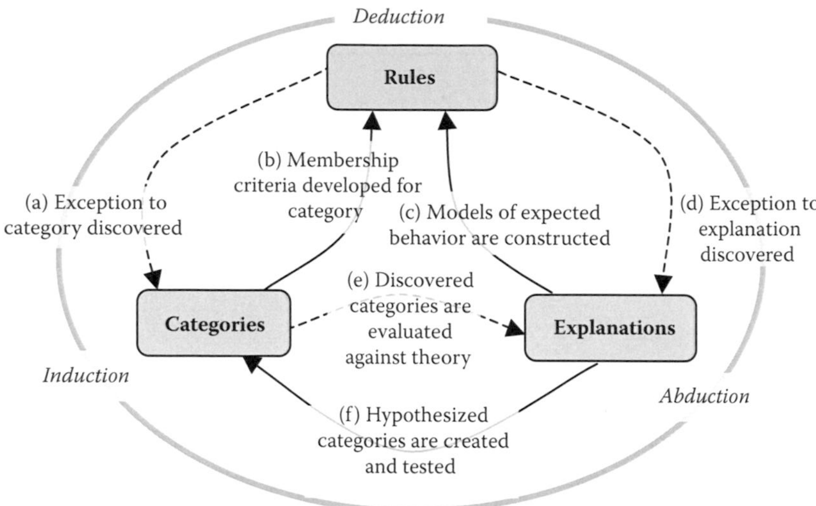

FIGURE 11.11 A roadmap for discovery-based science. Transformations such as these occur when normal science breaks down. They result in the creation, modification, and retirement of conceptual structures such as rules, categories, and explanations. Constructive activities are shown as solid arrows, revision activities as dashed arrows.

(d), or a mismatch between category and explanation (e). Such failures are usually the triggers for further conceptual changes to occur, as when rules, categories, and explanations are revised to address the problems uncovered. Arrows (b), (c), and (f) show the corresponding activities of rebuilding the rules, categories, and explanations brought into question by (a), (d), and (e). Thus, retraction triggers creation, which in turn triggers evaluation. A key question is therefore: How might visualization techniques support these mental activities better?

Local failures of conceptual structures are commonplace, as when a brushing operation in a visual display fails to separate the data into useful groups. Breakdowns deep within the structure can have wide reverberations, such as when an old law is disproved or a new theory becomes dominant. A good example of a new development with substantial impacts on conceptual structures is the theory of plate tectonics, which revolutionized much geoscientific understanding in the 1930s through the 1960s (Shrager and Langley, 1990). The interaction between categories and theories, shown by arrows (e) and (f), represents the struggle to match the emergent properties of the data under investigation with the established wisdom of a discipline. There is sometimes a healthy tension between these two ends of a problem, exacerbated by the fact that each geographical setting is unique, and thus established categories and theory may not perfectly fit any current situation.

Finally, one should not forget that the model described in this section is itself a hypothesis — an explanation put forward to help understand a complex and opaque process, built on lower-level conceptual structures, experience, and literature, and presented via diagrams and rhetoric. It too will be replaced in time by a model that is shown to be stronger.

11.4 COMBINING THE TECHNIQUES WITH THE PHILOSOPHY

The production of a visual display implies the creation of certain visual stimuli that are predisposed toward some aspects of the underlying data, but not others. As we have seen, this bias can be supplied by the user, the system, or the data itself. However, in all cases, the task of recognizing these stimuli rests entirely with the user and all modes of reasoning fail completely if target patterns are not observed. It is perhaps this single aspect that separates visual approaches from analytical approaches, since the *observation* of a stimulus cannot be modeled as a function of the data (except in a probabilistic sense based on a post-hoc study). In most simple analytic systems, the same data arrangement will produce the same response. The predisposition and nonsystematic behavior of the human analyst make it impossible to ensure that the most significant trends in the data are always the most visually striking, to guarantee that a bias-free perspective on the data is presented, or even to make certain that all the underlying data are actually explored.

As described previously, many exploratory visualization techniques rely heavily on one or more predefined hypotheses by which targets might be uncovered. When only pre-defined hypotheses are presented to the user, one could argue that the mode of inference is primarily deduction (the user is recognizing a pattern known by the creator of the visual display) and the display is restricted from providing further insights. However (and as noted previously), the degree of discovery depends largely on the knowledge, engagement, and imagination of the analyst; it is therefore difficult to predict in advance.

Putting these insights together with the description of exploratory techniques given previously, we begin to see that to be effective, techniques must support very specific cognitive tasks, with quite definable goals. We can therefore start to understand visualization techniques according to: (1) how a stimulus is defined and presented, (2) whether the set of targets is pre-defined (closed) or open ended, (3) if the space of possible targets is searched, and if so, how, and (4) whether explicit support for knowledge construction can be provided by the system. Thus, it is possible to propose a number of specific *approaches* to exploratory visualization, which can be differentiated by the roles that the system and the user adopt, and particularly how the stimulus is constructed, and whether by the system or the user. These roles influence the scientific process and can encourage or prohibit different ways of thinking.

1. *Static Exploratory* — User as observer: The exploratory mode is considered *static* if the visual display is fixed. Display properties and behavior are pre-defined, with the user playing the role of a passive observer who is presented with an array of stimuli to interpret. Web mapping applications are often of this type, as are the pixel-based techniques described previously. Graph-based and hierarchical methods that do not support interactive activities, such as zooming and brushing, would also be grouped here.
2. *Dynamic Exploratory* — User as initiator: The exploratory mode is considered *dynamic* if the visual variables can be interactively changed to shift the focus of attention to different kinds of patterns in the data. This implies some user control over the formation of a visual stimulus, by being

able to change the appearance of data. Map-based methods and some compositional techniques fall into this category, as do graph and hierarchical methods that support interaction. The U.S. Census FactFinder mapping application falls into this category (http://factfinder.census.gov/servlet/ThematicMapFramesetServlet?_bm=y&-_lang=en).

3. *Adaptive Exploratory* — Data or task as initiator: Instead of being pre-defined, scene construction is influenced by the data under consideration, as well as possibly some declared purpose. Languages and task-specific approaches for exploratory visualization have been proposed by Duclos and Grave (1993), Spitaleri (1993) and Gahegan and O'Brien (1997), and typically relate the data to suitable representational structures according to the statistical scale of the data (nominal, ordinal, interval, ratio*) and the notion of generic visual tasks, such as to "explore" or to "contrast" the data. Task-based approaches might use a variety of visualization techniques; they are differentiated here because of the way they automate the process of scene construction rather than any underlying difference in visualization metaphor. The resulting scene is likely to depend heavily on the type of data being displayed (i.e., data-driven), and possibly the task to be undertaken, so is not pre-defined as in static approaches. The projection techniques defined previously fall into this category because their appearance is ultimately a function of the data used.
4. *Knowledge Construction* — Knowledge construction requires the further step of extracting relevant structures or relationships from the scene and *learning* them, that is, representing them in some machine-based or human-readable form for later use. Knowledge discovery can be either expert-led or machine-led. It is fair to say that to date, no system perfectly addresses the challenge of knowledge construction, whereby identified structures in the data can be represented, retained, retrieved, compared, and revised. To date we do have systems that can remember (record) specific visual patterns (Chen, MacEachren, and Guo, 2006), connect patterns together into evidence chains (Yang et al., 2007), and represent the process of concept formation (Pike and Gahegan, 2007) — all useful steps in this direction.

The major question here concerns how humans and machines might best collaborate in discovery and knowledge construction. The richest source of knowledge constructs in geography is clearly still the human experts because they are able to create, organize, and reason with concepts and relationships that extend far beyond the simple schemas of geographic databases and ontologies. This observation strongly supports the inclusion of the human expert in the abductive, concept formation stage and in the interpretative/evaluative stage of the scientific process. The reasoning processes that underpin the exploratory visualization approaches described before

* Most approaches treat interval and ratio data identically, under the heading of quantitative data, the reason being that the presence or absence of a real zero point is largely irrelevant to the perceptual task of observing quantity.

are clearly different. Understanding this is crucial because modes of inference dictate the tasks that the system and the observer are asked to carry out, and implicitly define the limits on what might be observed and how it might be explained.

11.5 CONCLUSIONS AND CHALLENGES FOR GEOVISUALIZATION

There is no "magic bullet" for data exploration and knowledge discovery, no single algorithm or approach can be guaranteed as the best. Successful techniques will be those that can make use of a wide variety of strategies, that can involve human experts, and that can "learn" or adapt to a particular problem over time. Knowledge discovery is complex because it forces us to think deeply about how concepts and relationships are identified, defined, and represented and poses new challenges related to data modeling, semantics, and ontology. The benefits are clear though — insight into complex datasets and a better understanding of the process by which knowledge is created.

For the present, exploring geographic datasets remains a complex task and, despite efforts to the contrary, still requires the skills and experience of a geographer. In recognition of this fact, here the role of the computer is to provide insight into the structure of the data via visual display. Visualization is chosen as the medium because of its capacity to depict large quantities of data in a variety of new ways that can increase accessibility and promote understanding. This flexibility differs from the objective approach taken by GIS in that there are more degrees of freedom relating to representation and categorization. While it is possible now to show results that demonstrate some success, it remains to be seen whether this extra freedom can, in general, be used constructively or whether it acts to further confound the problem. Basic research is required here to firmly establish proof of concept.

The various visualization techniques described previously employ a number of implicit mechanisms to construct a scene. Some use pre-defined mappings to automate the process; others attempt to define mappings from objects identified by the user or from trends observed in the data. What is not clear yet is how these different modes of inference (deduction, induction, and abduction) can be used together for maximum effect, making exploration as flexible, robust, and repeatable as possible. Specific research is required here to establish the suitability of different approaches and techniques for distinct application areas and datasets.

There are many caveats to visual analysis. The processing activity becomes largely the responsibility of the human visual system, which can be easily misled, does not behave linearly (or even straightforwardly) for many basic tasks, and is adept only under certain paradigms (e.g., Gordon, 1989). In addition, embracing a visual approach to analysis does not remove the representational question, but it does change it. Instead of attempting to build a formal model of the data in a language that a machine can understand, we must instead build a visual model of the data in a form that a human can understand. Cartographic theory needs further extension into the science of visualization to meet this challenge (as argued by MacEachren, 1995).

Three specific sets of challenges follow, relating to the themes introduced in this chapter.

1. **The need for deeper representation of the discovery process.** The major focus of this chapter, and the challenge for exploratory geovisualization, is that the activities comprising the process of discovery and the connections between them (as shown in Figure 11.11) are what our systems must support if they are to be truly effective and intuitive to use.

   As described previously, some of the basic functionality to build more effective visual displays, to support the analyst moving back and forth between different inference tasks, to record what is done, and to pass findings back into a knowledge representation system, is available — but currently in various disjoint and experimental systems. This functionality does not yet connect together. Thus far, with few exceptions we attempt to engage the analyst's deep conceptual understanding, but do not record the process or the findings. As examples of what we might aspire to, imagine a system that makes explicit the conceptual structures developed during exploration, with the facility to import the stable findings into domain ontologies, while checking for consistency. One might also imagine a system that responds to user interactions by creating and retracting appropriate conceptual structures automatically. For example, if the analyst creates a discrete color scheme and imposes it on a set of observations (a kind of visual classification), the system could respond by creating a category for each of the groups so defined. If the analyst abandons this scheme in favor of a new one, these categories are destroyed; otherwise, they are retained, at least for the time being. Table 11.2 suggests some immediate goals related to each of the inferential modes described before.
2. **The reliability of the search for structure.** As with all searches, visual or computational, you only find useful structures where you look! There are two aspects to this challenge, the first being that it is the nature of the human visual system to find patterns, even where they do not exist. Our visual

TABLE 11.2
Some Ideas for Supporting Closer Ties between the Process of Discovery and Its Support in Visualization Systems, Organized by Inference Mode

	Functionality to Support this Mode of Inference in Visual Displays
Deduction	Visually applying and evaluating rules, e.g., "brushing" rules
	Conditioning displays based on rules
	Locating and emphasizing exceptions to rules
Induction	Collecting learning examples of some category or process
	Learning a general model from such examples
	Evaluating and exploring the categories so formed
Abduction	Representing arguments and chains of reasoning visually
	Importing constructed knowledge directly into concept maps or ontologies
	Communicating and challenging arguments, finding alternative explanations

systems are designed to impose meaning on even the most bizarre of visual displays, so that the absence of any significant patterns in the data does not mean we will not discover some. The second aspect involves the completeness of any search for structure. If we think of a *search space* comprising all the ways in which we might visualize the data we have, then our problem is to find instances in this space that are useful, and that tell us things about the data that we do not yet know. We also want to avoid searching in parts of this space where the probability of success is low. The mathematics and machine learning communities have developed sophisticated techniques to explore such large "combinatorial landscapes" of possibilities (Reidys and Stadler, 2002) and to address the problem of overlooking one part of a search space, or spending too much effort on another. However, with visualization systems, there are no such tight controls or optimizations. Now we rely wholly upon the analysts to judge where to look, what visual techniques to use, and for how long. What guarantees do we have that they will be diligent, they know where best to look, and alternative explanations for discoveries will be sought? The answer thus far is "none," though as mentioned previously, there are some projection techniques developed to systematically explore large attribute spaces visually, such as the grand tour. Some key challenges are therefore

- Providing better tools for systematic exploration, rather than the very *ad hoc* and unstructured approaches we currently employ.
- Conveying to the user the combinatorial landscape of possibilities, and which regions have been explored — this equates to how hard we have looked for alternative explanations for our findings.
- Providing some kind of assessment of the strength of our findings, based on the diligence employed in finding them.

3. **Paying attention to the user.** Finally, much current work in visualization is focused on providing large quantities of information in graphs or other chart-like structures, with consistent and perceptually defensible use of color and symbolization, and often using multiple, coordinated displays. This would seem to be a logical, structured way to proceed. However, we could put the task another way: what is it that grabs and holds the visual attention of humans? The answer might vary from person to person, and culture to culture, but nevertheless there may be some universal principles we can uncover; a disengaged user is not likely to discover much!

   Relevant questions here might be

   - How is it that we can engage and extend attention spans so that the analyst is drawn into visual exploratory tasks with relish?
   - Do multiple, but separate, displays truly aid cognition? Alternatively, do they confuse the analyst with too complex a visual task?
   - What are the cognitive costs of interpreting these displays? Can they be lowered?

At present, I suspect we rely too much on low-level research concerning visual variables to determine how we build our systems, and not enough on the visual allure of the fine arts (Landa, 2004).

ACKNOWLEDGMENT

All of the visualization examples used here stem from research projects conducted at the GeoVISTA center at Penn State University, by the author and many colleagues from 1998 to 2007. Worthy of special mention for their work in this regard are: Ritesh Agrawal, Cindy Brewer, Xiping Dai, Frank Hardisty, Junyan Luo, Alan MacEachren, James Macgill, Masa Takatsuka, Wenhui Wang, Steve Weaver, and Mike Wheeler. Further details, examples, related papers, and software for download are available to interested readers from the GeoVISTA Website: http://www.geovista.psu.edu/.

REFERENCES

Agrawal, R., Imielinski, T., and Swami, A. (1993). Mining association rules between sets of items in large databases. *ACM SIGMOD*, pp. 207–216.

Andrienko, G.L. and Andrienko, N.V. (1999). Interactive maps for visual data exploration. *International Journal of Geographic Information Science*, **13**(4), 355–374.

Andrienko, N. and Andrienko, G. (2005). *Exploratory Analysis of Spatial and Temporal Data: A Systematic Approach*, Springer, Berlin.

Andrienko, N., Andrienko, G., and Gatalsky, P. (2003). Exploratory spatio-temporal visualization: an analytical review. *Journal of Visual Languages & Computing,* **14**, 503–541.

Asimov, D. (1985). The grand tour: a tool for viewing multidimensional data. *SIAM Journal of Science and Statistical Computing*, **6**, 28–143.

Baker, V. R. (1999). Geosemiosis. *GSA Bulletin*, **111**(5), 633–645.

Bertin, J. (1985). *Graphical Semiology*. University of Wisconsin Press, Madison, Wisconsin.

Bryson, S., Kenwright, D., Cox, M., Ellsworth, D., and Haimes, R. (1999). Visually exploring gigabyte data sets in real time. *Communications of the ACM*, **42**(8), 83–90.

Buja, A., McDonald, J.A., Michalak, J., and Stuetzle, W. (1991). Interactive data visualization focusing and linking. Proc. *IEEE Conference on Visualization (Visualization '91)*, San Diego, CA, IEEE Computer Society, pp. 156–163.

Chen, J., MacEachren, A.M., and Guo, D., (2006). Visual Inquiry Toolkit – An integrated approach for exploring and interpreting space-time, multivariate patterns. *Proc. AutoCarto 2006*, Vancouver, WA, June 26–28, 2006.

Chernoff, H. (1973). The use of faces to represent points in k-dimensional space graphically. *Journal of the American Statistical Association*, **68**, 361–336.

Chernoff, H. (1978). Graphical representations as a discipline. In: *Graphical Representations of Multivariate Data*, Wang, P.C.C., Ed., Academic Press, New York.

Cleveland, W.S. and McGill, M.E. (1988). *Dynamic Graphics for Statistics*, Wadsworth & Brookes/Cole, Belmont, CA.

Compieta, P., Di Martino, S., Bertolotto, M., Ferrucci, F., and Kechadi, T. (2007). Exploratory spatio-temporal data mining and visualization, *Journal of Visual Languages and Computing*, **18**(3), 255–279.

Cook, D., Buja, A., Cabrera, J., and Hurley, C. (1995). Grand tour and projection pursuit. *Computational and Graphical Statistics*, **4**(3), 155–172.

de Oliveira, M. and Levkowitz, H. (2003). From visual data exploration to visual data mining: A survey, *IEEE Transactions on Visualization and Computer Graphics,* **9**(3), 378–394.

DiBiase, D., MacEachren, A., Krygier, J., and Reeves, C. (1992). Animation and the role of map design in scientific visualization. *Cartography and Geographic Information Systems*, **19**(4), 204–214.

Dorling, D. (1994). Cartograms for human geography. In: *Visualization in Geographical Information Systems*, Hearnshaw, H.M. and Unwin, D.J., Eds., Wiley, Chichester, England, pp. 85–102.

Duclos, A.M. and Grave, M. (1993). Reference models and formal specification for scientific visualization. In: *Scientific Visualization: Advanced Software Techniques*, Palamidese, P., Ed., Ellis Harwood Ltd., Chichester, England, pp. 3–14.

Dykes, J.A. (1997). Exploring spatial data representation with dynamic graphics. *Computers & Geosciences*, **23**(4), 345–370.

Dykes, J.A., MacEachren, A.M., and Kraak, M.J., Eds. (2005). *Exploring Geovisualization.* Elsevier, Amsterdam.

Edsall, R.M., Harrower, M., and Mennis, J. (1999). Visualizing properties of temporal and spatial periodicity in geographic data. *Computers & Geosciences*, **26**(1), 109–118.

Ester, M., Kriegel, H.-P., and Sander, J. (1998). Algorithms for characterization and trend detection in spatial databases. *Proc. 4th International Conference on Knowledge Discovery and Data Mining (KDD'98)*, New York, pp. 44–50.

Ester M., Kriegel, H.-P., Sander, J., and Xu, X. (1996) A density based algorithm for discovering clusters in large spatial databases with noise. *Proc. 2nd International Conference on Knowledge Discovery and Data Mining (KDD-96)*, pp. 226–231.

Fayyad, U. and Grinstein, G. (2001). Introduction. In: *Information Visualization in Data Mining and Knowledge Discovery,* Fayyad, U., Grinstein, G., and Wierse, A., Eds., Morgan Kaufmann Publishers, San Francisco, CA.

Fayyad, U., Grinstein, G., and Wierse, A., Eds. (2001). *Information Visualization in Data Mining and Knowledge Discovery*, Morgan Kaufmann, San Francisco, CA.

Fayyad, U., Piatetsky-Shapiro, G., and Smyth, P. (1996). The KDD process for extracting useful knowledge from volumes of data. *Communications of the ACM*, **39**(11), 27–34.

Gahegan, M. (1999). Four barriers to the development of effective exploratory visualization tools for the geosciences. *International Journal of Geographic Information Science*, **13**(4), 289–310.

Gahegan, M.N. (1998). Scatterplots and scenes: Visualization techniques for exploratory spatial analysis. *Computers, Environment and Urban Systems*, **21**(1), 43–56.

Gahegan, M.N. (2003). Is inductive machine learning just another wild goose (or might it lay the golden egg?). *International Journal of Geographical Information Science* **17**, 69–92.

Gahegan, M.N. (2005). Beyond tools: Visual support for the entire process of GIScience. In *Exploring Geovisualization,* Dykes, J.A., MacEachren, A.M., and Kraak, M.J., Eds., Elsevier, Amsterdam, pp. 83–99.

Gahegan, M. and Brodaric, B. (2002). Computational and visual support for geographical knowledge construction: Filling in the gaps between exploration and explanation. In: Proceedings, Spatial Data Handling 2002, July 9–12, Ottawa, Canada.

Gahegan, M., Wachowicz, M., Harrower, M., and Rhyne, T. (2001). The integration of geographic visualization with knowledge discovery in databases and geocomputation. *Cartography and Geographic Information Science*, **28**(1), 29–44.

Gordon, I.E. (1989). *Theories of Visual Perception.* John Wiley & Sons, Chichester, U.K.

Grinstein, G. and Levkowitz, H., Eds. (1995). *Perceptual Issues in Visualization*, Springer-Verlag, Berlin, Germany.

Han, J. (1999). Characteristic rules. In: *Handbook of Data Mining and Knowledge Discovery*, Kloegen, W. and Zytkow, J., Eds., Oxford University Press, Oxford, U.K.

Haslett, J., Bradley, R., Craig, P., Unwin, A., and Wills, G. (1991). Dynamic graphics for exploring spatial data with application to locating global and local anomalies. *The American Statistician*, **45**(3), 234–242.

Hinneburg, A., Keim, D., and Wawryniuk, M. (1999). HD-Eye: Visual mining of high dimensional data. *IEEE Computer Graphics and Applications*, September/October, 22–31.

Inselberg, A. (1997). Multidimensional detective. *Proc. IEEE Conference on Visualization (Visualization '97)*, Los Alamitos, CA, IEEE Computer Society, pp. 100–107.

Jerding, D.F. and Stasko, J.T. (1998). The information mural: A technique for displaying and navigating large information spaces. *IEEE Transactions on Visualization and Computer Graphics*, **4**(3), 257–271.

Johansson, J., Treloar, R., and Jern, M. (2004) Integration of unsupervised clustering, interaction and parallel coordinates for the exploration of large multivariate data. *Proceedings. Eighth International Conference Information Visualization, IV 2004*, pp. 52–57.

Keim, D.A. (1996). Pixel-oriented database visualizations. *Sigmod Record*, Special Issue on Information Visualization, Dec. 1996.

Keim, D. and Kriegel, H.-P. (1994). VisDB: Database exploration using multidimensional visualization. *Computer Graphics and Applications*, September, 44–49.

Keim, D. and Kriegel, H.-P. (1996). Visualization techniques for mining large databases: A comparison. *IEEE Transactions on Knowledge and Data Engineering* (special issue on data mining).

Keim, D.A., Mansmann, F., Schneidewind, J., and Ziegler, H. (2006). Challenges in visual data analysis, *Proc. Information Visualization 2006* (IV'06), IEEE, pp. 9–16.

Keim, D.A., Panse, C., Sips, M., and North, S. (2004). Visual data mining in large geospatial point sets. *IEEE Computer Graphics and Applications*, **24**(5), 36–44.

Koperski, K. Han, J., and Adhikary, J. (1998). Mining knowledge in geographic data. Comm. ACM, 41(12), 47–66. URL: http://db.cs.sfu.ca/sections/publication/kdd/kdd.html.

Kovalerchuk, B. and Schwing, J. (2004). *Visual and Spatial Analysis: Advances in Data Mining, Reasoning, and Problem Solving*. Springer, Berlin.

Kraak, M.-J. and MacEachren, A.M., Eds. (1999). *International Journal of Geographic Information Science*, special issue on exploratory cartographic visualization, **13**(4).

Kuhn, T.S. (1962). *The Structure of Scientific Revolutions*. University of Chicago Press, Chicago, IL.

Landa, R. (2004). *Advertising by Design: Creating Visual Communications with Graphic Impact*. John Wiley & Sons, New York.

Landgrebe, D. (1999). Information extraction principles and methods for multispectral and hyperspectral image data. In: *Information Processing for Remote Sensing*, Chen, C. H., Ed., World Scientific, River Edge, NJ.

Langley, P. (2000). The computational support of scientific discovery. *International Journal of Human-Computer Studies*, **53**, 393–410.

Lee, H.Y. and Ong, H.L. (1996). Visualization support for data mining. *IEEE Expert Intelligent Systems and their Applications*, **11**(5), 69–75.

Leedy, P.D. (1993). *Practical Research: Planning and Design*, 5th ed., MacMillan, New York.

Luger, G.F. and Stubblefield, W.A. (1998). *Artificial Intelligence: Structures and Strategies for Complex Problem Solving*. Addison-Wesley, Reading, MA.

MacDougall, E.B. (1992). Exploratory analysis, dynamic statistical visualization and geographic information systems. *Cartography and Geographical Information Systems*, **19**(4), 237–246.

MacEachren, A.M. (1995). *How Maps Work*. Guilford Press, New York.

MacEachren, A., Dai, X., Hardisty, F., Guo, D., and Lengerich, G. (2003). Exploring high-D spaces with multiform matrices and small multiples. *Proceedings of the IEEE Symposium on Information Visualization 2003* (INFOVIS'03), URL: http://ieeexplore.ieee.org/iel5/8837/27965/01249006.pdf?arnumber=1249006.

MacEachren, A.M., Gahegan, M., and Pike, W. (2004). Geovisualization for knowledge construction and decision support. *IEEE Computer Graphics and Applications*, **24**(1), 13–17.

MacEachren, A.M. and Kraak, M.-J., Eds. (1997). Exploratory cartographic visualization. *Computers & Geosciences* (special issue on exploratory cartographic visualization), **23**(4).

MacEachren, A.M., Wachowitz, M., Edsall, R., Haug, D., and Masters, R. (1999). Constructing knowledge from multivariate spatio-temporal data: Integrating geographical visualization with knowledge discovery in database methods. *International Journal of Geographic Information Science*, **13**(4), 311–334.

Mackinlay J.D. (1986). Automating the design of graphical presentations of relational information. *ACM Transactions on Graphics*, **5**(2), 110–141.

Mark, D.M., Freska, C., Hirtle, S.C., Lloyd, R., and Tversky, B. (1999). Cognitive models of geographical space. *International Journal of Geographic Information Science*, **13**(8), 747–774.

Mitchell, T.M. (1997). *Machine Learning*, McGraw-Hill, New York.

Mutton, P. and Golbeck, J. (2003). Visualization of semantic metadata and ontologies. *Seventh International Conference on Information Visualization (IV03)*, London, pp. 300–305.

Pal, S. and Mitra, P. (2004). *Pattern Recognition Algorithms for Data Mining*. CRC Press, Boca Raton, FL.

Peirce, C.S. (1891). The architecture of theories. *The Monist*, **1**, 161–176.

Peterson, M.P. (1999). Active legends for interactive cartographic animation. *International Journal of Geographic Information Science*, **13**(4), 375–384.

Piatetsky-Shapiro, G., Fayyad, U., and Smith, P. (1996). From data mining to knowledge discovery: An overview. In: *Advances in Knowledge Discovery and Data Mining*, Fayyad, U., Piatetsky-Shapiro, G., Smith, P., and Uthurusamy, R., Eds., AAAI/MIT Press, Boston, MA, pp. 1–35.

Pickett, R.M. and Grinstein, G.G. (1988). Iconographic displays for visualizing multidimensional data. *Proc. IEEE Conference on Systems, Man and Cybernetics*, IEEE Press, Piscataway, NJ, pp. 514–519.

Pike, W. and Gahegan, M. (2007). Beyond ontologies: Toward situated representations of scientific knowledge. *International Journal of Human-Computer Studies,* 65(7), 674–688.

Popper, K. (1959). *The Logic of Scientific Discovery*, Basic Books, New York.

Psillos, S. (2000). Abduction: Between conceptual richness and computational complexity. In: *Abduction and Induction*, Flach, P.A. and Kakas, A.C. Eds., Kluwer, Dordrecht, pp. 59–74.

Reidys, C.M. and Stadler, P.F. (2002). Combinatorial landscapes. *SIAM Review*, **44**, 3–54.

Ribarsky, W., Katz, J., and Holland, A. (1999). Discovery visualization using fast clustering. *IEEE Computer Graphics and Applications*, September/October, 32–39.

Robertson, G.G. (1991). Cone trees: Animated 3D visualization of hierarchical information. *Proc. ACM CHI'91*, ACM Press, New Orleans, LA, pp. 189–194.

Robertson, P.K. (1997). Visualizing spatial data: The problem of paradigms. *International Journal of Pattern recognition and Artificial Intelligence*, **11**(2), 263–273.

Roddick, J.F. and Spiliopoulou, M. (1999). A bibliography of temporal, spatial and spatio-temporal data mining research. *SIGKDD Explorations*. **1**(1).

Shrager, J. and Langley, P., Eds. (1990). *Computational Models of Scientific Discovery and Theory Formation*, Morgan Kaufman, San Mateo, CA.

Slocum, T.A., Blok, C., Jiang, B., Koussoulakou, A., Montello, D., Fuhrmann, S., and Hedley, N.R. (2001). Cognitive and usability issues in geovisualization. *Cartography and Geographic Information Science* **28**, 61–75.

Sowa, J.F. (1999). *Knowledge Representation: Logical, Philosophical, and Computational Foundations*, Brooks/Cole, Pacific Grove, CA.

Sowa, J. and Majumdar, A. (2003). Analogical reasoning. In: *Conceptual Structures for Knowledge Creation and Communication*, de Moor, A., Lex, W., and Ganter, B., Eds., Springer-Verlag, Berlin, pp. 16–36. URL: http://www.jfsowa.com/pubs/analog.htm.

Spitaleri, R. (1993). Reference models for computational visual simulations. In: *Scientific Visualization: Advanced Software Techniques*, Palamidese, P., Ed., Ellis-Horwood, Chichester, pp. 3–14.

Stolte, C. (2003). Query, Analysis, and Visualization of Multidimensional Databases, Ph.D. dissertation, Stanford University, Stanford, CA. URL: http://graphics.stanford.edu/papers/cstolte_thesis/thesis.pdf.

Tang, Q. (1992). A Personal Visualization System for Visual Analysis of Area-Based Spatial Data: Proc. GIS/LIS' 92, Vol. 2, American Society for Photogrammetry and Remote Sensing, Bethesda, MD, pp. 767–776.

Thomas J. and Cook, K., Eds. (2005). *Illuminating the Path: The Research and Development Agenda for Visual Analytics*, IEEE CS Press, Piscataway, NJ.

Treinish, L.A. (1999). Task-specific visualization design. *IEEE Computer Graphics and Applications*, September/October, 72–77.

Treisman, A. (1986). Features and objects in visual processing. *Scientific American*, **255**(5), 114B–125.

Tufte, E.R. (1990). *Envisioning Information*, Graphics Press, Cheshire, CT.

Tukey, J.W. (1977). *Exploratory Data Analysis*. Addison-Wesley, Reading, MA.

Ware, C. (2000). Information Visualization: Perception for Design, Morgan Kaufmann, Los Altos, CA.

Weaver, C. (2004). Building highly coordinated visualizations in Improvise. *Proceedings of the IEEE Symposium on Information Visualization 2004*, Austin, TX, pp. 159–166.

Wilson, E.O. (1998). *Consilience: The Unity of Knowledge*. Knopf, New York.

Wise, S., Haining, R., and Signoretta, P. (1998). The role of visualization for exploratory spatial data analysis of area-based data. Proc. *Fourth International Conference on Geocomputation (GeoComputation'98)*, Bristol, U.K.

Yang, D., Xie, Z., Rundensteiner, E., and Ward, M. (2007). Managing discoveries in the visual analytics process, *SIGKDD Explorations*, **9**(2), 22–29.

12 Multivariate Spatial Clustering and Geovisualization*

Diansheng Guo

CONTENTS

12.1 INTRODUCTION

The study and understanding of complex geographic phenomena often depends on the analysis of multivariate spatial data to discover complex structures and gain new knowledge. Figure 12.1 shows a conceptual representation of a typical data set that contains multiple variables and geographic information, which can be viewed as a spatial data matrix (Haining 2003), lattice data (Cressie 1991), or a "map cube" [a simple case of the map cube model introduced in Shekhar et al. (2001) and Chapter

* This research is in part supported by National Science Foundation Grant BCS-0748813.

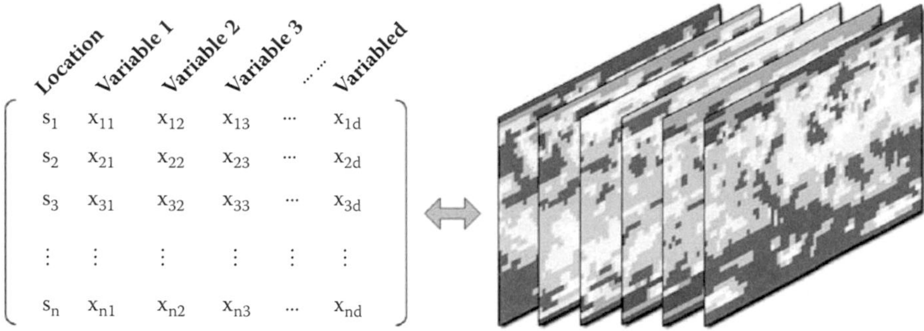

FIGURE 12.1 A multivariate spatial data set is often conceptually represented as a spatial data matrix (left) or a map cube (right), which can be transformed from one to the other.

4 of this book]. Such data sets are commonly encountered in various spatial research fields such as socioeconomic analysis, public health, climatology, and environmental studies, among others. For example, to study global climate patterns and their change over time, we not only examine temporal trends or patterns of climate variables (e.g., temperature) at a specific location, but we are also interested in the geographic variation of such trends or patterns. It is a challenging task to explore large multivariate spatial data sets and tease out complex (and often unexpected) patterns, which may take various forms (linear or nonlinear) and involve multiple spaces (e.g., multivariate space and geographic space) (National Research Council 2003).

To detect the unexpected and understand the data in its entirety, it is important to support an exploratory analysis process and let the data speak for themselves (Gould 1981; Gahegan 2003). Existing methods for exploratory spatial analysis and spatial data mining span across three main groups: *computational*, *statistical*, and *visual approaches*. Computational approaches resort to computer algorithms to search large volumes of data for specific types of patterns such as spatial clusters (Han, Kamber, and Tung 2001), spatial association rules (Han, Koperski, and Stefanovic 1997), homogeneous regions (Openshaw and Rao 1995; Assunção et al. 2006; Guo 2008), co-location patterns (Huang, Shekhar, and Xiong 2004), and spatial outliers (Shekhar, Lu, and Zhang 2003). Statistical approaches include spatial scan statistics or models (Openshaw, Cross, and Charlton 1990; Jona-Lasinio 2001; Kulldorff 2006), geographically weighted regression (Fotheringham, Brunsdon, and Charlton 2002), multivariate lattice models (Saina and Cressie 2007), and spatial association tests (Getis and Ord 1992; Anselin 1995). Visualization-based methods for multivariate spatial analysis include, for example, multivariate mapping (Chernoff and Rizvi 1975; Zhang and Pazner 2004), conditioned choropleth map (Carr, White, and MacEachren 2005), spatial statistics graphics (Anselin 1999), and other geovisualization techniques (MacEachren et al. 1999; Andrienko, Andrienko, and Gatalsky 2003; Dykes, MacEachren, and Kraak 2005).

Different methods have their own strengths and weaknesses. In general, computational methods are able to search for structures in large datasets with great efficiency but lack the ability to interpret and attach meaning to patterns. Statistical methods are rigorous and verifiable but often assume *a priori* model and relation form. Visual methods can facilitate the discovery and understanding of complex patterns by presenting data visually to allow human experts to interact directly with the data. However, visualization-based methods alone usually can only handle relatively small data sets and primarily rely on users to pick up patterns, which can be very time consuming (to visually sort and summarize massive amounts of information across multiple dimensions) and sometime biased (e.g., users may only see what they expect or what the visual representation allows). Given the increasingly large volume and complexity of multivariate spatial data sets, it is not likely that any individual method can adequately support an exploratory process to detect, interpret, and present complex information lurking in the data (Figure 12.1). To leverage the power of different analysis approaches, there are recent research efforts that focus on the integration of visualization with statistical and/or computational methods (for example, Swayne et al. 2003; Guo et al. 2005, 2006).

Built upon several recent developments in geovisualization and computational approaches (Guo et al. 2005, 2006; Guo 2008), this chapter introduces an integrated approach to multivariate analysis and geovisualization, which couples a suite of methods that are either complementary or competitive to each other. Complementary methods examine the data from different perspectives and together present an overview of complex patterns. On the opposite side, competitive methods focus on the same perspective or analysis task and their results can validate and crosscheck each other. The integrated approach couples a self-organizing map (SOM, which is a multivariate clustering and projection method), a regionalization method (which is based on spatially constrained hierarchical clustering), a multidimensional visualization component, and a multivariate mapping component. On one hand, the SOM and the regionalization method are competitors as they both seek clusters (with different cluster formulation algorithms and under different constraints). On the other hand, they both are complemented by the multivariate visualization and mapping components so that information can be examined and understood. The integrated approach not only supports user interactions and multiple linked views but also merges or overlays results of different components (views) into a single view (overview). Such an overview facilitates an overall understanding of major patterns and efficiently guides user interactions toward the hot spots that warrant closer attention.

The remainder of the chapter is organized as follows. Related work is briefly reviewed in the next section, primarily focusing on clustering, regionalization, and geovisualization. Section 12.3 elaborates on the methodologies adopted in the research, including SOM-based clustering and color encoding; regionalization based on spatially constrained hierarchical clustering; and their integration with multivariate visualization and mapping techniques. Section 12.1.4 presents an application of the approach for global climate change analysis. Finally, there is discussion and conclusion. The implemented software package is available at www.SpatialDataMining.org.

12.2 RELATED WORK

12.2.1 Cluster Analysis

Cluster analysis is a widely used data analysis approach, which organizes a set of data items into groups (or clusters) so that items in the same group are similar to each other and different from those in other groups (Jain and Dubes 1988; Gordon

1996; Jain, Murty, and Flynn 1999). Many different clustering methods have been developed in various research fields such as statistics, pattern recognition, data mining, machine learning, and spatial analysis. Different methods, or even the same method with different parameter configurations, can give quite different clustering results. Clustering methods may differ in many ways, including (1) the definition of distance (or dissimilarity) between data items (and between clusters), (2) the definition of "cluster," (3) the strategy to group or divide data items into clusters, (4) the data types that can be analyzed (e.g., numerical, categorical, and spatial), and (5) application-specific context and constraints. It is beyond the scope of this chapter to provide a comprehensive review of clustering methods. Readers are referred to the surveys provided by Jain et al. (1999, 2000), Han et al. (2001) and Chapter 7 of this book. Below is a brief review of relevant methods for this chapter.

Clustering methods can be broadly classified into two groups: *partitioning clustering* and *hierarchical clustering*. Partitioning clustering methods, such as K-means and maximum likelihood estimation (MLE), divide a set of data items into a number of nonoverlapping clusters. A data item is assigned to the "closest" cluster based on a proximity or dissimilarity measure. Hierarchical clustering, on the other hand, organizes data items into a hierarchy with a sequence of nested partitions or groupings. Hierarchical clustering can be represented with dendrograms, which consist of a hierarchy of nodes, each of which represents a cluster at a certain level (Jain and Dubes 1988). Commonly used hierarchical clustering methods include the Ward's method (Ward 1963), single-linkage clustering, average-linkage clustering, and complete-linkage clustering (Jain and Dubes 1988; Gordon 1996).

SOM (Kohonen 1997, 2001) is a unique partitioning clustering method, which not only segments data into clusters but also orders the clusters in a two-dimensional layout so that nearby clusters are similar to each other. Therefore, SOM is also considered both a visualization method and a dimension reduction technique that projects multidimensional data to a 2-D space. SOMs are widely used in various research fields and application areas (see Kaski, Kangas, and Kohonen 1998; Oja, Kaski, and Kohonen 2003 for comprehensive reviews). There are also numerous applications of SOM in geographic analysis, for example, the visualization of census data (Skupin and Hagelman 2003), spatialization of nonspatial information (Skupin and Fabrikant 2003), and exploration of health survey data (Koua and Kraak 2004).

Clustering in general can be used to summarize or compress large data sets by aggregating similar data items into clusters while preserving the overall data distribution and patterns. However, clusters (especially those derived with multivariate data) are not easy to interpret and understand unless they can be examined in the original multivariate data space and the geographic space (when the data also contains geographic information). Guo et al. (2005, 2006) present an approach that couples the SOM and geovisualization methods through color encoding and multiple linked views. This approach is adopted in this research as part of the integrated system (see Section 3.1).

12.2.2 Regionalization

General-purpose clustering methods, as those introduced in the previous section, normally do not consider geographic information or spatial constraints. Therefore, data items in a cluster are not necessarily close or contiguous in the geographic space. In contrast, regionalization is a special form of clustering that seeks to group data items (or spatial objects) into spatially contiguous clusters (i.e., regions) while optimizing an objective function (e.g., a homogeneity measure based on multivariate similarities within regions). Regionalization has long been an important analysis task for a large spectrum of research and application domains, for example, climatic zoning (Fovell and Fovell 1993), eco-region analysis (Handcock and Csillag 2004), map generalization (Tobler 1969), census reengineering (Openshaw and Rao 1995), and public health analysis (Haining, Wise, and Blake 1994; Osnes 1999).

Regionalization is a combinatorial problem. Given a large set of spatial objects, it is not feasible to enumerate all possible partitions or groupings to find the best set of regions (according to the objective function). Existing regionalization methods can be classified into four groups: (1) optimization through a trial-and-error search, (2) multivariate (nonspatial) clustering followed by spatial processing, (3) clustering with a spatially weighted dissimilarity measure, and (4) contiguity constrained clustering and partitioning. The first group, represented by the Automatic Zoning Procedure (AZP) method (Openshaw 1977; Openshaw and Rao 1995), starts with a random regionalization and iteratively improves the solution by switching boundary objects between neighboring regions. The second group uses a general clustering method to derive clusters based on multivariate similarity and then divides or merges the clusters to form regions (Fovell and Fovell 1993; Haining et al. 1994). The third type of regionalization method incorporates spatial information explicitly in the similarity measure for a general clustering method (e.g., K-means; Wise, Haining, and Ma 1997). The latest group of methods, represented by SKATER (Assunção et al. 2006), a scale-space method (Mu and Wang 2008), and REDCAP (Guo 2008), explicitly considers spatial contiguity constraints (rather than spatial similarities) in a hierarchical clustering process. Particularly, both SKATER and REDCAP can optimize an objective function while partitioning the cluster hierarchy to obtain a given number of regions.

As the latest development, REDCAP is a family of six regionalization methods, which respectively extend the single-linkage, average-linkage, and complete-linkage hierarchical clustering methods to enforce spatial contiguity constraints during the clustering process (Guo 2008). These six methods are similar in that they all iteratively merge clusters (which must be spatial neighbors) until all data items are in one cluster. However, they differ in their definitions of "similarity" between two clusters. Among the six methods, the one based on the complete-linkage clustering [named *Full-Order-CLK*, see Guo (2008) for algorithms and evaluations] consistently produces better regionalization results than existing methods in the literature. This regionalization method is included in the approach presented in this chapter (see Section 12.3.4).

12.2.3 Geovisualization

Geovisualization concerns the development of theory, methods, and tools for the visual analysis and presentation of geographic data (i.e., any data with geographic information). As an emerging domain, geovisualization has drawn interests from various cognate fields and evolved along a diverse set of research directions, as evidenced in a recent edited volume on geovisualization by Dykes, MacEachren, and Kraak (2005). To analyze multivariate spatial data, geovisualization research often draws upon approaches from related disciplines such as cartography, information visualization, and exploratory data analysis.

In the literature of information visualization, many techniques have been developed to visualize multivariate data, for example, stacked histograms and charts (Harris 1999), scatterplot matrices (Andrews 1972), glyphs (Pickett et al. 1995), pixel-oriented approaches (Keim and Kriegel 1996), and parallel coordinate plots (PCP) (Inselberg 1985). One of the major challenges for multidimensional (or multivariate) visualization is related to the number of dimensions (variables). The more dimensions a dataset has, the more challenging to visualize the data and the more difficult for the analyst to recognize patterns across dimensions. To alleviate this problem, multivariate data are often projected to a lower dimensional space using dimensional reduction techniques such as multidimensional scaling (Wong and Bergeron 1997; Williams and Munzner 2004), principle component analysis (PCA), RadViz spring visualization (Bertini, Aquila, and Santucci 2005), SOM (Kohonen 1997, 2001), or other projection pursuit methods (Cook et al. 1995; Wong and Bergeron 1997).

In addition to high dimensionality, large data volume can also cause serious problems for most visualization techniques. A large number of data items often lead to a cluttered visual display (e.g., points overlapping in a scatter plot or line segments overlapping in a PCP) and thus make it very difficult (if possible at all) for the analyst to visually perceive patterns (Keim and Kriegel 1996). Several solutions to such overlapping problems have been proposed. One type of solution is to resolve the overlapping by data sampling, density mapping (Johansson et al. 2005), or re-positioning (or shifting) data points (Keim et al. 2004). A second type of solution relies on user interactions to dynamically filter, select, zoom, and adjust detail levels in the visualization (reference). A third type of solution is to aggregate data items to a relative small number of clusters, visualize the clusters instead of data items, and then provide details (data items) for each cluster upon user request (Guo et al. 2003, 2005; Ward 2004). This latter solution — data abstraction with drill down — is used in this chapter.

Maps are essential for visualizing geographic patterns. Multivariate mapping has long been an interesting and challenging research problem. One of the best-known approaches for multivariate mapping is the Chernoff face (Chernoff and Rizvi 1975), which visualizes multivariate data by relating different variables to different facial features to form a face icon for each data item and then draw each face icon on a map. Generally, multivariate mapping methods can be classified into three types. The first type, with the Chernoff face being an example, depicts each dimension (variable) independently through some attribute of the display and then integrates all variable depictions into one map using composite glyphs, attributes of color, or other methods (Grinstein et al. 1992; DiBiase et al. 1994; Wittenbrink et al. 1995; Gahegan 1998; Zhang and Pazner 2004). The second type uses multiple linked views (or maps) that show one (or more) variables per view (Monmonier 1989; Dykes 1998; MacEachren et al. 1999; Andrienko and Andrienko 2001). For example, Carr et al. (2005) proposed the conditioned choropleth maps (CCmaps) that uses a two-way layout of maps (arranged by two potential explanatory variables) to facilitate the exploration of potential associations between a dependent variable (as represented in colors) and the two explanatory variables. The third type projects data to a lower-dimensional (normally 1D or 2D) space through clustering, encodes the clusters with colors, and then uses the colors to generate a multivariate map (Guo et al. 2003, 2005).

There are three major challenges for multivariate mapping, including large data volume, high dimensionality, and the understandability (or perception) of complex patterns. The first type of mapping (e.g., composite glyphs and Chernoff faces) cannot deal with either large datasets or many variables. The second type (e.g., multiple linked views) alone is not very effective in presenting an overview of major patterns as it relies on human interactions to visually detect connections and structures. The third type (with clustering, projection, coloring, and mapping) uses computational algorithms to summarize patterns and thus it can handle larger datasets and more variables. However, it needs visualization techniques to help understand the patterns.

12.3 AN INTEGRATED APPROACH TO MULTIVARIATE CLUSTERING AND GEOVISUALIZATION

12.3.1 A Theoretical Framework

To build an integrated approach, it is necessary to examine the relations between different methods, which can be either *complementary* or *competitive*. Complementary methods usually analyze the data from different perspectives and help each other overcome weaknesses. For example, multidimensional visualization can be complemented by a cartographic map to explore multivariate spatial data interactively (Andrienko and Andrienko 2001). Computational and visual approaches are usually complementary to each other as the former process and summarize large data sets while the latter can help present and understand the findings (Guo et al. 2003). In contrast, competitive methods usually focus on the same analysis task (e.g., clustering). For example, two different clustering methods often produce different clusters from the same data due to different searching strategies or underlying constraints. It would be useful and often critical to be able to compare the results of such competitive methods, find commonalities, examine differences, crosscheck each other's validity, and thus better understand the data and patterns. Although there are considerable efforts on integrating complementary methods, few have focused on competitive approaches.

The strategies to couple different methods may be classified into four groups (see Table 12.1):

1. *Different time and different view*, for example, independently applying different methods to analyze the same data and examine their results separately
2. *Different time and same view*, for example, one method's output being another's input so that the final output is a joint outcome of the two methods
3. *Same time and different view*, for example, simultaneously feeding the data into multiple linked views and examining their results side by side through brushing and linking

TABLE 12.1
A Scheme for Classifying Integrated Approaches

	Different View	Same View
Different Time	Independent application of different methods and examine their results separately. *Example:* Performing two cluster analyses (separately) with two different methods. *Relation:* Competitive	Sequential application of two different methods and one method's output is the other method's input. *Example:* Principle Component Analysis → using principle components to render a scatterplot matrix. *Relation:* Complementary
Same Time	Parallel application of different methods and comparing results in different views with user interactions such as brushing and linking. *Example:* Multiple linked views (e.g., PCP + Scatterplot + Map + user interactions) *Relation:* Complementary and/or competitive	Parallel application of different methods and immersing or overlaying results in a unified overview. *Example:* Multivariate patterns are coded with colors and shown in a map, which also overlays a regionalization result (polygons). *Relation:* Complementary and/or competitive

4. *Same time and same view*, for example, immersing or overlaying (Roberts 2005) the results of different methods in a unified view to facilitate a holistic understanding across *all* perspectives (for complementary methods) and to support precise comparison and crosschecking (for competitive methods)

This classification scheme is borrowed from the research in collaborative decision making or group work, which distinguishes four different types of collaboration among people according to location and time (Jankowski et al. 1997).

The primary difference between the third group and fourth group is that the former has to rely on human interactions (e.g., brushing and linking) to perceive the connection between multiple views, while the latter merges different results into the same view so that one can perceive an overview of major patterns even without human interactions. However, interactive exploration and multiple linked views remain important when it comes to concise understanding and detailed inspection of specific patterns.

The integrated approach presented in this section (1) couples both complimentary approaches and competitive methods, and (2) supports both the same-time-different-view and same-time-same-view coupling strategies. The design framework for the approach is illustrated in Figure 12.2. A SOM is extended to perform multivariate clustering, dimension reduction, and pattern encoding with colors. The clusters (with association to data items) and their colors are passed on to a *multivariate mapping* component and a *multivariate visualization* component. The multivariate map uses the colors (assigned by the SOM) to represent multivariate information and the multivariate visualization signifies the meaning (i.e., multivariate characteristic) of

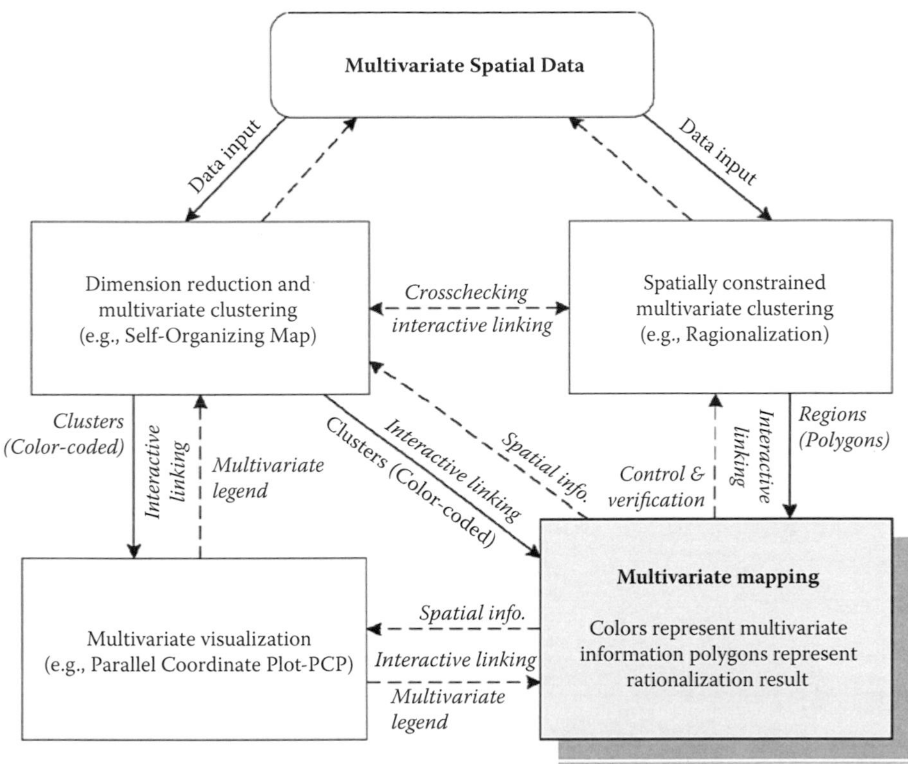

FIGURE 12.2 The framework for the integrated approach, which can be naturally expanded by adding new components or replacing current components.

those colors. In other words, the multivariate visualization component serves as the "legend" for the multivariate map. A *regionalization* method, as a competitor to the SOM, takes the same data input and derives homogeneous regions, which are overlaid in the multivariate map. Thus, one can easily perceive an overall picture of how the multivariate patterns (colors) change over the geographic space and how the two clustering results (polygons and colors) agree with or differ from each other. With interactions (e.g., brushing and linking across different views), the analyst may focus on specific patterns (e.g., a cluster, a region, or a specific data item) and examine details (e.g., showing all the data items contained in a cluster).

The following subsections focus on individual components in the framework, that is, clustering with SOM, multivariate visualization and mapping, and regionalization. User interactions will be demonstrated through the global climate analysis presented in Section 12.4.

12.3.2 Multivariate Clustering and Pattern Encoding

A SOM is a special clustering method that seeks clusters in multivariate data and orders the clusters in a two-dimensional layout so that nearby clusters are similar in terms of multivariate characteristics (Kohonen 2001). Each cluster (node) is associated with a multivariate vector (or codebook vector), which represents the centroid of the cluster in the multivariate space. A SOM first arranges a user-specified number of nodes in a regularly spaced grid and then initializes each node by assigning its codebook vector randomly (or using any specific initialization method). During the learning process, the SOM iteratively adjusts each codebook vector according to the data items falling inside the node (cluster) and the codebook vectors of its neighboring nodes. Once the learning is complete, each node has a position in the multivariate space and all the nodes in the SOM form a nonlinear surface in the multivariate space. Then data items are projected onto the surface by assigning each item to its nearest cluster (whose codebook vector is most similar to the data item). Although SOM nodes are equally spaced in the two-dimensional layout, their codebook vectors are not equally spaced in the multivariate space. Rather, the distribution of nodes adapts to the actual data density — dense areas (in the multivariate space) tend to have more clusters.

Figure 12.3 (top left) shows the implementation of a SOM in this research. Each SOM node (cluster) is represented with a circle, whose size (area) is linearly scaled

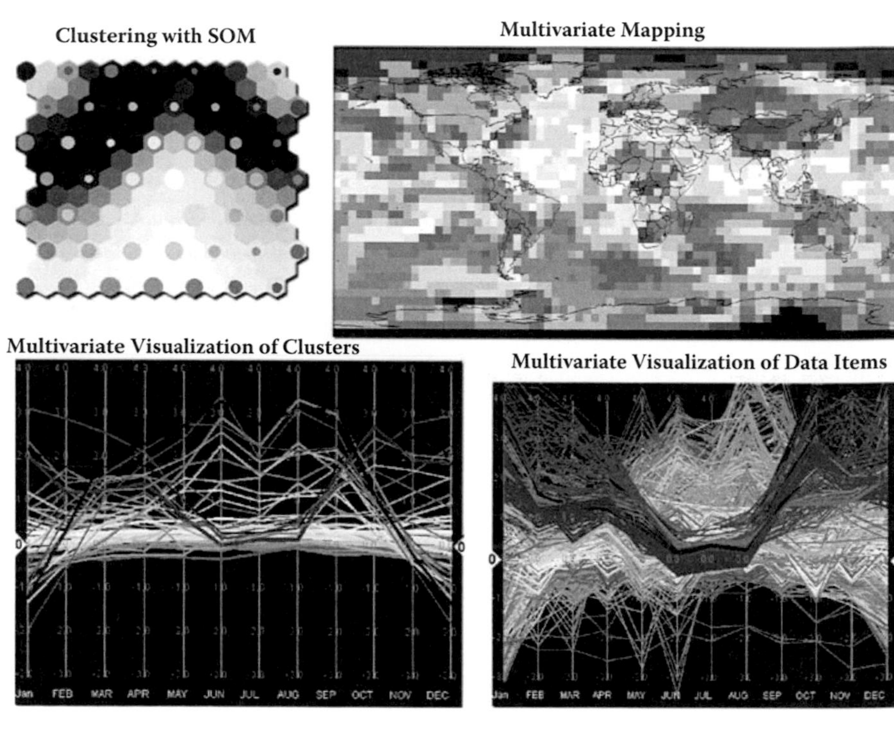

FIGURE 12.3 This is the overview of multivariate (seasonal) and spatial patterns of global climate (temperature) change. The multivariate mapping component (top right) is the central view, while other views can assist the understanding of the map. From such an integrated overview, one can easily perceive the major patterns in the data even without user interactions. See color insert after page 148.

according to the number of data items that it contains. Nodes (circles) are equally spaced in a two-dimensional space. Behind the nodes (circles), there is a layer of hexagons, which are shaded to show the multivariate dissimilarity between neighboring nodes — darker tones represent greater dissimilarity. Such a view is called the U-matrix (Kohonen 2001), which can reveal natural clusters and data distributions in the multivariate space. Clusters (nodes) in a brighter area are more similar to each other than those in a darker area are. Dark ridges in the U-matrix signify natural divides in the data. For example, the SOM in Figure 12.3 has 49 nodes and the U-matrix indicates that there are three high-level clusters: the top-left corner (in green colors), top-right corner (in red colors), and the bottom triangular area (in blue, purple, and white colors).

With the U-matrix, a SOM is also considered a visualization method that can show inherent structures in the data. However, as is true with most multivariate clustering or dimension reduction methods, it is not easy to find out the meaning (i.e., multivariate characteristics) of each cluster (node). Traditionally each SOM node is labeled (using item names or keywords) to indicate the kind of data items that are covered by the node. However, such a labeling strategy becomes less useful when the data set is large or it is not convenient to summarize data items with keywords. This research adopts the strategy proposed by Guo et al. (2005), which uses a two-dimensional color scheme to assign a unique color to each SOM node so that nearby (and therefore similar) clusters have similar colors (see Figure 12.3, top left). In other words, the SOM surface is folded onto a color surface so that colors represent the multivariate patterns (clusters). Colors are then passed on to other visualization components where multivariate meanings of colors (and thus clusters) can be understood.

12.3.3 Multivariate Visualization and Mapping

To reveal the multivariate meaning of SOM clusters (which are now represented by colors), an extended version of a PCP (Guo et al. 2005, 2006) is used. The PCP can visualize the data at two different levels: the cluster level or the data item level. At the cluster level, the PCP shows each cluster as a single entity (string) and thus partially avoids the overlapping problem. Each string (which represents a cluster with its mean vector) has the same color as the cluster does in the SOM. Figure 12.3 (bottom left) shows a PCP at the cluster level. The thickness of each string is proportional to the cluster size. At the data item level, each string in the PCP represents an individual data item (see the bottom-right plot in Figure 12.3). Each string (data item) has the same color as that of its containing cluster. Evidently, the colors dramatically improve the visual effectiveness of the PCP in presenting multivariate patterns. Without colors, it would be very difficult to track each string across many dimensions (variables). By comparing the PCP at the two different detail levels, we can also see that the aggregation of data items into clusters significantly helps accentuate major patterns.

With the clusters (and colors) derived by the SOM, it is straightforward to create a multivariate map (see Figure 12.3, top right), which is similar to a univariate choropleth map except that colors now represent multivariate information. This map is an overview of the spatial distribution of multivariate patterns. The PCP introduced previously serves as the legend for interpreting the meaning of colors (e.g., what does a green color represent in the map?).

The three visual components (i.e., SOM, PCP, and map) allow a variety of user interactions such as selection-based brushing and linking. A selection made in one component will be highlighted in all other components simultaneously. A selection can be progressively refined by, for example, adding or subtracting new selections. The user may select at either the cluster level or the data item level. For example, one may show data at the item level in the PCP and select a single data item to read its exact variable values. To respond to this single item selection, the SOM will highlight the cluster that contains the selected data item and change the circle to a wedge accordingly if that item is not the only one in that cluster. Some of these interactive features will be demonstrated in Section 12.4.

12.3.4 Regionalization with Spatially Constrained Hierarchical Clustering

Regionalization is a unique multivariate spatial analysis task that seeks to group data items (or spatial objects) into spatially contiguous regions while optimizing an objective function. Among the six regionalization methods in the REDCAP family (Guo, 2008), this research adopts the full-order-CLK regionalization method, which extends the traditional complete linkage clustering method to consider geographic contiguity constraints. Readers are referred to Guo (2008) for detailed algorithms and evaluations. In the following is a brief and conceptual explanation of the method.

The complete linkage clustering (CLK) defines the distance (or dissimilarity) between two clusters as the longest distance (dissimilarity) between all possible pairs of data items across the two clusters. Therefore, CLK considers two clusters similar only if all the observations in the two clusters are similar to each other. At the beginning of the clustering, each individual data item is a cluster by itself. With an iterative merging process, the most similar pair of clusters is merged into one cluster at each step. This merging process repeats until all data items are in the same cluster.

To extend the CLK clustering to consider spatial contiguity constraints, the full-order-CLK regionalization method requires that two clusters cannot be merged if they are not spatially contiguous. At each iteration step, the most similar pair of *spatially neighboring* clusters will be merged. A contiguity matrix of clusters is maintained and updated during the clustering process. With this extended CLK algorithm, it is guaranteed that each cluster (at any level in the hierarchy) occupies a spatially contiguous region. Although this may seem simple conceptually, it demands a rather complicated algorithm to achieve an acceptable efficiency. The algorithm proposed in Guo (2008) achieves a time complexity of $O(n^2 \log n)$ and space (i.e., RAM memory) complexity of $O(n^2)$.

Deriving a hierarchy of spatially contiguous clusters is only halfway to the final regionalization result. The hierarchy of clusters will be re-partitioned from the top down to obtain a number of regions while optimizing an objective function. This partitioning step is necessary especially when the objective function includes other criteria (e.g., region size limit) besides multivariate similarity within regions. In this research, the objective function is to minimize the overall heterogeneity and (optionally) to satisfy a minimum region size threshold (i.e., a region cannot be smaller than the threshold size). The heterogeneity of a region is defined as the sum of squared differences between each data item in the region and the region mean vector. The overall heterogeneity for a regionalization result is the total of the heterogeneity values of all regions. The partitioning process is iterative — one region is selected (according to the objective function) to cut into two at each step until a specified number of regions is obtained. The detail of this partitioning algorithm is available in (Guo 2008).

12.4 APPLICATION IN GLOBAL CLIMATE CHANGE ANALYSIS

12.4.1 Data Source and Preprocessing

The climate data being analyzed is a spatio-temporal data set of monthly mean surface air temperature for 60 years (Jan. 1948–Dec. 2007). It has a global coverage based on a matrix of 2.5° latitude by 2.5° longitude grid, ranging from 90N to 90S and from 0E to 357.5E. This data set is part of the NCEP/NCAR re-analysis data archive (Kalnay et al. 1996), which is available at: ftp://ftp.cdc.noaa.gov/Datasets/ncep.reanalysis.derived/. The analysis task is to discover global climate change patterns with a focus on surface air temperature and global warming.

To analyze temperature changes (i.e., warming or cooling), an anomaly value is calculated for each month and each grid cell following three steps. First, a 40-year average temperature (1948–1987) is calculated for each grid cell and month. Then the 10-year average temperature for 1998–2007 (i.e., the most recent 10 years) is calculated for each grid cell and each month. Finally, an *anomaly* value is derived by taking the difference between the 40-year average (1948–1987) and the 10-year average (1998–2007). A positive anomaly value represents a temperature increase (warming) for a specific month and grid cell during the last 10 years (compared to the average temperature during the 40-year period, for the same month and location). To smooth the data and remove some noise, the grid cells are aggregated to a 5° × 5° resolution by merging each 2 × 2 block of original cells. Thus, the original data is transformed to a 12 × 72 × 37 cube. In other words, the data set has 2664 spatial objects (grid cells) and 12 variables (monthly anomalies). For the regionalization method, two grid cells are considered neighbors if they share a side (i.e., the rook type of contiguity).

12.4.2 Discovering Climate Change Patterns

There have been extensive research efforts on global climate change (Houghton et al. 2001). To understand global temperature changes (warming or cooling), existing analyses usually examine the data from a selected perspective, for example, analyzing temporal trends of global annual temperatures (ignoring geographic and seasonal variations) or mapping spatial distributions of annual temperature anomalies (ignoring their seasonal variations). The integrated approach introduced here offers a more comprehensive methodology that can examine seasonal trends and geographic variations simultaneously, which enables a more effective and efficient exploratory process for more complex patterns.

Figure 12.3 shows the overview of patterns in the data (without using the regionalization method). The SOM derives 49 clusters and the shaded U-matrix suggests that there are three high-level groups of clusters: the green/light-green nodes at the top left corner, the red-reddish nodes at the top right corner, and the rest (blue/white/purple) nodes at the bottom and the center. The PCPs and map can then help in understanding the meaning of these groups. The green and light green clusters, which are mainly in the Antarctic area, had high positive anomaly values for winter months (April–September) and negative values for summer months (January, February, November, and December). In other words, for the last 10 years the Antarctic area became much warmer in winter but cooler in summer. In contrast, the red/reddish clusters, located primarily in the Arctic area, had very high positive values for winter and around zero for summer (June, July, and August). This means that, for the last 10 years, the Arctic areas were much warmer in winter but relatively stable for summer. The difference in patterns for the Arctic and Antarctic areas for the last 10 years is very interesting — both showed a warming trend in winter but the Antarctic area actually was cooler than before during winter. This may potentially and partially explain why the ice cap in the Arctic diminishes while the Antarctic ice cap recently grew to a historical maximum since 1979 (see http://arctic.atmos.uiuc.edu/cryosphere/, which updates snow and ice extent for both hemispheres daily).

One may notice that in the cluster-level PCP (Figure 12.3, lower left), a red cluster represents the highest positive values (i.e., the most severe warming trend) in January and December. To examine this cluster in detail, we select the string of that cluster in the PCP or select the node in the SOM (see Figure 12.4). The PCPs (one at the cluster level and the other at the data item level), SOM, and the map all highlight the selected cluster/items by hiding other clusters and items. Then we can examine all the data items covered by this cluster in the item-level PCP and understand that those places,

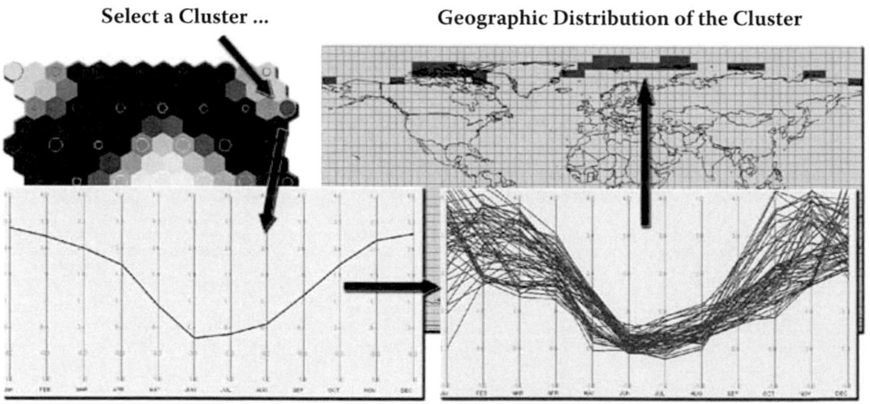

FIGURE 12.4 Human interaction with multiple linked views. A cluster is selected in the SOM and other components respond by highlighting the cluster or data items are selected.

which have experienced the most dramatic warming effect in winter, are exclusively in the coastal area along the Arctic Ocean. Guided by the overview and facilitated by user interactions, one can efficiently obtain a variety of new knowledge from the data.

12.4.3 Comparing Regionalization and SOM Results

The regionalization method is also applied to the data to search regions of similar climate change patterns. Figure 12.5 shows two 5-region results: one has no region size limit (bottom right map) and the other result requires that each region be larger than 200 grid cells. The plot at the lower left corner shows the relationship between the number of regions and the overall heterogeneity value. For the first two cuts (to get two or three regions), the heterogeneity value drops dramatically. Further cuts (i.e., more regions), however, cannot significantly reduce the heterogeneity value. This suggests that there are three natural regions, which are the Arctic (region 1 in the top map), the Antarctic (region 5 in the top map), and the rest of the world.

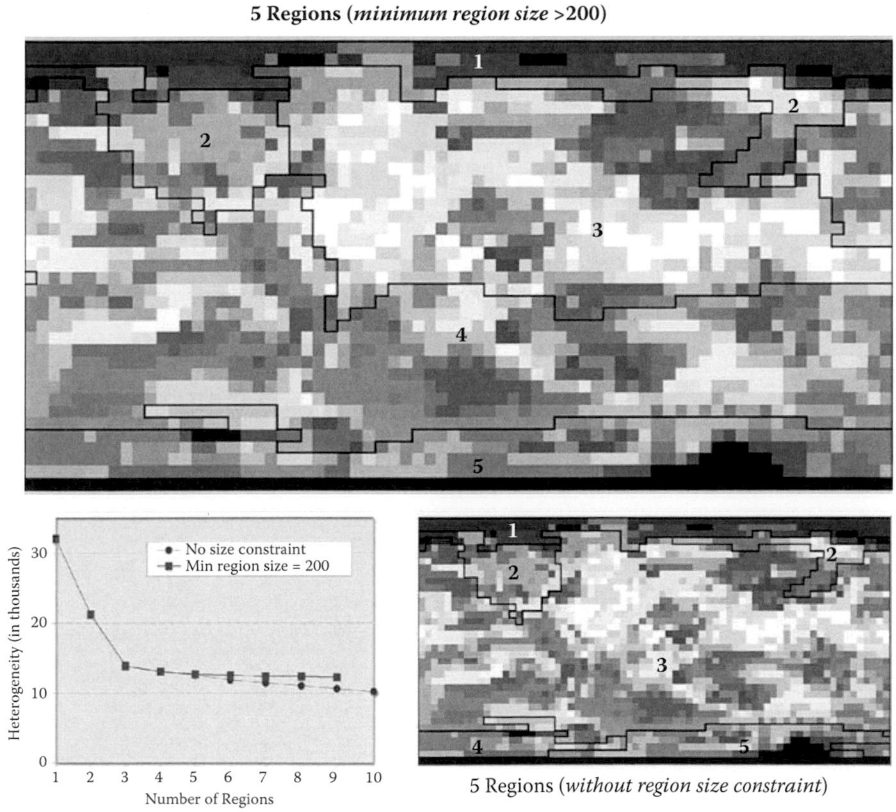

FIGURE 12.5 Two regionalization results. The small map shows five regions derived without a size constraint while the top (larger) map shows five regions under a size constraint (i.e., a region must be larger than 200 grid cells). See color insert after page 148.

With a size constraint, the regionalization method will not divide a region if one of the resultant regions is smaller than the size constraint. For example, if the size constraint is 200, the method can only produce at most nine regions because having more regions will result in a region that is smaller than the size limit (see the plot in the lower left corner in Figure 12.5).

By examining how the regions match the colors in the map, we can make two interesting observations. On one hand, a region tends to contain similar colors and region boundaries approximately follow color discrepancies (i.e., where colors change dramatically). This indicates that the regionalization and the SOM clustering are in good agreement. On the other hand, when there is no size constraint (see the smaller map in Figure 12.4), the regionalization method divides the Antarctic area into two regions (regions 4 and 5 in the small map) before partitioning region 3 (which is visually more heterogeneous — in terms of colors — than the Antarctic region). This reflects the methodological differences between the SOM and the CLK clustering. As explained in Section 12.3.2, the SOM can adapt to the data distribution and tend have more nodes (and thus more colors) in a dense (but similar) data area. The complete linkage clustering, however, does not adapt to data distribution and strictly follows the distance measure. Thus, it will produce large clusters in dense areas and many smaller clusters in sparse areas.

Through the integration of SOM and regionalization, one can better understand both the methods and their results. The integrated approach makes it possible for the user to inspect in detail how a regionalization method works, understand the meaning of regions, and gain insights. Figure 12.6 shows a selection of the Antarctic region (as derived by the regionalization method) to take a closer look at its multivariate

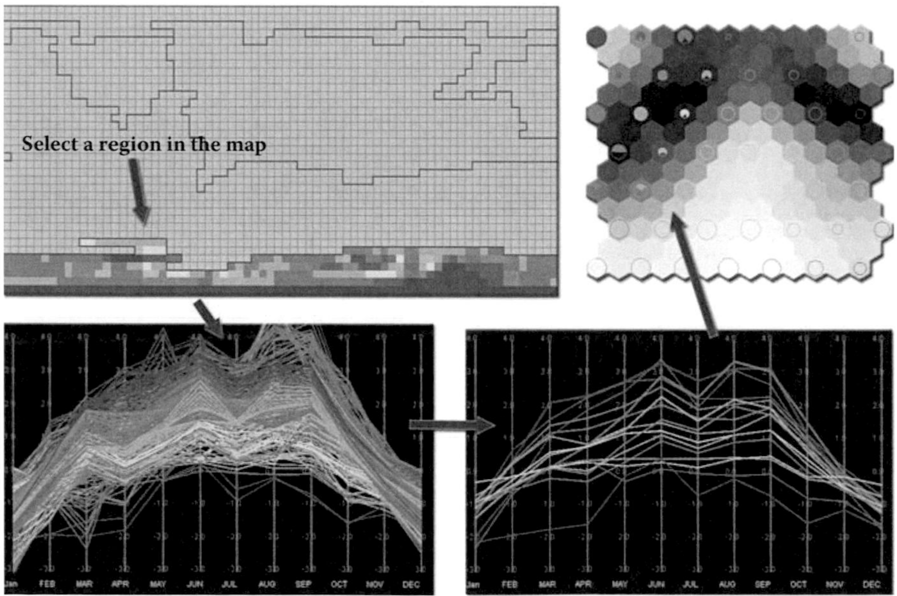

FIGURE 12.6 An example of user interactions with regions and multivariate information.

characteristic and internal variation. The data items (i.e., grid cells) contained in this region belong to several SOM clusters and some of the clusters are only being partially selected (shown as wedges, proportional to the percentage of items being selected in that cluster). The cluster-level PCP (Figure 12.6, lower right) will re-calculate the mean vector for each partially selected cluster based on selected items only. From the PCPs and the map, we can see how the multivariate (seasonal) patterns change within the region. From the SOM, we can see that the data items contained in the selected region are also recognized by the SOM as a similar group (which occupies the top left corner of the SOM — see Figure 12.6, top right).

12.5 CONCLUSION AND DISCUSSION

This chapter introduces an integrated approach to multivariate clustering and geovisualization, which builds upon the synergy of multiple computational and visual methods. As demonstrated with its application for climate data analysis, the approach can effectively facilitate the exploration and understanding of complex patterns in multivariate spatial data sets. Its unique strength is evident in several aspects. First, by leveraging the power of computational methods (e.g., clustering), it can handle larger datasets and more variables than would be possible with visual methods alone. Second, it effectively synthesizes different perspective information (e.g., multivariate relations and geographic variations) to enable an overview of complex patterns across multiple spaces. Third, its component-based design provides flexibility to extend the system by adding new components or replacing current components.

Through the static linking (via colors and overlay) and user interactions (via brushing and linking), complex relationships can be perceived and understood easily. The consistent use of colors across all visual components significantly improves the presentation of connections between different views. However, there is a limitation on the color configuration — it is not always possible to assign a cognitively meaningful color to each cluster. The color surface must be smooth and continuous, meaning that the user may not always be able to assign a desirable color to a cluster, for example, a red color to represent a warming trend and a blue color to represent a cooling trend. The software allows the user to rotate or flip the two-dimensional color surface to find a reasonable match between clusters and colors. Another potentially confusing factor is that when the user interactively changes the size of the SOM (i.e., the number of nodes), the resultant clusters are different, and thus the multivariate meaning of colors will change. In other words, the same color may represent two different groups of data items for different runs of the SOM (with different parameters).

Spatial analysis tasks are often application-dependent and the data may vary dramatically. The set of methods introduced in this chapter is only suitable for the analysis of lattice data, which contains a set of spatial objects and a multivariate vector for each object. It is not applicable, for example, to the analysis of point patterns or network-based phenomena. However, the general concept and framework for the integration of different methods will remain a viable direction for addressing other complex spatial analysis problems.

REFERENCES

Andrews, D.F., 1972, Plots of high-dimensional data. *Biometrics*, **29,** 125–136.

Andrienko, G. and N. Andrienko, 2001, Exploring spatial data with dominant attribute map and parallel coordinates. *Computers, Environment and Urban Systems*, **25,** 5–15.

Andrienko, N., G. Andrienko and P. Gatalsky, 2003, Exploratory spatio-temporal visualization: an analytical review. *Journal of Visual Languages & Computing*, **14,** 503–541.

Anselin, L., 1995, Local indicators of spatial association—LISA. *Geographical Analysis*, **27,** 93–115.

Anselin, L., 1999, Interactive techniques and exploratory spatial data analysis. In: *Geographical Information Systems—Principles and Technical Issues*. P.A. Longley, M.F. Goodchild, D.J. Maguire, and D.W. Rhind, Eds., New York: John Wiley & Sons, **1**, 253–266.

Assunção, R.M., M.C. Neves, G. Câmara and C.D.C. Freitas, 2006, Efficient regionalization techniques for socio-economic geographical units using minimum spanning trees. *International Journal of Geographical Information Science*, **20,** 797–811.

Bertini, E.D., L. Aquila and G. Santucci, 2005, SpringView: Cooperation of Radviz and Parallel Coordinates for View Optimization and Clutter Reduction. *Proceedings, the Third International Conference on Coordinated and Multiple Views in Exploratory Visualization* (CMV 2005), 22–29.

Carr, D.B., D. White and A.M. MacEachren, 2005, Conditioned choropleth maps and hypothesis generation. *Annals of the Association of American Geographers*, **95,** 32–53.

Chernoff, H. and M.H. Rizvi, 1975, Effect on classification error of random permutations of features in representing multivariate data by faces. *Journal of American Statistical Association*, **70,** 548–554.

Cook, D., A. Buja, J. Cabrera and C. Hurley, 1995, Grand tour and projection pursuit. *Journal of Computational and Graphical Statistics*, **4,** 155–172.

Cressie, N., 1991, *Statistics for Spatial Data*, Chichester: John Wiley & Sons.

DiBiase, D., C. Reeves, J. Krygier, A.M. MacEachren, M.V. Weiss, J. Sloan and M. Detweiller, 1994, Multivariate display of geographic data: Applications in earth system science. In: *Visualization in Modern Cartography*. A.M. MacEachren and D.R.F. Taylor, Eds., Oxford, U.K.: Pergamo, 287–312.

Dykes, J., 1998, Cartographic visualization: Exploratory spatial data analysis with local indicators of spatial association using Tcl/Tk and cdv'. *The Statistician*, **47,** 485–497.

Dykes, J., A.M. MacEachren and M.-J. Kraak, Eds., 2005, *Exploring Geovisualization*, Amsterdam: Elsevier.

Fotheringham, A.S., C. Brunsdon and M. Charlton, 2002, *Geographically Weighted Regression: The Analysis of Spatially Varying Relationships*, New York: John Wiley & Sons.

Fovell, R.G. and M.-Y.C. Fovell, 1993, Climate zones of the conterminous United States defined using cluster analysis. *Journal of Climate*, **6,** 2103–2135.

Gahegan, M., 1998, Scatterplots and scenes: visualization techniques for exploratory spatial analysis. *Computers, Environment and Urban Systems*, **22,** 43–56.

Gahegan, M., 2003, Is inductive machine learning just another wild goose (or might it lay the golden egg)? *International Journal of Geographical Information Science*, **17,** 69–92.

Getis, A. and J.K. Ord, 1992, The analysis of spatial association by use of distance statistics. *Geographical Analysis*, **24,** 189–206.

Gordon, A.D., 1996, Hierarchical classification. In: *Clustering and Classification*. P. Arabie, L.J. Hubert and G.D. Soete, Eds., River Edge, NJ: World Scientific Publisher, 65–122.

Gould, P.R., 1981, Letting the data speak for themselves. *Annals of the Association of American Geographers*, **71,** 166–176.

Grinstein, G., J.C.J. Sieg, S. Smith and M.G. Williams, 1992, Visualization for knowledge discovery. *International Journal of Intelligent Systems*, **7,** 637–648.

Guo, D., 2008, Regionalization with dynamically constrained agglomerative clustering and partitioning (REDCAP). *International Journal of Geographical Information Science*, **22**(7), pp. 801–823.

Guo, D., J. Chen, A.M. MacEachren and K. Liao, 2006, A visualization system for space-time and multivariate patterns (VIS-STAMP). *IEEE Transactions on Visualization and Computer Graphics*, **12,** 1461–1474.

Guo, D., M. Gahegan, A.M. MacEachren and B. Zhou, 2005, Multivariate analysis and geovisualization with an integrated geographic knowledge discovery approach. *Cartography and Geographic Information Science*, **32,** 113–132.

Guo, D., D. Peuquet and M. Gahegan, 2003, ICEAGE: Interactive clustering and exploration of large and high-dimensional geodata. *GeoInformatica*, **7,** 229–253.

Haining, R., 2003, Spatial Data Analysis—Theory and Practice, Cambridge, U.K.

Haining, R.P., S.M. Wise and M. Blake, 1994, Constructing regions for small area analysis: material deprivation and colorectal cancer. *Journal of Public Health Medicine*, **16,** 429–438.

Han, J., M. Kamber and A.K.H. Tung, 2001, Spatial clustering methods in data mining: A survey. In: *Geographic Data Mining and Knowledge Discovery*. H.J. Miller and J. Han, Eds., London: Taylor & Francis, 33–50.

Han, J., K. Koperski and N. Stefanovic, 1997, GeoMiner: A System Prototype for Spatial Data Mining. ACM SIGMOD International Conference on Management of Data, Tucson, AZ, 553–556.

Handcock, R. and F. Csillag, 2004, Spatio-temporal analysis using a multiscale hierarchical ecoregionalization. *Photogrammetric Engineering and Remote Sensing*, **70,** 101–110.

Harris, R.L., 1999, *Information Graphics: A Comprehensive Illustrated Reference*. Oxford, U.K.: Oxford Press.

Houghton, J.T., Y. Ding, D.J. Griggs, M. Noguer, P.J.V.D. Linden, X. Dai, K. Maskell and C.A. Johnson, Eds., 2001, *Climate Change 2001: The Scientific Basis*. Cambridge, England: Cambridge University Press.

Huang, Y., S. Shekhar and H. Xiong, 2004, Discovering colocation patterns from spatial data sets: A general approach. *IEEE Transactions on Knowledge and Data Engineering*, **16,** 1472–1485.

Inselberg, A., 1985, The plane with parallel coordinates. *The Visual Computer*, **1,** 69–97.

Jain, A.K. and R.C. Dubes, 1988, *Algorithms for Clustering Data*. Englewood Cliffs, NJ: Prentice Hall.

Jain, A.K., R.P.W. Duin and J.C. Mao, 2000, Statistical pattern recognition: A review. *IEEE Transactions on Pattern Analysis and Machine Intelligence*, **22,** 4–37.

Jain, A.K., M.N. Murty and P.J. Flynn, 1999, Data clustering: a review. *ACM Computing Surveys (CSUR)*, **31,** 264–323.

Jankowski, P., T.L. Nyerges, A. Smith, T.J. Moore and E. Horvath, 1997, Spatial group choice: A SDSS tool for collaborative spatial decision making. *International Journal of Geographical Information Systems*, **11,** 577–602.

Johansson, J., P. Ljung, M. Jern and M. Cooper, 2005, Revealing Structure within Clustered Parallel Coordinates Displays. *Proceedings, IEEE Symposium on Information Visualization*, Minneapolis, MN, Oct. 23–25, IEEE Computer Society, 125–132.

Jona-Lasinio, G., 2001, Modeling and exploring multivariate spatial variation: A test procedure for isotropy of multivariate spatial data. *Journal of Multivariate Analysis*, **77,** 295–317.

Kalnay, E., M. Kanamitsu, R. Kistler, W. Collins, D. Deaven, L. Gandin, M. Iredell, S. Saha, G. White, J. Woollen, Y. Zhu, M. Chelliah, W. Ebisuzaki, W. Higgins, J. Janowiak, K.C. Mo, C. Ropelewski, J. Wang, A. Leetmaa, R. Reynolds, R. Jenne and D. Joseph, 1996, The NCEP/NCAR 40-year reanalysis project. *Bulletin of the American Meteorological Society,* **77,** 437–470.

Kaski, S., J. Kangas and T. Kohonen, 1998, Bibliography of self-organizing map (SOM) papers: 1981–1997. *Neural Computing Surveys*, **1,** 102–350.

Keim, D.A. and H.P. Kriegel, 1996, Visualization techniques for mining large databases: A comparison. *IEEE Transaction on Knowledge and Data Engineering*, **8**.

Keim, D.A., C. Panse, M. Sips and S.C. North, 2004, Visual data mining in large geospatial point sets. *IEEE Computer Graphics and Applications*, **24,** 36–44.

Kohonen, T., 1997, *Self-Organizing Maps*, Berlin: Springer.

Kohonen, T., 2001, *Self-Organizing Maps*, 2nd ed., Berlin: Springer.

Koua, E.L. and M.-J. Kraak, 2004, Geovisualization to support the exploration of large health and demographic survey data. *International Journal of Health Geographics*, **3**.

Kulldorff, M., 2006, Tests for spatial randomness adjusting for an underlying inhomogeneity: A general framework. *Journal of the American Statistical Association*, **101,** 1289–1305.

MacEachren, A.M., M. Wachowicz, R. Edsall, D. Haug and R. Masters, 1999, Constructing knowledge from multivariate spatiotemporal data: Integrating geographical visualization with knowledge discovery in database methods. *International Journal of Geographical Information Science*, **13,** 311–334.

Monmonier, M., 1989, Geographic brushing: Enhancing exploratory analysis of the scatterplot matrix. *Geographical Analysis*, **21,** 81–84.

Mu, L. and F. Wang, 2008, A scale-space clustering method: Mitigating the effect of scale in the analysis of zone-based data. *Annals of the Association of American Geographers*, **98,** 85–101.

National Research Council, 2003, *IT Roadmap to a Geospatial Future*, Washington, D.C.: National Academy Press.

Oja, M., S. Kaski and T. Kohonen, 2003, Bibliography of self-organizing map (SOM) papers: 1998–2001 addendum. *Neural Computing Surveys*, **3,** 1–156.

Openshaw, S., 1977, A geographical solution to scale and aggregation problems in region-building, partitioning, and spatial modelling. *Transactions of the Institute of British Geographers*, **NS 2,** 459–472.

Openshaw, S., A. Cross and M. Charlton, 1990, Building a prototype geographical correlates exploration machine. *International Journal of Geographical Information Systems*, **4,** 297–311.

Openshaw, S. and L. Rao, 1995, Algorithms for reengineering 1991 census geography. *Environment & Planning A*, **27,** 425–446.

Osnes, K., 1999, Iterative random aggregation of small units using regional measures of spatial autocorrelation for cluster localization. *Statistics in Medicine*, **18,** 707–725.

Pickett, R.M., G. Grinstein, H. Levkowitz and S. Smith, 1995, Harnessing preattentive perceptual processes in visualization. In: *Perceptual Issues in Visualization*. G. Grinstein and H. Levkowitz, Eds., New York: Springer, 33–45.

Roberts, J.C., 2005, Exploratory visualization with multiple linked views. In: *Exploring Geovisualization*. J. Dykes, A.M. MacEachren and M.-J. Kraak, Eds., Amsterdam: Elsevier, 158–180.

Saina, S.R. and N. Cressie, 2007, A spatial model for multivariate lattice data. *Journal of Econometrics*, **140,** 226–259.

Shekhar, S., C.-T. Lu and P. Zhang, 2003, A unified approach to detecting spatial outliers. *GeoInformatica*, **7,** 139–166.

Shekhar, S., C.T. Lu, X. Tan, S. Chawla and R.R. Vatsavai, 2001, Map cube: a visualization tool for spatial data warehouses. In: *Geographic Data Mining and Knowledge Discovery*. H.J. Miller and J. Han, Eds., London: Taylor & Francis, 74–109.

Skupin, A. and S. Fabrikant, 2003, Spatialization methods: A cartographic research agenda for nongeographic information visualization. *Cartography and Geographic Information Science*, **30,** 99–119.

Skupin, A. and R. Hagelman, 2003, Attribute Space Visualization of Demographic Change. *Proceedings of the Eleventh ACM International Symposium on Advances in Geographic Information Systems*, New Orleans, LA, ACM Press, 56–62.

Swayne, D.F., D.T. Lang, A. Buja and D. Cook, 2003, GGobi: Evolving from XGobi into an extensible framework for interactive data visualization. *Computational Statistics and Data Analysis*, **43,** 423–444.

Tobler, W.R., 1969, Geographical filters and their inverses. *Geographical Analysis*, **1,** 234–253.

Ward, J.H., 1963, Hierarchical grouping to optimise an objective function. *Journal of the American Statistic Association*, **58,** 236–244.

Ward, M.O., 2004, Finding needles in large-scale multivariate data haystacks. *Computer Graphics and Applications*, **24,** 16–19.

Williams, M. and T. Munzner, 2004, Steerable, Progressive Multidimensional Scaling. *IEEE Symposium on Information Visualization*, 57–64.

Wise, S.M., R.P. Haining and J. Ma, 1997, Regionalization tools for the exploratory spatial analysis of health data. In: *Recent Developments in Spatial Analysis: Spatial Statistics, Behavioural Modelling and Neuro-Computing*. M. Fischer and A. Getis, Eds., Berlin: Springer-Verlag.

Wittenbrink, C.M., E. Saxon, J.J. Furman, A. Pang and S. Lodha, 1995, Glyphs for visualizing uncertainty in environmental vector fields. *IEEE Transactions on Visualization and Computer Graphics*, **2,** 266–279.

Wong, P.C. and R.D. Bergeron, 1997, Multivariate Visualization Using Metric Scaling. *Proceedings of the 8th IEEE Visualization '97 Conference,* Phoenix, AZ, ACM Press, 111–118.

Zhang, X. and M. Pazner, 2004, The icon imagemap technique for multivariate geospatial data visualization: approach and software system. *Cartography and Geographic Information Science*, **31,** 29–41.

13 Toward Knowledge Discovery about Geographic Dynamics in Spatiotemporal Databases

May Yuan

CONTENTS

13.1 INTRODUCTION

This chapter proposes spatiotemporal constructs and a conceptual framework to lead geospatial knowledge discovery beyond what is directly recorded in databases. While knowledge discovery is fundamentally a data-driven approach to elicit novel, previously unknown patterns from massive, heterogeneous data, what we can discover from a database is constrained by what can be conceptualized and therefore represented in the database. Just as analytical possibilities for the data depend on the chosen representation schemes (Miller and Wentz 2003), the knowledge that can be discovered from data records is limited to patterns and rules of the data objects represented in the employed data models. Conventional geospatial data models adopt space-centric representations. Geospatial facts are recorded based on geometry and location, and, therefore, the knowledge that can be discovered is constrained to patterns and relationships derived from geometry and location.

Going beyond what is directly recorded in spatiotemporal databases, geospatial knowledge discovery seeks geographic constructs at a higher level of conceptualization than location and geometry. In addition to discerning spatiotemporal clustering of points representing locations of infected cases, for example, higher-level concepts, such as the spread of infectious disease and movements of infected individuals in relation to the spatiotemporal behavior of the spread, can provide a richer field for knowledge discovery to facilitate understanding and decision making.

Geographic dynamics are considered here as the umbrella concept of high-level spatiotemporal constructs. Dynamics is a common term used to characterize the work of forces that drive the behavior of a system and the system's components individually and collectively. Dynamics in an ecological system, for example, are reflected by interactions among species and habitats, and their feedback mechanisms across spatial and temporal scales. Dynamics result in change. Changes to one system component may trigger adjustments to the behavior and interactions of the other components and, ultimately, the system's dynamics. Likewise, geographic dynamics give rise to the spatiotemporal behavior of components and their interactions and consequently produce changes in a geographic system.

Nevertheless, geographic dynamics are a complex concept, and it is challenging to categorize spatiotemporal constructs that are universal to serve the knowledge needs for all domains. The premise here is that activities, events, and processes are general spatiotemporal constructs of geographic dynamics. Attempts to understand geographic dynamics can be achieved by formulating knowledge about activities, events, or processes involved in the domain of interest. Therefore, discovery of knowledge about geographic dynamics eventually aims at synthesizing information about activities, events, or processes, and through this synthesis to obtain patterns and rules about their behaviors, interactions, and effects.

While there are no unified definitions for activities, events, and processes among disciplines, the next section (Section 13.2) discusses the key distinctions and concepts associated with each spatiotemporal construct with attempts to highlight the needed information for each construct. With the spatiotemporal constructs, the concept of geographic dynamics are extended from geographic space in a general GIS analysis framework (Section 13.3). Considerations are elaborated to ensure a meaningful assemblage of these constructs (Section 13.4), and a conceptual framework is proposed to guide the elicitation of geographic dynamics from spatiotemporal databases (Section 13.5). The concluding section synthesizes key ideas, contributions, and research challenges (Section 13.6).

13.2 SPATIOTEMPORAL CONSTRUCTS OF GEOGRAPHIC DYNAMICS: ACTIVITY, EVENT, AND PROCESS

Philosophical, conceptual, cognitive, and methodological fundamentals of space and time have been discussed broadly in temporal GIS literature (Langran 1992; Egenhofer and Golledge 1998; Peuquet 2002). Important concepts are examined extensively, such as the continuous, discrete, fuzzy, linear, cyclic intermittent, instantaneous, and persistent nature of space and time. While the current GIS technology lacks robust representations and functions to adequately handle the full spectrum of geographic dynamics, many researchers have proposed spatiotemporal data models over the last two decades to capture information about activities (Wang and Cheng 2001; Frihida et al. 2002), events (Peuquet and Duan 1995; Worboys and Hornsby 2004; Worboys 2005; Beard 2007), and processes (Yuan 2001; Pang and Shi 2002; Reitsma and Albrecht 2007). Most publications, however, only address one of the three spatiotemporal constructs. This section attempts to discuss the three constructs in the context of geographic dynamics exploration.

A *geospatial activity* is an action taken by one or more agents, such as going to school, shopping, or jogging. Agents may be any active beings, including entities or phenomena. Humans, animals, volcanoes, rivers, and cultural groups are a few examples. Information about an activity provides a summary of an agent and actions that invoke the activity. Key information to characterize an activity may include *activity type, agent, action, when, duration, frequency, where,* and *outcome.* Many organizations or clubs provide templates for members to keep activity logs for time management analysis or other applications (Figure 13.1). With the combined popularity of global positioning systems (GPS), GIS services, and agent-based modeling, activity-based research has grown significantly over the last five years by tracking the movement of individuals, including animal movements (Bennett and Tang 2006). Activity-based analysis is also particularly common in travel pattern analysis and transportation research (Kwan 2004; Charypar and Nagel 2005; Chen et al. 2006; Davidson et al. 2007).

Some geographic activities are stationary, but others may be mobile. Stationary activities may be held simultaneously at multiple locations. For example, votes may be cast simultaneously at multiple polling stations. Other stationary activities are doing homework, sewing, and cooking. Points or a group of points can serve as the spatial representation of stationary activities. For mobile activities, like shopping or jogging, chains of points or point groups will be needed to denote locations over time. However, whether an activity is stationary or mobile depends on the scale of observation. Cleaning a house is a stationary activity at the scale of houses but is a

DEPARTMENT OF
HOME LAND SECURITY
U.S. COAST GUARD
ANSC-7029 (2-05)

U.S. COAST GUARD AUXILIARY

MEMBER ACTIVITY LOG

Division ___ Flotilla ___

AIXDATA USE ONLY

SUBMIT AT LEAST MONTHLY-USE THIS FORM FOR ACTIVITIES NOT REPORTED ON OTHER AUXDATA REPORTS.

SECTION I–MEMBER INFORMATION

MEMBER ID	LAST NAME AND INITIALS

Hours: ___

Activity: UMS Mission: 99

SECTION II–ACTIVITY INFORMATION

	DATE DDMMM	TYPE & LOCATION OF ACTIVITY	HOURS ACTIVITY	PREP	TRAVEL	TOTAL
1						
2						
3						

FIGURE 13.1 An example of activity logs.

mobile activity at the scale of rooms. Likewise, the temporality of an activity depends upon the scale of observation. Daily activities may be overlooked when data are collected monthly, for example. Knowledge about activities emphasizes the agents, their actions, and outcomes of their actions. Exploring activities taken by an individual can reveal the whatabouts and whereabouts of the individual over a period. Besides tracking applications, results can help individuals with time management. Exploring activities among a group provides information about the location and frequency of the group activities, as well as opportunities for additional group activities.

A geospatial event is a notable occurrence when certain conditions have been met. Notability depends upon the scale of observation in space and time. These conditions may be environmental and activity-oriented. To hold a meeting, we need conditions in which participants can exchange ideas and discuss issues; a game requires conditions to enable competitors to participate in a contest; or a traffic accident happens when conditions trigger a vehicle collision. Information about an event provides a spatiotemporal summary of what has happened. The summary may include the preconditions, the components involved, and the consequences. As an example, Figure 13.2 shows a tornado event report archived at the U.S. National Climate Data Center (http://www.ncdc.noaa.gov).

Key information to characterize an event may include *event type*, *cause* or *trigger*, *participants*, *when*, *duration*, *periodicity*, *where*, and *consequence*. Event-based

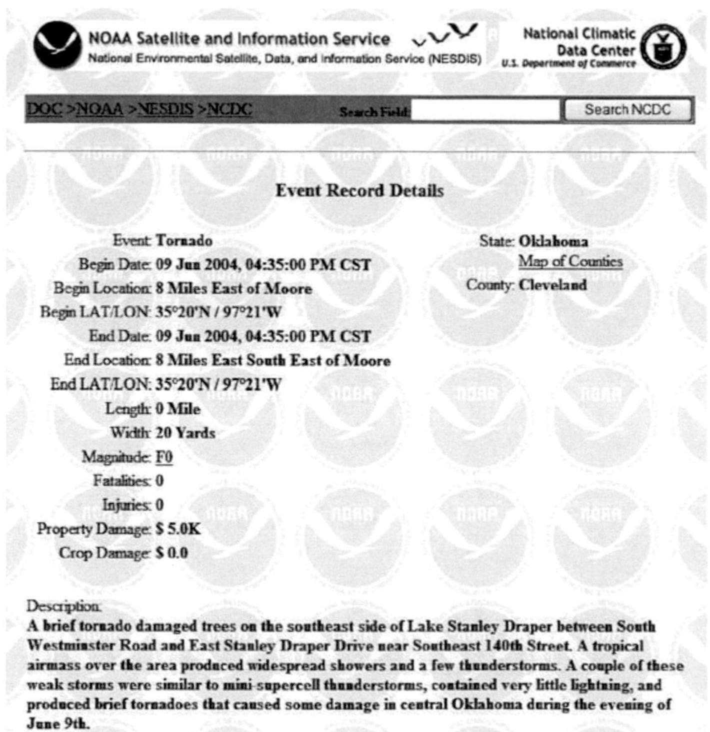

FIGURE 13.2 An example of event reports.

approaches are common in geosciences (Guo and Adams 1998; Minocha et al. 2004), operation research (Glen 1996), computer architecture and programming (Galdámez et al. 1999; Hermanns et al. 2005; Belli et al. 2007), linguistics and media analysis (Lund 2000; Zelnik-Manor and Irani 2001; Ou et al. 2006), political science and history (McLaughlin et al. 1998; Boehmke 2005), and many other fields.

While an activity requires one or more active beings to act, events take place when conditions permit. Exploring events seeks notable happenings that generate outcomes. While outcomes may not be apparent right after an event, outcomes are as important as the event itself. Like activities, events can be stationary or mobile; events can be intermittent or continuous; and events can reoccur in space and time. These spatiotemporal characteristics are also applicable to the consequences of an event or interactions of events. Discovery of knowledge about events can help us understand what happened, where it happened, how often it happened, what initial or boundary conditions afford the happening, and what outcome was produced.

A geospatial process is a progression of continuous or discrete states and phases of a dynamic system. A carbon cycle process, for example, is driven by primary production, respiration, and fossil fuel emission. Carbon fluxes cycle through several carbon reservoirs in plants, soils, fossil fuels, atmosphere, and ocean where deforestation, land use change, and industrial combustion provide the forcing necessary to carry out the cycle (Figure 13.3). Being carried out by a series of actions or events, a geospatial process may transform the system continuously from one state-phase to another with smooth transitions. In contrast, a geospatial process may exhibit stepwise development or stochastic behavior. Information about a process needs to capture every sequence of state-phases and transitions over space and time to elaborate on the sequential changes of states and phases in a geographic system.

Key information to characterize a process may include *process type*, *drivers*, *state*, *phase*, *phase duration*, *phase transition*, *when*, *where*, and *becoming*. Understanding

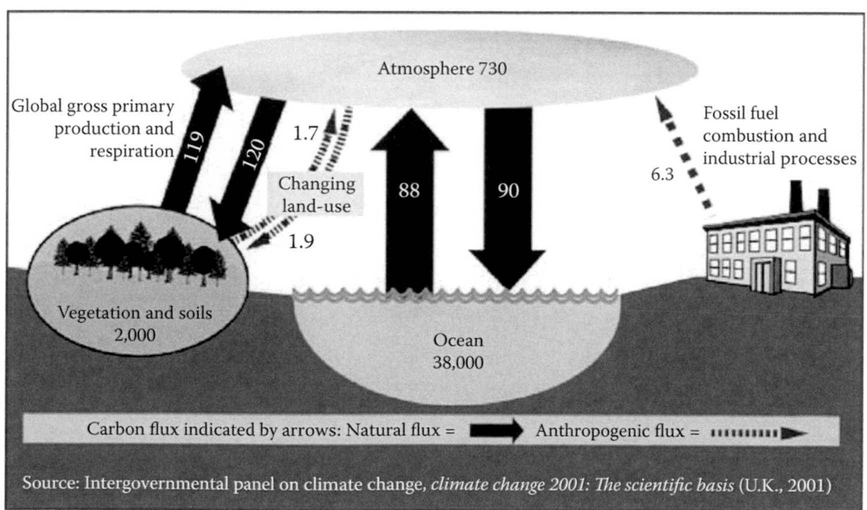

FIGURE 13.3 An example of geospatial processes.

processes is arguably one of the key motivators to all scientific inquiries. Process-based modeling is fundamental to all natural (Casalí et al. 2003), biological and ecological (Friend et al. 1997), social and political (Szayna 2002), and behavioral sciences (Bloch et al. 2005). In addition, sociologists and historians also aim to understand historical processes (Clemens 2007; Dreassi and Gottard 2007; Johnson 2007), and business managers strive to understand production and sales processes (Moreno and Roberto 1995; Greasley 2003).

Compared to activities and events, processes emphasize the path of becoming. Along the path, there are states and phases that transition from one to the next. The path can be linear (e.g., lifelines), cyclic (e.g., carbon cycle), or spiral (e.g., community planning). States summarize the characteristics of a process at a given time or period. Phases indicate the developmental stages of distinct periods. The distinction may be determined by significant changes in states or functions of the process.

While research usually addresses activities, events, or processes separately, the three spatiotemporal constructs are scale-dependent in the context of geographic dynamics (Yuan and Hornsby 2007). Activities, events, and processes can be considered as different levels of abstraction and they interact. For example, jogging is considered as an activity when we emphasize the action (e.g., we went jogging two hours ago). Jogging can be an event, if the emphasis is on happening (e.g., someone jogged across the street). Likewise, jogging can be considered a process with an emphasis on stages of jogging (e.g., Mary has a good jogging technique. She begins at a slow pace, raises her knees with each step, lands on her heel and pushes off for the next step with the ball of her foot, keeping her arms relaxed, bent at no less than a 90° angle, and swings them gently with each step, breaths easily and deeply, and keeps her head up). A jogging event (such as a marathon) involves many joggers doing jogging activities, and each jogger may practice different processes of jogging techniques. In addition, a process may consist of many events that drive phase transitions in the process, such as a community planning process that is driven by several stakeholder meetings and activities within each and between each meeting. On the other hand, an event may consist of many processes and activities such as a holiday parade that has different stages of floats with activities to entertain the bystanders.

With the three spatiotemporal constructs, knowledge discovery of geographic dynamics can inquire as to what happened (event), how did it happen (process), and what were the agents and actions involved (activities). For example,

- What are the geospatial activities, events, and processes in the geographic system of interest? How do they relate?
- What are the geographic components that may be influenced by these activities, events, and processes? How do the influences manifest themselves in space and time? How can the influences be observed and measured quantitatively or qualitatively?
- What are similar activities? What are the common settings in which similar activities take place? (Similar questions to events and processes.)
- In retrospect, what synthesis of geospatial activities, events, and processes is needed to improve understanding of their behavior and relationships?

A conceptual model integrating activities, events, and processes will free any implicit or explicit pre-assumptions of what geographic dynamics are to be discerned from a spatiotemporal database. Depending upon one's emphasis in understanding geographic dynamics, discovery of knowledge about geographic dynamics can explore activities, events that trigger these activities, and processes in which these activities evolve. Alternatively, one can center on processes to explore events that perturb these processes and activities that system components take to adjust the perturbations. The following section incorporates activities, events, and processes into an existing conceptual framework for spatial analysis of geographic space and suggests a logical extension of the conceptual framework to geographic dynamics.

13.3 EXTENDING GEOGRAPHIC SPACE TO GEOGRAPHIC DYNAMICS IN A GIS ANALYSIS FRAMEWORK

Similar to the inherent tension between spatial analysis and representation (Miller and Wentz 2003), knowledge discovery is also restricted by the selective approximation of reality captured by the chosen representation scheme. Expanding upon the representation framework proposed by Miller and Wentz (2003), Figure 13.4 illustrates a conceptual framework for geographic dynamics. The geographic measurement framework provides locations and spatial measurements based on geographic observations. When summarizing locations and measurements at locations over attributes, higher-level spatial constructs emerge as forms (or features) and relationships that give meaning to geography. The expanded conceptual framework of geographic dynamics takes what is identified in geographic space as project activities, events, and processes through sequencing footprints (e.g., forms and relationships in geographic space) to recognize change and movement over space and time. Activities, events, and processes drive change and movement. Change and movement meanwhile serve as observables of geographic dynamics. Nevertheless, observations of change and movement result from comparisons of differences in footprints over time.

As the geographic measurement framework supports geographic space where geographic forms and relationships emerge, geographic space supports geographic dynamics where geographic footprints of forms and relationship change (or move) over time and suggest driving forces motivated by activities, events, or processes. Implicitly, the support of the geographic measurement framework to geographic space relies on prior identification of phenomena or entities and associated sets of measurable attributes. There is a rich set of GIScience literature that provides in-depth discussion on geographic phenomena and entities (Couclelis 1992; Goodchild 1992; Burrough and Frank 1996; Peuquet et al. 1998; Smith and Mark 1998; Mark et al. 1999; Bennett 2001). For the support of geographic space to geographic dynamics, activities, events, and processes provide the necessary links from spatial to dynamic constructs.

Activities, events, and processes create footprints in space and time, and by sequencing these footprints, the space–time path and evolution of the geographic dynamics can be reconstructed. For activities, footprints indicate the locations of an agent, and sequencing these footprints therefore follows the agent. For an event,

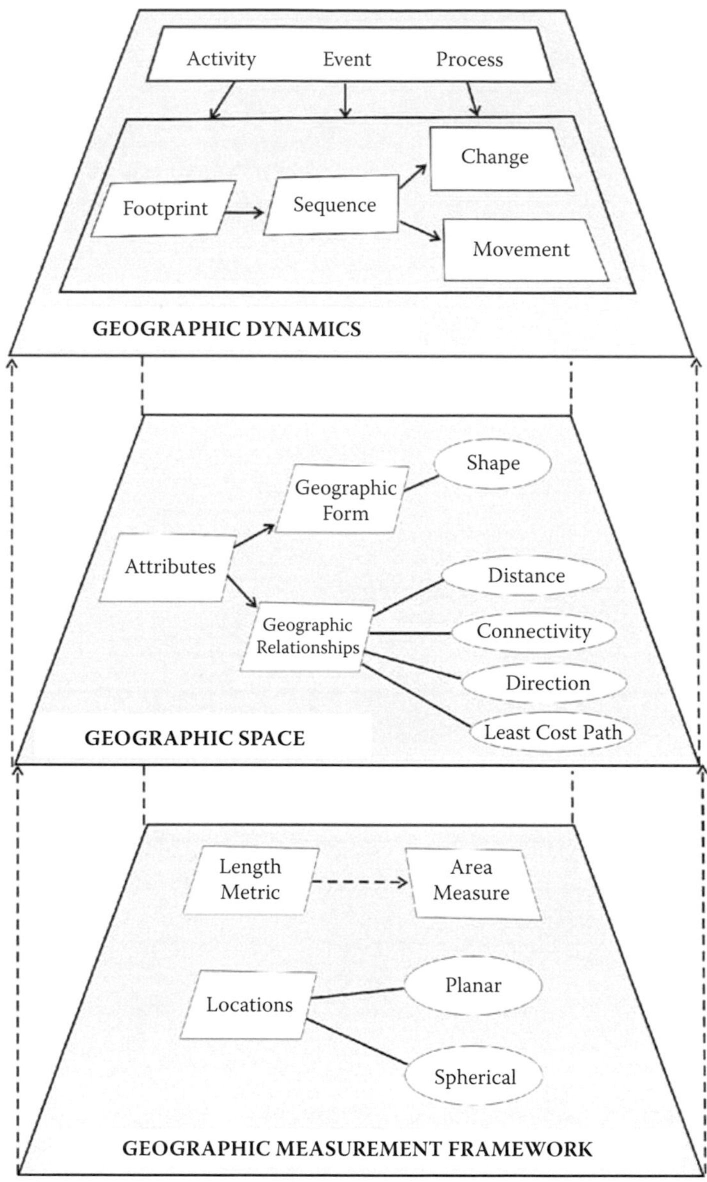

FIGURE 13.4 A general GIS analysis framework for geographic dynamics and geographic space (expanded from Miller and Wentz 2003).

footprints are aggregates of all locations where the event occurred. Aggregated footprints provide a space–time composite view of the happening, and summarize its outcomes and consequences in the environment. As to processes, footprints represent states. Footprints of a process should be sequenced based on the spatiotemporal resolution of the data and spatiotemporal characteristics of the process. A two-stage strategy may be useful: first, connect sets of similar footprints to form phases, and second, connect phases of footprints to reconstruct the development of the process.

Once footprints are established and sequenced, information about change and movement can be synthesized to characterize the geographic dynamics. Change has been a key concept in temporal GIS research. Key categories of change include no change, appearance, disappearance, expansion, contraction, deformation, split, merge, displacement, rotation, convergence, and divergence. These categorization schemes serve as the foundation for change analysis. A next step is to address how to sequence footprints to facilitate the recognition of changes resulting from activities, events, and processes and investigation of their behaviors and interactions.

Nevertheless, determination of footprints from the same geographic dynamics is a nontrivial task. The next section suggests considerations useful to sequence footprints for the development of knowledge discovery algorithms. Effective algorithms for knowledge discovery of geographic dynamics will first probe footprints resulting from the same geographic dynamics and order them in proper sequences to assemble spatiotemporal traces of the geographic dynamics. Next, the conceptual framework will provide guidelines to characterize spatiotemporal characteristics of the sequential footprints and to discern connections of geographic dynamics of different kinds in an automatic means.

13.4 SEQUENCING CONSIDERATIONS IN ASSEMBLING SPATIOTEMPORAL CONSTRUCTS OF GEOGRAPHIC DYNAMICS

There are three broad dimensions in the consideration of sequencing legibility: property, existence, and structure. The three dimensions assume predefined geographic systems or system components with which property, existence, and structure can be associated. Geographic dynamics are perceived and interpreted through recognition of changes to any of the dimensions. Therefore, legible sequencing footprints are critical to identify change. As geographic observations and measurements are taken at a discrete space and time, knowledge construction about geographic dynamics builds upon internal and external changes to property, existence, and structure to develop computational solutions for discovery of geographic dynamics knowledge.

The three board dimensions of property, existence, and structure are used here to determine footprint sequential legibility. Each of the dimensions has internal and external components. Internal components are essential to identifying the continuation of an activity, an event, or a process. External components, on the other hand, provide situational characteristics of the focal geographic dynamics and its environment. For example, some *internal properties* of a hurricane include a well-defined surface circulation and sustained winds exceeding 119 km/h (74 mph). Change to internal properties requires a definition change, which should only be led by domain experts. Internal properties are critical to the identification of footprints for the geographic dynamics of interest. Based on the internal properties listed here, footprints of a hurricane can be identified based on delineations of wind fields from weather balloon data, radar data, Geostationary Satellite Server (GEOS), and other satellite images. The speed constraint of a hurricane offers a logical basis for linking hurricane footprints to form a sequence of the hurricane and show its extent over space and time.

Hurricane forecasts, hurricane path, and socio-economic damage are examples of *external properties.* Change to external properties has no effect on the definition and identification of the focal geographic dynamics. External properties characterize the environment and effects external to the dynamics of interest. For example, timely and accurate hurricane forecasts are critical to preparedness and emergency planning. However, changing hurricane forecasts only reflects a change in the forecaster's judgment but does not change the hurricane being forecasted. When following forecasts along the development of a hurricane, lessons can be learned to improve hurricane forecasts as well as emergency responses to forecasts. Similarly, hurricane paths and socio-economic damage data are important for hurricane impact analysis. Both are outcomes of a hurricane, not what defines the hurricane.

An *internal existence* specifies the location and time of possible existence of the geographic dynamics of interest. For example, hurricanes occur during the summer or early fall over the North Atlantic Ocean, the Caribbean Sea, or the Gulf of Mexico. When a hurricane moves toward higher latitudes or over land, colder air, less moisture, and greater wind shear can weaken and eventually dissipate the hurricane. Internal existence sets spatial and temporal bounds for geographic dynamics. Derivation of the bounds may be empirical or theoretical. Hurricanes are examples of geographic dynamics with theoretically defined bounds. In contrast, many habitats or forge ranges of species are mostly empirically defined.

An *external existence* denotes the development of geographic dynamics such that a hurricane starts with a tropical disturbance with building rain clouds, grows to a tropical depression with a circular pattern, develops into a tropical storm with winds exceeding 61 km/h (38 mph), and eventually becomes a hurricane. In other words, internal existence characterizes the spatial and temporal constraints for the geographic dynamics of interest to operate. External existence highlights the spatiotemporal context in which the focused geographic dynamics exist as part of a greater existence of other geographic dynamics.

An eye, circular bands of dense clouds, and counter-clockwise rotation in the Northern Hemisphere and clockwise rotation in the Southern Hemisphere characterize the *internal structure* of a hurricane. In addition, internal structure addresses temporal relationships among the spatial components. When the eye of a hurricane widens, the winds decrease and the hurricane becomes less organized. In addition to spatial and temporal relations among components, internal structures also consider spatial and temporal variances within geographic dynamics. For example, the spatial variances of precipitation and winds within a hurricane are part of the internal structure of the hurricane. Increased spatial variances reflect the hurricane becoming less organized and therefore weakening.

An *external structure* depicts spatial or temporal connections between geographic dynamics and their environment. For example, a hurricane can cause damaging winds, flooding, mudslides, and storm surges. External structures are particularly important for chain effects, which are critical to multihazard modeling. Similar to hurricanes, wildfires can cause consequent damage to the environment by removing vegetation cover and leaving land vulnerable to surface erosion. External structures also include atmospheric teleconnections associated with multiple types of geographic dynamics to form a large-scale, lasting pattern of highly positively or negatively correlated regions in temperature and pressure that reflect changes in the atmospheric wave and jet stream patterns due to pressure and circulation anomalies (Schneider 1996). The concept of teleconnections was developed to describe how patterns that last for several weeks or months may become prominent for several consecutive years and contribute to intra-annual and inter-decadal climate variability. Much potential remains to apply the concept of teleconnections to address social and economic dynamics.

Internal property, existence, and structure provide the basis for sequencing footprints of geographic dynamics by facilitating identification of footprints from the same geographic dynamics. Every activity, event, and process is defined according to domain knowledge. While the same geographic dynamics may have distinct definitions among different domains, internal properties may vary according to domain needs. Hence, spatiotemporal data may be re-purposed in knowledge discovery based on the domain interest. Snow intensity for a severe snowstorm in Buffalo, New York will be much higher than what is considered as a severe snowstorm in Norman, Oklahoma. On the other hand, definitions of traffic accidents or shopping activities are likely to be comparable anywhere.

While internal properties determine if the data meet the requirements to be considered in a footprint, internal existence ensures that the footprint is within the reasonable realm of consideration. Shopping activities can only take place when stores are open. Two people can only meet if their space–time prisms overlap (Miller 1991). Furthermore, the internal structure within a footprint suggests the development of geographic dynamics. A certain degree of continuity or systematic discrete patterns can be assumed as the internal structure of geographic dynamics holds all components together. Component features and their structure can be used to signal the state or phase taking place. When the internal structure can no longer hold, the activity, event, or process will end consequently.

TABLE 13.1
Internal and External Investigations to Discover Knowledge about Geographic Dynamics

	Property	Existence	Structure	Facility to KDD
Internal	Definitions	Spatiotemporal constraints	Components, variances, and connections	Identity, footprints, and sequences
External	Consequences and impacts	Developments and associated geographic dynamics	Other geographic dynamics across spatial and temporal scales	Connections and interactions among geographic dynamics

External property, existence, and structure, on the other hand, support connections to other activities, events, and processes. No geographic dynamics exist in isolation. Activities, events, and processes evolve with each other and with the environment. Geographic dynamics may leave spatiotemporal footprints in physical, biological, social, economic, historical, or cultural environments. These footprints may last long after the geographic dynamics have ended and therefore leave clues for retrospective investigations of the responsible activities, events, and processes. External existence exhibits the support and context of other geographic dynamics to sustain the focal activity, event, or process. Geographic dynamics at a large scale set the boundary conditions for dynamics at a small scale, such that climate processes bound weather events. On the other hand, small-scale dynamics collectively sustain large-scale dynamics. Daily activities afford lifelong accomplishments. Moreover, external structure broadly connects geographic dynamics at one space–time location that are positively or negatively correlated with dynamics at another space–time location.

The internal and external characterizations of property, existence, and structure address both intrinsic and environmental characteristics of geographic dynamics to determine the sequencing legibility of footprints. Such a conceptual approach facilitates knowledge discovery by identifying and sequencing footprints from the same geographic dynamics, then situating the dynamics in physical and human environments, and making connections to other geographic dynamics. Subsequent logical and computational implementations need to incorporate domain definitions and knowledge of the geographic dynamics of interest to construct property, existence, and structure frames of activities, events, and processes.

13.5 FROM SPATIOTEMPORAL DATABASES TO KNOWLEDGE OF GEOGRAPHIC DYNAMICS

The previous sections discussed two central ideas in the chapter: (1) activity, event, and process are three basic types of geographic dynamics constructs; and (2) each of the geographic dynamics constructs can be recognized from spatiotemporal databases and furthermore analyzed by their property, existence, and structure both internally and externally. Based on these two ideas, discovery of knowledge about geographic dynamics from spatiotemporal databases builds on procedures to identify activities, events, and processes and discern their behavior, interactions, and effects. This section outlines an approach to synthesize the two ideas for developing such procedures.

Figure 13.5 illustrates an approach that builds geographic dynamics constructs from spatiotemporal databases. Internal and external characterizations of property, existence, and structure guide the construction of activities, events, and processes from spatiotemporal data. Ontology of the geographic dynamics constructs of interest should be determined *a priori* so that property, existence, and structure can be determined according to existing domain knowledge.

Internal characterization is weighted heavier than external characterization in data subsetting that potentially include footprints of the activities, events, and

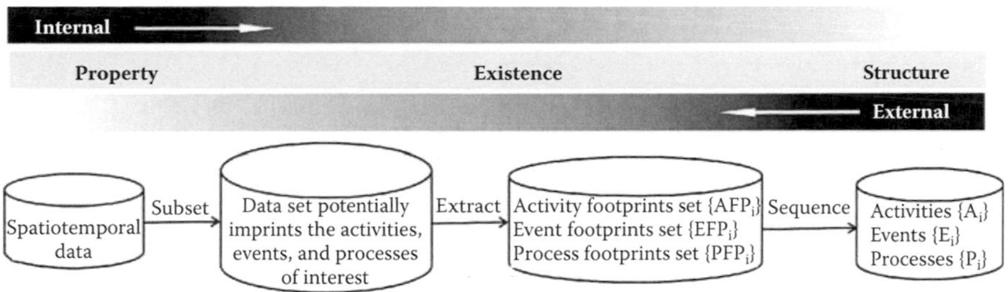

FIGURE 13.5 A conceptual framework to elicit geographic dynamics constructs from spatiotemporal databases.

processes of interest. In particular, specifications of internal properties and internal existence largely determine the potential data range for the activities, events, and processes. When a data set is compiled for specific activities, such as GPS points for the movements of individuals, data subsetting is not necessary. In other cases, data subsetting can be as easy as selecting an area and period that meet the identified bounding conditions of the focal geographic dynamics or as challenging as including all possible data sets when the bounding conditions are uncertain or very broad. In the challenging situation, data may need to be subset into similar areas and periods based on factors of the bounding conditions, e.g., spatiotemporal regions of possible existence. In the case of crime events, data sets may be subset based on social and economic variables in space and time. Crime events identified from each subset are likely to exhibit distinct internal or external structures, but all identified crime events will be incorporated into the final event data set ($\{E_i\}$ in Figure 13.5) for further analysis.

Once the potential data set has been identified, specifications of internal property, existence, and structure are used to extract footprints of the chosen activities, events, and processes. Identifying footprints should be a two-step procedure. First, internal property and existence are used to identify features that meet the property and existence criteria. Internal structural specifications should be applied to determine whether these features are part of the activities, events, or processes. Exceptions are point-based footprints to which there is no internal structure. Second, external property, existence, and structure should be used to refine these footprints. Refinements may include repositioning or reshaping footprints. For example, automobile trips should be along a road, and therefore, an off-road point should be repositioned to the nearest road. All confirmed footprints corresponding to the same type of geographic dynamics would be extracted and organized into footprint sets according to the *types* of activities, events, and processes. For example, there may be fundraising footprint sets, parade footprint sets, ice storm footprint sets, and others.

Within each footprint set, internal and external specifications should be considered to sequence footprints. Internal specifications ensure the dynamics consistency. Nevertheless, external specifications have heavier weights than internal specifications in determining whether footprints result from the same *instance* of geographic dynamics. External properties describe environmental conditions and effects, external existence depicts paths and developmental stages, and external structure makes connections to other geographic dynamics. Footprints can be attributed to the same instance of geographic dynamics when they share similar external property, existence, and structure or are in a logical transition based on the external specifications. Once determined, the outcome is a dataset with different types of activities, events, and processes, each of which consists of instances of activities, events, or processes. Each instance is a sequence of footprints. For activities of jogging, an instance will include a sequence of points to show the jogging path. For events of an oil spill, an instance will include a sequence of areas of contamination, clean-up, and broader impacts. For processes of urban sprawl, an instance will include a sequence of urbanized and suburban areas and the physical infrastructure and socioeconomic compositions within the areas.

As a result, the proposed conceptual framework represents all activities, events, and processes as sequences of footprints. As discussed in Section 13.2, activities, events, and processes emphasize distinct aspects of geographic dynamics. Their spatiotemporal constructs exhibit a progression of complexity from stationary points to mobile fields (Galton 2004; Goodchild et al. 2007), in which a stationary point can be considered as the simplest case of a mobile field with spatial and temporal extents condensed to a point. Moving objects or geospatial lifelines are commonly represented by sequences of points (Kwan 1998; Mark and Egenhofer 1998; Martin et al. 1999; Kwan 2004; Sinha and Mark 2005). Analytically, the most complicated form of spatiotemporal constructs will be a moving field with changing boundaries, capable of splitting, merging, and reincarnation, and in need of maintaining a spatial and temporal structure (which may vary over space and time) with other entities or dynamics in a large environment. Rainstorms are examples of such complicated cases (Yuan 2001). In fact, most geographic processes fall in this category. Migration, desertification, erosion, and many other geographic processes exhibit fields of spatially varied properties and progress in space and time. The idea of footprint sequences has been used to develop representation and computation solutions to analyze and mine such complicated processes (Martin et al. 1999; Stefanidis et al. 2003; McIntosh and Yuan 2005a, 2005b).

Individual sequences of footprints can be used to analyze the spatiotemporal behaviors of the chosen instance of geographic dynamics. Analysis of footprint sequences of instances from the same type of geographic dynamics can reveal trends and re-occurring patterns. Interactions among geographic dynamics of different types can be discerned from spatial and temporal correlations of their footprint sequences or spatiotemporal buffers of their footprint sequences. Furthermore, their interdependence with environmental factors can be examined along the footprint sequence (Mennis and Liu 2005; Sinha and Mark 2005).

13.6 CONCLUDING REMARKS

This chapter advocates for a step toward discovery knowledge about geographic dynamics beyond what is directly recorded in spatiotemporal databases. In addition to mining spatiotemporal clusters or other interesting patterns, knowledge discovery should move to extract high-level concepts that help us understand geographic dynamics.

Two main ideas are proposed to making the step forward. First, activities, events, and processes are basic spatiotemporal constructs of geographic dynamics. To enable discovery of knowledge about geographic dynamics, these geographic dynamics constructs need to be extracted from spatiotemporal databases and serve as the basis for spatiotemporal analysis and mining. While definitions for activities, events, and processes are diverse, this chapter outlines distinctions among the three types of geographic dynamics constructs. Activities emphasize agents and actions taken intentionally, events emphasize occurrences, and processes emphasize transitions and operations. Spatiotemporal representation of geographic dynamics depends on the scale of observation and emphases, just as the spatial representation of a geographic feature is scale-dependent. A vehicle collision can be considered as an event with the emphasis on happening. On the other hand, it can be considered as a process with the emphasis on how it happened. Another example of jogging is discussed in Section 13.2. The idea of activities, events, and processes can extend Miller and Wentz's (2003) conceptual model of geographic space to geographic dynamics.

The second idea is the use of internal and external characterizations of activities, events, and processes to identify and sequence footprints in building the constructs of geographic dynamics from spatiotemporal databases. Both internal and external characterizations consider the property, existence, and structure of geographic dynamics. Internal characterization specifies the essential properties, boundary conditions of existence, and internal structure of component features that define the chosen geographic dynamics. Specifications of internal characterization are determined by domain knowledge. External characterization denotes environmental properties, developmental stages, and connections to other geographic dynamics in space and time. Internal characterization facilitates mostly the determination of footprints from a particular type of geographic dynamics. Complementarily, external characterization assists largely in the identification of footprints from a specific instance of geographic dynamics. Composed of both internal and external characterization, spatiotemporal data sets can be transformed into various types of activities, events, and processes as well as instances within each type of these geographic dynamics.

The types and instances of activities, events, and processes serve as the spatiotemporal constructs for discovery knowledge about geographic dynamics. Sequences of footprints at varying degrees of complexity are common structures of activities, events, and processes. Several researchers have developed analytical or visual methods to explore sequences of footprints in the form of lifelines (Mark and Egenhofer 1998), trajectories (Pfoser et al. 2003), helixes (Stefanidis et al. 2003), and field-objects (McIntosh and Yuan 2005b). Knowledge discovery of these footprint sequences can be considered at four levels: (1) mining instances to reveal spatiotemporal behavior, (2) mining types to uncover trends and re-occurring patterns in space and time, (3) mining multiple types of geographic dynamics to identify their interactions, and (4) mining instances with environmental variables to recognize interdependence between the type of geographic dynamics and the environment.

While the proposed framework is only one of many possible approaches to advancing knowledge discovery of geographic dynamics, the approach is based on a general consideration of activities, events, and processes, expands upon an established conceptual framework of geographic space, and extends it to geographic dynamics. Specific frameworks of activities, events, and processes are necessary to satisfy particular domain applications. However, the proposed framework is intended to provide a common conceptual basis for most, if not all, domain applications to support large-scale data integration of all types of geographic dynamics. Subsequently, analytical or computational methods can be developed to mine relationships and interactions among multiple types of geographic dynamics. Nevertheless, the proposed conceptual framework is only one step toward advancing discovering knowledge of geographic dynamics. Research challenges remain to test and refine the conceptual framework, compile existing methods, and develop new algorithms to analyze and mine footprint sequences.

REFERENCES

Beard, K., Ed. (2007). Modeling change in space and time: an event-based approach. *Dynamic and Mobile GIS: Investigating Changes in Space and Time*. Boca Raton, FL: CRC/Taylor & Francis.

Belli, F., A. Hollmann, et al. (2007). Modeling, analysis and testing of safety issues — an event-based approach and case study. *Lecture Notes in Computer Science* (4680): 276–282.

Bennett, B. (2001). Space, time, matter and things. *The International Conference on Formal Ontology in Information Systems*. Ogunquit, ME, ACM Press, 105–116.

Bennett, D.A. and W. Tang (2006). Modelling adaptive, spatially aware, and mobile agents: Elk migration in Yellowstone. *International Journal of Geographical Information Science* 20(9): 1039–1066.

Bloch, A., D. Krob, et al. (2005). Modeling commercial processes and customer behaviors to estimate the diffusion rate of new products. *Journal of Systems Science and Systems Engineering* 14(4): 436–453.

Boehmke, F.J. (2005). Event history modeling: A guide for social scientists. *Perspectives on Politics* 3(2): 366–368.

Burrough, P.A. and A.U. Frank, Eds. (1996). *Geographic Objects with Indeterminate Boundaries*. London: Taylor & Francis.

Casalí, J., J.J. López, et al. (2003). A process-based model for channel degradation: Application to ephemeral gully erosion. *Catena* 50(2–4): 435–447.

Charypar, D. and K. Nagel (2005). Generating complete all-day activity plans with genetic algorithms. *Transportation* 32(4): 369–397.

Chen, X., J.W. Meaker, et al. (2006). Agent-based modeling and analysis of hurricane evacuation procedures for the Florida Keys. *Natural Hazards* 38(3): 321–338.

Clemens, E.S. (2007). Toward a historicized sociology: Theorizing events, processes, and emergence. *Annual Review of Sociology* 33(1): 527–549.

Couclelis, H. (1992). People manipulate objects (but cultivate fields): Beyond the raster-vector debate in GIS. *Theories and methods of spatio-temporal reasoning in geographic space*. A.U. Frank, I. Campari, and U. Formentini, Eds. Berlin: Springer Verlag, 65–77.

Davidson, W., J. Freedman, et al. (2007). Synthesis of first practices and operational research approaches in activity-based travel demand modeling. *Transportation Research Part A: Policy and Practice* 41(5): 464–488.

Dreassi, E. and A. Gottard (2007). A Bayesian approach to model interdependent event histories by graphical models. *Statistical Methods and Applications* 16(1): 39–49.

Egenhofer, M.J. and R.G. Golledge, Eds. (1998). *Spatial and Temporal Reasoning in Geographic Information Systems*. New York: Oxford University Press.

Friend, A.D., A.K. Stevens, et al. (1997). A process-based, terrestrial biosphere model of ecosystem dynamics (Hybrid v3.0). *Ecological Modelling* 95(2–3): 249–287.

Frihida, A., M. Thériault, et al. (2002). Spatio-temporal object-oriented data model for disaggregate travel behavior. *Transactions in GIS* 6(3): 277–294.

Galdámez, P., D. Murphy, et al. (1999). Event-based techniques to debug an object request broker. *Journal of Supercomputing* 13(2): 133.

Galton, A. (2004). Fields and objects in space, time, and space-time. *Spatial Cognition and Computation*(1): 39–68.

Glen, D.E. (1996). An event-based approach to modelling intermodal freight systems. *International Journal of Physical Distribution & Logistics Management* 26(6): 4.

Goodchild, M.F. (1992). Geographical data modeling. *Computers & Geosciences*. 18**:** 401–408.

Goodchild, M.F., M. Yuan, et al. (2007). Towards a general theory of geographic representation in GIS. *International Journal of Geographic Information Science* 21(3): 239–260.

Greasley, A. (2003). Using business-process simulation within a business-process reengineering approach. *Business Process Management Journal* 9(4): 408–420.

Guo, Y. and B. J. Adams (1998). Surface water and climate — hydrologic analysis of urban catchments with event-based probabilistic models, 2, Peak discharge rate (Paper 98WR02448). *Water Resources Research* 34(12): 3433.

Hermanns, M.A., B. Mohr, et al. (2005). Event-based measurement and analysis of one-sided communication. *Lecture Notes in Computer Science*(3648): 156–165.

Johnson, D. (2007). Connectivity in antiquity: Globalization as long-term historical process. *American Anthropologist* 109(1): 217–218.

Kwan, M.-P. (1998). Space-time and integral measures of individual accessibility: a comparative analysis using a point-based framework. *Geographical Analysis* 30(3): 191.

Kwan, M.-P. (2004). GIS methods in time-geographic research: geocomputation and geovisualization of human activity patterns. *Geografiska Annaler, Series B: Human Geography 86,* 4: 267–280.

Langran, G. (1992). *Time in Geographic Information Systems*. London: Taylor & Francis.

Lund, P.D.N.J. (2000). Assessment of language structure: from syntax to event-based analysis. *Seminars in Speech and Language* 21(3): 267.

Mark, D.M. and M.J. Egenhofer (1998). Geospatial lifelines. *Integrating Spatial and Temporal Databases*. O. Gunther, T. Sellis and B. Theodoulidis, Eds. Germany, Schloos Dagstuhl.

Mark, D.M., B. Smith, et al. (1999). Ontology and geographic objects: An empirical study of cognitive categorization. *Spatial Information Theory: A Theoretical Basis for GIS*. C. Freksa, and D.M. Mark, Eds. Berlin: Springer-Verlag.

Martin, E., G.T. Ralf Hartmut, et al. (1999). Spatio-temporal data types: an approach to modeling and querying moving objects in databases. *GeoInformatica* V3(3): 269–296.

McIntosh, J. and M. Yuan (2005a). A framework to enhance semantic flexibility for analysis of distributed phenomena. *International Journal of Geographic Information Science* 19(10): 999–1018.

McIntosh, J. and M. Yuan (2005b). Assessing similarity of geographic processes and events. *Transactions in GIS* 9(2): 223–245.

McLaughlin, S., S. Gates, et al. (1998). Timing the changes in political structures: A new polity database. *Journal of Conflict Resolution* 42(2): 231–242.

Mennis, J. and J.W. Liu (2005). Mining association rules in spatio-temporal data: an analysis of urban socioeconomic and land cover change. *Transactions in GIS 9,* 1: 5–17.

Miller, H. (1991). Modeling accessiblity using space-time prism concepts within a GIS. *International Journal of Geographical Information Systems* 5(3): 287–301.

Miller, H.J. and E.A. Wentz (2003). Representation and spatial analysis in geographic information systems. *Annals of the Association of American Geographers* 93(3): 574–594.

Minocha, V.K., A. Jain, et al. (2004). Discussion of "comparative analysis of event-based rainfall-runoff modeling techniques — Deterministic, statistical, and artificial neural networks" by Ashu Jain and S.K.V. Prasad Indurthy. *Journal of Hydrologic Engineering* 9(6): 550–553.

Moreno, M. and P. Roberto (1995). A process-based view for customer satisfaction. *International Journal of Quality & Reliability Management* 12(9): 154.

Ou, S., C.S.G. Khoo, et al. (2006). Multidocument summarization of news articles using an event-based framework. *Aslib Proceedings* 58(3): 197–217.

Pang, M.Y.C. and W. Shi (2002). Development of a process-based model for dynamic interaction in spatio-temporal GIS. *GeoInformatica* 6(4): 323–344.

Peuquet, D.J. (2002). *Representation of Space and Time*. New York: The Guilford Press.

Peuquet, D.J. and N. Duan (1995). An event-based spatiotemporal data model (ESTDM) for temporal analysis of geographical data. *International Journal of Geographical Information Systems* 9(1): 7–24.

Peuquet, D.J., B. Smith, et al. (1998). The ontology of fields: Report of a Specialist Meeting Held under the Auspices of the Varenius Project. Bar Harbor, Maine, the National Center for Geographic Information and Analysis (NCGIA)**:** 42.

Pfoser, D., Y. Theodoridis, et al. (2003). Generating semantics-based trajectories of moving objects. Building temporal topology in a GIS database to study the land-use changes in a rural-urban environment. *Computers, Environment and Urban Systems* 27(3): 243–263.

Reitsma, F. and J. Albrecht, Eds. (2007). nen, A Process-Oriented Data Model. *Dynamic and Mobile GIS: Investigating Changes in Space and Time*. Boca Raton, FL: CRC/Taylor & Francis.

Schneider, S.H., Ed. (1996). *Encyclopedia of Climate and Weather*. New York: Oxford University Press.

Sinha, G. and D.M. Mark (2005). Measuring similarity between geospatial lifelines in studies of environmental health. *Journal of Geographical Systems* 7(1): 115–136.

Smith, B. and D.M. Mark (1998). Ontology and geographic kinds. 8th International Symposium on Spatial Data Handling (SDH'98), Vancouver, Canada, International Geographical Union.

Stefanidis, A., K. Eickhorst, et al. (2003). Modeling and comparing change using spatiotemporal helixes. The 11th ACM International Symposium on Advances in Geographic Information Systems, New Orleans, Louisiana, ACM Press.

Szayna, T.S. (2002). Identifying potential ethnic conflict: application of a process model. *Peace Research Abstracts* 39(2): 155–306.

Wang, D. and T. Cheng (2001). A spatio-temporal data model for activity-based transport demand modelling. *International Journal of Geographical Information Science* 15(6): 561–585.

Worboys, M. (2005). Event-oriented approaches to geographic phenomena. *International Journal of Geographical Information Science 19,* 1: 1–28.

Worboys, M. and K. Hornsby (2004). *From Objects to Events: GEM, the Geospatial Event Model. Geographic Information Science 2004,* Adelphi, MD: Springer.

Yuan, M. (2001). Representing complex geographic phenomena with both object- and field-like properties. *Cartography and Geographic Information Science* 28(2): 83–96.

Yuan, M. and K.S. Hornsby (2007). *Visualization and Computation for the Understanding of Dynamics in Geographic Domains.* Boca Raton, FL, CRC/Taylor & Francis.

Zelnik-Manor, L. and M. Irani (2001). Event-based analysis of video. *Proceedings* 2(2001): 11–130.

14 The Role of a Multitier Ontological Framework in Reasoning to Discover Meaningful Patterns of Sustainable Mobility

Monica Wachowicz

Jose Macedo

Chiara Renso

Arend Ligtenberg

CONTENTS

14.1 INTRODUCTION

Geographical knowledge of nature must come from a systemic interpretation of *patterns* as well as from the exploration of different data sources that contain empirical observations in time and space (Tiezzi 2004). Currently, a geographical knowledge discovery process is supported by computer-based environments that provide a narrow view of patterns by allowing users to be interactive, and automated information processing with many decisions made by the users about how to uncover the meaning of these patterns, or how to determine the interesting patterns from different data sources.

It is important to recognize that the discovery of a vast number of unknown patterns alone do not explain the meaning of interesting trends, relationships, and dynamics of empirical observations over space and time (Gendlin 1995). On the contrary, it is a *metaphor*, and only after it makes sense can an unknown set of patterns from a GKDD process be interpreted and understood by an expert of an application domain. In general, metaphors have been proposed as artifacts of understanding, specifically understanding one kind of conceptual domain in terms of another. They are not just a pattern or a logical form. Johnson (1987) proposes metaphors as a "concrete and dynamic, embodied imaginative schemata," which are surely not just logical patterns, images, or diagrams. Moreover, Lakoff (1987) argues that metaphors are something "nonpropositional," which should not be thought of as if they were commonalities, classes, structures, or image schemata, although we might be interested to formulate those.

In a GKDD context, metaphors can help the comprehension of what makes one pattern structurally and meaningfully different from another. Ideally, metaphors would be inferred classes by a domain expert, having a high-level or abstract reason that makes sense within a geographical knowledge domain. Metaphors will lead to the "discovery" and explanation of interesting higher-level abstraction concepts, entities, relationships, or processes within some application domain of interest. Poore and Chrisman (2006) draw attention to the fact that *information metaphors* do not relate directly to reality, but instead they are more successful when they can have the effect of structuring reality to fit a hypothesis. For example, in GIS, the *landscape-as-layer* metaphor has structured the landscape into a set of layers and, nowadays, software packages encourage organizations to collect their data according to layers. Although researchers have proposed new information metaphors such as objects (Wachowicz 1999, Peuquet 2002) and agents (Deadman 1999, Ligtenberg et al. 2004), numerous practitioners are locked into the layer metaphor for geographical representation and reasoning.

What will be the information metaphors of a GKDD process for a geographical knowledge domain as sustainable mobility? The *movement-as-trajectories* metaphor is already being used to structure the history of the past and current locations of mobile entities. Pfoser and Jensen (2001) have implemented the metaphor of trajectory as polylines consisting of connected line segments, which can be grouped accordingly to two movement scenarios, termed as unconstrained movement (vessels at sea) and constrained movement (cars and pedestrians). Another example is given by the account of the *movement-as-balance* metaphor that provides an interpretive artifact of a *balance scale* for analyzing the traffic flow of cars in the presence of transportation problems (Richmond 1998). A transportation system that operates under the conditions of free-flow will be in balance. On the contrary, if the components such as road and rail are in wrong proportions, they are out of balance, having as a result traffic that is unbearable with a need to remove the load from the roads.

The research challenge lies in mapping the discovered patterns with information metaphors of movement. For example, how movement metaphors can be used to explain the discovered patterns, in such a way so that "discovered" clusters could be understood as representing those patterns that occurred in low-density fringe growth in urban developments, which show the reduced effectiveness of public transport and increased reliance on the private car. It is still to be proven that information metaphors will enhance the likelihood that experts will "see" not only the discovered patterns of movement, but instead, they will understand their meaning, and as a result, they will "see" the interesting ones as well. However, it is already clear that information metaphors can generate a chain of commonalities and differences, not a single pattern. A better account of the role of information metaphors in a GKDD process will allow the experts to form and operate on semantic concepts of space and time, and not only on the GKDD steps (i.e., data cleaning, data mining, and interpretation of the results).

The complex relationship between information metaphors of movement and the discovered patterns of a GKDD process will remain a topic for further research. In this chapter, we outline our first effort to understand such a relationship by developing a *multitier ontological framework* that consists of three tiers for bringing together movement metaphors and patterns of a GKDD process. It is important to emphasize the role of metaphors in clarifying, naming, and structuring what might otherwise be vague and inapplicable patterns within the context of an application domain. Therefore, reasoning becomes an integral part of the discovery process, and we propose that discovery and reasoning should be studied together. This will facilitate not only the extraction of patterns from very large databases, but also exploration of the use of metaphors to infer geographical knowledge from these patterns.

Using movement metaphors to infer geographical knowledge from patterns in the context of a multitier ontological framework means that each tier should have a formal representation and a formal logical theory that support an inference task. A formal representation of concepts requires a "specification of a conceptualization," according to Gruber's definition of ontology (Gruber 1993). Formal ontologies have been studied extensively in the last decades especially in artificial intelligence (Guarino 1998, Sowa 1999). Recently, they have been exploited in several other research fields such as semantic web, bioinformatics, GIS, and many others (Fonseca and Egenhofer 1999, Smith 1998, Yuan et al. 2000). Moreover, a formal ontology is based on "a conceptualization of objects, concepts and other entities that exist in a given domain and their relationships" (Gruber 2007). They rely upon a logical theory that defines the basic concepts along with their relations, and formal axioms to constrain the possible interpretation of terms. Concepts are defined as taxonomic hierarchies of classes; thus, the most important type of relation is the *subsumption,* commonly known as an *is-a* relationship. An ontology may also include *individuals*, the ground facts or concrete objects that are instances of a class. The set of individuals represents the *knowledge base* of a formal ontology. A logical theory supporting ontological definitions by means of classes, relationships, and axioms typically also provides a reasoning engine to perform inferences.

Previous research on spatio-temporal reasoning has primarily dealt with hierarchical concepts based on static and well-defined closed environments, and unfortunately, without having them associated to a geographical knowledge discovery process. Some examples include the spatio-temporal granularity description of spatial regions (Stell 2003), and the concept of perceptual hierarchical spatial units for representing people behavior in urban environments (Reginster and Edwards 2001). The dominant view has been that these representations are hierarchically organized (McNamara, Hardy and Hirtle 1989), and the locations, objects, circumstances, and factors may be perceived and understood in separate representations, which are required accordingly to a particular situation or task (Huttenlocher et al. 1991). Most of the studies have been conducted at a specific scale level by building scenarios on the variations in urban form characteristics such as urban morphology, transportation network, availability of facilities, and density of a city and the relative location of neighborhoods. The reasoning task involved has been of deriving the most likely explanations of the known facts and assumptions about urban form characteristics, and their influence on travel behavior. Such explanations have usually pointed out to four major factors that have explained such an influence on a specific scale. They are density of development, land use mix, transport networks, and layout development (Stead and Marshall 2001).

This chapter proposes a multitier ontological framework as a geographical knowledge base for the integration of movement metaphors, reasoning tasks, and discovered patterns within a GKDD process. Section 14.2 describes the ontological foundation of a GKDD process. The multitier ontological framework is also described using the domain, application, and data ontology tiers. The formalization of the data ontology tier is described in more detail using the movement-as-trajectory metaphor. Section 14.3 provides an overview of the framework architecture and the description of the data ontology tier implementation. Section 14.4 illustrates the deductive reasoning on movement-as-trajectory metaphor by focusing on a querying reasoning task. Section 14.5 concludes this chapter and points to our future research work.

14.2 THE ONTOLOGICAL FOUNDATION FOR A GKDD PROCESS

Many different styles of GKDD processes have been proposed in the literature (see Roddick and Spiliopoulou, 1999 for a wide-ranging bibliography). The GKDD methods currently available concentrate on pre-processing and mining techniques to discover categories, clusters, outliers, and other kinds of patterns that might occur in data, rather than on mapping these patterns into knowledge structures such as information metaphors proposed in this chapter. The role of geographical knowledge representation has already been pointed out as one of support for the iterative and interactive nature of the GKDD processes (Mennis and Peuquet 2003). Parallel efforts have emphasized the complex aspects of different forms of inference (i.e., abductive, deductive, and inductive inference modes) that are required to perform the steps of a GKDD process such as pre-processing, data mining, and interpretation of the results (Gahegan et al. 2001, Wachowicz et al. 2008, Zimmerman 2000). The consensus is when patterns are uncovered it is impossible to represent that geographic knowledge formally because there is nowhere to put them after a GKDD process is finished.

This section describes the conceptualization of a multitier ontological framework to formally represent and infer geographic knowledge with a GKDD process. A GKDD process constructed from an ontological perspective aims to integrate different reasoning tasks in a framework by mapping the complex relationship between movement metaphors and discovered patterns. The main advantage of this proposed framework is the possibility to focus on the spatial and temporal semantics of a geographical knowledge domain rather than on the GKDD steps. Geographic knowledge discovery is not a trivial process, it requires the examination of similarities and differences, interrelations, behavior, and evolution of what experts believe the world is like. This will lead to uncovering new and innovative patterns that can be understood by experts through the conceptualization of movement metaphors, which in turn can be used to infer a chain of commonalities and differences between these discovered patterns. Therefore, a multitier ontological framework plays an important role in the support of different geographical knowledge backgrounds as well as the integration of different inference modes (i.e., abductive, deductive, and inductive inference modes).

14.2.1 The GKDD Ontological Tiers

The formal ontologies counterpart of the multitier vision is the so-called *ontological layers*. Indeed, the definition of a formal ontology can be given at different abstraction levels and typically, more of them can coexist in a single ontology. The *domain ontology* represents the canonical descriptions of a given domain, or the associated classification theories in domains such as sustainable mobility, spatial planning, hydrology, or any other domain of interest. Domain ontology represents all the concepts of interest in a given domain, abstracting away from implementation and data details. What is commonly called *application (or task) ontology* represents a vocabulary or classification system that describes concepts operating in a given domain through definitions that are sufficiently detailed for capturing the semantics of that domain. GIS software designers can build systems that are tailored to users' needs based on appropriate application ontologies. For example, all the concepts representing kinds of car traffic behavior related to a specific urban area can be considered as an application ontology, and they can be used to build up explicit services to a mobility agency of such an urban area. In several applications, a formal ontology is built on top of a database or a given dataset. In this case, the ontology represents a conceptual abstraction of a data structure, analogously to a database conceptual model. These kinds of ontologies are called *data ontologies* because they are more dependent on data with respect to other ontology types. Figure 14.1 illustrates these three ontological levels of abstractions.

Another abstraction level that can possibly be on top of the domain ontology is the *upper ontology* (or foundation ontology) that models the common objects that are generally applicable across a wide range of domain ontologies. There are several standardized upper ontologies such as Dublin Core, GFO, OpenCyc/ResearchCyc, SUMO, and DOLCE (Wikipedia).

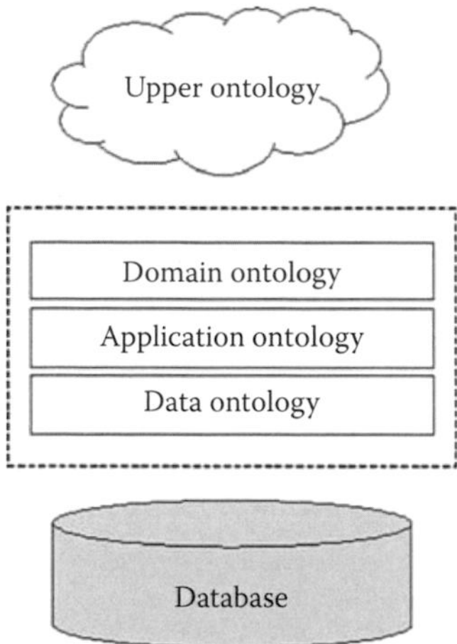

FIGURE 14.1 The three ontological tiers of a GKDD process.

Therefore, the proposed multitier ontological framework consists of layers of space and time semantic abstractions that can be described as one of the following:

- Domain Ontology Tier is based upon information metaphors as possibly defined in spatial theory.
- Application Ontology Tier is based upon information metaphors as different notions of spatial, temporal, and spatio-temporal continuity of geographical phenomena.
- Data Ontology Tier is based upon the information metaphors that time is constant, distance is relative, and there are observed states of which the only true state is that which an empirical observation is possible.

These layers also represent abstraction levels, from concrete data concepts up to more abstract conceptual notions. It is worth noticing that the border between each layer is far from clear because the same concept can be intended at different abstraction layers. Therefore, these layers are not to be intended as built one upon the other, but they represent the ontological abstraction view of a GKDD process. Indeed, a concept definition in a given layer may result from a logical expression that involves concepts belonging to other layers. For example, let us consider a traffic management application, where data include GPS positions collected from cars and a road network. A data ontology over these data may represent basic concepts such as the trajectory concept (see Section 14.2.2), the road concept, and the relations between them, i.e., a trajectory passes through a set of roads. An application ontology may represent specific concepts and definitions related to a particular traffic management application, such as the definition of traffic jam or home–work movements. The domain ontology represents all concepts typical from a traffic management domain, such as concepts representing accessibility and network infrastructure.

The main research challenge is how to formalize this multitier ontological framework for the integration of metaphors, reasoning tasks, and patterns between these layers and, in turn, to expand a GKDD process into the realms of human knowledge, information systems, and the geospatial world. Our first attempt was to establish the types of concepts, relations, and axioms at the data ontology tier, from which the mapping between the complex relationship between movement metaphors and patterns can be defined within a GKDD process. Next, we describe in detail the ontological formalism that allows representing in a uniform way this data ontology tier. The chosen formalism for representing the formal ontology is OWL (2004), which is a well-known standard that has arisen from the semantic Web. Another interesting feature of OWL is that it relies upon a family of languages known as description logics (DL), which provides an inference system based on formal well-founded semantics (Baader et al. 2003). The basic components of DL are concepts, roles, and individuals. Concepts describe the common properties of a collection of individuals and roles are binary relations between objects. Furthermore, a number of language constructs, such as intersection, union, and role quantification, can be used to define new concepts and roles. The main reasoning tasks are *classification and satisfiability*, *subsumption,* and *instance checking. Classification* is the computation of a concept hierarchy based on subsumption, whereas *instance checking* verifies that an individual is an instance of a concept. Finally, DL programs are split in two parts, the terminological box (TBox) and the assertional box (ABox). The TBox contains sentences describing concept hierarchies and relations, whereas the ABox contains the ground facts, for example, the individuals and their relations with concepts. Indeed, some inferences are related to Tbox or Abox only.

14.2.2 The Formalization of the Data Ontology Tier

A question that comes up is "Why not put all database data inside an ontology (i.e., as ABox individuals)?" This question may be answered by looking from the perspective of reasoning. Data complexity of reasoning in DL estimates the performance of reasoning algorithms measured in the size of the ABox only. Even for the very expressive DL SHIQ, satisfiability checking is data complete for Nonpolynomial (NP). Very expressive DLs such as SHIQ are interesting mainly due to their high expressivity combined with the clearly defined model-theoretic semantics and known formal properties, such as the computational complexity of reasoning. In particular, for applications with large ABoxes, the combined complexity of checking satisfiability of a SHIQ knowledge base (KB) is EXPTIME-complete in size of KB. EXPTIME-completeness is a rather discouraging result since trajectory applications have large |KB| in practice (Calvanase et al. 2007).

Before explaining the data ontology tier, it is important to notice that this tier is based on a local-as-view-based query processing approach. More specifically, we extend the approach proposed by Calvanase et al. 2007, which addresses the problem of establishing sound mechanisms for linking existing data to the instances of the concepts and the roles in the ontology. In this paper, Calvanese et al. (2007) propose a new DL, called DL-LiteA^{+}, which is the largest fragment of DL that allows for answering complex queries (namely, conjunctive queries, i.e., SQL select-project join queries, and unions of conjunctive queries) in LOGSPACE with respect to data complexity (i.e., the complexity measured only with regard to the size of the data). More importantly, they allow for delegating query processing, after a preprocessing phase, which is independent of the data, to the relational Data Base Management System (DBMS) managing the data layer.

Our proposal differentiates from the Calvanese et al. (2007) approach in two aspects. First, we map an ontology to a conceptual data model, instead of mapping to a logical relational model. Second, we do not translate an ontology query to an SQL query; by contrast, we map an ontology query into a conceptual query. Both conceptual schema and conceptual language are based on the Modelisation de Donnees Pour Applications Spatio-Temporelles (MADS) model (Parent et al. 2006a), which is a spatio-temporal conceptual model that provides algebra operators taking into account spatial, temporal, and multirepresentation features. Basically, our idea is to use a query rewriting mechanism that permits us to translate query over an ontology (i.e., which uses concepts defined within the ontology) into a conceptual query, which is a query that uses the conceptual schema vocabulary and takes advantage of spatial and temporal operators already defined. Thus, this idea permits us to reduce the semantic gap between database query language (e.g., SQL) and ontology language (e.g., OWL) since a conceptual data language is closer to the user's mental model and an ontology language is the generalization of both the object-oriented data model (e.g., UML) and the extended entity-relationship (EER) semantic data model.

14.2.2.1 The Movement-as-Trajectory Metaphor within the Data Ontology Tier

While developing a rich body of work for managing moving objects, the database research community has shown very little interest in the ontological viewpoint on moving objects, i.e., providing support for the movement metaphor of trajectories. From the ontological point of view, trajectory is the most important metaphor within the data ontology tier, and as a result, it should be manipulated as first class citizens within databases.

We propose representing trajectories from a spatio-semantic abstraction point of view (Figure 14.2). Since traveling objects do not necessarily continuously move during a trajectory (this is the case, for example, of traffic management applications),

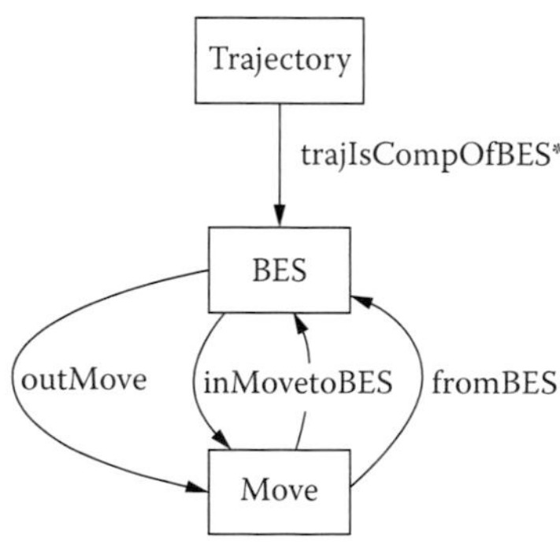

FIGURE 14.2 The formalization of movement-as-trajectory within the data ontology tier.

trajectories may themselves be semantically segmented by defining a temporal sequence of time sub-intervals where alternatively the object position changes and stays fixed. The former is called the moves and the latter is called the stops. Thus, a trajectory can be viewed as a sequence of moves going from one stop to the next (or as a sequence of stops separating the moves). For example, a car will stop in some locations along the city due to traffic congestion or traffic lights.

Definition 14.1 (Trajectory): A trajectory is the user-defined record of the evolution of the position (perceived as a point) of an object that is moving in space during a given time interval in order to achieve a given goal. trajectory : [tbegin , tend] $\rightarrow$ space

Definition 14.2 (Stop): A stop is a part of a trajectory, such that

- The user has explicitly defined this part of the trajectory ([tbeginstopx, tendstopx]) to represent a stop.
- The temporal extent [tbeginstopx, tendstopx] is a nonempty time interval.
- The traveling object does not move (as far as the application view of this trajectory is concerned), i.e., the spatial range of the trajectory for the [tbeginstopx, tendstopx] interval is a single point. All stops are temporally disjoint, i.e., the temporal extents of two stops are always disjoint.

Definition 14.3 (Move): A move is a part of a trajectory, such that

- The part is delimited by two extremities that represent either two consecutive stops, or tbegin and the first stop, or the last stop and tend, or [tbegin, tend].
- The temporal extent [tbeginmovex, tendmovex] is a nonempty time interval.
- The spatial range of the trajectory for the [tbeginmovex, tendmovex] interval is the spatio-temporal line (not a point) defined by the trajectory function (in fact, it is the polyline built upon the sample points in the [tbeginmovex, tendmovex] interval).

The formalization of such concept classes for the inference of movement-as-trajectory metaphor can be used to deduce some *a priori* knowledge from patterns, such as density clusters of points in space and time that represent stops with a very short duration of some minutes due to traffic light or stop signs. On the other hand, time-windows can be deduced from clustering the point data that represents moves, and a sequential temporal snapshot can be inferred using the linear patterns of move points. Figure 14.3 illustrates the concepts of stops and moves of the movement-as-trajectory metaphor.

14.3 ARCHITECTURE OVERVIEW OF THE DATA ONTOLOGY TIER

Before describing the framework architecture, the following section will introduce the conceptual database model used for the implementation of the data ontology tier.

14.3.1 Conceptual Database Modeling for Trajectory

Since conceptual database models are adequate for representing concepts that are closer to the data ontology tier, they are the best tools for specifying applications

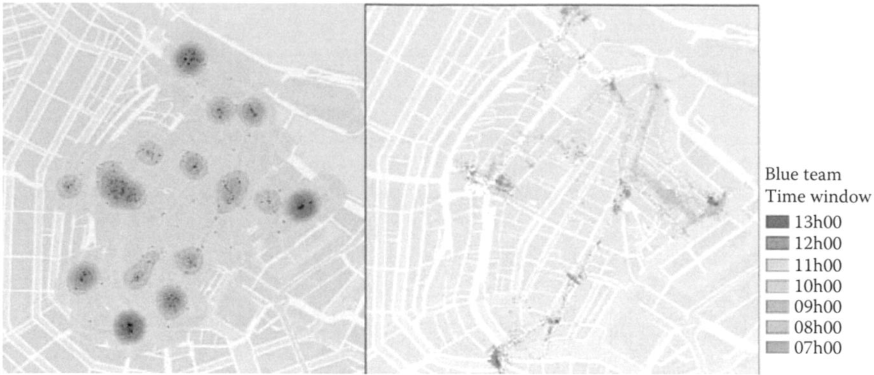

FIGURE 14.3 The visualization of (a) spatial clusters of stops and (b) the temporal clusters of moves of a group of 50 people (Blue Team) moving around the city of Amsterdam. (Data Source: Waag Society, Netherlands.) See color insert after page 148.

requirements in terms of data. However, considering complex domain applications (e.g., spatio-temporal applications), the creation of conceptual schemas becomes a hard task due to the limit expressive power of generic conceptual models. Thus, specialized conceptual models are required in order to give more comfort to the user in specifying those conceptual schemas. The MADS model (Parent et al. 2006b) is an example of this attempt in creating a model adequate for representing spatio-temporal data.

MADS is an object+relationship spatio-temporal conceptual data model. In this model, the real world of interest that is to be represented in the database is composed of complex objects and relationships between them; both characterized by properties (attributes and methods), and both may be involved in a generalization hierarchy (is-a links). Data structuring capabilities of MADS are orthogonally complemented with space and time modeling concepts, i.e., spatiality and temporality may be associated at the various structural levels: object, attribute, and relationship. The spatiality of an object conveys information about its location and its extent; the temporality describes its lifecycle. A set of predefined spatial and temporal abstract data types is used to describe the spatial and temporal extents of data.

The abstract data types are organized in a generalization hierarchy where generic data types are used to describe domains whose values may be of different, more specific types, e.g., small rivers may be described as lines, bigger ones as areas; hence, their domain should be of the generic type Geo. Attributes may also be space- or time-varying, supporting in this way the continuous view of space. Relationships are either classical n-ary relationships among individual objects or n-ary relationships among sets of objects (multiassociation). Relationships may be enhanced with one or several specific semantics, such as aggregation, topological, synchronization, and inter-representation semantics. Topological and synchronization semantics define constraints between spatial and temporal objects, respectively.

Multirepresentation has been added in MADS as an additional orthogonal dimension. Multirepresentation allows the definition in the same schema of several representations for the same real world objects. Those multiple representations may be the consequence of diverging requirements during the database design phase or, in the particular context of spatial data, of the description of data at various levels of detail. The MADS multirepresentation feature may also be used in the context of spatial database integration where the full integration, possibly based on different levels of detail, is not possible.

Therefore, MADS, as a spatio-temporal conceptual data model, is a natural candidate for supporting the movement-as-trajectory metaphor at the data ontology tier. Two alternative approaches for trajectory modeling were developed based on data types and design patterns. The former is inspired by MADS conceptual model extension using a new kind of data type. The latter resumes an old idea of creating a predefined schema that can be instantiated according to some rules. The two approaches may be used together if the application requirements suggest that both are useful in a specific reasoning task. Figure 14.4 illustrates these two different approaches.

The modeling goal of the first solution (Figure 14.4a) is to hide as much trajectory data as possible into a dedicated TrajectoryType data type, equipped with methods providing access to trajectory components (stops, moves, begin, and end). The definition of dedicated data types is a well-known technique to extend a data model to take into account new types of data. It is the technique we have used to embed support of spatial and temporal features into MADS.

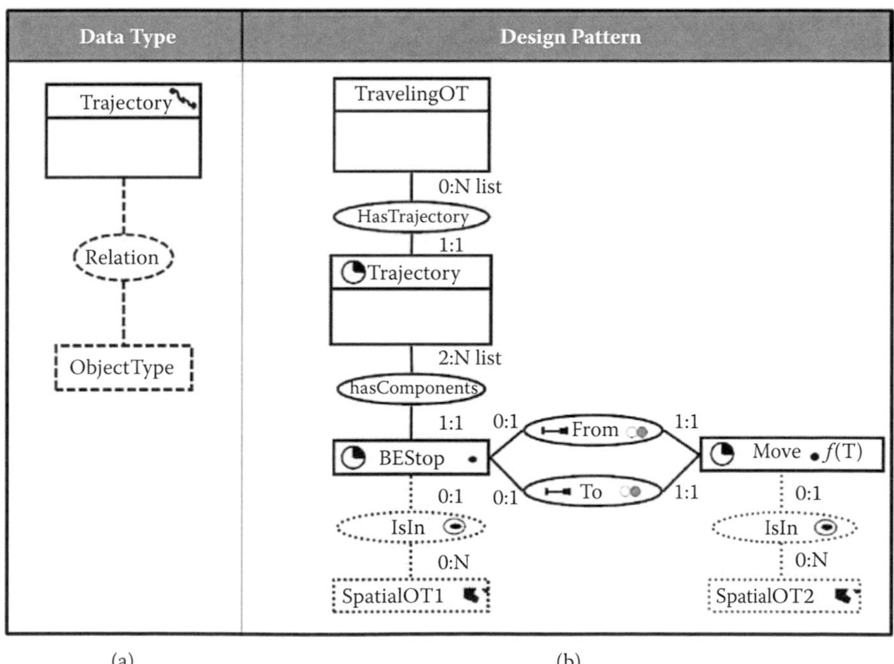

FIGURE 14.4 Trajectory conceptual database modeling approaches.

The trajectory design pattern approach (Figure 14.4b) holds object types for representing trajectories and their begin, end, stops, and moves. In the design pattern approach, it is proposed an object type group's begin, end, and stop (B.E.S) objects as instances of the same type because of their similar features. Each B.E.S object has a lifecycle, which is a simple time interval, and a geometry, which is a point. A relationship type hasComponents relates trajectories to their components. Its cardinalities enforce that each component belongs to a single trajectory, and each trajectory has at least two components (begin and end). The object type Move has a lifecycle, which is a simple time interval, and a geometry, which is a time-varying point. The two object types B.E.S and Move are related by two relationship types, From and To, materializing the fact that each move starts and ends in a stop. Both From and To bear a topological (adjacency) and a synchronization (meet) constraint enforcing that each move is linked to stops that are adjacent to it in both space and time. The lifecycle of Trajectory objects is a time interval that is inferred from the lifecycles of the first and last instances of B.E.S to which they are linked (i.e., the instants of its begin and end).

In addition to expressing the internal structure of a trajectory, the pattern includes the hooks used for its connection to application objects. In Figure 14.4, the names of the hooks are written in italics. Trajectories are linked to a hook object type TravelingOT that represents the traveling objects covering the trajectories. B.E.S may be linked to a hook spatial object type SpatialOT1 that represents the corresponding location in terms of application objects. The IsIn relationship bears a topological inside constraint. As this hook is optional, it is drawn with dotted lines in Figure 14.4. Similarly, moves may be related to a hook object type SpatialOT2 by another topological inside relationship, called IsIn, which may be used to model network-constrained trajectories.

14.3.2 Implementation

Figure 14.5 illustrates the main components of our proposed architecture for combining the data ontology tier and a spatio-temporal database. This architecture is composed of four components identified by letters A, B, C, and D. The component A represents any ontology management system (e.g., Protégé) that has a query interface for end users. This component stores and manages the data ontology tier, which describes the concepts of the movement-as-trajectory metaphor, for instance, in a traffic management application; the instance of the stop concept could be a "traffic jam." In this architecture, every concept defined in the data ontology tier must be based on the movement-as-metaphor. This is a pre-requisite for using our framework because the movement-as-trajectory metaphor has a pre-defined mapping to a MADS constructor (e.g., concepts, relationships, etc.).

The component B is responsible for translating an ontology query written in OWL-DL syntax to a MADS query. Since this component is conscious about the MADS model, it knows how to translate each spatial, temporal, or spatio-temporal concept into the MADS concept. In addition, it knows how to build a MADS query according to its internal mapping. Clearly, this component is a quite simple component because the MADS model contains each constructor defined in MADS

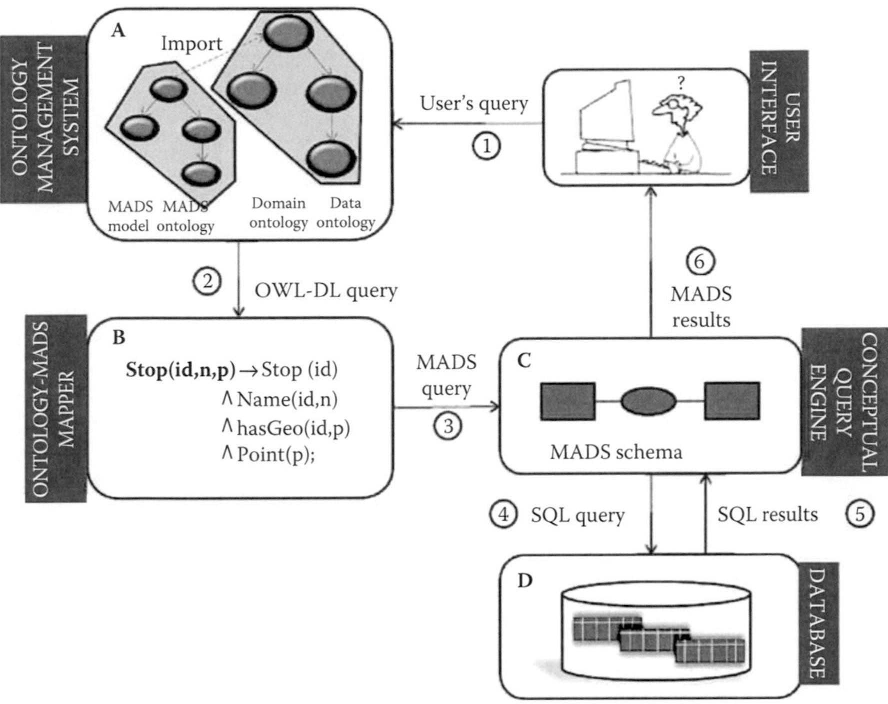

FIGURE 14.5 Framework for combining ontology and spatio-temporal database.

conceptual language. It is worth noting that the MADS model also includes new concepts for semantic trajectory specification because our main objective is to support applications that manipulate trajectory data. The concepts defined for the movement-as-trajectory metaphor (i.e., trajectory, stops, and moves) are defined also using MADS concepts. The component C stands for the conceptual query engine. This component receives a conceptual query, which is written using MADS-compliant query language syntax that translates it into a MADS query and submits it to the underlying database (component D). Then, the results are collected and sent to the user.

Figure 14.5 also illustrates a typical querying scenario where an end-user sends a query using a graphical interface. This query is sent to an ontology management system being translated to a specific ontology query language (e.g., OWL DL Query). Then, this ontology query is dispatched to the ontology-MADS Mapper where the ontology query is translated to a conceptual MADS query and sent to a MADS query engine. Then, the underlying relational database executes an SQL query returning the results, which are processed by a MADS query engine and returned to the user interface. We could mention many applications that would be benefit from our framework. Analysts and decision makers in the field, as well as end users would benefit from writing sophisticated queries without worrying about underlying database schema.

14.3.2.1 Mapping Assertions

Now we turn our attention to the problem of mapping objects in the ontology to the conceptual view of data in a database. Our approach mainly differentiates from the Calvanese et al. (2007) approach by the fact that we are mapping an ontology into MADS objects and relationships, instead of mapping to a relational database. Clearly, we will map queries using MADS algebraic operators, which implement spatial, temporal, and multirepresentation features. The mappings are composed by a set of assertions, called mapping assertions, each one of the form F → Y where F is a MADS query over conceptual schema of arity n, and Y is a DL-LiteA+ conjunctive query over DL-LiteA+ knowledge base of arity n.

A conjunctive query (CQ) q over a knowledge base K is an expression of the form q(x) ← ∃y.conj (x, y), where x are the so-called distinguished variables, y are existentially quantified variables called the nondistinguished variables, and conj (x, y) is a conjunction of atoms of the form A(x), D(x), P(x, y), UC(x, y), UR(x, y, z), or x = y, where x, y, z are either variables in x or y or constants in an alphabet of constants. A union of conjunctive queries (UCQ) is a query of the form q(x) ← Wi $\exists y_i$.conj (x, y_i).

Table 14.1 illustrates four mappings, which model the situation where every tuple (i, s, d) c trajectory, which is an object defined in the conceptual schema presented in Section 14.2.1. This tuple corresponds to a trajectory of a car whose identification is id, whose source is s, and whose destination is d. Besides, we denote that this trajectory is identified into the ontology by its id. To assist the creation of an identity for trajectory objects, a domain of object identifiers is considered that is built starting from data values, in particular as logic terms over data items. In other words, object terms are constructed by function symbols applied to data value constants. In our example, we use a function symbol trj, whereas trj(id), which is called a variable term, defines the identity of a trajectory.

Similarly, we model every tuple (id, name, geo) that corresponds to a building described by an identification, a name, and a geometry, which is a point. Following,

TABLE 14.1
Some Examples of Mapping Assertions

	F → Y	
	MADS Query Over Conceptual Schema	**Conjunctive Query Over DL-LiteA+ KB**
M1	projection [id, s, d] Trajectory	Trajectory(trj(id), s, d)
M2	projection [id, n, geo] Building	Building(bd(id), pt(geo)), hasName(bd(id), n), Point(pt(geo), geo),
M3	projection [m] selection [∃m ∈ Move m.stopsVarying = t.stopsVarying] Trajectory	Trajectory(trj(id), s, d), Move(mv(m)), has_Move(trj(id),mv(m))
M4	projection [x, y] selection [∃x, y ∈ Geometry x.geometry.distance(y.geometry) > z]	has_distance_greater(x, y, z)

concerning the mapping assertion M3, we describe a MADS query that corresponds to the moves of a trajectory. This MADS query represents a set of tuples (t, m) whose trajectory is t, and whose move is m. Note that we use the method stopsVarying of trajectory concept, which returns a set of stops belonging to the trajectory. This method permits us to compare if a move has the same set of stops of one trajectory. Finally, we describe the mapping 4 that models every tuple (x, y) where x and y are geometries. In this case, this mapping corresponds to each pair of spatial objects that distances more than z units from each other do. The corresponding DL conjunctive query is represented by a role attribute. This last mapping shows clearly that this approach can separate the semantic of concept or role in the ontology from its computation.

14.4 REASONING AND QUERYING

The reasoning task has been focused on query answering. Basic classical inferences that an ontological formalism should provide are *consistency* and *querying knowledge*. Consistency means to check whether a given concept C subsumes a concept D. Querying knowledge means to check whether a given individual is an instance to a given class. Formally, given a knowledge base K and a union of conjunctive queries q(x) over K, return the certain answers to q(x) over K, i.e., all tuples t of elements of q such that for every model I of K. First, the query answering method consists of compiling the TBox into a finite reformulation of the query that is evaluated afterward over the minimal model db (A) of the ABox. The technique for query answering over a DL-Lite+A ontology with mappings is implemented as follows:

- Each mapping assertion $F \rightarrow Y$ is split into several assertions of the form $F \rightarrow p$, one for each atom p in Y.
- The atoms in the query q are unified in all possible ways to be evaluated with the right-hand side atoms of the (split) mappings, thus obtaining a (bigger) union of conjunctive queries containing variable object terms.
- Then we unfold each atom with the corresponding left-hand side-mapping query. Observe that, after unfolding, we obtain an SQL query that can be evaluated over DB, and possibly return terms built from values extracted from DB.

For example, refer to the previous example and consider now the following query over the TBox, asking for the trajectories within a distance of 1 km from any hospital:

$$\begin{aligned}&\mathrm{Q(id)} \leftarrow \mathrm{Move(m), Hospital(h, geo), has\_distance\_greater(m, geo, 1000),}\\&\quad \mathrm{Trajectory(id, s, d), has\_Move(id, m)}\end{aligned}$$

The first step is to expand the previous query into a union of queries containing all possible concepts, roles, and role attributes, according to the knowledge specified in ontology ABox, that could provide objects to satisfy this query.

$$\begin{aligned}&\mathrm{Q(id)} \leftarrow \mathrm{Trajectory(id, s, d)}\\&\mathrm{Q(id)} \leftarrow \mathrm{has\_Move(id, m)}\\&\mathrm{Q(id)} \leftarrow \mathrm{has\_Stop(id, m)}\\&\mathrm{Q(id)} \leftarrow \cdots\end{aligned}$$

TABLE 14.2
DB Query Mapping Assertions

Atom From Mappings	Corresponding DB Query
Trajectory(trj(id),s,d)	Q1(id,s,d)
Building(bd(id),pt(geo)	Q2(id,n,geo)
hasName(bd(id),n)	Q2(id,n,geo)
Point(pt(geo),geo)	Q2(id,n,geo)
Trajectory(trj(id),s,d),	Q3(m)
Move(mv(m))	Q3(m)
has_Move(trj(id),mv(m))	Q3(m)
mv(m)	Q3(m)
has_distance_greater(x,y,z)	Q4(x,y)

These expanded queries will be used further for describing the union of all possible sets of individuals defined by tuples corresponding to some database query answering.

The second step concerns splitting each mapping assertion $\varphi \rightarrow \psi$ into several assertions of the form φ p, one for each atom p in ψ. Table 14.2 illustrates some of the mapping assertions that undergo the splitting process.

In the third step, the expanded queries showed in Table 14.2 are joined in all possible ways with the left-hand side of (split) mapping assertions presented in Table 14.2. Then, a new query over ontology TBox is generated by unifying the atoms in the query with the left-hand side atoms in the split mapping, thus obtaining:

$$\begin{aligned}&Q(id) \leftarrow Move(mv(m)), Hospital(bd(h), pt(geo)),\\ &has\_distance\_greater(m, geo, 1000), Trajectory(trj(id), s, d),\\ &has\_Move(trj(id), mv(m))\end{aligned}$$

Then, we unfold each atom with the corresponding left-hand side of the mapping query, and obtain the MADS query:

Projection[t]
selection[$\exists$m $\in$ Move $\exists$h $\in$ Hospital
(m.geometry.distance(h.geometry) > 1000) $\wedge$ [$\exists$t $\in$ Trajectory (m.stopsVarying
$\subseteq$ t.stopsVarying)]

which can be simply evaluated over the database to get certain answers to q. Table 14.3 summarizes the process of rewriting a conjunctive query over a DL-Lite+A ontology into a MADS query using mappings. The former query is translated into an intermediate query that permits its translation to a MADS query.

TABLE 14.3
Examples of Conjunctive Queries

	Queries
QUERY	Return the trajectories within a distance of 1 km from any hospital
DL CONJUNCTIVE QUERY	Q(id) ← Move(m), Hospital(h,geo), has_distance_greater(m,geo,1000), Trajectory(id,s,d), has_Move(id,m)
Rewrote Query	Q(id) ← Move(mv(m)), Hospital(bd(h),pt(geo)), has_distance_greater(m,geo,1000), Trajectory(trj(id),s,d), has_Move(trj(id),mv(m))
MADS QUERY	Projection[t] selection[∃m ∈ Move ∃h ∈ Hospital (m.geometry.distance(h.geometry)>1000) ∧ [∃t ∈ Trajectory (m.stopsVarying ⊆ t.stopsVarying)]

14.5 CONCLUSIONS

The development of a multitier ontological framework for a GKDD process is a research challenge because it addresses the integration of movement metaphors, reasoning tasks, and data management using different perspectives. In other words, the challenge is in integrating knowledge representation and data representation. The former is oriented to the semantic organization of data and powerful reasoning tasks. The latter focuses on efficient data storage and access. In particular, a GKDD process needs to cope with these two perspectives of data management because of the following reasons. First, large data sets of positioning data delivered by GPS devices must pass through an extracting, transforming, and loading process in order to remove data error and uncertainty, which will permit us to make these data available for use. Second, due to the lack of semantics in positioning raw data, which is represented by timestamp and position coordinates, there is a need for a semantic enrichment of this data. Enriching data means, for instance, associating them with existing data available at legacy systems (e.g., GIS, thematic information system, etc.). Third, a GKDD process is fundamental for extrapolating the limits imposed by the query answer process of traditional databases. To this end, positioning data must be annotated semantically using a geographical, application domain (e.g., traffic management) and background knowledge (e.g., sustainable mobility). The semantic enrichment by association and annotation of collected data is the key to supporting both efficient querying answering and reasoning tasks. Thus, a GKDD process requires efficient mechanisms for integrating knowledge representation and data representation in a seamless way.

The multitier ontological framework described in this chapter is our first attempt toward supporting both efficient querying answering and reasoning tasks on the movement-as-trajectory metaphor. Ontology languages, such as DLs, are the state-of-the-art of knowledge representation with application domains, while relational database technology is nowadays the best technology for efficient management of very large quantities of data. In addition, relational databases are the habitual destination for positioning raw data; thus, combining ontological representation with database querying mechanisms is fundamental for the use of ontologies in GKDD processes. The data ontological tier has been used to illustrate our proposed approach.

Although, we may find in the literature some solutions for combining ontologies with database (Calvanase et al. 2007; Poggi 2006), none of them cope with the problem of representing and reasoning on spatio-temporal data. More specifically, there is no logic language in the literature that provides a good expressiveness for representing spatio-temporal concepts/relationships and provides query answering over it in LogSpace with respect to data complexity. Consequently, there is a need for developing methods to permit seamless integration with spatio-temporal ontologies (i.e., domain, application, and data ontology tiers) and databases. Our approach proposes to exploit an ontological formalism that allows representing and reasoning on movement in a uniform way within different tiers, such as the data, application, and domain ontology tiers. Furthermore, adopting a logical formalism, such as DLs, allows us to perform inferences on these amalgamated data with an improved expressive power.

Our proposal focuses on the combination of spatio-temporal databases and a multitier ontology framework. In this respect, this chapter describes the results of the integration of the data ontology tier and a spatio-temporal conceptual schema. Our approach relies on two well-known formalisms: description logics and conceptual models. Spatio-temporal conceptual schemas to be integrated are specified using the MADS conceptual data model (Parent et al. 2008), which can represent rich spatio-temporal semantics. DLs are then used to model the intentional knowledge needed in real-world applications. Thus, reasoning that deals with the data (e.g., ABox reasoning) is delegated to a DBMS via a particular query answering service, while DL reasoning is used for the terminological part of the data ontology (e.g., TBox reasoning).

In summary, the main contribution of our approach is twofold, which aims at (1) reducing the semantic gap between ontological and database representation by using mappings between an ontology and a conceptual database model, and (2) permitting definition of spatio-temporal relationships within an ontology, which can be translated to spatio-temporal conceptual queries.

The conceptualization of a multitier ontological framework is meant to formally represent and infer geographic knowledge with a GKDD process. A GKDD process constructed from an ontological perspective aims to integrate different reasoning tasks in a framework by mapping the complex relationship between movement metaphors and discovered patterns. The main advantage of this proposed framework is the possibility to focus on the spatial and temporal semantics of a geographical knowledge domain rather than on the GKDD steps. Therefore, our further research will be focused on the application ontology tier where the reasoning task is characterized by inferring descriptive knowledge such as the trajectory characteristics (e.g., work-home and recreational trajectories), the geographical context where the trajectory occurs (i.e., landscape structure, transportation mode), the topological relations between the trajectories (e.g., branching, converging), and the association between the trajectories and specific features of a landscape. Such information can be a set of properties that are associated with individual trajectories themselves, or a pre-defined group of trajectories. The overall goal will be to help the experts to deduce the consequences for the existence of linear patterns of the movement of the trajectories.

ACKNOWLEDGMENTS

This research work has been funded by the FP6 STREP Programme under the GeoPKDD project, and the BSIK RGI Programme under the People in Motion Project. This research is part of the the strategic research program "Sustainable spatial development of ecosystems, landscapes, seas and regions" which is funded by the Dutch Ministry of Agriculture, Nature Conservation and Food Quality. We would also like to thank Waag Society for providing the GPS data set, Amsterdam, Netherlands.

REFERENCES

Baader, F., Calvanese, D., McGuinness, D.L., Nardi, D., and Patel-Schneider, P.F. 2003. *The Description Logic Handbook: Theory, Implementation, and Applications*, Cambridge University Press, Cambridge, U.K.

Baader, F., Küsters, R. and Molitor, R. 2000. Rewriting concepts using terminologies. Proceedings of the Seventh International Conference on Knowledge Representation and Reasoning (KR2000), Breckenridge, CO.

Calvanese, D., De Giacomo, G., Lembo, D., Lenzerini, M., Poggi, A., and Rosati. R. 2007. Ontology-based database access (Extended Abstract). *Proceedings of the 15th Italian Conference on Database Systems (SEBD 2007).*

Calvanese, D., De Giacomo, G., Lembo, D., Lenzerini, M., and Rosati, R. 2006. Data complexity of query answering in description logics. *Proceedings of the 11th International Conference on Principles of Knowledge Representation and Reasoning.*

Deadman, P. 1999. Modelling individual behaviour and group performance in an intelligent agent-based simulation of the tragedy of the commons. *Journal of Environment Management,* 56,159–172.

Fonseca, F. and Egenhofer, M. 1999. Ontology-driven geographic information systems. *Proceedings 7th ACM Symposium on Advances in Geographic Information Systems*, Kansas City, MO, pp. 14–19.

Gahegan M., Wachowicz, M., Harrower, M., and Rhyne, T.M. 2001. The integration of geographic visualization with knowledge discovery in databases and geocomputation. *Cartography and Geographic Information Systems* (special issue on the ICA research agenda), 28(1), 29–44.

Gendlin, E.T. 1995. Crossing and dipping: some terms for approaching the interface between natural understanding and logical formulation. *Minds and Machines*, 1995, 5(4), 547–560.

Gruber, T.R. 1993. A translation approach to portable ontologies. *Knowledge Acquisition*, 5(2),199–220.

Gruber, T.R. 2007. Ontology. In: *Encyclopedia of Database Systems*, Liu, L. and Tamer Özsu, M. (Eds.), Springer-Verlag, Berlin, http://tomgruber.org/writing/ontology-definition-2007.htm.

Guarino, N. 1998. Formal ontology and information systems. In: *Formal Ontology in Information Systems*, N. Guarino (Ed.). IOS Press, Amsterdam, The Netherlands, pp. 3–15.

Hustadt, U., Motik, B., and Sattler, U. 2005. Data complexity of reasoning in very expressive description logics. *Proceedings of the 20th International Joint Conference on Artificial Intelligence* (IJCAI 2005), pp. 466–471.

Huttenlocher, J., Hedges, L.V., and Duncan, S. 1991. Categories and particulars: Prototype effects in estimating spatial location. *Psychological Review*, 98, 352–376.

Johnson, M. 1987. *The Body in the Mind*. University of Chicago Press, Chicago, IL.

Lakoff, G. 1987. *Women, Fire, and Dangerous Things*. University of Chicago Press, Chicago, IL.

Ligtenberg, A., Wachowicz, M., Bregt, A., Beulens, A., and Kettenis, D.L. 2004. A design and application of a multiagent system for simulation of multifactor spatial planning. *Journal of Environment Management*, 72, 43–55.

MASTRO-I: Efficient integration of relational data through DL ontologies.

McNamara, T.P., Hardy, J.K., and Hirtle, S.C. 1989. Subjective hierarchies in spatial memory. *Journal of Experimental Psychology: Learning, Memory, and Cognition*, 15, 211–227.

Mennis, J. and Peuquet, D. 2003. The role of knowledge representation in geographical knowledge discovery: A case study. *Transactions on GIS* 7(3), 371–391.

Ortiz, M.M., Calvanese, D., and Eiter, T. 2006. Characterizing data complexity for conjunctive query answering in expressive description logics. *Proceedings of the 21st National Conference on Artificial Intelligence* (AAAI 2006).

OWL 2004. OWL Web Ontology Language Guide: W3C Recommendation 10 February 2004. W3C (2004-02-10).

Parent, C., Spaccapietra, S., and Zimanyi, E. 2006a. Conceptual Modeling for Traditional and Spatio-Temporal Applications — The MADS Approach, Springer, Berlin.

Parent, C., Spaccapietra, S., and Zimanyi, E. 2006b. The MADS query and manipulation languages. In: *Conceptual Modeling for Traditional and Spatio-Temporal Applications — The MADS Approach*, Springer, Berlin.

Peuquet, D.J. 2002. *Representations of Space and Time*. New York: The Guilford Press.

Pfoser, D. and Jensen, C.S. 2001. Querying the trajectories of on-line mobile objects. Proceedings of the 2nd ACM International Workshop on Data Engineering for Wireless and Mobile Access, Santa Barbara, CA, pp. 66–73.

Poggi, A. 2006. Structured and Semi-Structured Data Integration. PhD Dissertation. Joint work between the University of Rome "La Sapienza" (Italy) and the University of Paris-Sud (France).

Poore, B.S. and Chrisman, N.R. 2006. Order from noise: Toward a social theory of geographic information. *Annals of the Association of American Geographers,* 96(3), 508–523.

Reginster, I. and Edwards, G. 2001. The concept and implementation of perceptual regions as hierarchical spatial units for evaluating environmental sensitivity. *URISA Journal*, 13(1), 5–16.

Richmond, J.E.D. 1998. Simplicity and complexity in design for transportation systems and urban forms. *Journal of Planning Education and Research*, 17, 220–230.

Roddick, J.F. and Spiliopoulou, M. 1999. A bibliography of temporal, spatial and spatio-temporal data mining research. *SIGKDD Explorations Newsletter*. 1(1), 34–38.

Smith, B. 1998. An introduction to ontology. In: Peuquet, D., Smith, B. and Brogaard, B. (Eds.). *The Ontology of Fields*. Technical Report, National Center for Geographic Information and Analysis, University of California, Santa Barbara, pp. 10–14.

Sowa, J.F. 1999. Knowledge Representation: Logical, Philosophical, and Computational Foundations. Brooks/Cole, Pacific Grove, CA.

Spaccapietra, S., Parent, C., Damiani, M., De Macedo, J., Porto, F., Vangenot, C. 2008. A conceptual view on trajectories. Data & Knowledge Engineering, V. 65, pp. 126–146.

Stead, D. and Marshall, S. 2001. The relationships between urban form and travel patterns: An international review and evaluation. *European Journal of Transport and Infrastructure Research*, 1(2), 113–141.

Stell, J.G. 2003. Qualitative extents for spatio-temporal granularity. *Spatial Cognition and Computation*, 3(2&3), 199–136.

Tiezzi, E. 2004. *Beauty and Science*, WIT Press, Southamptom, U.K.

Wachowicz, M. 1999. *Object-Oriented Design for Temporal GIS*. Taylor & Francis, London.

Wachowicz, M., Ligtenberg, A., Renso, C., and Gurses, S. 2008. Characterising mobile applications through a privacy-aware geographic knowledge discovery process. In: *Mobility, Data Mining and Privacy, and Geography: Geographic Knowledge Discovery*. F. Giannoni and D. Pedreski (Eds.). Springer-Verlag, Berlin.

Wikipedia, the free encyclopedia, http://www.wikipedia.com.

Yuan, M., Buttenfield, B., Gahegan, M., and Miller, H. 2000. Geospatial Data Mining and Knowledge Discovery. A UCGIS White Paper on Emergent Research Themes. URISA Journal, 12(2).

Zimmerman, C. 2000. The development of scientific reasoning skills. *Developmental Review*, 20, 99–149.

15 Periodic Pattern Discovery from Trajectories of Moving Objects

Huiping Cao

Nikos Mamoulis

David W. Cheung

CONTENTS

15.1 INTRODUCTION

The rapid advances in telecommunications (e.g., GPS, cellular networks, etc.) facilitate the collection of large amounts of trajectory data. To make good use of such data, effective and efficient methods are needed to manage and analyze them.

Management and analysis of moving object trajectories is challenging due to the vast amount of collected data and novel types of queries and analysis tasks. The relational model cannot directly be deployed for managing trajectory data because of its insufficient representation ability. Because of this, the efficient management of such data has gained a lot of interest during the past few years [19, 23, 11, 22].

In the knowledge discovery and data mining (KDD) area, we face a similar problem: methods that work well for traditional relational and transactional data have their limitations in analyzing trajectory data. Furthermore, the lack of a consistent theoretical framework in the KDD field renders the trajectory data analysis harder. These challenges triggered a lot of work (see [18, 17, 13, 5, 8]) on analyzing trajectory data. In this chapter, we focus our discussion on one of these exploratory efforts; the discovery of trajectory periodic patterns, which capture regular movement of objects in space and time.

This chapter is organized as follows. We overview some efforts in mining trajectory data in Section 15.2. In Section 15.3, we present the model of trajectory periodic patterns and algorithms to discover them. Section 15.4 discusses some open issues in mining trajectory data. Finally, the chapter is concluded in Section 15.5.

15.2 LITERATURE OF MINING TRAJECTORY DATA

Clustering is a long-standing technique to group objects showing similar behavior and differentiate objects performing differently. For identifying trajectories of similar shapes, Gaffney and Smyth [7] propose a probabilistic mixture regression model so that routes in one cluster follow the trends of a core representative trajectory and vary a little from it. The same objective motivated Vlachos et al. [24] to formalize a similarity function for trajectories based on the longest common subsequence (LCSS). Besides these efforts, Li et al. [17] and Kalnis et al. [13] have studied the moving cluster model. Reference [17] focuses on maintaining moving micro-clusters, each of which contains objects moving close in a time period, whereas Reference [13] finds clusters with unchangeable density during their lifetime while the objects in a group may vary (with some objects leaving it and others entering it).

Pattern discovery is another intensively studied topic in trajectory data mining. Frequent patterns reflect the regular behavior of moving objects. Mamoulis et al. [18] enrich the time-series periodic pattern model to capture periodic movements in space; for example, somebody regularly visits place A and then place B at 8:00 A.M. and 9:00 A.M., respectively, every day. References [2] and [8] present different models and techniques for extracting sequential trajectory patterns. One example pattern of Reference [2] is a region list $r_1 \ldots r_m$, where each automatically discovered region r_i contains sub-trajectories with similar shapes. In Reference [8], patterns model regular routes by tourists (e.g., Railway Station $\xrightarrow{15min}$ Castle Square $\xrightarrow{2h15min}$ Museum), where the time intervals between sequential regions are captured. Discovery of motion similarity (e.g., flock, leadership, convergence, and encounter patterns) has been studied in [16, 9]. With different interestingness measurements, References [26], [25], [3], and [5] have extended the spatial collocation model [20] to include temporal aspects such that the discovered patterns represent object types whose instances move close to each other in the history.

The discovered clusters or patterns could help to predict the future locations of objects. Yavas et al. [27] present a three-stage algorithm to find historical mobility patterns and predict future positions of mobile users from them. To improve location-based services, Karimi and Liu [14] propose a model to predict locations in the road-level granularity. Laasonen [15] has devised online algorithms to learn routes (clusters of cell sequences) between important locations. From these clusters, destination probabilities can be derived to help predict a user's future location. References [10] and [21] have investigated the prediction of objects number (density) in an area.

15.3 PERIODIC PATTERNS

In many applications, the movements obey periodic patterns; that is, the objects follow the same routes (approximately) over regular time intervals. Such objects following approximate periodic patterns include transportation vehicles (buses, boats, airplanes, trains, etc.), animals, mobile phone users, etc. For example, Bob wakes up at the same time and then follows, more or less, the same route to his work every day.

The problem of discovering periodic patterns from historical object movements is very challenging. Usually, the patterns are not explicitly specified, but have to be discovered from the data. The patterns can be thought of as (possibly noncontiguous) sequences of object locations that reappear in the movement history periodically. In addition, since an object is not expected to visit *exactly* the same location at every time instant of each period, the patterns are not rigid but differ slightly from one occurrence to the next. The approximate nature of patterns in the spatio-temporal domain increases the complexity of the mining tasks. The proposed method needs to discover, along with the patterns, a flexible description of how they variate in space and time.

In what follows, Section 15.3.1 motivates the necessity to propose a new model to represent periodic patterns. Section 15.3.2 models formally the concept of periodic pattern. The algorithms that discover such patterns are presented in Section 15.3.3. Finally, Section 15.3.4 briefly shows two variants of this problem.

15.3.1 MOTIVATING EXAMPLE AND NAIVE METHOD

This section motivates the research on periodic pattern discovery by showing that previous work on event sequences is not expected to perform well when applied to trajectories.

Definition 15.1 ***An object trajectory*** *is an n-length sequence* $\mathcal{S}$ *of spatial locations, one for each timestamp in the history, of the form* $\{(l_0, t_0), (l_1, t_1), \ldots, (l_{n-1}, t_{n-1})\}$, *where* l_i *is the object's location at time* t_i *and is expressed in terms of spatial coordinates. If the difference between consecutive timestamps is fixed,* $\mathcal{S}$ *is simplified to* $\{l_0, l_1, \ldots, l_{n-1}\}$.

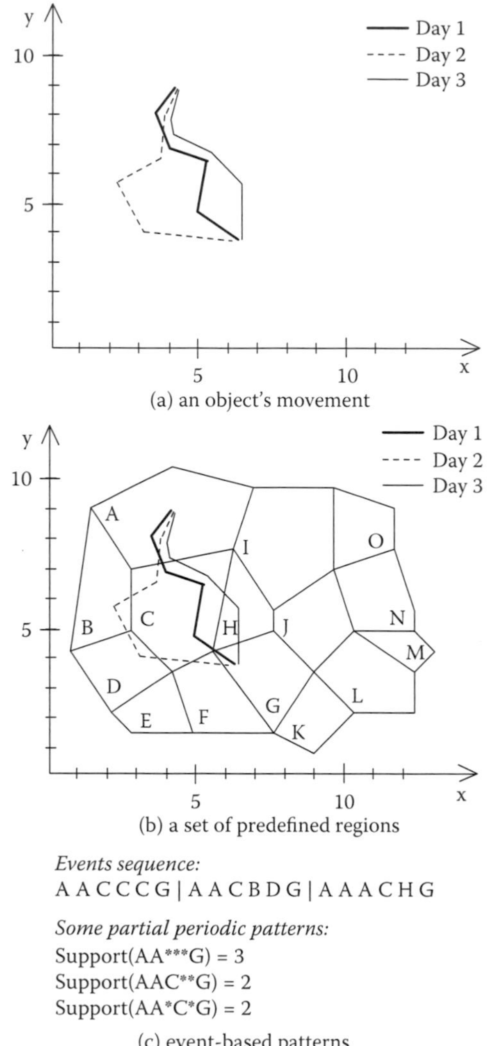

FIGURE 15.1 Periodic patterns with respect to pre-defined spatial regions.

Figure 15.1a, for example, illustrates the movement of an object in three consecutive days (assuming that it is tracked only during specific hours, e.g., working hours). It can be modeled with a sequence $\mathcal{S} = \{<4, 9>, <3.5, 8>, \ldots\}$.

Given a sequence $\mathcal{S}$, a minimum support *min_sup* ($0 < min\_sup \leq 1$), and an integer T, called *period*, our problem is to discover movement patterns that repeat themselves every T timestamps. A discovered pattern P is a T-length sequence of the form $r_0 r_1 \ldots r_{T-1}$, where r_i is a *spatial region* or the special character $*$, indicating the whole spatial universe. For instance, pattern `AB*C**` implies that at the beginning of the cycle the object is in region `A`, at the next timestamp it is found in region

B, then it moves irregularly (it can be anywhere), then it goes to region C, and after that it can go anywhere, until the beginning of the next cycle, when it can be found again in region A. The patterns are required to be followed by the object in at least $\alpha(\alpha = min\_sup \cdot \lfloor \frac{n}{T} \rfloor)$ periodic intervals in $\mathcal{S}$.

Existing algorithms for mining periodic patterns (e.g., [12]) operate on event sequences and discover patterns of the above form. However, in this case, the elements r_i of a pattern are events (or sets of events). As a result, these techniques cannot directly be applied to our problem, unless the exact locations l_i's are treated as discrete categorical values. Nevertheless, it is highly unlikely that an object repeats an identical sequence of $<x, y>$ locations precisely either because of the imprecise spatial routes or due to unsynchronized location transmissions. Thus, the object cannot reach the same location at the same time every day, and as a result the sampled locations at specific timestamps (e.g., at 9:00 A.M. sharp, every day), are different. In Figure 15.1a, for example, the first daily locations of the object are very close to each other; however, they are treated differently by a straightforward mining algorithm.

One way to handle the noise in object movement is to replace the exact locations by the regions (e.g., districts, mobile communication cells, etc.) which contain them. Figure 15.1b shows an example of an area's division into such regions. Sequence {A, A, C, C, C, G, A,...} can now summarize the object's movement, and periodic pattern mining algorithms, like Reference [12], can directly be applied. Figure 15.1c shows three (closed) discovered patterns for $T = 6$, and $min\_sup = \frac{2}{3}$. A disadvantage of this approach is that the discovered patterns may not be very descriptive if the space division is not very detailed. For example, regions A and C are too large to capture in detail the first three positions of the object in each periodic instance. On the other hand, with detailed space divisions, the same (approximate) object location may span more than one different region. For example, in Figure 15.1b, observe that the third object positions for the three days are close to each other, however, they fall into different regions (A and C) at different days. Therefore, we should aim at the *automated* discovering of patterns *and their descriptive regions*.

15.3.2 Model

Let $\mathcal{S} = \{l_0, l_1, \ldots, l_{n-1}\}$ represent the movement of an object. Let $T \ll n$ be a user-specified integer called *period* (e.g., day, week, month). A *periodic segment s* is defined by a subsequence $l_i l_{i+1} \ldots l_{i+T-1}$ of $\mathcal{S}$, such that i modulo $T = 0$. Thus, segments start at positions $0, T, \ldots, (\lfloor \frac{n}{T} \rfloor - 1) \cdot T$, and there are exactly $m = \lfloor \frac{n}{T} \rfloor$ periodic segments in $\mathcal{S}$.[1] Let s^j denote the segment starting at position $l_{j \cdot T}$ of $\mathcal{S}$, for $0 \le j < m$, and let $s_i^j = l_{j \cdot T + i}$, for $0 \le i < T$.

Definition 15.2 ***A* periodic pattern** *P is defined by a sequence* $r_0 r_1 \ldots r_{T-1}$ *of length T, such that* r_i *is either a spatial region or* $*$. *The* length *of a pattern P is the number of non-*$*$ *regions in P.*

---

[1] If n is not a multiple of T, then the last n modulo T locations are truncated, and the length n of sequence S is reduced accordingly.

A segment s^j is said to *comply with P*, if for each $r_i \in P$, $r_i = *$ or s^j_i is *inside* region r_i.

Definition 15.3 *The* **support** $|P|$ *of a pattern P in* $\mathcal{S}$ *is defined by the number of periodic segments in* $\mathcal{S}$ *that comply with P.*

In the rest of the discussion, P may refer to a pattern or the set of segments that comply with it. Let *min_sup* be a fraction in the range (0, 1] (*minimum support*). A pattern P is *frequent* if its support is larger than $min\_sup \cdot m$.

A problem with the definition above is that it has no control over the density of the pattern region r_i. In other words, if the pattern regions are too relaxed (e.g., each r_i is the whole map), the pattern may always be frequent. Therefore, an additional constraint is imposed as follows. Let $\mathcal{S}^P$ be the set of segments that comply with a pattern P. Each region r_i of P is *valid* if the set of locations $R^P_i := \{s^j_i \mid s^j \in \mathcal{S}^P\}$ form a *dense cluster*, which is initially defined in Reference [6] using two parameters $\in$ and *MinPts*. A point p in the spatial dataset R^P_i is a *core* point if the circular range centered at p with radius $\in$ contains at least *MinPts* points. If a point q is within distance $\in$ from a core point p, it is assigned in the same cluster as p. If q is a core point itself, then all points within distance $\in$ from q are assigned in the same cluster as p and q. If R^P_i forms a single, dense cluster with respect to some values of $\in$ and *MinPts*, then region r_i is valid. If all non-$*$ regions of P are valid, then P is a valid pattern.

Figure 15.2a shows an example of a valid pattern, if $\in = 1.5$ and $MinPts = 4$. Each region at positions 1, 2, and 3 forms a single, dense cluster and is therefore a dense region. Notice, however, that it is possible that two valid patterns P and P' of the same length (1) have the same $*$ positions, (2) every segment that complies with P', complies with P, and (3) $|P'| < |P|$. In other words, *P implies P'*. For example, the pattern of Figure 15.2a implies the one of Figure 15.2b (denoted by the three circles). A frequent pattern P' is *redundant* if it is implied by some other frequent pattern P.

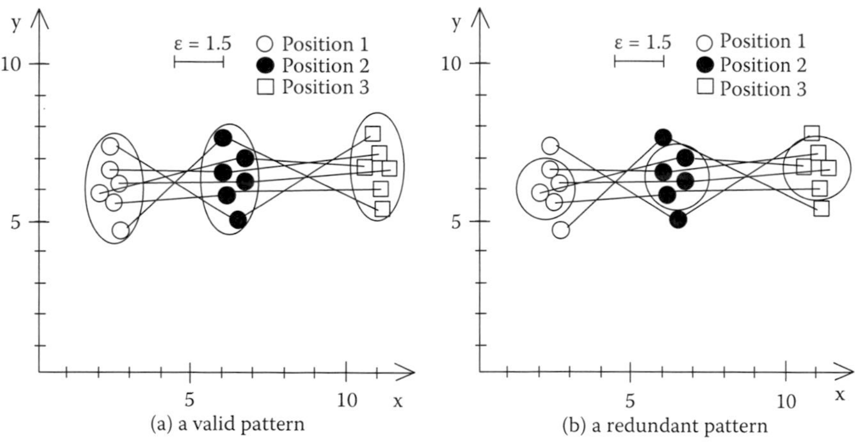

FIGURE 15.2 Redundancy of patterns.

Definition 15.4 *The* **mining periodic patterns problem** *searches for all valid periodic patterns P ("frequent pattern" for short) in $\mathcal{S}$, which are frequent and nonredundant with respect to a minimum support min_sup.*

15.3.3 Algorithms

This section presents techniques for mining frequent periodic patterns and their associated regions in object trajectories. First, we discuss how to find frequent 1-patterns (i.e., of length 1). Then, three algorithms that identify longer patterns; a bottom-up, level-wise technique, denoted by STPMine1 (SpatioTemporal periodic Pattern Min(e)ing 1), a faster top-down approach, referred to as STPMine2, and a simplified version of the top-down approach STPMine2-V2, which solves the problem approximately but efficiently.

15.3.3.1 Obtaining Frequent 1-Patterns

To discover frequent 1-patterns, the following methodology is applied. The sequence $\mathcal{S}$ is divided into T spatial datasets, one for each offset of the period T. In other words, locations $\{l_i, l_{i+T}, \ldots, l_{i+(m-1)\cdot T}\}$ go to set R_i, for each $0 \leq i < T$. Each location is tagged by the id $j \in [0, \ldots, m-1]$ of the segment that contains it. Figure 15.3a shows the spatial datasets obtained after decomposing the object trajectory of Figure 15.1a.

Observe that a dense cluster r in dataset R_i corresponds to a frequent pattern, having r at position i and $*$ at all the other positions. Figure 15.3b shows examples of five clusters discovered in datasets R_1, R_2, R_3, R_4, and R_6. These correspond to five 1-patterns (i.e., $r_{11}{*}{*}{*}{*}{*}$, ${*}r_{21}{*}{*}{*}{*}$, etc.). In order to identify the dense clusters for each R_i, a density-based clustering algorithm like DBSCAN [6] can be applied. Clusters with less than α ($\alpha = min\_sup \cdot m$) points are discarded, since they are not frequent 1-patterns according to our definition. The original DBSCAN algorithm has quadratic cost to the number of clustered points; therefore, a hash-based method is utilized to reduce its cost (see [4] for details).

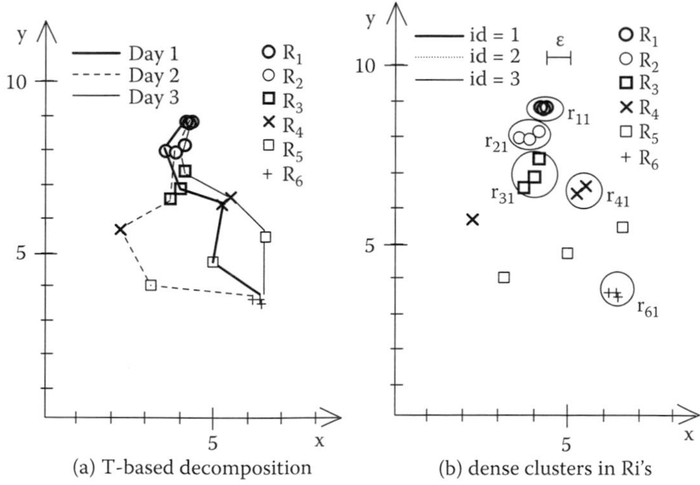

FIGURE 15.3 Locations and regions per periodic offset.

15.3.3.2 A Level-Wise, Bottom-Up Approach

Starting from 1-patterns (i.e., clusters for each R_i), the bottom-up approach applies a variant of the level-wise a *priori*-TID algorithm [1] to discover longer ones, as shown in Algorithm STPMine1. The input of the algorithm is a collection $\mathcal{L}_1$ of frequent 1-patterns, discovered as described in Section 15.3.3.1. Pairs $<P_1,P_2>$ of $(k-1)$-patterns in $\mathcal{L}_{k-1}$, with their first $k-2$ non-∗ regions in the same position and different $(k-1)$-th non-∗ position create candidate k-patterns (lines 4 to 6). For each candidate P_{cand}, a segment-id join is performed between P_1 and P_2, and if the number of segments that comply with both patterns is at least $min\_sup \cdot m$, a pattern validation function checks whether the regions of P_{cand} are still clusters. After the patterns of length k have been discovered, the patterns at the next level are found, until there are no more patterns at the current level, or there are no more levels.

Algorithm STPMine1($\mathcal{L}_1$, T, *min_sup*);
1). k:=2;
2). **while** ($\mathcal{L}_{k-1} \neq \emptyset \wedge k < T$)
3). $\mathcal{L}_k := \emptyset$;
4). **for each** pair of patterns $(P_1,P_2) \in \mathcal{L}_{k-1}$
5). such that P_1 and P_2 agree on the first $k-2$
6). and have different $(k-1)$-th non-∗ position
7). P_{cand} := **candidate_gen**(P_1, P_2);
8). **if** ($P_{cand} \neq null$) then
9). $P_{cand} := P_1 \bowtie_{P_1 \cdot s_{id} = P_2 \cdot s_{id}} P_2$; //segment-id join
10). **if** ($|P_{cand}| \geq min\_sup \cdot m$) then
11). **validate-pattern**(P_{cand}, $\mathcal{L}_k$, *min_sup*);
12). k:=k + 1;
13). return $\mathcal{P} := \cup \mathcal{L}_k$, $\forall 1 \leq k < T$;

To facilitate fast and effective candidate generation, STPMine1 uses the MBRs (i.e., *minimum bounding rectangles*) of pattern regions. For each common non-∗ position i, the intersection of the MBRs of the regions for P_1 and P_2 must be nonempty; otherwise, a valid superpattern cannot exist. Function **validate_pattern** computes a number of actual k-patterns from P_{cand}. The rationale is that the points at all non-∗ positions of P_{cand} may no longer form a cluster after the join of P_1 and P_2. Thus, for each non-∗ position of P_{cand} the points are re-clustered. If for some position the points can be grouped to more than one cluster, a new candidate pattern is created and validated for each cluster. Note that, from a candidate pattern P_{cand}, it is possible to generate more than one actual pattern eventually. If no position of P_{cand} is split to multiple clusters, the non-∗ positions of P_{cand} may need to be re-clustered, since some points (and segment-ids) may be eliminated during clustering at some position.

EXAMPLE 15.1

Consider the 2-patterns $P_1 = r_{1x}r_{2y}*$ and $P_2 = r_{1w}*r_{3z}$ of Figure 15.4a. Assume that *MinPts* = 4 and $\in$ =1.5. The two patterns have common first non-∗ position

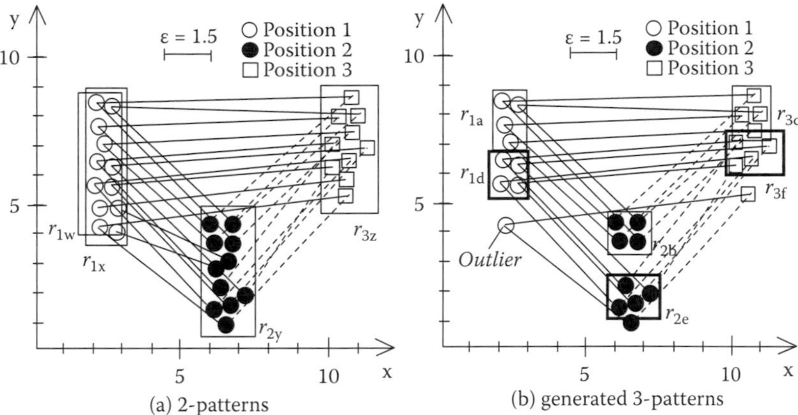

FIGURE 15.4 Example of STPMine1.

and $MBR(r_{1x})$ overlaps $MBR(r_{1w})$. Therefore, a candidate 3-pattern P_{cand} is generated. During candidate pruning, STPMine1 verifies that there is a 2-pattern with non-$*$ positions 2 and 3 which is in $\mathcal{L}_2$. Indeed, such a pattern can be spotted at the figure (see the dashed lines). After joining the segment-ids in P_1 and P_2 at line 9 of STPMine1, P_{cand} contains the trajectories shown in Figure 15.4b. Notice that the locations of the segment-ids in the intersection may no longer form clusters at some positions of P_{cand}. This is why **validate_pattern** has to be called, in order to identify the valid patterns included in P_{cand}. Observe that the segment-id corresponding to the lowermost location of the first position is eliminated from the cluster as an outlier. Then, while clustering at position 2, two dense clusters are identified, which define the final patterns $r_{1a}r_{2b}r_{3c}$ and $r_{1d}r_{2e}r_{3f}$.

15.3.3.3 A Two-Phase, Top-Down Algorithm

Although STPMine1 can find all periodic patterns correctly, it can be very slow due to the huge number of region combinations to be joined. If the actual patterns are long, all their subpatterns have to be computed and validated. So, a top-down method (STPMine2) is proposed to discover long patterns more efficiently.

After applying clustering on each R_i (Section 15.3.3.1), the frequent 1-patterns with their segment-ids are discovered. The first phase of STPMine2 algorithm replaces each location in $\mathcal{S}$ with the cluster-id it belongs to or with an "empty" value (e.g., $*$) if the location belongs to no cluster. For example, assume that the following clusters have been discovered: $\{r_{11}, r_{12}\}$ at position 1, $\{r_{21}\}$ at position 2, and $\{r_{31}, r_{32}\}$ at position 3. A segment $\{l_1, l_2, l_3\}$, such that $l_1 \in r_{12}$, $l_2 \notin r_{21}$, and $l_3 \in r_{31}$ is transformed to subsequence $\{r_{12}{*}r_{31}\}$. Therefore, the original trajectory $\mathcal{S}$ is transformed to a symbol sequence $\mathcal{S}'$.

The algorithm in Reference [12] can be used to discover quickly all frequent patterns of the form $r_0r_1\ldots r_{T-1}$, where each r_i is a cluster in R_i or $*$. However, it is unknown whether the results of the sequence-based algorithm are actual patterns, since the contents of each non-$*$ position may not form a cluster. For example, $\{r_{12} * r_{31}\}$ may be frequent, however if the algorithm considers only the segment-ids that qualify

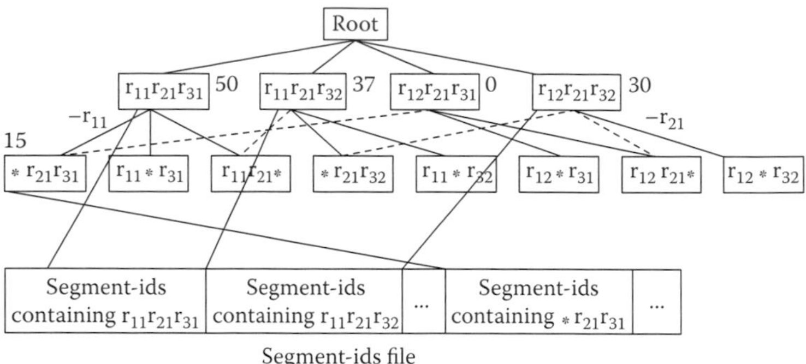

FIGURE 15.5 Example of a max-subpattern tree.

this pattern, r_{12} may no longer be a cluster or may form different actual clusters (as illustrated in Figure 15.4). The patterns P' which can be discovered by the algorithm of Reference [12] are called *pseudo-patterns*, since they may not be valid.

To discover the actual patterns, some changes are applied in the original algorithm of Reference [12]. While creating the max-subpattern tree, each tree node is stored with the segment-ids that correspond to the pseudo-pattern of the node. In this way, one segment-id goes to exactly one node of the tree. However, $\mathcal{S}$ could be too large to manage in memory. In order to alleviate this problem, while scanning $\mathcal{S}$, for every segment s, the following operations are performed.

- First, the segment is inserted to the max-subpattern tree, as in Reference [12], increasing the counter of the candidate pseudo-pattern P' that s corresponds to after the transformation. An example of such a tree is shown in Figure 15.5. This node can be found by finding the (first) maximal pseudo-pattern that is a superpattern of P' and following its children, recursively. If the node corresponding to P' does not exist, it is created (together with any nonexistent ancestors). Notice that the dotted lines are not implemented and not followed during insertion (thus, STPMine2 materializes the tree instead of a lattice). For instance, for segment with $P' = \{*r_{21}r_{31}\}$, STPMine2 increases the counter of the corresponding node at the second level of the tree.
- Second, an entry $<P'.id, s.sid>$ is inserted to a file F , where $P'.id$ is the node id that corresponds to a pseudo-pattern P' and $s.sid$ is the id of segment s. At the end, file F is sorted on $P'.id$ to bring together segment-ids that comply with the same (maximal) pseudo-pattern. For each pseudo-pattern with at least one segment, STPMine2 inserts a pointer to the file position, where the first segment-id is located.

Instead of finding frequent patterns in a bottom-up fashion, the tree is traversed in a top-down, breadth-first order. For every pseudo-pattern with at least $min\_sup \cdot m$ segment-ids, the **validate_pattern** function is applied to recover potential valid patterns. All segment-ids that belong to a discovered pattern are removed from the current pseudo-pattern because the interested patterns are those not spatially contained in some superpattern. The segment-ids that are not included in a pattern are used to verify its subpatterns.

Thus, after scanning the first level of the lattice, some patterns may be discovered, and segment-id lists of the pseudo-patterns may be shrunk. Then, STPMine2 moves to the next level. The support of a pseudo-pattern P' at each level is the recorded support of P' plus the supports of all its superpatterns (recall that a segment-id is assigned to the *maximal* pattern it complies with). The supports of the superpatterns can be immediately accessed from the lattice. If the total support of the candidate is at least $min\_sup \cdot m$, then the segment-ids have to be loaded for application of **validate_pattern**. The segment-ids of a superpattern may already be in memory from previous level executions. If not, they are loaded from the file F. After validation, only the disqualified segment-ids are kept to be used at lower level patterns.

The fact that segment-ids are clustered in F according to the breadth-first traversal of the lattice minimizes random accesses and restricts the number of loaded blocks to memory. The segment-ids for a superpattern remain in memory to be used at lower-level validations. If the algorithm runs out of memory, the segment-ids of the uppermost lattice levels are rewritten to disk, but this time possibly to a smaller file if there were some deletions.

```
Algorithm STPMine2(ℒ1, T , min_sup);
build max-subpattern tree 𝒯 and pattern-file F;
sort F on P'.id and connect it to the nodes of 𝒯;
for k:=T down to 2
    for each pattern P' at level k of 𝒯
        |P'|:=P'.counter +Σ_{P''⊃P',length(P'') = k+1} |P''|;
        if (|P'|≥ min_sup · m) then
            P_cand := ∪_{P''⊇P'} P''.sids;
            validate-pattern(P_cand, ℒ, min_sup);
            if (𝒫 has changed) then
                remove from P' those sids in new patterns of 𝒫;
                if (unassigned sids less than min_sup · m) then
                    return 𝒫;
return 𝒫;
```

A pseudocode for STPMine2 (top-down pattern mining) is shown in Algorithm STPMine2. Initially, the tree and the segment-ids file are created and linked. Then for each level, it finds the support of a pseudo-pattern $|P'|$ at level k by accessing only the supports of its superpatterns $P'' \supset P$ at level $k + 1$, since the tree is accessed in breadth-first order. If $|P'| \geq min\_sup\ .\ m$, the pattern is validated as in STPMine1, and if some pattern is discovered, STPMine2 removes from P' all those segment-ids that comply with the discovered pattern. Thus, the number of segment-ids decreases as STPMine2 goes down the levels of the tree, until it is not possible to discover any more patterns, or there are no more levels. Notice that the patterns discovered here are only maximal, as opposed to STPMine1, which discovers *all* frequent patterns.

In real applications, these maximal patterns are more useful, compared to the huge set of all patterns.

15.3.3.4 A Simplified Algorithm: STPMine2-V2

A pattern P is valid if (1) its frequency exceeds $min\_sup \cdot m$; and (2) the locations in R_i^P form a single dense cluster for all non-$*$ positions in P. Property (2) incurs a high computational burden to the mining algorithms, because it requires repetitive applications of the clustering algorithm and maintenance of the segment-ids that comply with each tree node.

A simplified version of the mining algorithm is proposed to consider the second property only in the discovery of frequent 1-patterns. After computing the dense regions at each R_i, no clusters are re-validated. As a result, the segment-id lists are not used at each node, only the node counters are considered to measure the pattern frequency. This mining technique is identical to STPMine2, excluding the validation and re-assignment of segment-ids, thus it is denoted by STPMine2-V2.

STPMine2-V2 is less accurate than STPMine2, since it may discover patterns that are invalid according to the strict definition. In addition, the regions that define the patterns discovered by STPMine2-V2 are identical to the regions of the clusters forming frequent 1-patterns (e.g., the region refinement of the example in Figure 15.4 is not performed). On the other hand, STPMine2-V2 is expected to be significantly faster than STPMine2.

15.3.3.5 Effectiveness

To demonstrate the effectiveness of STPMine1, STPMine2, and STPMine2-V2, a short trajectory with $n = 1000$ locations is generated. T is set to 20, and the object follows a single periodic pattern P at 39 out of 50 segments, whereas the movement is irregular in 11 segments. Figure 15.6a shows the object trajectory, where the periodic movement can roughly be observed. For this dataset $\ell = 10$, i.e., there are 10 non-$*$ positions in P. Figure 15.6b shows the maximal frequent pattern P of length 10, successfully discovered by STPMine1 and STPMine2, when $min\_sup = 0.6$. The object's movement is plotted, interpolated using only the non-$*$ positions. The discovered pattern is identical to the generated one. The dense regions are successfully detected by the clustering module, and the spatial extents of the pattern are minimal.

A *grid-based* mining method is developed to directly apply the mining algorithm in Reference [12]. The space is divided using a regular $M \times M$ grid. Each location of S is transformed to the cell-id which encloses the location. Then, the algorithm of Reference [12] is used to find partial patterns of cell-ids. Figure 15.6c shows a maximal pattern P' discovered by this grid-based technique, when using a 10×10 grid. P' has the largest length among all discovered patterns, however it is only 4 (whereas the actual pattern P has 10 non-$*$ positions). With a grid with fine granularity, frequent regions that span multiple cells cannot be identified (e.g., the cluster r[19][1] is split between cells c_{47} and C_{57} and neither of these cells has higher support than $min\_sup \cdot m$), whereas with a grid of low granularity the patterns are formed by very large regions.

STPMine2-V2 also finds the maximal pattern with length 10 shown in Figure 15.6d. The region for each non-$*$ position is represented with the MBR of

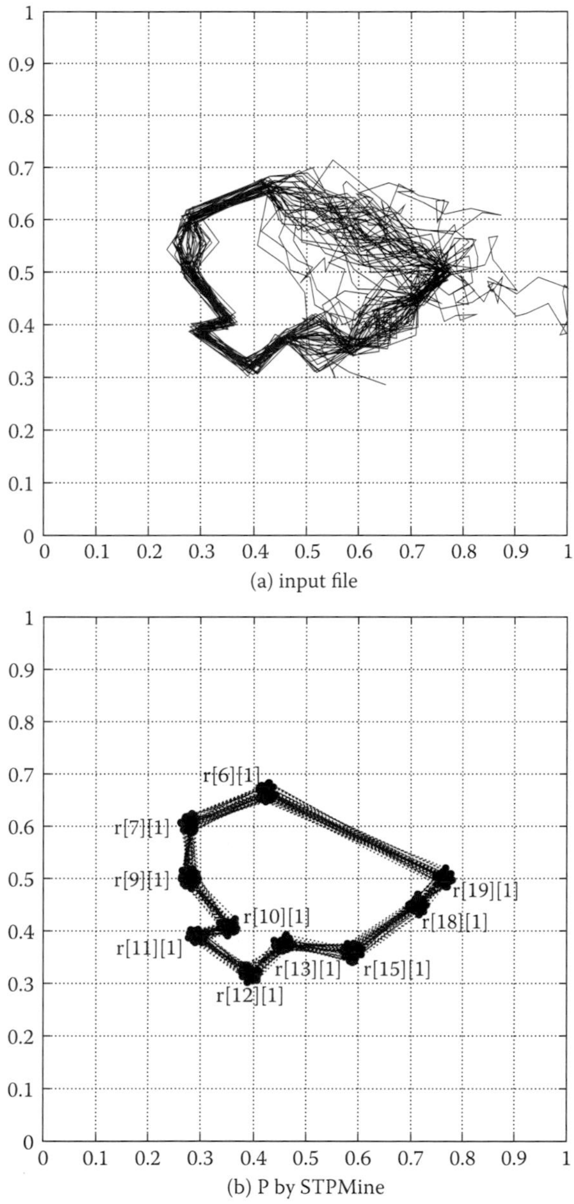

FIGURE 15.6 Effectiveness comparison.

its associated initial cluster. Thus, STPMine2-V2 retrieves comparative results to STPMine1 and STPMine2 in finding the descriptive regions and patterns, except that the non-$*$ regions are a little larger (i.e., a little less descriptive) than the ones in the pattern of Figure 15.6b.

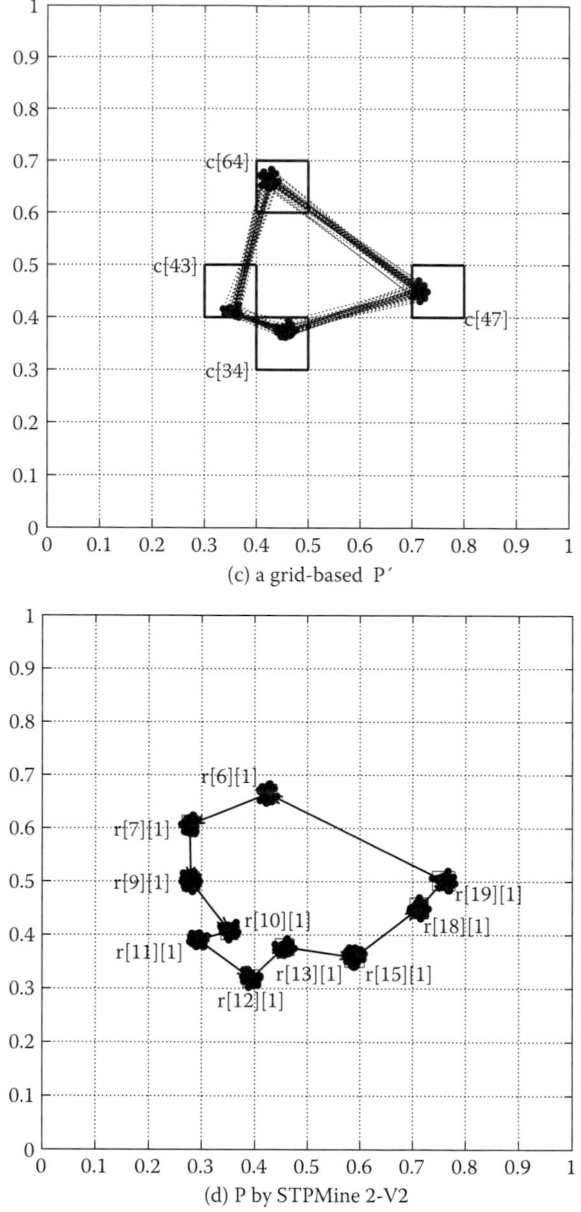

FIGURE 15.6 (Continued).

15.3.4 Variants

In practice, the model presented thus far may fail to identify some noncrisp periodic patterns. This section discusses two variants of this model that relax the definition of frequent periodic patterns and allow for the discovery of additional trends in the data. The first variant captures periodic patterns that may not be frequent in the whole sequence. For instance, assume that Bob changes his route to work after being transferred from department A to department B. In this case, his route to department A is frequent only during the time interval he works there. This variant mines frequent patterns and their *validity eras*; that is, the (maximal) time ranges (*eras*) during which these patterns are frequent.

The second variant discovers patterns whose instances may be shifted or distorted. For example, if Bob wakes up late on a certain day, the movement to his work is shifted on that day (e.g., for 10 minutes). Or, Bob gets up at the usual time, but arrives at the company a little late due to traffic congestion. Although Bob follows the same route (pattern) to the company in the above two cases, the corresponding pattern instances are shifted and/or distorted. In both cases, in the counting of a pattern's frequency, its shifted and/or distorted instances should be included.

This section defines the above two variants and briefly describes appropriate techniques for their discovery.

15.3.4.1 Patterns with Validity Eras

Given a trajectory $\mathcal{S}$, a period T, and a segment s^j starting from $l_{j \cdot T}$ and spanning T consecutive locations in $\mathcal{S}$, an **era** $[b, e]$ is the subsequence of $\mathcal{S}$, from the beginning of segment s^b until the end of s^e. The **time span** of the era $[b, e]$ is $e - b + 1$. An era $[b, e]$ is a *superset* of the era $[b', e']$ if $b \leq b'$ and $e \geq e'$; accordingly, $[b', e']$ is a *subset* of $[b, e]$.

A periodic pattern with a validity era, **era pattern** for short, refers to a periodic pattern associated with some era, $P = r_0 r_1 \ldots r_{T-1}[b, e]$. In Figure 15.1c, the era of subsequence AACBDG AAACHG is [1,2] ($T = 6$), whereas the era of the whole sequence is [0, 2]. Examples of era patterns are AA***G[0, 2] and AAC**G[0, 1].

Recall that $\mathcal{S}^P$ is the set of segments that comply with a pattern P. Let b_{min} and e_{max} be the minimum and maximum segment-ids in $\mathcal{S}^P$, respectively. Given $[b, e]$, a subset of $[b_{min}, e_{max}]$, let $\mathcal{S}^P_{[b,e]}$ contain all the segments in $\mathcal{S}^P$ with segment-ids in $[b, e]$, and $|\mathcal{S}^P_{[b,e]}|$ denote the number of segments in $\mathcal{S}^P_{[b,e]}$.

Definition 15.5 *Given min_sup and MinPts, the era pattern P [b, e] is a* **valid era pattern** *if* $|S^P_{[b,e]}| \geq$ **Min***Pts and* $\frac{|s^P_{[b,e]}|}{e-b+1} \geq min\_sup$.

Consider two era patterns $P = r_0 r_1 \ldots r_{T-1}[b, e]$ and $P' = r'_0 r'_1 \ldots r'_{[T-1]}[b', e']$. P is a **super-pattern** of P' if (1) $r_i = r'_i$ or $r'_i = *$ for $0 \leq i < T$ and (2) $[b, e]$ is superset of $[b', e']$. An era pattern P is *maximal* if it has no proper valid superpattern.

Definition 15.6 *The* **mining era patterns problem** *aims to find all the* maximal valid era patterns, *given a sequence $\mathcal{S}$, a period T, a minimum support min_sup* ($0 <$ *min_sup* ≤ 1)*, and cluster parameters* $\in$ *and MinPts.*

The discovery of era patterns involves two steps: detecting valid 1-patterns and discovering the patterns with longer length. In finding valid 1-patterns, the era pattern mining algorithm (EP-Mine) first employs the procedure in Section 15.3.3.1 to find candidate 1-patterns. Then, for each candidate P, it computes the maximal era that makes P valid.

For each candidate pattern P, its initial era is $[b_{min}, e_{max}]$. The maximal valid era is a subset $[b, e]$ of $[b_{min}, e_{max}]$ such that no superset of $[b, e]$ makes P valid. This can be computed by recursively checking the subsets of $[b_{min}, e_{max}]$ from eras with longer length to those of shorter length. The first reported valid era is the maximum one.

For finding longer era patterns, STPMine2-V2 is adapted. A max-subpattern P_{max} is formed by combining the valid 1-patterns, and its era is the *union* of the eras from the 1-patterns that define P. For example, the union of a set of eras $\{[b_1, e_1], [b_2, e_2],\ldots [b_k, e_k]\}$ is defined by $[\min_{i=1}^{k} b_i, \max_{i=1}^{k} e_i]$. In addition, the eras of the 1-patterns that form a max-subpattern P have nonempty intersection; otherwise, there can be no valid instance of P. From the candidate max-patterns, EPMine builds the max-subpattern tree for each of them. The max-subpattern trees are traversed in a breadth-first order and longer patterns with valid maximal eras are reported.

15.3.4.2 Shifted and Distorted Patterns

Recall that s^j denotes a segment starting at position $j \cdot T$. Given a tolerance integer $\tau (0 \le \tau \le \lfloor T/2 \rfloor)$, a segment starting at position $j \cdot T + d$, $-\tau \le d \le \tau$, is denoted by $s^j[d]$ (note that $s^j[0] = s^j$).

Definition 15.7 *Given a sequence $\mathcal{S}$ and an integer τ, a segment $s^j[d]$, $-\tau \le d \le \tau$, is a* **shifted pattern instance** *of a pattern P if it complies with P, i.e., P's occurrence in $\mathcal{S}$ is shifted at most τ timestamps forward or backward from its expected position $j \cdot T$.*

EXAMPLE 15.2

Let $T = 5$ and $\mathcal{S}' = r_0r_1r_2r_3r_4\ r_0r_0r_1r_4r_3\ r_2r_0r_1r_3r_3$ be the transformed sequence, after replacing the locations in $\mathcal{S}$ by spatial regions. The pattern $r_0r_1 * r_3 *$ has one nonshifted instance, s^0, starting at position 0, and two shifted pattern instances, $s^1[1]$ and $s^2[1]$, starting at positions 6 $(1 \cdot T + 1)$ and 11 $(2 \cdot T + 1)$.

Definition 15.8 *A segment $s^j[d]$, $-\tau \le d \le \tau$, is a* **distorted instance** *for a pattern $P = r_0r_1\ldots r_{T-1}$ with length P_{len}, with respect to τ, if there exist P_{len} ordered locations in $s^j[d]$ such that (1) these locations follow the order of non-$*$ elements in P; and (2) for every non-$*$ element in P, its period offset differs at most τ from the period offset of its related location in $s^j[d]$.*

EXAMPLE 15.3

Consider a segment $s^0 = l_0l_1l_2l_3l_4$ and let $\tau = 1$. If $l_1 \in r_0$, $l_2 \in r_2$, and $l_4 \in r_3$, s^0 is a distorted instance of pattern $P = r_0 * r_2r_3*$.

Two pattern instances (segments) *overlap* if they have some locations in common. For example, $s^0[1] = \{l_1, l_2, l_3, l_4, l_5\}$ overlaps with $s^1 = \{l_5, l_6, l_7, l_8, l_9\}$ since they have l_5 in common.

Definition 15.9 *If a pattern P has more than min_sup.m (shifted/distorted) pattern instances in $\mathcal{S}$, such that no two instances overlap, then P is a frequent pattern with shifted/distorted instances. Given a sequence $\mathcal{S}$, minimum support min_sup* ($0 < min\_sup \leq 1$)*, cluster parameters* $\in$ *and MinPts, and the maximum shifting/distortion parameter* $\tau (0 \leq \tau \leq \lfloor T/2 \rfloor)$*, the* **problem of discovering shifted/distorted patterns** *aims at finding all frequent patterns with shifted/distorted instances from $\mathcal{S}$.*

The discovery of frequent patterns with shifted/distorted instances also involves two stages. In the first stage, a similar procedure in Section 15.3.3.1 is applied to find 1-patterns. Different from Section 15.3.3.1, which generates a single point in the corresponding dataset R_i and applies clustering to each R_i, this stage generates points for an object location at offset position i at all τ-neighbor positions $R_{(i-\tau) \bmod T}$, $R_{(i-\tau+1) \bmod T}, \ldots, R_{(i+\tau) \bmod T}$. Consider the 5th position of day 1 in Figure 15.3a and assume that $\tau = 1$. Instead of generating a single '□' point at that location, the algorithm generates one '□' point (to file R_5), one '+' point (to file R_4), and one '×' point (to file R_6). In the second stage, STPMine2-V2 is adopted to facilitate counting of longer (shifted/distorted) pattern instances. The algorithm works in a top-down manner by starting the pattern validation from the max-subpattern P_{max}. The subpatterns of P_{max} are examined level-by-level. For each candidate subpattern P at a level, the algorithm sequentially fetches one location from the segment-id set for every non-* position. If the set of locations forms a valid shifted/distorted pattern instance, these locations contribute one to the support of P. The support of P is the total number of nonoverlapping location sets, which intrinsically form pattern instances. The details of finding nonoverlapping location sets can be found in Reference [4].

EXAMPLE 15.4

*Assume two clusters $r_0 = \{l_0, l_5, l_{10}\}$ and $r_1 = \{l_1, l_5, l_6, l_7, l_{11}\}$. Consider a candidate pattern $P = r_0 r_1$ *** and let $\tau = 1$. (l_0, l_1) is the first point-id pair from the two sets, falling into the same segment, so they contribute 1 to $|P|$. Then, (l_5, l_6) forms another segment and adds 1 to P's frequency. The final contributing pair is (l_{10}, l_{11}).*

15.4 OPEN ISSUES

The current research on trajectory analysis is still at a starting stage due to the lack of a consistent theoretical model for representing and accessing such data. Therefore data mining on such data is ad-hoc; analysis tasks are defined from application requirements and appropriate algorithms are designed for them. In the following, we outline some open issues for future research in this field.

- A fundamental theory for modeling trajectory data and their access/analysis should be defined. Modeling is not an easy process; therefore, it should be done gradually. To begin with, a set of typical analysis tasks should be defined and benchmark data should be provided for them.
- Similar to the management of other types of complex data (e.g., spatial, temporal, multimedia, etc.), it is necessary to develop a systematic

framework that combines the dominant methods in managing and analyzing trajectory data.

- The current trajectory data mining algorithms require users to provide values for several parameters. This is neither desirable nor applicable in many cases. In addition, different parameters may cause instability to the system. Due to these reasons, for typical tasks, some heuristics or models for setting and tuning parameters are required. Benchmark datasets and mining tasks should be provided to allow for the testing of such models.
- Real applications impose additional requirements to data trajectory analysis. For example, the locations of objects may not be certain due to translation delay or collection granularity. It is desirable to consider uncertainty in analyzing trajectory data.

15.5 CONCLUSION

In this chapter, we discussed the problem of periodic pattern discovery from trajectory data. Such patterns capture movement regularities for objects that periodically visit specific places at specific times. We presented three algorithms, which not only discover the patterns but also find the pattern regions automatically. Two variants of the basic periodic patterns definition are discussed. In the first variant, the objective is to find the patterns and the maximal time interval in which the patterns are of interest, while patterns in the second variant get supported by shifted/distored instances of the core pattern description. Finally, we discussed open issues in trajectory data mining, which will hopefully stimulate future research in the field.

REFERENCES

[1] R. Agrawal and R. Srikant. Fast algorithms for mining association rules. In *Proc. of Very Large Data Bases Conference*, pp. 487–499, 1994.

[2] H. Cao, N. Mamoulis, and D. W. Cheung. Mining frequent spatio-temporal sequential patterns. In *Proc. of IEEE International Conference on Data Mining*, pp. 82–89, 2005.

[3] H. Cao, N. Mamoulis, and D. W. Cheung. Discovery of collocation episodes in spatiotemporal data. In *Proc. of IEEE International Conference on Data Mining*, pp. 823–827, 2006.

[4] H. Cao, N. Mamoulis, and D. W. Cheung. Discovery of periodic patterns in spatiotemporal sequences. *IEEE Trans. Knowl. Data Eng.*, 19(4):453–467, 2007.

[5] M. Celik, S. Shekhar, J. P. Rogers, J. A. Shine, and J. S. Yoo. Mixed-drove spatiotemporal cooccurence pattern mining: A summary of results. In *Proc. of IEEE International Conference on Data Mining*, pp. 119–128, 2006.

[6] M. Ester, H. P. Kriegel, J. Sander, and X. Xu. A density-based algorithm for discovering clusters in large spatial databases with noise. In *Proc. of International Conference on Knowledge Discovery and Data Mining*, pp. 226–231, 1996.

[7] S. Gaffney and P. Smyth. Trajectory clustering with mixtures of regression models. In *Proc. of International Conference on Knowledge Discovery and Data Mining*, pp. 63–72, 1999.

[8] F. Giannotti, M. Nanni, F. Pinelli, and D. Pedreschi. Trajectory pattern mining. In *Proc. of International Conference on Knowledge Discovery and Data Mining*, pp. 330–339, 2007.

[9] J. Gudmundsson, M. J. van Kreveld, and B. Speckmann. Efficient detection of motion patterns in spatio-temporal data sets. In *ACM International Workshop on Geographic Information Systems (GIS)*, pp. 250–257, 2004.

[10] M. Hadjieleftheriou, G. Kollios, D. Gunopulos, and V. J. Tsotras. On-line discovery of dense areas in spatio-temporal databases. In *Proc. of Symposium on Advances in Spatial and Temporal Databases*, pp. 306–324, 2003.

[11] M. Hadjieleftheriou, G. Kollios, V. J. Tsotras, and D. Gunopulos. Efficient indexing of spatiotemporal objects. In *Proc. of Extending Database Technology Conference*, pp. 251– 268, 2002.

[12] J. Han, G. Dong, and Y. Yin. Efficient mining of partial periodic patterns in time series database. In *Proc. of International Conference on Data Engineering*, pp. 106–115, 1999.

[13] P. Kalnis, N. Mamoulis, and S. Bakiras. On discovering moving clusters in spatio-temporal data. In *Proc. of Symposium on Advances in Spatial and Temporal Databases*, pp. 364–381, 2005.

[14] H. A. Karimi and X. Liu. A predictive location model for location-based services. In *ACM International Workshop on Geographic Information Systems (GIS)*, pp. 126–133, 2003.

[15] K. Laasonen. Clustering and prediction of mobile user routes from cellular data. In *European Conference on Principles and Practice of Knowledge Discovery in Databases*, pp. 569–576, 2005.

[16] P. Laube and S. Imfeld. Analyzing relative motion within groups of trackable moving point objects. In *Proc. of International Conference on Geographic Information Science (GIScience)*, pp. 132–144, 2002.

[17] Y. Li, J. Han, and J. Yang. Clustering moving objects. In *Proc. of International Conference on Knowledge Discovery and Data Mining*, pp. 617–622, 2004.

[18] N. Mamoulis, H. Cao, G. Kollios, M. Hadjieleftheriou, Y. Tao, and D. W. Cheung. Mining, indexing, and querying historical spatiotemporal data. In *Proc. of International Conference on Knowledge Discovery and Data Mining*, pp. 236–245, 2004.

[19] D. Pfoser, C. S. Jensen, and Y. Theodoridis. Novel approaches in query processing for moving object trajectories. In *Proc. of Very Large Data Bases Conference*, pp. 395–406, 2000.

[20] S. Shekhar and Y. Huang. Discovering spatial co-location patterns: A summary of results. In *Proc. of Symposium on Advances in Spatial and Temporal Databases*, pp. 236–256, 2001.

[21] J. Sun, D. Papadias, Y. Tao, and B. Liu. Querying about the past, the present, and the future in spatio-temporal. In *Proc. of International Conference on Data Engineering*, pp. 202–213, 2004.

[22] Y. Tao, G. Kollios, J. Considine, F. Li, and D. Papadias. Spatio-temporal aggregation using sketches. In *Proc. of International Conference on Data Engineering*, pp. 449–460, 2004.

[23] Y. Tao and D. Papadias. MV3R–tree: A spatio-temporal access method for timestamp and interval queries. In *Proc. of Very Large Data Bases Conference*, pp. 431–440, 2001.

[24] M. Vlachos, G. Kollios, and D. Gunopulos. Discovering similar multidimensional trajectories. In *Proc. of International Conference on Data Engineering*, pp. 673–684, 2002.

[25] J. Wang, W. Hsu, and M.-L. Lee. A framework for mining topological patterns in spatiotemporal databases. In *Proc. of Conference on Information and Knowledge Management*, pp. 429–436, 2005.

[26] H. Yang, S. Parthasarathy, and S. Mehta. A generalized framework for mining spatio-temporal patterns in scientific data. In *Proc. of International Conference on Knowledge Discovery and Data Mining*, pp. 716–721, 2005.

[27] G. Yavas, D. Katsaros, O. Ulusoy, and Y. Manolopoulos. A data mining approach for location prediction in mobile environments. *Data and Knowledge Engineering*, 54(2):121–146, 2005.

16 Decentralized Spatial Data Mining for Geosensor Networks

Patrick Laube

Matt Duckham

CONTENTS

16.1 INTRODUCTION

Conventional approaches to geographic knowledge discovery and spatial data mining are founded on powerful, centralized algorithms that screen large data sets for interesting patterns and rules. Such global algorithms allow fast detection of interesting patterns if centralized access to the whole data set can be guaranteed.

However, new technologies for distributed spatial data capture and processing, such as geosensor networks, present new challenges to conventional knowledge discovery and data mining algorithms. Increasingly, access to the whole data set cannot be guaranteed; instead, multiple computing units, none of which possess global knowledge, must cooperate in knowledge discovery.

This chapter investigates the structure and design of *decentralized* algorithms for spatial data mining. Our increasing ability to collect data at finer and finer spatiotemporal granularities has the potential to generate such overwhelming volumes of data that the paradigm of central processing is no longer practicable (Kargupta and Chan 2000). Instead, knowledge discovery must descend into the network, detecting patterns as the spatial data is captured.

Specifically, the chapter contributes to the theory of geographic knowledge discovery and spatial data mining by

- identifying spatial data mining and knowledge discovery as a crucial application layer for geosensor networks, the latest technology for spatial data capture,
- exploring the notion of *decentralized spatial data mining* (DSDM) for geographic knowledge discovery, and
- presenting an overview of techniques for DSDM, including an investigation of the potential of DSDM for classical spatial data mining applications, such as clustering.

The remainder of this chapter is organized as follows. Section 16.2 surveys the relevant background literature for DSDM. Section 16.3 presents the concept of decentralized spatial data mining and proposes a set of generic strategies for DSDM algorithms. The chapter then investigates a specific case study of different decentralized algorithms for spatial clustering (Sections 16.4 and 16.5). Finally, the chapter concludes with a discussion of the results in Section 16.6 and the formulation of a research agenda for DSDM (Section 16.7).

16.2 BACKGROUND

16.2.1 Distributed and Decentralized Spatial Computing

A *distributed system* is defined as a collection of multiple information system units that synchronously cooperate via a communication network to complete some computing task (Worboys and Duckham 2004). A *wireless sensor network* (WSN, ad-hoc wireless networks of sensor-enabled miniature computing platforms, Zhao and Guibas 2004) is a form of distributed system, where individual sensor nodes cooperate to ensure the network as a whole can meet the requirements of the specific application. Applications of WSN in the spatial domain include environmental monitoring (Duckham et al. 2005, Werner-Allen et al. 2006), smart farming (Wark et al. 2007), traffic management (Kellerer et al. 2001), and robotics (Correll and Martinoli 2006). Considerable recent research activity in the area of WSN has focused on the issues surrounding the establishment and maintenance of the communication networks necessary for distributed computing (e.g., Braginsky and Estrin 2002, Cheng and Heinzelman 2005), including many ingenious techniques using the spatial characteristics of the network for that purpose (e.g., Karp and Kung 2000, Mauve et al. 2001, Yu et al. 2001, Xu et al. 2001).

In many systems that are commonly referred to as "distributed," the cooperating information systems each take responsibility for logically or functionally distinct sub-tasks. For example, the architecture of distributed client-server systems is typically founded on a clear delineation of the distinct services provided and consumed by different logical units (e.g., the classic three-tier client-server architecture of Web browser, Web server, and spatial database server used in Web mapping applications, Worboys and Duckham 2004). However, in some distributed systems, such as peer-to-peer networks, there is no such partitioning of sub-tasks; multiple units in the distributed system may have similar or equivalent responsibilities. In such systems, specific processing tasks may be distributed throughout the network, each individual unit performing a small part of the required processing. Here we reserve the term *decentralized* for describing these distributed systems and algorithms, where the processing task itself is distributed throughout the network and no component of the distributed system "knows" the entire system state (Lynch 1996).

A *geosensor network* is defined by Nittel et al. (2004) as a WSN that monitors phenomena in geographic space. A geosensor network is, therefore, also a type of distributed system. There are four main reasons why decentralized algorithms are important in geosensor networks:

- *Energy resources*: WSN are highly resource-constrained systems, especially with respect to sensor node energy resources (Zhao and Guibas 2004). Wireless communication is one of the most energy-intensive activities of a sensor node, so continually relaying data to a central system can dramatically shorten the useful lifetime of a WSN.
- *Information overload*: The fine-grained detail becoming available from larger sensor networks means that individual data items become less and less meaningful. Transmitting all data can lead to high levels of redundancy and ultimately information overload (Rabiner et al. 1999, Datta et al. 2006a).
- *Scalability*: As networks scale from tens to thousands to millions of nodes, effective centralized control of the network becomes impossible. The issue of scalability is especially important in geosensor networks, which must by definition contain a large number of nodes in order to provide enough spatial detail to monitor geographic phenomena (Estrin et al. 1999).
- *Sensor/actuator networks*: The results of the analysis of sensor network data are often required by the network itself in order to modify the behavior

of the network (e.g., activate or deactivate sensors to adapt the granularity of monitoring of important events, Duckham et al. 2005). Removing information from the network, processing it centrally, and then returning it to the network is an inefficient drain on network resources.

The key challenge of decentralized, in-network processing is to use "decentralized coordination with local decision making to achieve the intended global goal" (Estrin et al. 2000, p. 40); in other words, to generate *global* knowledge using *local* processes (local in this context refers to a node and its immediate neighborhood or locality). Thus, in decentralized spatial computing we are interested in developing algorithms that can operate using purely local knowledge, but are still able to monitor geographic phenomena with global extents. This is a very different approach from conventional spatial computing paradigms (exemplified by GIS) where processes (e.g., spatial analysis routines) operate upon entire data sets (e.g., stored in a spatial database).

16.2.2 Centralized (Geographic) Knowledge Discovery and Data Mining

Conventional *knowledge discovery in databases* (KDD) and its most prominent step, *data mining*, rely on data available at a single location. Association rule mining, for example, is based on global counts of frequent item sets in order to compute support for and confidence in a rule (Gidofalvi and Pedersen 2005). Similarly, many point pattern measures used for clustering purposes depend on globally fixed criteria such as "nearest neighbor" or "neighbors within 50m" (O'Sullivan and Unwin 2003). Even though classical spatial data mining patterns fundamentally depend on local spatial relations (Shekhar et al. 2003), the vast majority of current algorithms for detecting these patterns rely on global data structures and algorithms.

For example, clustering is a classic spatial data mining technique that organizes observations into coherent and contrasted groups. Clustering approaches are normally classified into two categories: *hierarchical* and *partitional* clustering. Hierarchical clustering techniques establish a nested hierarchy of clusters by successively building new clusters based on previously merged leaves in the clustering tree (O'Sullivan and Unwin 2003). At every step, merging the closest clusters requires finding the smallest distance of any pair in the distance matrix for the whole data set, and hence relies on global knowledge. The most common *partitional* algorithm is k-means clustering. The algorithm is based on an initial assignment of all observation to k randomly seeded cluster heads followed by successive improvement of the partitioning by iterative reassessment of computed mean centers of the partitions. In its conventional variants, k-means clustering assumes global knowledge (however, see Section 16.2.3).

Spatial and spatiotemporal clustering often takes a different perspective and first asks *if* there are clusters at all in some given point distribution. In the field of spatial statistics, a series of techniques have been developed to quantify the randomness of a point distribution, including density-based methods (quadrat count, kernel estimation) and distance-based methods (nearest neighbor, distance functions, O'Sullivan and Unwin 2003). The locations of potential clusters are then the focus of a second stage. A common application would be the identification of crime hot spots or the origin of an infectious disease (Shekhar et al. 2003). Again, these centralized algorithms require global access to data to operate.

In summary, conventional data mining approaches allow efficient screening for patterns and rules, given the proviso that all data is available at a single location, data structures are centralized, and algorithms are omniscient.

16.2.3 Distributed Data Mining

The emergence of network-based distributed computing environments has added a new dimension to knowledge discovery in databases and data mining. Distributed data mining (DDM) has evolved over the last decade in an attempt to develop distributed versions of many standard data mining algorithms (Datta et al. 2006a). DDM embraces the growing trend of merging computation with communication. DDM aims at finding patterns and rules from distributed and heterogeneous data using minimal communication (Kargupta and Chan 2000). Privacy concerns, as well as bandwidth and resource constraints in distributed systems, often dictate that data collected at different nodes be analyzed in a decentralized fashion, without collecting everything to a central site (Datta et al. 2006a). The limitations of using purely local knowledge means distributed data mining often focuses on *approximate* algorithms that may not always match the *exact* answers provided by conventional centralized data mining algorithms (Datta et al. 2006a). The goal remains to derive new and useful information, but potentially to sacrifice a small degree of certainty for substantial computational gains.

Taking for instance clustering, there exists ample research on distributed clustering algorithms (see Bandyopadhyay et al. 2006 for an introductory overview). For example, clustering of sensor nodes can be used for communication load balancing in *ad hoc* sensor networks (Younis and Fahmy 2004). Other authors have explicitly focused on distributed clustering for data mining purposes (Bandyopadhyay et al. 2006) and even dynamic distributed networks (Datta et al. 2006b). However, most distributed clustering approaches thus far focus on partitional clustering, assigning the nodes of a network to a given number of k cluster heads. Apart from clustering there also exists work on distributed in-network association rule mining (Wolff and Schuster 2004) and outlier detection (Branch et al. 2006).

The distributed data mining field shows a growing interest in distributed partitional clustering. However, decentralized spatial data mining, for example decentralized cluster detection and localization in spatial point distributions, remains an open research task.

16.3 DECENTRALIZED SPATIAL DATA MINING (DSDM)

This section discusses the concept of *decentralized spatial data mining* (DSDM). Many of the patterns and rules of interest in conventional spatial data mining are defined based on *local* inter-object relationships, including density clusters and colocation patterns. This spatial locality is exploited by decentralized algorithms for DSDM. The section first defines the problem addressed by DSDM, then proposes a series of general strategies for DSDM.

16.3.1 Problem Definition

DDM is the attempt to develop distributed versions of standard data mining algorithms (Datta et al. 2006a). Similarly, the aim of DSDM is the introduction of decentralized algorithms for spatial data mining. Like spatial data mining, DSDM gains its strength from exploiting the special characteristics of spatial information (primarily, spatial autocorrelation). Hence, DSDM is not about *functionally* distributing a complex task among cooperating sub-systems. Instead, DSDM attempts to *spatially* distribute a global task throughout a decentralized network, each individual computing unit relying on local data and processes to operate.

Consequently, DSDM can be defined as the process of discovering new spatial patterns within a distributed system using decentralized algorithms with no central coordination operating upon locally defined spatial data.

For example, consider the simple spatial cluster detection illustration in Figure 16.1. The black observations lie in two distinctive density clusters. The task of DSDM is to detect these clusters using decentralized data mining algorithms that operate in computing nodes themselves. Each node is expected to be able to perceive only local information about its own and its immediate neighbors' geographic environment, but there exists no node that can perceive the entire geographic space.

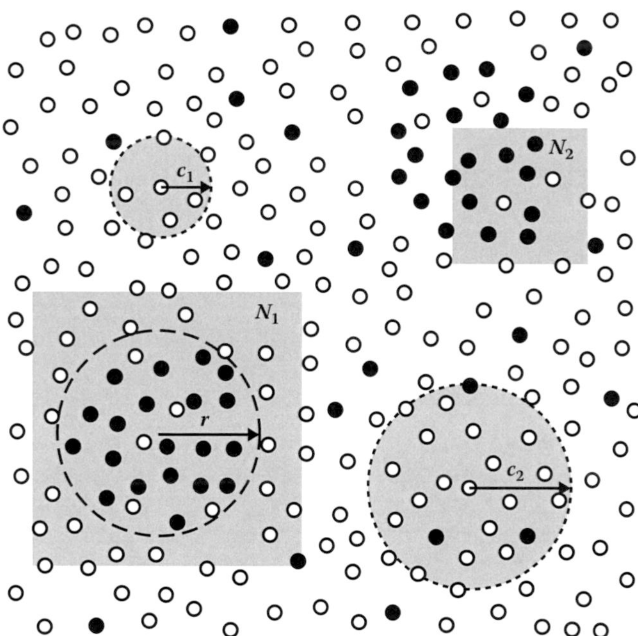

FIGURE 16.1 Decentralized clustering. A set of V observations features two clusters of hot nodes. Detecting such a clustering is easier given a larger neighborhood N_1 covering the whole cluster, but more difficult in a smaller neighborhood N_2 that only covers parts of the cluster. Applied to a geosensor network scenario, where the observations represent the nodes, the DSDM task is the decentralized detection of the clusters.

Environmental monitoring using a geosensor network would be one specific application of the previous example. The observations represent sensor nodes in a geosensor network, physically distributed in space, and monitoring some environmental variable, such as temperature. Given some threshold t, then the clusters of "hot" nodes might represent temperature "hot spots," where the temperature is above t. Each individual sensor node is expected to possess only partial spatial knowledge about its own temperature and the temperature of its immediate neighbors. The DSDM task for this geosensor network is then to detect the temperature hot spots using only its local knowledge about its own and its neighbors' observations.

Obviously, the size of the neighborhood is a central parameter in this example. The larger the neighborhood, the more the problem resembles a conventional centralized, global spatial data mining problem. Neighborhood N_1 , for example, covers the entire cluster bottom left and it should potentially be straightforward for a node with only local knowledge of that neighborhood still to detect the pattern. By contrast, neighborhood N_2 only covers a fraction of the whole cluster top right, making decentralized detection of this cluster much harder. In the context of a geosensor network, the size of the neighborhood will be determined to a large extent by the communication range of individual nodes. The technical and physical limitations of communication in geosensor networks means that in general it is to be expected that each sensor node can only communicate with a tiny fraction of the nodes in the entire network: those in its immediate spatial vicinity.

16.3.2 Formal Problem Definition

In this section we more precisely specify the problem outlined before using a formal model of geosensor networks. For simplicity we assume only a static geosensor network, where nodes are immobile. However, later sections indicate how this model can be extended to deal with dynamism.

16.3.2.1 Geosensor Networks

As indicated previously, the key features of a geosensor network are the nodes and short-range radio frequency (RF) communication links between nearby nodes. The most commonly used model of such a network is a graph, where vertices in the graph model nodes in the geosensor network and edges in the graph model the potential for communication between neighboring nodes. Such a graph is static (nodes do not move and edges are fixed), and can be formally defined as in Definition 16.3.2.1.

Definition 16.3.2.1 *A* geosensor network *may be modeled as a graph $G = (V, E)$, where V is the set of vertices (sensor nodes) and $E \subseteq V \times V$ is the set of edges (communication links) between neighboring nodes. For a node $v \in V$, its neighborhood $\{v' \in V \mid \{v, v'\} \in E\}$ is written $nbr(v)$.*

Note that by adopting an *undirected* graph to model a geosensor network (as in Definition 16.3.2.1), we are implicitly assuming symmetric bidirectional communication: If node a can communicate with node b, then node b can communicate with node a. While this is a natural and common simplifying assumption in geosensor networks, in actuality it does not always hold (Min and Chandrakasan 2003). In more sophisticated situations, a directed graph might be required to model any communication asymmetry.

We can further model the location of a sensor node as a *locator* function (Definition 16.3.2.2).

Definition 16.3.2.2 *A (static)* locator *is a function* $l : V \rightarrow \mathbb{R}^n$, *where for any vertex* $v \in V$, $l(v)$ *maps to the coordinate location of that node (where n is 2 or 3). The distance function* $\delta : V \times V \rightarrow \mathbb{R}$ *is the usual metric for Euclidean distance between nodes.*

An implicit assumption is commonly made that no two nodes occupy the same location (i.e., the locator function is an injection). Because the communication links between nodes are constrained by the physical limitations of RF communication, the locator function can be used to generate the set of edges E for a particular set of vertices V assuming a maximum communication range c. For example, the *unit distance graph* (UDG) is the graph formed when all nodes that are within communication range may potentially communicate (Definition 16.3.2.3).

Definition 16.3.2.3 *Given a maximum communication distance c, the UDG is the geosensor network* $G = (V, E)$ *where* $E = \{(u, v) \in V \times V | 0 < d(u,v) \leq c\}$.

Note that in addition to assuming symmetric, bi-directional communication, the UDG also assumes a constant communication distance across the entire network. Again, the actual situation may in practice be more complex.

Often in spatial applications it is more useful to assume that only a subset of the communication links in the UDG are available. In particular, subsets of the UDG that form planar graphs (such as triangulations) are commonly used in specific spatial applications (e.g., Karp and Kung 2000, Worboys and Duckham 2006). Common planar subsets of the UDG include the relative neighborhood graph (RNG) and the Gabriel graph (GG) (Zhao and Guibas 2004).

It is important to note that although the underlying communication graph is usually constructed with reference to a locator function l, we do not necessarily assume that an individual node is location-aware (i.e., has access to knowledge about its own location). In some geosensor networks, all nodes may be location aware. However, the technical limitations of achieving high precision and accuracy location of nodes in a geosensor network means that where possible it is safer to assume nodes are only able to determine their qualitative location in terms of the nodes in their immediate neighborhood. Despite this limitation, it is possible to generate many interesting spatial properties and behaviors using only such qualitative location information, as we will see later in Section 16.4.

Finally, the geosensor network is assumed to be monitoring some environmental variable in space using its sensors. The environmental variable may itself be highly structured and complex; however, in this chapter we assume the most simple domain for an environmental variable: Boolean values.

Definition 16.3.2.4 *The sensor data for the set of nodes V can be represented using a (static) sensor function* $s: V \rightarrow D$, *where D is the domain for some environmental*

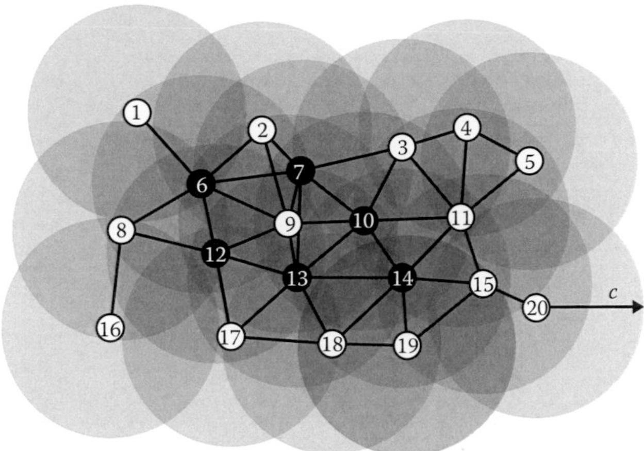

FIGURE 16.2 Summary of formal model. Geosensor network $G = (V, E)$, where $V = \{v_1, \ldots, v_{20}\}$ and $E = \{\{v_1, v_2\}, \ldots\}$ is the UDG based on communication distance c. Hot nodes, where $s(v) = 1$, are shown in black; cold nodes are shown in white.

variable. In this chapter we assume a Boolean domain $D = \{0,1\}$. *Any node* v *where* $s(v) = 1$ *(i.e., that can detect the environmental variable) is termed a "hot" node; any node* v *where* $s(v) = 0$ *(i.e., that cannot detect the environmental variable) is termed a "cold" node.*

Figure 16.2 summarizes the formal model of the geosensor network, showing the UDG for a small group of hot (black) and cold (white) nodes.

16.3.2.2 Clusters

The decentralized algorithms introduced in later sections are designed to find *clusters*, such as those in Figure 16.1. Here we adopt a simple definition of a cluster as a set of at least n related observations that lie within a circle of radius r spatial region (Definition 16.3.2.5).

Definition 16.3.2.5 *Given a geosensor network* $G = (V, E)$, *a locator function* $l : V \rightarrow \mathbb{R}^2$, *and a sensor function* $s : V \rightarrow \{0,1\}$, *an (nr) cluster is defined as a set of nodes* $V' \subseteq V$ *such that* $|V'| \geq n$; *there exists some circle* e_r *of radius r such that for all* $v \in V$, $l(v)$ *is spatially contained within* e_r*; and for all* $v \in V'$, $s(v) = 1$.

Note that this definition concerns only the *static* scenarios outlined before, although a more sophisticated dynamic clustering definition can also be defined based on the previous definitions.

The problem facing a DSDM algorithm is how to detect a cluster of n nodes within a circle of radius r, when individual nodes $v \in V$ can only communicate with their immediate neighbors, v' such that $\{v,v'\} \in E$. If we assume the neighborhood of a node is defined by the UDG, then the neighborhood of v depends on the communication radius c. Assuming c is substantially larger than r, then finding a cluster is relatively straightforward: each node can locally look for a cluster only in its immediate neighborhood. However, when c and r are similar, or c is smaller than r, more sophisticated strategies are required.

16.3.3 Compensation Strategies

Based on the previous problem definition, it is possible to describe four classes of strategies for decentralized spatial data mining as follows.

16.3.3.1 Local Extrapolation

In local extrapolation, nodes infer knowledge about patterns that extend their local knowledge range. For example, if $c = \frac{1}{2}r$ a node v can expect to receive knowledge only on approximately $\frac{1}{2}^2 = \frac{1}{4}$ of the entire pattern extent from its immediate neighbors $nbr(v)$. Thus, if a node detects $\frac{1}{4}$ of the required cluster (i.e., for all $v' \in nbr(v)$, $s(v') = 1$ but $|nbr(v)| \approx \frac{1}{4}n$), it may locally infer that the cluster has been detected. For example, the clusters in Figure 16.3 consist of 20 hot nodes. In the top left case, the node with communication range c has 5 hot neighbors and hence naively assumes that it has detected a cluster.

16.3.3.2 Local Absorption

Nodes can reach beyond the limits of their communication ranges by absorbing their neighbors' knowledge. In other words, nodes propagate their knowledge by locally restricted flooding (Zhao and Guibas 2004). If $c = \frac{1}{2}r$, nodes may reach out to the edges of patterns by relaying knowledge about their neighbors to their other neighbors (termed two-hop communication). If the central node in the example top right in Figure 16.3 receives knowledge from neighbors up to two hops away, it will be able to locally infer the presence of the clustering pattern. The logical extreme of local absorption is to flood knowledge throughout the network (multihop communication). In such a case, every node could possess knowledge about the state of all other nodes in the network. However, as already discussed, the physical and technical limitations of sensor networks makes such an approach unscalable and impractical. Hence, local absorption must typically be limited to only a few hops.

16.3.3.3 Selective Collaboration

A third compensation strategy is to invoke more targeted communication between nodes only if some pre-defined condition is met. Such conditions are similar to *certificates* used for kinetic data structures (Guibas 2002). In that context, data structures for dynamic systems only update when some local certificate (that is, some elementary relations among the objects involved) is violated. In a decentralized data mining context, nodes do not communicate until they have good reason to believe that they might be involved in a pattern. Then nodes select other nodes (typically close neighbors) to solve collaboratively the task at hand. In Figure 16.3 (bottom right) the central node has detected a certain number of hot nodes within communication range and established a collaboration with six nearby nodes in order to cover the pattern. Since the selected nodes may not necessarily be in the immediate neighborhood of a node (i.e., $n' \notin nbr(n)$), selective collaboration may require more sophisticated routing protocols

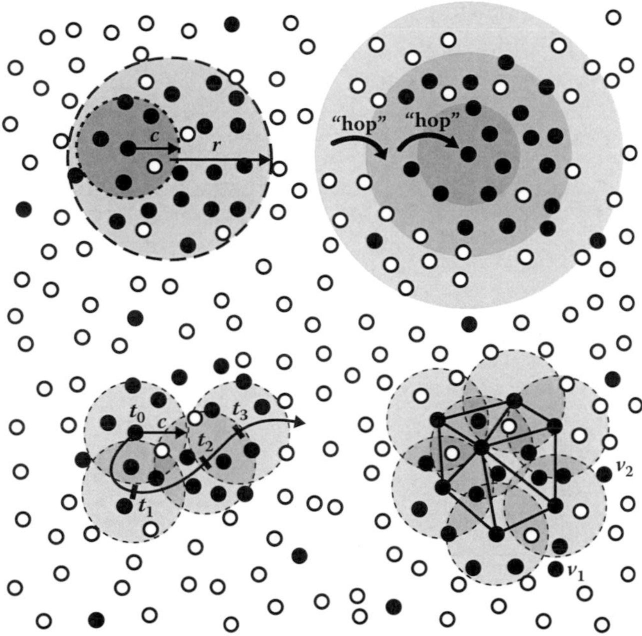

FIGURE 16.3 Four compensation strategies. Clockwise from top left: local extrapolation, local absorption, selective collaboration, node mobility.

to organize communication between remote nodes. Schemes for such routing protocols are legion in the literature, so in this chapter we do not consider this issue further.

16.3.3.4 Node Mobility

In more dynamic situations than considered thus far, an important possibility is for nodes to extend their spatially limited communication range through mobility. As mobile nodes move around the geographic space, they "see" different parts of the geographic area and can potentially communicate with different neighbors.

Given a set T of discrete, totally ordered times $\{t_1, \ldots, t_n\}$, we can extend the formal definitions presented in Section 16.3.2 to model mobility. Assuming the number of nodes in the geosensor network is constant, the mobility of nodes can be modeled with a dynamic locator function, $l : V \times \mathrm{T} \to \mathbb{R}^n$, where for any vertex $v \in V$ and time $t \in T$, $l(v, t)$ maps to the coordinate location of that node at time t. The dynamic neighborhoods can be modeled as a dynamic graph, where the set of edges changes over time. For example, given a maximum communication distance c, the *dynamic* UDG can be defined as $G(t) = (V, E(t))$ where $E(t) = \{(u, v) \in \mathrm{V} \times \mathrm{V} | 0 < \delta(l(u, t), l(v, t)) \leq c\}$. Similarly, the changing environmental variables sensed by a node can be modeled as a dynamic sensor function $s : V \times T \to \{0, 1\}$.

Node mobility opens at least two options for exploiting mobility for knowledge discovery (Grossglauser and Vetterli 2006, Grossglauser and Tse 2002). First, nodes might "graze" information while moving and store it in a constantly updated memory, termed *mobility memory*. Formally, an individual node v tracks its sensed values $s(v, t)$ and potentially its location $l(v, t)$ over a range of times t, and combines that knowledge in its pattern detection algorithm. Figure 16.3 (bottom left) illustrates a node passing through a pattern and thereby collecting enough information to reason about the presence of a pattern.

Mobility memory can operate even without any communication between nodes. However, knowledge discovery can clearly be improved by additionally enabling nodes to exchange information with their constantly changing neighbors while moving, termed *mobility diffusion*. Formally, an individual node v may communicate information with its neighbors $\{v'|\{v, v'\} \in E(t)\}$ over a range of times t. In some senses, mobility diffusion can be regarded as inexpensive variant of local absorption, since mobility (rather than multihop communication) is used to move information around the system beyond a node's immediate neighbors at a particular time (Grossglauser and Tse 2002).

Combinations of these individual strategies can be used, and indeed are expected to be more effective than strategies used in isolation. An obvious combination is the use of local extrapolation as a preliminary for other strategies. Local extrapolation can be used to establish a local state of belief about the presence of a pattern. If this state of belief reaches some threshold, it may trigger one of the other more involved methods in order to derive further information.

16.4 DECENTRALIZED SPATIAL CLUSTERING ALGORITHMS

In this section we present two algorithms implementing two of the previously mentioned compensation strategies, local extrapolation and local absorption. Although work is ongoing on examples of algorithms in all four categories, in this chapter we restrict the discussion to these two cases because they are relatively easy to grasp and representative of the issues faced in DSDM.

In both algorithms, nodes process locally collected knowledge about their neighborhood in order to develop a state of belief as to whether they have detected a cluster. As we like to refer to agents that have detected a pattern as "happy," our algorithms are termed *happiness extrapolation clustering* (HEC) and *happiness absorption clustering* (HAC), with a preliminary base-case algorithm termed *naive clustering*.

16.4.1 NAIVIE CLUSTERING (NC)

As discussed previously, where the communication range c is substantially greater than the cluster size r, it is potentially possible for a node to locally detect a cluster without any need for compensation strategies. Algorithm 16.1 presents such a naive base-case algorithm, where each node simply examines those nodes in its immediate neighborhood to determine whether it can locally detect a cluster. The algorithm cycles through every node (line 1.1); checks whether enough hot nodes to form a cluster are within its neighborhood (line 1.3); and if so whether they lie inside a circle of radius r (line 1.5).

Algorithm 16.1: *NC:* Naive cluster algorithm to check for each node whether it can sense a cluster in its local neighborhood

---

Data: Geosensor network graph $G = (V, E)$; locator function $l : V \rightarrow E$; sensor function
$s : V \rightarrow \{1, 0\}$; cluster radius r; cluster size n

foreach $v \in V$ **do**
    $X_v = \{v' \in V | s(v') = 1 \text{ and } v' \in nbr(v)\}$;
    if $|X_v| \geq n$ **then**
        $d = \max_{v1, v2 \in nbr(v)} (\delta(l(v_1), l(v_2)))$;
        if $d \geq r$ **then**
            Node v has detected a cluster of radius r and size n;

Several points are worth noting about this algorithm. First, the NC algorithm is obviously expected to fail in cases where $c < r$ or even $c \approx r$. Second, in Algorithm 16.1 several nodes may potentially detect the same cluster. The distinction between cases where a *particular* node detects a phenomenon, and where *some* node detects a phenomenon is an important one in DSDM. Normally in DSDM we are interested primarily in the latter situation, where *some* node detects a phenomenon because individual node behavior is not as important as the overall global network behavior. In other words, an algorithm can be regarded as successful as long as some node detects a phenomenon, although it may not matter whether any particular node detects it.

Unfortunately, even Algorithm 16.1 is not guaranteed to detect a cluster in all cases. For larger communication ranges, certain configurations of nodes could result in clusters with a relatively small radius r being missed. A more sophisticated algorithm to solve this would need to perform the computationally intensive process of checking the distances between different permutations of subsets of the entire set of neighbors. For simplicity, and in recognition of the limited processing power of sensor nodes, this more simple algorithm has been preferred here. As already noted, decentralized algorithms are often approximate, in this case potentially missing some clusters (error of omission), although never incorrectly identifying a cluster (error of commission).

It is also important to note that the algorithm also relies on some quantitative information about each node's location (location-awareness), or at the very least quantitative information about distances between nodes (e.g., using range-finding techniques). As already intimated, in many practical situations for geosensor networks, location-awareness may be unreliable or unavailable, limiting the applicability of such an algorithm.

16.4.2 Happiness Extrapolation Clustering (HEC)

HEC implements a basic form of local extrapolation for decentralized cluster detection. HEC is based purely on instantaneous and local neighbor counts. Every node in

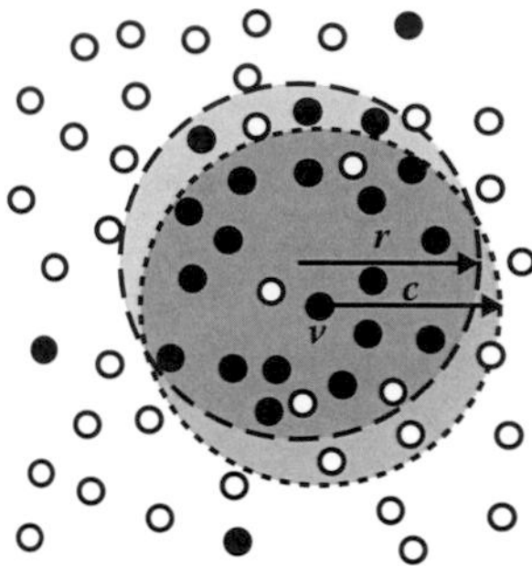

FIGURE 16.4 Limits of naive clustering. Nodes randomly distributed within a cluster are unlikely to be in a position to observe a cluster using the naive cluster algorithm for $c \approx r$. Even though placed centrally, node v misses three hot nodes for $c = r$.

parallel extrapolates its local knowledge and computes its belief in having detected a cluster, or in "being happy" (as illustrated earlier in Figure 16.1).

In performing the extrapolation, an important observation is that it is unlikely that a node will be located exactly in the middle of a cluster. Thus, even when $c \approx r$ the naive clustering algorithm is likely to fail. In such cases although it would be *possible* for a single node to observe the entire cluster, there is no *a priori* reason for expecting any node to be so conveniently located. The problem is illustrated in Figure 16.4, where although $c = r$, there exists no node that can be located in such a way to be able to detect the cluster of radius r.

Algorithm 16.2 addresses this problem using a threshold t (line 2.1) that adjusts a node's expectation of what proportion of a cluster size n it should see, given the known ratio between the communication and cluster areas (c^2/r^2). Figure 16.5 illustrates the threshold t in a graph, plotting the ratio between communication and cluster areas

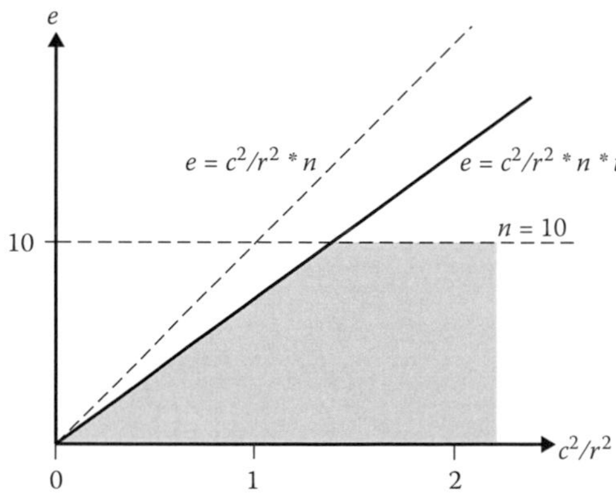

FIGURE 16.5 Expected numbers of hot neighbors. Graph of the ratio between communication and cluster areas against e, the expected number of hot neighbors, for a cluster where $n = 10$.

against the expected number of hot neighbors a node should see in a cluster where $n = 10$. In practice, t can be empirically determined. For example, for a cluster size of about $n = 10$ a t-value of approximately 0.8 provides the required adjustment (i.e., when communication range and cluster radius are the same, a node would expect to see at least 80% of the cluster).

If this adjusted expected number of hot neighbors is greater than the cluster size, then Algorithm 16.2 (line 2.3) resorts to the naive cluster algorithm (since we expect some node to be able to observe the entire cluster). Otherwise, Algorithm 16.2 cycles through each node, checking whether it has enough hot nodes in its neighborhood to justify a belief that it can see a cluster and hence be happy (lines 2.5 through 2.8).

Algorithm 16.2: *HEC*: Local extrapolation algorithm to check for each node if it can locally infer a cluster from its neighborhood

Data: Geosensor network graph $G = (V, E)$; locator function $l : V \rightarrow E$; sensor function
$s : V \rightarrow \{1, 0\}$; cluster radius r; cluster size n

```
2.1    Set x ← n * t * c²/r²;
2.2    if x ≥ n then
2.3        Use NC algorithm (naive cluster) to determine whether cluster is detected;
2.4    else
2.5        foreach v ∈ V do
2.6            X_v = {v′ ∈ V | s(v′) = 1 and v′ ∈ nbr(v)};
2.7            if |X_v| ≥ x ≥ 2 then
2.8                Node v has detected a cluster of radius r and size n;
```

Clearly, the smaller the communication range (and so the smaller the number of hot neighbors must be detected for a cluster), the more likely it becomes that a node misidentifies a random constellation of nodes as a cluster. Consequently, for small communication ranges the number of errors of commission (false positives) is expected to increase. The final threshold for HEC algorithms is $n = 2$ (line 2.8), since a "group" of one node provides no rationale to believe there are any other hot nodes nearby.

16.4.3 Happiness Absorption Clustering (HAC)

Local absorption aims at extending a node's limited communication range by absorbing knowledge from its neighbors. In the HAC algorithm, this knowledge is again simple neighbor counts, that is, the number of hot nodes within communication range. This time, however, nodes pass their local counts on to their neighbors, and after a limited number of hops, the aggregated knowledge is analyzed and used to compute expectations about the presence or absence of clusters and a node's happiness, respectively.

Algorithm 16.3 begins as for the HEC algorithm, resorting to the naive cluster algorithm if the communication range is high enough to enable nodes to expect to see entire clusters (line 3.1 to 3.3). Otherwise, for each node v a new parameter z_v is initialized as that node's initial happy value (i.e., 1 if they are hot, 0 otherwise, line 3.6).

Algorithm 16.3: *HAC*: Local absorption algorithm to check for each node if it can locally infer a cluster from its multihop neighborhood

---

Data: Geosensor network graph $G = (V, E)$; locator function $l : V \rightarrow E$; sensor function
$s : V \rightarrow \{1, 0\}$; cluster radius r; cluster size n, discount function d

3.1 Set $x \leftarrow n * t * \frac{c^2}{r^2}$;
3.2 **if** $x \geq n$ **then**
3.3 Use *NC* algorithm (naive cluster) to determine whether cluster is detected;
3.4 **else**
3.5 **foreach** $v \in V$ **do**
3.6 $z_v \leftarrow s(v)$;
3.7 **for** $i = 1$ *to ceiling* $(\frac{r}{c})$ **do**
3.8 **foreach** $v \in V$ **do**
3.9 $z_v \leftarrow z_v + \Sigma_{v' \in nbr(v)} z_{v'}$;
3.10 **foreach** $v \in V$ **do**
3.11 **if** $z_v / d(c, r, n) > n$ **then**
3.12 Node v has detected a cluster of radius r and size n;

Next, nodes communicate and aggregate their neighbors' z_v values for a number of hops (lines 3.8 to 3.9). The number of hops depends solely on the ratio of communication range r and the cluster radius p. Each time the (multihop) communication range drops below cluster radius, an extra hop is added in order to make sure that again an area of at least the cluster extent $r^2\pi$ is used for information collection and for reasoning about presence or absence of a pattern. If, for example, $p > r > \frac{p}{2}$, one additional hop is added in order to cover the whole cluster (see Figure 16.6).

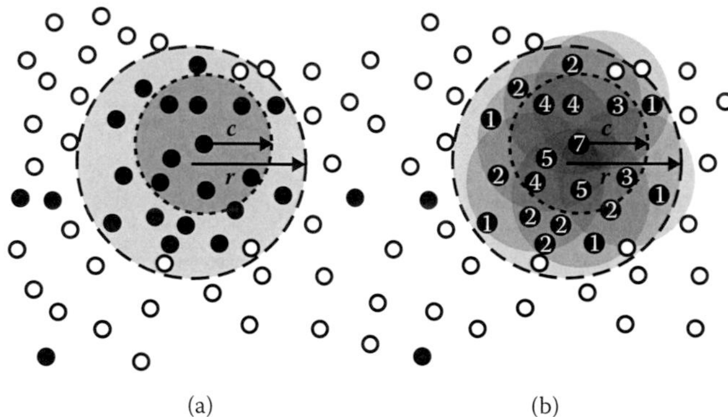

FIGURE 16.6 One additional absorption-hop in HAC. Additional hops result in overlapping communication ranges and hence multiple counting of found hot neighbors. The one-hop constellation in (b) illustrates for all hot nodes in the cluster their degree of over-estimation. A divisor has to be applied in order to discard redundant counts.

Using this happiness absorption method, nodes within clusters accumulate counts of hot neighbors, then neighbors' hot neighbors, and so on. Nodes not within clusters accumulate many fewer hot neighbors counts, if any. Because this procedure will count shared neighbor nodes more than once, the final node count needs to be discounted when deciding if a node has actually detected a cluster. For example, Figure 16.6 shows a two-hop constellation with several overlapping communication ranges. Instead of the actual 20 two-hop hot neighbors, double-counts mean the central node in fact observes 47 hot nodes. Consequently, a heuristic discount function d is used (line 3.11) to allow for the expected number of double counts. The discount function may depend on a number of factors, including the communication range c, and the cluster radius r and size n. As for the threshold t, the discount function can be empirically determined.

16.5 EXPERIMENTS

This section describes the results of experiments to compare the performance of the three algorithms, NC, HEC, and HAC, for decentralized detection of node clusters. The experiments were conducted using a popular free and open-source agent-based simulation and modeling toolkit, called Repast. Repast is implemented in several languages and features various libraries for simulation, visualization, and analysis. Each of the three algorithms was implemented in Repast, with sensor nodes modeled as agents (see Figure 16.7).

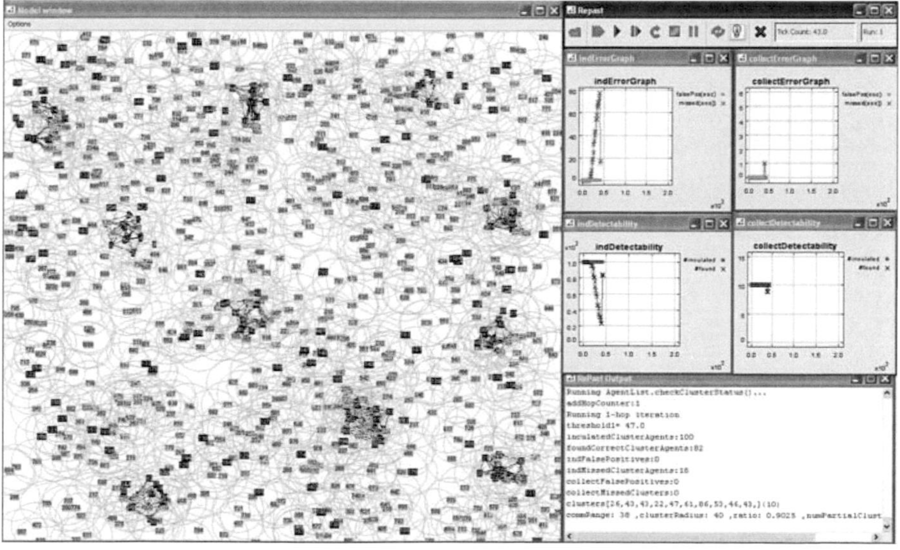

FIGURE 16.7 Implementation in Repast. The framework features a map view (communication ranges as gray circles, clustering nodes connected with edges), error plots (for errors of omission and commission), and a system log window.

16.5.1 Design

For each set of experiments, 1000 nodes were located in the square simulation space (set to have side length 1 unit). Ten nonoverlapping, but otherwise randomly located clusters of nodes were also generated. Each cluster consisted of 10 hot nodes with a cluster radius of 0.05 units. A further 100 hot nodes were also randomly distributed outside the clusters in order to reach a total of 200 hot nodes. Finally, a further 800 cold nodes were randomly distributed in the simulation space. Cold nodes were allowed to be located anywhere, including within existing hot clusters.

A set of experiments was then run, with each experiment varying the communication range r, starting from $c = 2r$ and decreasing step by step to 0. At each step the performance of the three algorithms was recorded. Performance was measured in terms of errors of omission (clusters that were placed in the simulation but not detected) and errors of commission (nonclusters that were incorrectly classified by a node as clusters) for each of the three algorithms. Errors of omission and commission were recorded against individual nodes (e.g., whether every node in a cluster correctly detected it was part of a cluster or not) as well as against individual clusters (e.g., whether *some* node in a cluster correctly detected it was part of a cluster). As discussed previously in Section 16.4.1, in the context of a distributed system it is the latter measure that is more important and so this measure is used in the following discussion of results.

16.5.2 Results

Figure 16.8 presents the results for the average performance of the NC, HEC, and HAC algorithms over several simulations. The x-axis represents the ratio of communication range c to cluster radius r, decreasing step-wise from $c = 2r$, through $c = r$, to $c = 0$. The y-axis shows error of commission (EOC, false positives) and error of omission (EOO, false negatives) expressed in number of clusters. As a consequence, when algorithms are performing well they will have zero or low corresponding values on the y-axis, and conversely high values when performing badly.

Figure 16.8 shows that the NC algorithm performs near-perfectly when the communication range is strictly larger than the cluster radius, in the range of 2 to 1.25 for the c/r ratio. However, below 1.25 for the c/r ratio, the performance of the NC algorithm degrades rapidly. This result is to be expected since at such high communication radii individual nodes can be expected to be able to detect entire clusters in their immediate neighborhood.

Since the HEC and HAC algorithms revert to the NC algorithm for larger c/r ratios, these algorithms similarly perform well in the 2 to 1.25 c/r ratio range. However, the HEC algorithm exhibits a clear improvement in performance over the NC algorithm, exhibiting on average less than two errors of omission or commission, in the 1.25 to 0.75 c/r ratio range. The local extrapolation adopted by HEC helps to extend its operating range beyond that of the NC algorithm. Below the 0.75 c/r ratio range, HEC algorithm performance also degrades rapidly, mirroring the fact that with decreasing c/r ratio, there is a greater chance that small groups of two or three hot nodes can falsely trigger cluster detection. The HAC algorithm exhibits further improvement on the HEC algorithm, finally degrading at beyond about the 0.4 c/r ratio mark. The local absorption

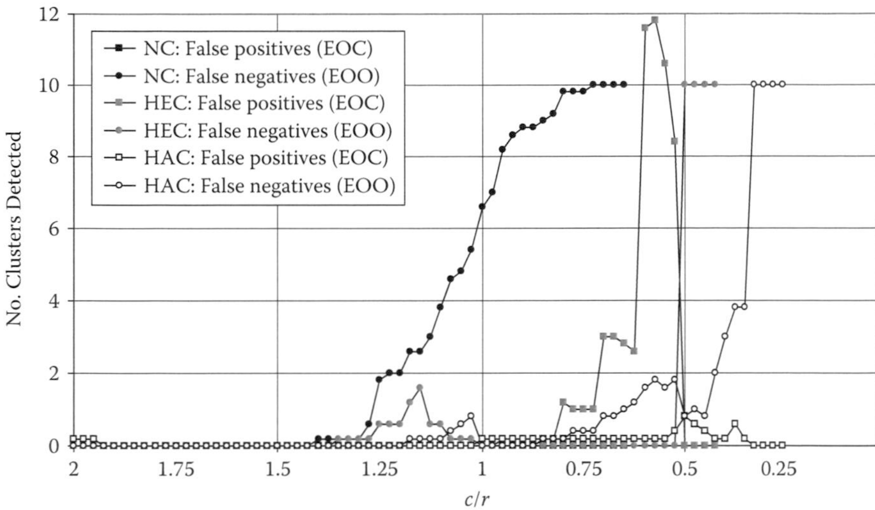

FIGURE 16.8 Experiment results. Errors of omission and commission for NC, HEC, and HAC algorithms.

used in HAC is able to extend the range of the cluster detection into the zone where individual nodes detect only a very small number of neighbors, and so where HEC fails.

16.6 DISCUSSION AND CONCLUSIONS

The experiments in the previous section provide specific examples of *decentralized* algorithms for spatial data mining. The three algorithms presented illustrate how increasingly sophisticated DSDM algorithms can be designed to deliver step improvements in performance. For the specific example of clustering, a naive decentralized algorithm is bettered by a local extrapolation algorithm, which in turn is outperformed by a local absorption algorithm. Current work is also investigating the further improvements that can be gained from using selective collaboration and, in cases where nodes are mobile, mobility memory and diffusion.

Because DSDM is often focused on efficient but approximate algorithms, (that may not match the solution generated using an exact centralized algorithm (Datta et al. 2006a), the performance of DSDM algorithms is primarily measured in terms of the certainty of its outcomes. Hence, we use the errors of omission and commission to assess our algorithms' performance. The results indicated the range of conditions under which the algorithms generate reliable results, and those where the algorithms' performance degrades. Different application domains may have different requirements for DSDM algorithm performance. For example, in safety-critical applications, like for example volcano monitoring, it may be vital never to miss a salient event (e.g., Werner-Allen et al. 2006). In such applications, approximate DSDM algorithms can still be useful if configured to guarantee no errors of omission, since a small number of errors of commission can be filtered out by additional scrutiny (e.g., human expertise).

The primary advantage of tolerating errors of omission and commission is computational. DSDM algorithms are computationally efficient and as a result highly scalable. Two of the three algorithms used in the previous experiments (NC and HEC) use only one one-hop communication, while the third (HAC) uses a small number of hops (up to three hops in practice). This dependence on local rather than global knowledge is what gives these algorithms scalability because their computational complexity depends not on the total number of nodes in the network, but rather on the number of neighbors a node has (which is expected to remain constant as *t*-size of the network increases, as long as node density remains constant). This contrasts strongly with centralized algorithms, where computational complexity typically increases with the number of observations (equivalent to nodes in the system). In moving from today's geosensor networks of tens or hundreds of nodes, to the predicted future networks of thousands or millions of nodes, scalability is paramount.

While centralized algorithms will long remain a core topic in knowledge discovery, DSDM represents a new approach to geographic knowledge discovery and, as we have shown, comes with new challenges beyond those posed by centralized algorithms. New technologies that blur the traditional separation between data capture and data processing (like geosensor networks) are driving the exploration of decentralized processing of spatial data. Longer term, the promise of these techniques is to contribute to the development of what is sometimes termed *ambient spatial intelligence*: spatial data capture, processing, and actuating capabilities embedded throughout our natural and built environment.

16.7 OUTLOOK: A DSDM RESEARCH AGENDA

To conclude, we identify four main research and development topics in DSDM:

1. The development of a *library* of fundamental DSDM algorithms for decentralized computation of classic spatial data mining tasks, including clustering, spatial outlier detection, co-location mining, and spatial association rule mining.
2. The investigation of *robust* and *fault-tolerant* methods for implementing DSDM in notoriously error-prone WSN environments.
3. The exploration of DSDM in *mobile* decentralized spatial computing systems, for example in the domain of traffic management or LBS.
4. The exploitation of decentralization as a means for providing geographic knowledge discovery at the same time as enhancing the *location privacy* of individuals in scenarios where nodes are associated with human users, for example in traffic or LBS applications.

ACKNOWLEDGMENTS

Patrick Laube and Matt Duckham's work is funded by the Australian Research Council (ARC), Discovery Grant DP0662906. Patrick Laube's research is additionally funded by the ARC Research Network on Intelligent Sensors, Sensor Networks and Information Processing (ISSNIP). Patrick Laube furthermore thanks organizers and attendees of the GADGET Workshop on Geometric Algorithms and Spatial Data Mining, funded by the Netherlands Organisation for Scientific Research (NWO) under BRICKS/FOCUS grant number 642.065.503, for an inspiring workshop.

REFERENCES

Bandyopadhyay, S., Giannella, C., Maulik, U., Kargupta, H., Liu, K., and Datta, S., 2006, Clustering distributed data streams in peer-to-peer environments. *Information Sciences*, **176**, 1952–1985.

Braginsky, D. and Estrin, D., 2002, Rumor routing algorithm for sensor networks, in *Proc. 1st ACM International Workshop on Wireless Sensor Networks and Applications*, (New York, NY: ACM) pp. 22–31.

Branch, J., Szymanski, B., Giannella, C., Wolff, R., and Kargupta, H., 2006, In-network outlier detection in wireless sensor networks, in *26th International Conference on Distributed Computing Systems (ICDCS), 2006.*

Cheng, Z. and Heinzelman, W. B., 2005, Flooding strategy for target discovery in wireless networks. *Wireless Networks*, **11**, 607–618.

Correll, N. and Martinoli, A., 2006, Collective inspection of regular structures using a swarm of miniature robots, in J. Ang, H. Marcelo, and O. Khatib (Eds.), *Experimental Robotics IX, The 9th International Symposium on Experimental Robotics (ISER), Singapore, June 18–21*, vol. 21 of *Springer Tracts in Advanced Robotics*, Heidelberg: (Springer) pp. 375–385.

Datta, S., Bhaduri, K., Giannella, C., Kargupta, H., and Wolff, R., 2006a, Distributed data mining in peer-to-peer networks. *IEEE Internet Computing*, **10**, 18–26.

Datta, S., Giannella, C., and Kargupta, H., 2006b, K-means clustering over a large, dynamic network, in *2006 SIAM Conf. Data Mining (SDM 06)* (SIAM Press), pp. 153–164.

Duckham, M., Nittel, S., and Worboys, M., 2005, Monitoring dynamic spatial fields using responsive geosensor networks, in *Proc. 13th ACM GIS 2005* (New York, NY: ACM).

Estrin, D., Govindan, R., and Heidemann, J., 2000, Embedding the Internet: Introduction. *Communications of the ACM*, **43**, 38–41.

Estrin, D., Govindan, R., Heidemann, J., and Kumar, S., 1999, Next century challenges: Scalable coordination in sensor networks, in *MobiCom '99: Proc. 5th annual ACM/IEEE International Conference on Mobile Computing and Networking* (New York, NY: ACM), pp. 263–270.

Gidofalvi, G. and Pedersen, T. B., 2005, Spatio-temporal rule mining: Issues and techniques, in *Data Warehousing and Knowledge Discovery, Proceedings*, vol. 3589 of *Lecture Notes in Computer Science*, (Heidelberg: Springer) pp. 275–284.

Grossglauser, M. and Tse, D.N.C., 2002, Mobility increases the capacity of ad hoc wireless networks. *IEEE/ACM Transactions on Networking*, **10**, 477–486.

Grossglauser, M. and Vetterli, M., 2006, Locating mobile nodes with ease: Learning efficient routes from encounter histories alone. *IEEE/ACM Transactions on Networking*, **14**, 457–469.

Guibas, L.J., 2002, Sensing, tracking and reasoning with relations. *IEEE Signal Processing Magazine*, **19**, 73–85.

Kargupta, H. and Chan, P., 2000, Distributed and parallel data mining: A brief infroduction, in H. Kargupta, and P. Chan (Eds.), *Advances in Distributed and Parallel Knowledge Discovery* (Menlo Park, CA: AAAI Press / The MIT Press). pp. xv–xxvi.

Karp, B. and Kung, H. T., 2000, GPSR: Greedy perimeter stateless routing for wireless networks, in *Proceedings of the 6th Annual International Conference on Mobile Computing and Networking* (New York, NY: ACM), pp. 243–254.

Kellerer, W., Bettstetter, C., Schwingenschlogl, C., Sties, P., and Steinberg, K. E., 2001, (Auto) mobile communication in a heterogeneous and converged world. *IEEE Personal Communications*, **8**, 41–47.

Lynch, N., 1996, *Distributed Algorithms* (San Mateo, CA: Morgan Kaufmann).

Mauve, M., Widmer, J., and Hartenstein, H., 2001, A survey on position-based routing in mobile ad-hoc networks. *IEEE Network*, **15**, 30–39.

Min, R. and Chandrakasan, A., 2003, Top five myths about the energy consumption of wireless communication. *SIGMOBILE Mobile Computing and Communications Review*, **7**, 65–67.

Nittel, S., Stefanidis, A., Cruz, I., Egenhofer, M., Goldin, D., Howard, A., Labrinidis, A., Madden, S., Voisard, A., and Worboys, M., 2004, Report from the First Workshop on Geo Sensor Networks. *ACM SIGMOD Record*, **33**.

O'Sullivan, D. and Unwin, D. J., 2003, *Geographic Information Analysis* (Hoboken, NJ: John Wiley & Sons).

Rabiner, W., Heinzelman, Kulik, J., and Balakrishnan, H., 1999, Adaptive protocols for information dissemination in wireless sensor networks, in *5th Annual ACM/IEEE International Conference on Mobile Computing and Networking* (New York, NY: ACM), pp. 174–185.

Shekhar, S., Zhang, P., Huang, Y., and Vatsavai, R. R., 2003, Trends in spatial data mining, in H. Kargupta, A. Joshi, K. Sivakumar, and Y. Yesha (Eds.), *Data Mining: Next Generation Challenges and Future Directions* (MIT/AAAI Press).

Wark, T., Corke, P., Sikka, P., Klingbeil, L., Guo, Y., Crossman, C., Valencia, P., Swain, D., and Bishop-Hurley, G., 2007, Transforming agriculture through pervasive wireless sensor networks. *Pervasive Computing, IEEE*, **6**, 50–57.

Werner-Allen, G., Lorinez, K., Welsh, M., Marcillo, O., Johnson, J., Ruiz, M., and Lees, J., 2006, Deploying a wireless sensor network on an active volcano. *IEEE Internet Computing*, **10**, 18–25.

Wolff, R. and Schuster, A., 2004, Association rule mining in peer-to-peer systems. *IEEE Transactions on Systems Man and Cybernetics Part B-Cybernetics*, **34**, 2426–2438.

Worboys, M. F. and Duckham, M., 2004, *GIS: A Computing Perspective*, 2nd ed. (Boca Raton, FL: CRC Press).

Worboys, M. F. and Duckham, M., 2006, Monitoring qualitative spatiotemporal change for geosensor networks. *International Journal of Geographic Information Science*, **20**, 1087–1108.

Xu, Y., Heidemann, J., and Estrin, D., 2001, Geography-informed energy conservation for ad hoc routing, in *Proc. ACM/IEEE International Conference on Mobile Computing and Networking*, pp. 70–84.

Younis, O. and Fahmy, S., 2004, Heed: A hybrid, energy-efficient, distributed clustering approach for ad hoc sensor networks. *IEEE Transactions on Mobile Computing*, **3**, 366–379.

Yu, Y., Govindan, R., and Estrin, D., 2001, *Geographical and Energy Aware Routing: A Recursive Data Dissemination Protocol for Wireless Sensor Networks*, Tech. Rep. UCLA/CSD-TR-01-0023, UCLA Computer Science Department.

Zhao, F. and Guibas, L. J., 2004, *Wireless Sensor Networks: An Information Processing Approach* (San Francisco, CA: Morgan Kaufmann).

17 Beyond Exploratory Visualization of Space–Time Paths

Menno-Jan Kraak

Otto Huisman

CONTENTS

17.1 INTRODUCTION

Natural disasters are a phenomenon of all times. However, if one considers recent events, such as tsunamis, earthquakes, and the (predicted) climate changes, as well as highly contagious and rapidly spreading diseases like SARS and avian bird flu, and if one takes into account the magnitude of economic globalization that has affected our world, it is clear that understanding and solving problems resulting from these events is of prime interest to society. The geosciences obviously have a prominent role to play, but the complexity of these problems is beyond the scope of a single discipline. From a geo-information perspective, a comprehensive understanding of these problems requires multidisciplinary approaches, collaboration among experts in order to bring together knowledge, and suitable tools for dealing with the large amounts of complex spatio-temporal data that analytical solutions to these problems would generate.

This chapter aims to draw together developments in various related fields, and to situate these within a broader discussion and illustration of geovisual analytics — the process of analytical reasoning using maps and other graphics. To do this, the chapter demonstrates use of the space–time cube (STC) as an interactive environment for the analysis and visualization of spatiotemporal data, drawing on two examples from the domain of human movement and activities. The first of these examines individual movement and the degree to which knowledge can be "discovered" by linking multiple attribute data to space-time movement data, and demonstrates how the STC can be deployed to query and investigate (individual-level) dynamic processes. The second example draws on the geometry of the STC as an environment for data mining through space–time query and analysis, illustrating work that deals with both individual and aggregate phenomena in the domain of tertiary education. These two examples form the basis of a broader discussion of the common elements of various disciplines and research areas concerned with moving object databases, dynamics, geocomputation, and geovisualization.

17.2 TRADITIONAL SOLUTIONS: FRAMEWORKS, TOOLS, AND TECHNIQUES

Both the tools and frameworks for understanding dynamic phenomena have been slow to develop since the "quantitative revolution" of the 1960s. While dynamics and the importance of space and time in the study of both human and physical processes are now widely accepted, they were largely ignored in traditional models, for a variety of conceptual and computational reasons. Conceptually, the fundamental notions underlying structure and functioning of transport systems, the spread of disease and, more generally, the representation of space and time in models of geographic phenomena differed considerably from those of today. The technology (hardware and software) for implementing enhanced concepts and models, particularly the software, was also slow to develop.

Recently, GIScience tools have had both fundamental and profound impacts on our ability to model, analyze, and visualize scenarios and phenomena. In the domain of human movement and activities, the significance of Hägerstrand's contribution to the modeling of spatiotemporal phenomena cannot be overlooked. Hägerstrand's original concepts formed the basis for the field now known as time-geography, by providing a series of simple geometric concepts. The time-geographic framework, as introduced by his seminal paper "What About People in Regional Science?" (Hägerstrand 1970), introduced the lifeline or space–time path (STP) of an object or entity as a continuous vector through space and time. This and further work introduced a range of other concepts, most notably the space-time prism (also known as the potential path area or PPA), and with these concepts, the notion that all movement and activities can be viewed inside an "aquarium" of space defined by time. Hägerstrand's original aim was to "invent a language which helps us to keep existents and events [...] together under a unifying perspective" (Hägerstrand 1982: 195), that of Newtonian or *absolute* space–time. One of the most notable features of these concepts is that they make use of time as the third dimension (z), collapsing space into an x,y plane. These initial concepts were simple and potentially powerful, but were unable to be made operational for significant sample size case studies due to a lack of computational tools. Aided by the "new" geographic information systems and other spatial modeling tools, researchers continued to apply and extend components of the time-geographic framework as deployed in Lenntorp's original PESASP simulation model (Lenntorp, 1978), and that of Burns (1979) on prism geometries. Early examples can be found in Forer and Kivell (1981) and Miller (1991), but more recently, better toolsets have become available to investigate and examine geographic phenomena in explicit spatial and temporal contexts (see Andrienko et al., 2007).

These developments have also initiated a wider trend of research into moving or mobile objects, with a significant focus on human behavior and movement (Forer 2002; Frihida, Marceau, and Theriault, 2004; Laube et al., 2005), as evidenced and supported by large-scale data collection exercises, including time budget surveys, and GPS-enabled tracking datasets (see Chapter 16 in this volume). Despite ongoing limitations in existing GIS data models, research has utilized various ways to cope with time, including attribute-based methods (see Peuquet, 1994; 2002), database approaches (Al-Taha et al., 1994; Varzigiannis and Wolfson, 2001), and hybrid rod-field data models (McDowall 2006). GIScience tools still only have limited ways to deal with time, and there is still no explicit spatio-temporal functionality in mainstream GIS software. One popular solution to this issue has been to utilize time as the third dimension (see Langran, 1992; Forer, 1998) in the same way as the space-time aquarium proposed by Hägerstrand. The examples in this chapter presented in the following sections illustrate the attractiveness of mapping movement and movement possibilities into an absolute space-time environment for both analysis and visualization. The STC — a computational version of the aquarium — facilitates *x,y,z* (plus attribute) investigation of the co-location of objects and events in space and time, and has been used successfully in a range of (geo-)visual and analytical studies (Kwan, 1999; Forer and Huisman, 2000; Kraak and Kousoulakou, 2004; Sinha and Mark, 2005; Ren and Kwan, 2007). One of the main reasons for its growing popularity is that it provides both a visual and analytical basis for linking objects and phenomena in space and time.

To summarize the trends just identified, Figure 17.1 presents a diagrammatic outline of the conceptual and operational developments in various diverse, yet related fields. In the 1970s and 1980s, although research was already informed by the notions of space and time, it was primarily application-driven, and little collaboration existed between these fields (Figure 17.1A). Furthermore, no specific "integrating technologies" existed, nor did any perspectives on how complementary aspects of some of these research fields might be aligned. At this stage, the notions of "visualization" and "analysis" were effectively two opposite ends of a spectrum of geo-information use. Several decades later (Figure 17.1B), we are witnessing a realignment of many disciplines and research fields, informed by one another and empowered by a common thread: toolsets that facilitate the combination of diverse data sets and provide a range of techniques for the investigation, summary, and analysis of spatial and, increasingly, spatio-temporal information.

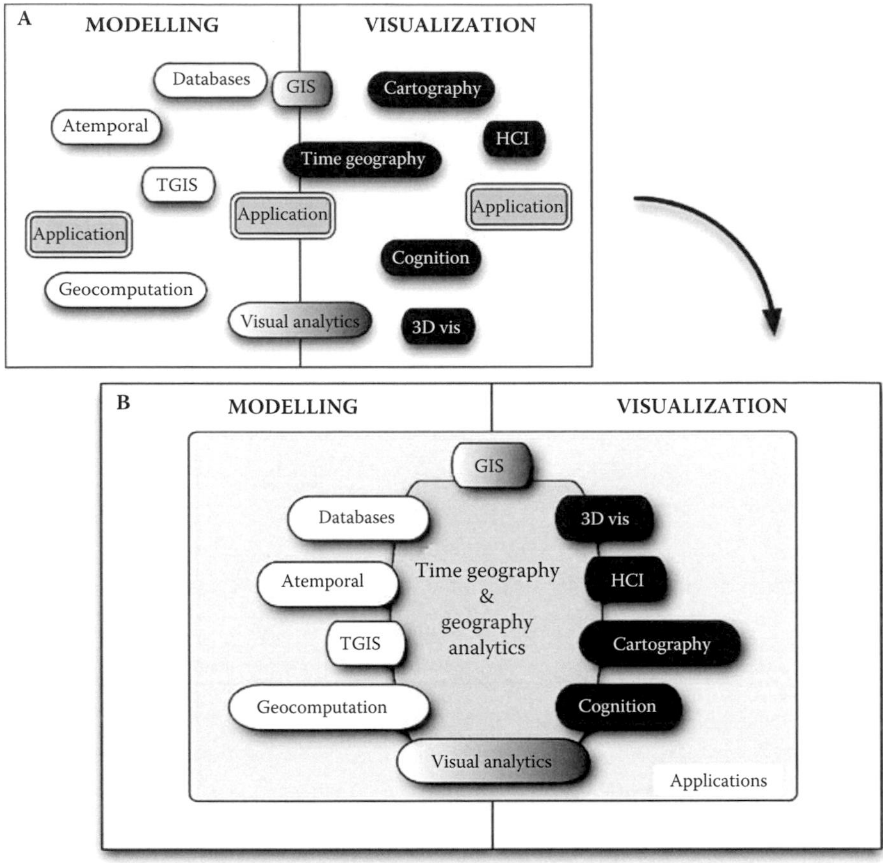

FIGURE 17.1 Developments around time-geography: Independent disciplines (A) either from the modelling or visualization domain each dealing with spatio-temporal data have over the years grown together (B) and can now offer an integrated approach to support solving geo-problems.

17.3 NEW AND IMPROVED: TIME-GEOGRAPHY AND GEOVISUAL ANALYTICS

17.3.1 Background

Over the last decades, cartography has developed considerably. One of the more prominent changes has been the introduction of the notion of geovisualization. In the book *Exploring Geovisualization* (Dykes, MacEachren and Kraak, 2005), it can be read that

> geovisualization can be described as a loosely bounded domain that addresses the visual exploration, analysis, synthesis, and presentation of geospatial data by integrating approaches from cartography with those from other information

representation and analysis disciplines, including scientific visualization, image analysis, information visualization, exploratory data analysis, and GIScience.

In a geovisualization context, maps are used to stimulate (visual) thinking about geospatial patterns, relationships, and trends, by offering interactive access to multiple alternative graphic representations of the data behind the map. As such, it supports knowledge construction, but this is not sufficient to deal with the global challenges mentioned before. Visualization has to be combined with analytics. The National Visualization and Analytics Center (NVAC) in the United States introduced the term *visual analytics,* which originates from their research agenda, "Illuminating the Path: The Research and Development Agenda for Visual Analytics" (Thomas and Cook, 2005) (http://nvac.pnl.gov/agenda.stm). In this book, it is described as "the science of analytical reasoning facilitated by interactive visual interfaces. The interfaces can be the maps as described in a geovisualization context. In more detail it should lead to synthesized information and derive insights from massive, dynamic, ambiguous, and often conflicting data," in other words "detect the expected and discover the unexpected."

According Thomas and Cook (2005), visual analytics helps solve problems because it offers methods and techniques that allow one to find, assimilate, and analyze continuously changing data about time-critical, evolving, real-world situations. For instance, for coastal protection one is interested in wind speed, direction and strength, water and wave heights, as well as the current situation of the dykes protecting the land. The findings of the analysis have to be communicated to a range of interest groups, including keepers of sluices and river barriers, but also to shipping control and local authorities, who will have to take necessary action. It is obvious that this is a process where different experts need to work together. In other words, it requires reasoning techniques that enable one to gain insight into the situation. Interactive visual representation and geocomputation techniques should be available depending on the situation — geocomputation for the number crunching (to process all weather data) and the human eye to explore and understand the resulting patterns. Maps and other graphics are there to offload memory. These might be annotated with thoughts and predictions supporting discussions and preparing decisions. Data integration functionality is crucial because the data will be obtained from all kinds of (different) sources. In many cases, this data will be both *incomplete* (a weather station fails due to the storms) and *uncertain* (not enough active sensors for trustworthy interpolations in both space and time).

The above trend made MacEachren (pers. comm., 2006) define the field of geovisual analytics as "the science of analytical reasoning and decision making with geospatial information, facilitated by interactive visual interfaces, computational methods, and knowledge construction, representation, and management strategies."

17.3.2 STC in Geovisual Analytics 1 — Linking Movement and Attribute Data

The first example discusses an elementary case of geovisual analytics to illustrate the process as such (see Figure 17.2). It is based on data of a single run. For data collection, a Garmin's Forerunner 305 was used. For visualization of the data, the Garmin Training Center software and our STC software were used. The device

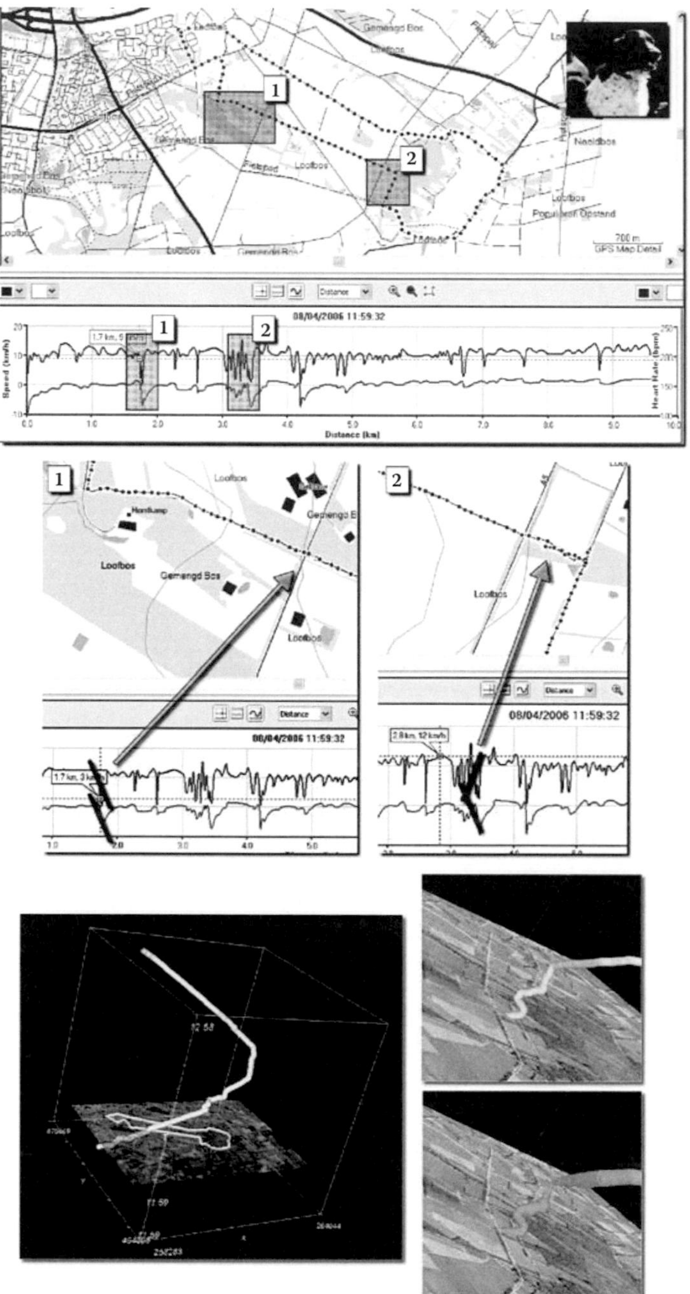

FIGURE 17.2 Geo-visual analysis of running data. The upper diagram shows geography and attribute data over time of a run. The analysis of the attribute data shows different patterns (1) and (2) which can be partly explained by the map and for the other part only with additional reasoning. The lower diagrams show the same data in the Space–Time Cube environment with the attributes visualized via the space–time path.

collects locations and heart rate values. From the first variable, others such as speed and pace are derived. One has to realize that the accuracy of the measurements is reasonable but the device is "sensitive" to noise, which should be filtered out before analysis.

The upper part of Figure 17.2 displays both speed and heart rate data for a short 8-km run in a map and graph. It can be observed that the heart rate values (the lower line in the graph) "follow" the speed, e.g., running faster will soon result in a higher heart rate value. This is illustrated in the left of the graph (see [1]) where two trend lines are plotted at a point where the speed reduces. These downward peaks can be recognized at several places in the graph. Is this a runner in bad shape who has to stop every so many meters, or is something else happening? Without particular knowledge of the capabilities of the runner, the linked map provides the answer. The location of the slowdown events seem to happen at crossings (see [1]) where the runner obviously watches for traffic before crossing. This seems to be a plausible reason.

While studying the graph in more detail with the above in mind, some anomalies can be observed. Around kilometer 2.8, the graph shows a high density in changes in both speed and heart rate (see [2]). However, if trend lines are plotted it can be seen that while the speed goes down, the heart rate increases. This is contradictory compared to earlier established trends. What goes on? Can the map assist? The map reveals no crossing around the 2.8 km and studying the track in more detail shows an up-and-down pattern along the road. With just common sense, it is not possible to find an explanation. More information, not available in the data collected, such as particular habits of the runner, is required. This is an example of the wide scope of geovisual analytics: one is often required to deal with incomplete data and the geo-expert often has to discuss or reason with other experts. In this particular example, the runner is accompanied by his dog. This might explain why at every crossing the runner slows down, but does not explain the above contradictory pattern. However, if we know the dog is a hunting dog and is running off-leash, and that at the location of the anomaly it observed and followed a rabbit, one will realize something different is going on. The runner slowed down, but his heart rate did not because he was yelling at the dog to follow him instead of chasing a rabbit.

Would an alternative view on the data have made the analytical reasoning simpler? In context of this chapter, would the display of the data in the STC be useful? The bottom of Figure 17.2 shows the result of these thoughts. Here the data is displayed in a STC. Here it should be clarified that the STC is not a stand-alone view but is spatially and attribute-linked to maps and graphs. On the left, the figure's full cube shows the run as a space-time path and its footprint is plotted on a photomap as well. This photomap can be moved up and down along the time axis of the cube, and one can zoom or pan. On the right, two details within the upper cube are an enlargement of the left cube. Here speed is represented by the path's angle; a vertical path segment means no movement. The lower right detail shows heart rate as an attribute of the path, represented by its thickness. This approach would, as with the basic Garmin software, not reveal a dog, but it does provide a different view of the same data where, because of the path speed, is better integrated into the "map." Seeing the anomalies might be easier, although one can argue that extra effort is required to interpret the three-dimensional scene when it comes to details.

17.3.3 STC in Geovisual Analytics 2 — Analyzing Potential Movement and Activities

This second example is drawn from the domain of tertiary education. Specifically, it considers the case of university students in Auckland, New Zealand. While as in the first example, movement and activities can be considered (modeled and visualized) in the form of space–time paths or timelines (*x,y,t* trajectories with attributes), the research from which the current example comes considers "potential" movement and activities rather than observed movement. Original concepts of the aquarium and the space–time prism are implemented using customized GIScience tools, and the data on which these examples are based derives from previous work investigating issues of access and interaction in the context of student learning (Forer and Huisman, 2000; Huisman, 2006). The details of generating individual time-geographies are quite complex, but the general procedures are illustrated in Figure 17.3:

- Modeling is based on actual lecture timetables and courses in which students were enrolled, and anonymized home locations. These are stored in database tables.
- Database tables are used in conjunction with an SQL-based scheduling algorithm to generate possible daily activity schedules for simulated individuals.
- Individual activity schedules are used in conjunction with a GIS-based multimodal transportation model and standard network-based shortest-path algorithms (implemented in ArcGIS workstation).
- Output from the above procedure is used to populate the STC using temporally referenced raster layers, and to assemble these into a 3D array of "taxels" (the space–time equivalent of voxels). This results in individual-based space–time volumes (prisms) and activities representing potential individual movement options over a day.

Figure 17.4 represents the output of the general procedures outlined above, and illustrates potential realizations of individual student days. Figures 17.4A, B, and C illustrate three unique individual volumes from a total sample of 2500 individual records in a database. These were generated by modeling each student's attendance at lectures, and modeling possible movement options between these fixed activities to and from a known (but anonymized) home and university campus location, using the concept of the STP. These three volumes (here termed "masks") represent the potential space–time locations that each student can potentially occupy given mandatory attendance at scheduled activities, and modeled transport options.

It is possible to see that in Figure 17.4A and Figure 17.4C, the students are able to "access" significantly larger space–time volumes than the student represented by

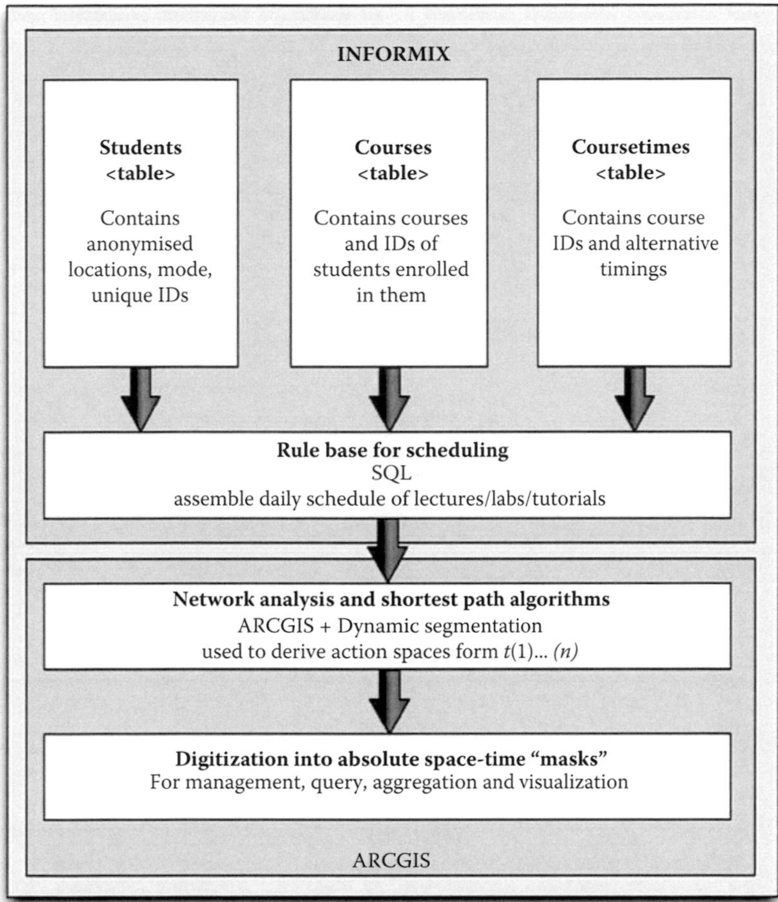

FIGURE 17.3 Generalized procedures for creating individual space-time *masks*.

Figure 17.4B. This is an illustration of the constraints for travel imposed by public transport, as the student in Figure 17.4B does not have a car at his or her disposal. Because of unique combinations of home location, transport mode availability, courses, and scheduled lecture times, each student mask is unique.

The STC can be used to combine any number of these masks, and various query and analysis functions have been developed specifically for their analysis and further aggregation to investigate resulting patterns (Forer and Huisman, 2000; Huisman, 2006). Figure 17.4D illustrates the three volumes in one single cube. From this figure, it is possible to identify common areas of space–time where the three students might meet. However, establishing exactly where and for how long these three individuals might meet requires further operations on this particular STC. As well as the linking of attribute information illustrated in the previous

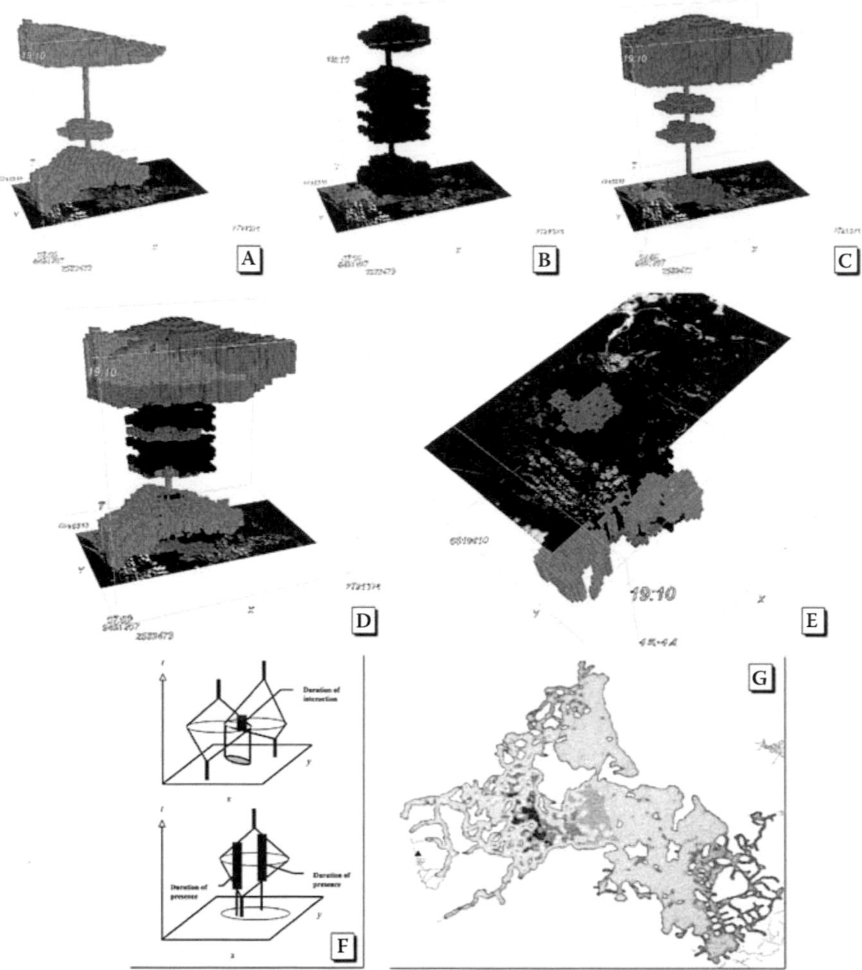

FIGURE 17.4 Diagrams (A), (B), and (C) illustrate three unique student "masks," each within its own space-time cube. These are intersected in (D) and (E). Examples of space-time query operations "drill" and "slice" are provided in (F). The map shown in (G) illustrates areas that the individual student in (A) might occupy for a given duration at 4:00 pm. The lightest areas are accessible for 10 min and dark areas in the middle of the figure for up to 3 h.

example (Figure 17.2), various other functions can also be applied here to derive new information, including:

- *Intersecting* the masks to identify areas of common space–time which the individuals can access.
- *Filtering* or thresholding of particular values to illustrate space–time clustering or potential space–time occupancy.
- *Drilling* down through the temporal axis to calculate durations.
- *Slicing* for examining patterns at specific temporal intervals.

The details of these functions are documented in Huisman (2006), but they all use the geometry of the space–time aquarium (the *x,y,t* data model) as their basis. Example outputs of intersecting masks and drilling down through the volumes are illustrated in Figure 17.4F and Figure 17.4G. Within the STC environment, transparency and filtering can be used to further segment the data and reveal hidden clusters or space–time patterns. Figure 17.4E illustrates the use of temporal clipping to view a specific time interval.

The functions just described can be combined in various ways to discover patterns and create knowledge. As an example of both *slicing* and *drilling*, Figure 17.4G depicts the places that an individual student can occupy as a result of his or her study timetable, and for how long. The color-ramp in this map ranges from light areas, representing areas where this individual can spend up to 10 min at this particular time of the day, to the dark areas, which represent places where the individual can spend up to 3 h of continuous free time. This output is generated through a simple time-interval query on a derivative mask of cells containing durations that the individual can be present at a location, which is itself generated through known constraints of modeled choices and movement options.

17.4 DISCUSSION AND CONCLUSION

The content of this book deals with data mining and knowledge discovery in a range of research fields. Our chapter has discussed these in the context of an investigation of the movement, and potential movement, of individuals as mobile objects in time and space. As noted here and in other chapters, space–time approaches are steadily growing in popularity, enabled by technology and driven by (research) demand. Hägerstrand's space–time aquarium provides a workable concept that is being deployed increasingly in various fields, including healthcare and the analysis of hazard and risk exposure (Loytonen, 1998; Forer and Huisman, 2000), as well as ongoing research into equity and accessibility (Kwan, 1999).

In attempting to draw together various approaches dealing with space–time visualization and analysis, this chapter has noted that in the past, the "visual" and the "analytical" represent points on opposite ends of a continuum. It has attempted to demonstrate that the aquarium (as implemented here using customized STC software) provides a flexible environment for the examination of space–time phenomena to support the study of moving objects and the dynamics of potential movement. The discovery of knowledge implies either accidental or planned confirmation of some kind of pattern or phenomenon revealed by (geo)data, as enabled by our tools, and (possibly) as expected or imagined by our minds. While some very powerful analytical and decision-support environments exist, and many more techniques have been developed for the analysis of large data sets, here we have emphasized the role of geovisual analytics in deriving new knowledge from data.

While there is a range of insights and analyses that could create new knowledge from detailed datasets, it should be noted that there are a number of issues relating to data quality and uncertainty to be dealt with in the context of human activities and behavior. For the purposes of this chapter, an acknowledgment of the complexity of modeling these is warranted. In the first example presented here, *observed* behavior was captured directly using GPS positioning. The key issue is to what degree *x,y,t* movement data can be used to "discover" knowledge. In the context of the discussion presented here, the answer depends on a range of factors including the quality of the input data, the ability to link a range of attribute data, and support for multivariate queries/analyses and visualization. A range of (geo)statistical tools and techniques can be used to aid in the investigation of results, visualize clustering for large movement data sets (Sinha and Mark, 2005), and inform hypothesis testing/generation algorithms or similar processes such as the detection of collective movement behavior in data using movement detection and generalization algorithms (see, for example, Laube et al., 2005). To enable wider insights, data that are more detailed might be required, and this could be linked to other visualization tools such as interactive 3D scatterplots (Kosara et al., 2004), for example, in order to be able to classify what types of people might take specific journeys at particular times of the day.

For *potential* behavior, there are processes at work that are even more complex. While these types of analyses steer clear of attempting to account for human agency or directly predicting behavior, achieving a degree of robustness at the daily or microscale requires significantly more detailed data (such as data describing the tasks to be undertaken during the day). For the student example presented in Section 17.3.3, data on space–time activities was readily available; however, in other cases, it needs to be "mined" from databases, such as time–budget surveys, and other approaches employed to generate realistic activity schedules. The mandatory lecture attendance assumed in this example is perhaps not directly representative of real-world events, but can be adjusted to reflect choice factors relatively easily, and results once again explored to determine possible outcomes.

In the wider domain of moving objects, as the scale of inquiry moves from examining individual-level to aggregate phenomena, there is an associated transition in research purpose from explaining and understanding behavior to the understanding and interpretation of *patterns* of behavior. This implies extended techniques for processing and generalization, which can be managed within the current environment of the STC for analysis and visualization.

REFERENCES

Al-Taha, K., R. Snodgrass, et al. 1994. Bibliography on spatiotemporal databases. *International Journal of Geographic Information Systems* 8(1): 95–103.

Andrienko, G., N. Andrienko, et al. 2007. Geovisual analytics for spatial decision support: Setting the research agenda. *International Journal of Geographic Information Science* 21(8): 839–857.

Burns, L. 1979. *Transportation, Temporal, and Spatial Components of Accessibility*. Lexington Books, Lexington, MA.

Dykes, J., A.M. MacEachren, and M. J. Kraak, Eds. 2005. *Exploring Geovisualization*. Elsevier, Amsterdam.

Forer, P.C. 1998. Geometric approaches to the nexus of time, space, and microprocess: Implementing a practical model for mundane socio-spatial systems. In: *Spatial and Temporal Reasoning in Geographic Information Systems*. M.J. Egenhofer and R.G. Golledge, Eds. Oxford University Press, New York, pp. 171–190.

Forer, P.C. 2002. Timelines, environments and issues of risk in health: The practical algebra of (x,y,t,a). In: *GIS for Emergency Preparedness and Health Risk Reduction*. D.J. Briggs, Ed. Kluwer Academic Publishers, The Netherlands, pp. 35–60.

Forer, P.C. and O. Huisman. 2000. Space, time and sequencing: Substitution at the physical/virtual interface. In: *Information, Place, and Cyberspace: Issues in accessibility*. D.G. Janelle and D. Hodge, Eds. Springer-Verlag, Berlin, pp. 73–90.

Forer, P.C. and H. Kivell. 1981. Space-time budgets, public transport, and spatial choice. *Environment and Planning A* 13: 497–509.

Frihida, A., D.J. Marceau, and M. Theriault. 2004. Extracting and visualising individual space-time paths: an integration of GIS and KDD in transport demand modelling. *Cartography and Geographic Information Science* 31(1): 19–28.

Hägerstrand, T. 1970. What about people in regional science? *Papers of the Regional Science Association* 24: 7–21.

Hägerstrand, T. 1982. Diorama, path and project. *Tijdschrift voor Economische en Sociale Geografie* 73(6): 323–339.

Huisman, O. 2006. The application of time-geographic concepts to urban micro-process. PhD Thesis, School of Geography and Environmental Science, University of Auckland, New Zealand.

Kosara, R., G.N. Sahling, et al. 2004. Linking scientific and information visualization with interactive 3D scatterplots. *Proceedings WSCG Communication (WSCG2004)*, Plzen, Czech Republic.

Kraak, M.J. and A. Kousoulakou. 2004. A visualization environment for the space-time-cube. In: *Developments in Spatial Data Handling: 11th International Symposium on Spatial Data Handling*. P.F. Fisher, Ed. Springer Verlag, Berlin, pp. 189–200.

Kwan, M.-P. 1999. Gender and individual access to urban opportunities: A study using space–time measures. *The Professional Geographer* 51(2): 211–227.

Langran, G. 1992, *Time in geographic information systems*. London: Taylor & Francis.

Laube, P., S. Imfeld, et al. 2005. Discovering relative motion patterns in groups of moving point objects. *International Journal of Geographical Information Science* 19(6): 639–668.

Lenntorp, B. 1978. A time-geographic model of individual activity programmes. In: *Human Activity and Time Geography*. T. Carlstein, D. N. Parkes, and N. J. Thrift, Eds. Edward Arnold, London, pp. 162–180.

Loytonen, M. 1998. GIS, time geography and health. In: *GIS and Health*. A.C. Gatrell and M. Loytonen. Taylor & Francis, London, pp. 97–110.

McDowall, C. 2006. Modelling vague dynamic entities within a GIS. PhD Thesis, School of Geography and Environmental Science, University of Auckland, New Zealand.

Miller, H.J. 1991. Modelling accessibility using space-time prism concepts within geographical information systems. *International Journal of Geographical Information Systems* 5(3): 287–301.

Peuquet, D. 1994. It's about time: A conceptual framework for the representation of temporal dynamics in geographic information systems. *Annals of the American Association of Geographers* 84 (3):441–461.

Peuquet, D.J. 2002. *Representations of Space and Time*. The Guilford Press, New York.

Ren, F. and M.-P. Kwan. 2007. Geovisualization of human hybrid activity-travel patterns. *Transactions in GIS* 11(5): 721–744.

Sinha, G. and D.M. Mark. 2005. Measuring similarity between geospatial lifelines in studies of environmental health. *Journal of Geographical Systems* 7(1): 115–136.

Thomas, J.J. and C.A. Cook, Eds. 2005. *Illuminating the Path: The Research and Development Agenda for Visual Analytics*. IEEE Press, Washington.

Vazirgiannis, M. and O. Wolfson. 2001. A spatiotemporal model and language for moving objects on road networks. Proceedings of the 7th International Symposium on Advances in Spatial and Temporal Databases, *Lecture Notes in Computer Science* 21: 20–35.

Index

A

B

E

H

I

N

O

P

Q

R

S

T

U

V

W

Y